Eugene Rosenberg/Yossi Loya (Eds.)

Coral Health and Disease

Springer

Berlin
Heidelberg
New York
Hong Kong
London
Milan
Paris
Tokyo

Eugene Rosenberg · Yossi Loya (Eds.)

Coral Health and Disease

With 143 Figures, 49 in Color

Springer

Professor Eugene Rosenberg
Department of Molecular Microbiology & Biotechnology
Tel Aviv University
Ramat Aviv
Israel 69978

e-mail: eros@post.tau.ac.il

Professor Yossi Loya
Department of Zoology
Tel Aviv University
Ramat Aviv
Israel 69978

e-mail: yossiloya@post.tau.ac.il

ISBN 3-540-20772-4 Springer-Verlag Berlin Heidelberg New York

Library of Congress Cataloging-in-Publication Data

Coral health and disease / Eugene Rosenberg, Yossi Loya, (eds).
p. cm.
Papers derived from an international meeting held in Eilat, Israel, April 2003.
Includes bibliographical references.
ISBN 3-540-20772-4 (alk. paper)
1. Coral reef ecology--Congresses. 2. Corals–Diseases--Congresses. I. Rosenberg, Eugene. II. Loya, Yossi.
QH541.5.C7C56 2004
577.7'89--dc22

Springer-Verlag is a part of Springer Science+Business Media
springeronline.com

Printed in Germany

Cover Photo: The coral reefs along the Gulf of Eilat/Aqaba are among the most diverse and spectacular reefs in the world (Photo Yossi Loya)
Cover design: design & production GmbH, 69121 Heidelberg
Typesetting: perform electronic publishing GmbH, Heidelberg
31/3150WI – 5 4 3 2 1 0 – Printed on acid-free paper

Preface

The major objective of this book is to present the current state of the art on coral disease – from observations in the water to laboratory research on the physiology, microbiology and molecular biology of these diseases. The book is divided into four parts: Part I describes the dynamics of coral diseases on coral reefs in different parts of the world. Part II presents experimental research on the microbial ecology and physiology of healthy and diseased corals. Part III deals separately with each of the major coral diseases, from observations in the field to experimental studies on the pathogens and modes of infection. Finally, Part IV emphasizes the key role of environmental stress in inducing coral bleaching and then presents different hypotheses to explain the phenomenon. The final chapter presents projections of future changes in coral reefs resulting from global warming and coral bleaching. Each of the chapters presents an overall view of the subject, the author's specific contribution and unresolved questions.

The chapters in this book were derived primarily from an international meeting on Coral Health and Disease which took place in Eilat, Israel, in April 2003. The meeting was sponsored by the Israel Center for Emerging Diseases (ICED). We thank the ICED for sponsoring the meeting and defraying the costs for the color photographs in this book and Dina Silverman and Ruth Rossing for their help in organizing the meeting and preparation of the manuscripts. It was a pleasure to work with Dr. Jutta Lindenborn of Springer-Verlag, who made our job much easier by providing all the necessary instructions. Finally, to the authors, thank you all.

Tel Aviv, March 2004

Eugene Rosenberg
Yossi Loya

Contents

Part I Regional Aspects of Coral Reef Health and Disease

List of Contributors

Baker, Andrew C. (e-mail: abaker@wcs.org)
Center for Environmental Research and Conservation, Columbia University, MC 5557, 1200 Amsterdam Avenue, New York, New York 10027, USA

Buddemeier, Robert W. (e-mail: buddrw@ku.edu)
Kansas Geological Survey, 1930 Constant Avenue, Lawrence, Kansas 66047, USA

Bythell, John (e-mail: j.c.bythell@ncl.ac.uk)
School of Biology, Newcastle University, Newcastle upon Tyne NE1 7RU, UK

Dinsdale, Elizabeth A.
Tropical Environment Studies and Geography, James Cook University, Townsville, Queensland 4811, Australia

Drew, Harvell C. (e-mail: cdh5@cornell.edu)
Department of Ecology and Evolutionary Biology, Cornell University, Ithaca, New York 14853, USA

Dubinsky, Zvy
Faculty of Life Sciences, Bar-Ilan University, Ramat-Gan 52900, Israel

Erez, Jonathan
The Hebrew University, Institute of Earth Sciences, Edmond Safra Campus, Givat Ram, Jerusalem, 91904, Israel

Fautin, Daphne G. (e-mail: fautin@ku.edu)
Division of Biological Sciences, University of Kansas, 1200 Sunnyside Avenue, Lawrence, Kansas 66045, USA

Fine, Maoz (e-mail: M.Fine@uq.edu.au)
Centre for Marine Studies, University of Queensland, St. Lucia, Queensland 4072, Australia

Garzón-Ferreira, Jaime (e-mail: jgarzon@invemar.org.co)
Instituto de Investigaciones Marinas y Costeras INVEMAR. P.O. Box 1016, Santa Marta, Colombia

Gil-Agudelo, Diego L. (e-mail: gilagude@msci.sc.edu)
University of South Carolina, Department of Marine Science, Columbia, South Carolina 29208, USA

Hoegh-Guldberg, Ove (e-mail: oveh@uq.edu.au)
Centre for Marine Studies, University of Queensland, St Lucia, Queensland 4072, Australia

Irikawa, Akiyuki
Department of Marine Sciences, University of the Ryukyus, Senbaru, Nishihara, Okinawa, 903–0213, Japan

Jacobs, J. Rebecca (e-mail: jacobs@biology.ucsc.edu)
Department of Ecology and Evolutionary Biology, University of California, Santa Cruz, California 95064, USA

Jokiel, Paul L. (e-mail: jokiel@hawaii.edu)
Hawaii Coral Reef Assessment and Monitoring Program (CRAMP), Hawaii Institute of Marine Biology, PO Box 1346, Kaneohe, Hawaii 96744, USA

Jordán-Dahlgren, Eric (e-mail: jordan@mar.icmyl.unam.mx)
Instituto de Ciencias del Mar y Limnología, U.N.A.M. Ap. Postal 1152, 77500 Cancún, Q. Roo, México

Kelley, Scott
Biology Department and the Center for Microbial Sciences, San Diego State University, 5500 Campanile Dr., San Diego, California 92182, USA

Kelman, Dovi (e-mail: dovi.kelman@weizmann.ac.il)
Department of Biological Chemistry, The Weizmann Institute of Science, 76100 Rehovot, Israel

Kramarsky-Winter, Esti (e-mail: wintere@post.tau.ac.il)
Department of Zoology, George S. Wise Faculty of Life Sciences, Tel Aviv University, Tel Aviv 69978, Israel

Kushmaro, Ariel (e-mail: arielkus@bgumail.bgu.ac.il)
Department of Biotechnology Engineering, Faculty of Engineering Sciences, Ben Gurion University, Beer Sheva, 84105, Israel

Lazar, Boaz
The Hebrew University, Institute of Earth Sciences, Edmond Safra Campus, Givat Ram, Jerusalem 91904, Israel

Loya, Yossi (e-mail: YossiLo@tauex.tau.ac.il)
Department of Zoology, George S. Wise Faculty of Life Sciences, Tel Aviv University, Tel Aviv 69978, Israel

McClanahan, Tim (e-mail: tmcclanahan@wcs.org)
Kibaki Flats #12, Bamburi, P.O. Box 99470, Mombasa, Kenya

Mullen, Kerri M.
Department of Ecology and Evolutionary Biology, Cornell University, Ithaca, New York 14853, USA

Page, Cathie A.
Australian Institute of Marine Science PMB No 3, Townsville MSO, Queensland 4810, Australia

Pantos, Olga
School of Biology, Newcastle University, Newcastle upon Tyne NE1 7RU, UK

Patterson Sutherland, Kathryn (e-mail: kathrynp@arches.uga.edu)
Department of Marine Sciences, University of Georgia, Athens, Georgia 30602, USA

Peters, Esther C.
Tetra Tech, Inc., 10306 Eaton Place, Suite 340, Fairfax, Virginia 22030, USA

Petersen, Dirk (e-mail: d.petersen@rotterdamzoo.nl)
Rotterdam Zoo, P.O. Box 532, 3000 AM Rotterdam, The Netherlands

Richardson, Laurie L. (e-mail: Laurie.Richardson@fiu.edu)
Department of Biological Sciences, Florida International University, Miami, Florida 33199, USA

Ritchie, Kim B. (e-mail: ritchie@mote.org)
Center for Coral Reef Research, Mote Marine Laboratory, 1600 Ken Thompson Parkway, Sarasota, Florida 34236, USA

Rodríguez-Martínez, Rosa E.
Instituto de Ciencias del Mar y Limnología, U.N.A.M. Ap. Postal 1152, 77500 Cancún, Q. Roo, México

Rohwer, Forest (e-mail: forest@sunstroke.sdsu.edu)
Biology Department and the Center for Microbial Sciences, San Diego State University, 5500 Campanile Dr., San Diego, California 92182, USA

Rosenberg, Eugene (e-mail: eros@post.tau.ac.il)
Department of Molecular Microbiology & Biotechnology, Tel-Aviv University, Ramat Aviv 69978, Israel

Rozenblat, Yael Ben-Haim (e-mail: yrozenblat@ucsd.edu)
Department of Molecular Microbiology and Biotechnology, Tel-Aviv University, Ramat Aviv 69978, Israel

Silverman, Jack (e-mail: jacobs@vms.huji.ac.il)
The Hebrew University, Institute of Earth Sciences, Edmond Safra Campus, Givat Ram, Jerusalem 91904, Israel

Smith, Garriet W. (e-mail:SmithRes@aiken.edu)
Department of Biology and Geology, University of South Carolina Aiken, Aiken, South Carolina 29801, USA

Stambler, Noga (e-mail: stambln@mail.biu.ac.il)
Faculty of Life Sciences, Bar-Ilan University, Ramat-Gan 52900, Israel

Van Woesik, Robert (e-mail: rvw@fit.edu)
Department of Biological Sciences, Florida Institute of Technology, 150 West University Boulevard, Melbourne, Florida 32901–6988, USA

Weil, Ernesto (e-mail: eweil@caribe.net)
Department of Marine Sciences, University of Puerto Rico, P.O. Box 3208, Lajas, Puerto Rico 00667, USA

Willis, Bette L. (e-mail: Bette.Willis@jcu.edu.au)
School of Marine Biology and Aquaculture, James Cook University, Townsville, Queensland 4811, Australia

1 The Coral Reefs of Eilat – Past, Present and Future: Three Decades of Coral Community Structure Studies

Yossi Loya

1.1 Introduction

Here, I shall present a brief review of ca. 35 years of our studies on changes in the coral species diversity and community structure at Eilat, Red Sea, at several scales in space and time. In the following, I shall: (1) summarize the geographical setting and the geological, physical and biological characteristics of the Gulf of Eilat/Aqaba, then point out the uniqueness of the coral reefs of Eilat, which are situated at the most northerly boundary of coral reef distribution, yet exhibit extraordinarily high within-habitat coral species diversity; (2) present the changes that took place in coral species diversity and community structure on the reef flats in the northern Gulf of Aqaba/Eilat (during the 1969–1980), due to natural disturbances (extreme midday low tides) and man-made perturbations (chronic oil spills); (3) discuss possible mechanisms that generate and maintain the high within-habitat coral diversity typifying pristine reefs in the Gulf of Eilat/Aqaba; (4) discuss the opposite mechanisms that caused a dramatic decrease in coral abundance and living cover at the Eilat Coral Nature Reserve (ECNR) during 1986–2000. I will also point out two major anthropogenic disturbances: first, eutrophication caused by Eilat's sewage discharge to the sea until 1995; and second, further eutrophication originating from intensive net pen mariculture off the northern coast of Eilat, which exponentially expanded activity from 1994–1995 to present times. The grave implications for the coral reefs of Eilat caused by this chronic eutrophication will be presented. Finally, (5) I conclude with a warning that, at present, the coral reefs of Eilat are severely damaged and subsist in a critical state. If eutrophication of the northern Gulf is not halted immediately, the final collapse and total destruction of the unique coral reefs of Eilat are certain. In their present fragile state, the only chance for the restoration of the Eilat reefs is extreme and instant protection measures against all man-made disturbances.

1.1.1 The Gulf of Eilat/Aqaba: Geographical Setting, Geological, Physical and Biological Characteristics

The Red Sea is a long body of water separating northeastern Africa from the Arabian Peninsula. Nearly 2000 km of water connects the south with the In-

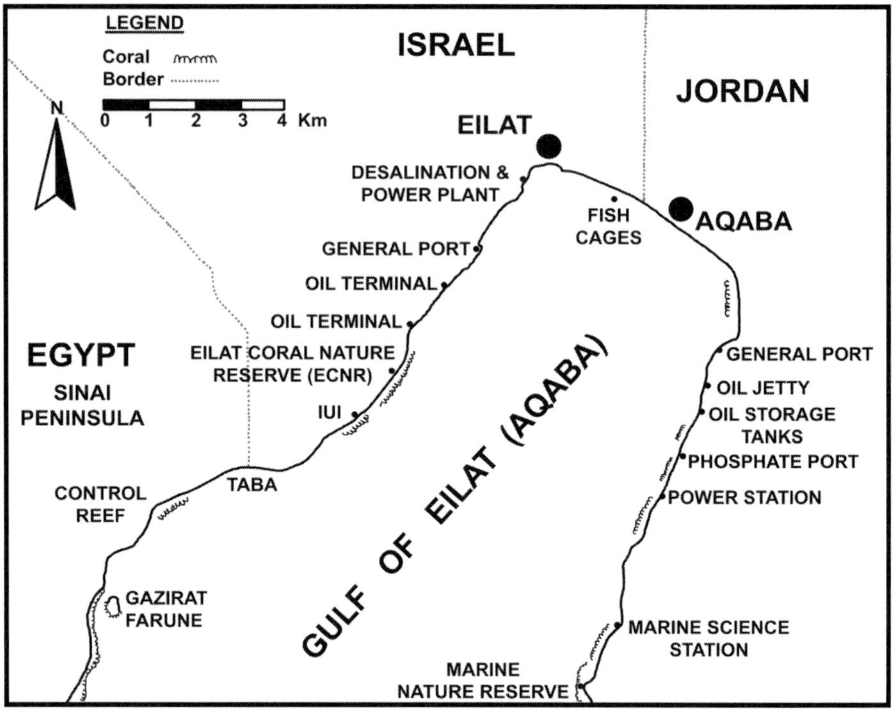

Fig. 1.1. General map of the northern Gulf of Eilat/Aqaba indicating the work sites: Eilat Coral Nature Reserve (ECNR), the Control Reef, the Interuniversity Institute (IUI) and the mariculture fish cages

dian Ocean and almost joins the Mediterranean Sea at the north of the Gulf of Suez. When Ferdinand de Lesseps completed the Suez Canal in 1869, the connection became direct. The Red Sea is connected to the Indian Ocean by the relatively shallow Bab el Mandeb Straits (270-m depth). Although it is relatively young, the Red Sea evolved to harbor a number of unique ecosystems. Its pelagic ecosystem is characterized by a photic zone (0–200 m depth) and an aphotic zone (200–2500 m depth).

The Gulf of Eilat/Aqaba (Figs. 1.1, 1.2A) is part of a major geological formation, the 1000-km-long Dead Sea rift, which is a portion of the 6000-km Syrian African rift that extends from Mozambique to Turkey. The rift was formed by the

Fig. 1.2. **A** Satellite view of the geographical setting. **B, C** Unpredictable extreme low tide at the northern Gulf of Eilat/Aqaba exposed corals during midday (September 1970) on the reef flats for 4 days (3–4 h every day) to high irradiance and air temperature of ca. 40 °C. The consequences were massive mortality of ca. 90% of the coral populations on the reef flats of the northern Gulf of Eilat/Aqaba. The coral community structure at the oil-polluted ECNR and a clean control reef (CR) 5 km further south were studied in detail by means of line transects, before, during and after the catastrophic low tide (see Fig. 1.4 for details). **D–F** During 1969–1980 an average of two to three oil spills per month covered the ECNR with crude oil ▶

A
Mediterranean Sea
Sinai Peninsula
Israel
Jordan
Egypt
Eilat
Aqaba
Gulf of Suez
Gulf of Eilat (Aqaba)
Saudi Arabia
Red Sea
B
C
D
E
F

opposing movement of the African and Arabian continents. The Gulf is the more eastern of the two northern horns of the Red Sea, which are separated by the Sinai Peninsula. It is surrounded by desert; hence, water inflow from land-based sources is negligible. The Gulf is ca. 180 km long, very steep-sided and deep (although it is only 5–26 km wide), reaching a maximum depth of over 1800 m near the eastern coast. The Gulf's photic zone is stratified, nutrient-depleted and has exceptionally crystal-clear surface waters (no river runoffs).

The climate in the region is arid with an average net evaporation of 1 m^3/day. Predominant northern winds enhance the evaporation. As a result, the Gulf's waters are extremely saline, ranging between 40 and 41.5 ‰. Temperature ranges from 20.5 (winter, northern Gulf) to 27.3 °C (summer, southern Gulf). Tides in the Gulf are minimal with a maximal range in the order of 1 m. Extreme midday low tides, which expose entire reef flats to the air are quite rare, but when they do occur, they may cause complete desiccation of the coral populations (Fig. 1.2B, C; see also Sect. 1.2.1).The general circulation pattern consists of an inflow of less saline Red Sea water in the upper layers through the straits of Tiran and an outflow of bottom heavier, more saline, Gulf water near sill depth. The coral reefs along the Gulf are of the fringing type and among the most spectacular and diverse in the world (Fig. 1.3A, B). Both the coral reefs and the mangroves are among the world's northern-most such ecosystems. The semi-isolation of the Red Sea from the main body of the Indian Ocean, as well as the semi-enclosed nature of the Gulf and the rather extreme oceanographic conditions resulted in the evolution of a high proportion of endemic species typical of the Red Sea flora and fauna (Sheppard et al. 1992).

Coral reef ecosystems are the most spectacular and diverse marine ecosystems. They form reservoirs of the highest marine biological diversity, including genetic resources and bioactive compounds. Unfortunately, coral reefs are also among the most heavily degraded marine ecosystems. Over the last two decades, coral reef communities have been experiencing increasingly stressful conditions due to a combination of natural and anthropogenic detrimental factors (Wilkinson 2000).

Fig. 1.3. The coral reefs of Eilat: past. **A, B** The ECNR 1969 (see Fig. 1.1). High within habitat diversity typified the ECNR in the past. **C–F** Urban development (Eilat) and pollution sources in the northern Gulf during the last three decades. **C** Phosphate pollution caused by the poor loading procedures used in Eilat port brought about eutrophication of the reef during the 1970s and 1980s. During the last decade, environmental law enforcement resulted in improved loading technology decreasing this pollution source almost to nil. **D** The Aqaba phosphate port has been another eutrophication source in the Gulf, but in the last decade has decreased substantially due to improved loading procedures. **E** Aerial view of the surroundings of the Inter University Institute at Eilat (IUI) in 1970. **F** Aerial view of the same surroundings in 2000 demonstrating the fast urban development of the area. **G** Discharge outlet of the desalination power plant of Eilat (operated during the early 70's). **H, I** Eilat's municipal sewage was flowing into the northern Gulf of Eilat until 1995. Since then, this source of nutrient pollution was stopped after a modern sewage treatment facility was built 6 km north of the city ▶

A
B
C
D
E
F
G
H
I

Scleractinian corals are the most important hermatypic (reef-building) organisms in the Gulf of Eilat (Loya and Slobodkin 1971). Hermatypic corals play a key role in forming the structure of coral reefs and in providing substrate and shelter for a wide variety of organisms. Acute damage to the corals may result in the collapse of the complex community of organisms that live in close association with them. Since the Gulf is a semi-enclosed basin, and the prevailing winds and currents are predominantly from the north, the Gulf is potentially vulnerable to pollution, particularly at its northern tip. The two cities located there, Eilat, in Israel, and Aqaba, in Jordan (Fig. 1.1), are both industrial and tourist centers to their respective countries. In the last three decades, the area has gone through rapid urban development (Loya 1995), increasing anthropogenic pressures on the reefs (Figs. 1.2D–F, 1.3C–I). Hence, the potential threat of pollution is very real, including possible spills from maritime activities and oil transport in the Gulf. In addition, chemical pollutants that may pollute the Gulf during transport and loading of phosphates (Fig. 1.3C, D), potash bromides and other cargoes, in both the commercial ports of Eilat and Aqaba (Loya 1995). Other major sources of human stresses to the reefs in the most northern section of the Gulf include eutrophication from Eilat's municipal wastewater (Fig. 1.3H, I), discharged in the past into the sea (stopped in 1995, see Sect. 1.4.), unregulated mariculture effluents in Eilat, which at present are the major source of coastal eutrophication of the northern Gulf of Aqaba/Eilat (Sect. 1.4.), as well as occasional ballast and bilge water spilled from various boat activities, occasional discharges of fuel, crude oil and detergents. In addition, due to tourism, physical damage to corals occurs, mainly by boat anchors, scuba divers and snorkelers.

1.1.2 The Coral Reefs of Eilat

Although situated in the most northern boundaries of coral reefs distribution, the pristine reefs of the northern Gulf of Aqaba (ca. 30°N), along the shorelines of the Sinai Peninsula (Fig. 1.1), exhibit extraordinary high within-habitat coral species diversity (sensu MacArthur 1972), among the highest in the world (Fig. 1.3A, B). Unfortunately, those reefs that have been chronically perturbed by anthropogenic activities, such as the coral reefs of Eilat, have severely deteriorated in the last three decades, especially in the last decade (Loya 1976a, 1990 and this chapter).

In 1968–1969, the community structure of the coral reefs in the northern Gulf of Eilat/Aqaba was studied in detail by means of 10-m-long line transects at the Eilat Coral Nature Reserve (ECNR, Loya 1972, 1975) and a site referred to as 'the control reef'(Figs. 1.1, 1.2C), 5 km further south. Both reefs were pristine at that time. The exact locations of the transects on both reefs were marked by stainless steel pegs, which enabled repeated long-term monitoring of the same transects in the following years (Loya et al. 1999). Any coral species that overlapped the line was recorded, and its projected length on the line was measured to the nearest centimeter. The line transects were surveyed from the reef flat to

30-m depth. The data provide estimates of the number of coral species, number of colonies, percentage of total living coverage, and diversity of corals. This large data set served as a baseline for continued long-term monitoring of the scleractinian coral populations of the northern Gulf until today.

In 1969, the average number of species per transect obtained at the two sites in the northern Gulf of Eilat/Aqaba (see below) was 13.5±3.8 (ECNR) and 15.7±3.4 (control reef, CR) species per 10-m transect (Loya 1972, 1976a). This diversity is very high for corals measured on reef flats (within-habitat diversity) in view of the low total number of species known from Eilat (ca. 100, Loya and Slobodkin 1971), and the fact that the coral reefs of Eilat are among the most northern reefs in the world. By contrast, about 360 coral species are known in the Great Barrier Reef (GBR, Veron 1986). Yet, within-habitat diversity on reef flats is higher in Eilat than on either the Australian inshore fringing reefs or the GBR proper.

A similar methodology suggested by Loya (1972) was used to study coral diversity on the GBR reef flats, with the only difference being the length of the transect lines (30 m). On Lizard Island, the average number of coral species per transect was 11.1±4.8 (data computed from Pichon and Morrissey 1981); on Heron Island, 9.8±4.1 (n=12 transects; Pichon, Weizman-Best and Loya, unpubl. data); and on an inshore reef at Magnetic Island, 4.8±3.7 (n=9 transects; Pichon, Loya and Bull, unpubl. data). Clearly, if the GBR data were transformed into 10-m transects, as performed in Eilat, the average number of species per transect would have decreased. Since information on community structure and species diversity of the GBR corals is limited and most probably regionally different (2500 km of reefs), any generalization on local patterns of diversity in the GBR would be premature.

Two major environmental disturbances dictated the changes in the coral community structure at Eilat for years to come. Firstly, an unexpected and extremely low tide that occurred in the northern part of the Gulf of Eilat/Aqaba in September 1970 killed almost all the corals on the reef flats (Fig. 1.2B, C; see Sect. 1.2.1), and secondly, the harmful effect of chronic oil spills that occurred during 1970–1980 in the vicinity of the ECNR to the coral populations (Fig. 1.2D–F, see Sect. 1.2.2).

1.2 Changes in Coral Community Structure at Eilat Due to Natural Perturbations and Human Disturbances During 1969–1982

1.2.1 Natural Catastrophes: Extreme Midday Low Tides

The Eilat Coral Nature Reserve (ECNR) is located 3 km south of the general port of Eilat, approximately 1 km south of two major oil terminals (Fig. 1.1). Oil tankers using the port of Eilat after 1970 caused two or three major oil spills every month, during which the ECNR was completely blackened by oil (Fig. 1.2D–F). This chronic oil pollution continued until ca.1979–1980. Since then, oil spills at

Eilat have become rare. A major reason for this remarkable improvement is the drastic decrease in the number of oil tankers arriving at Eilat (political reasons), in addition to the establishment of a Pollution Prevention Control Station by the Israeli Ministry of the Environment and the stiff fines imposed on the shipping companies.

Since the wind direction along the Gulf of Aqaba/Eilat is predominantly from the north or northeast, oil spilled around the terminals is carried by surface currents towards the ECNR. It does not, however, reach the 'control reef', a pristine coral reef located 5 km further south. The control reef (CR) has been termed so, since it has been found to be free of oil pollution (Loya 1975) and has served as a comparatively pristine site for studies on changes in the community structure of corals in the polluted ECNR.

An unexpected and extremely low tide occurred in the northern part of the Gulf from 15–20 September 1970 (Fig. 1.2B, C; Loya 1972). The water level fell approximately 20–25 cm below the reef flats. As a result, they were completely exposed to air and sun for 3–4 h during the hottest time of the day. The air temperature ranged from a minimum of 34 °C on 16 September to a maximum of 38.4 °C on 18 September. The immediate consequence of the low tide was the death of approximately 80–85% of the hermatypic corals along the northern part of the Gulf (Loya 1975, 1976a). The high temperatures and desiccation were probably the direct causes of the rapid coral mortality. The community structure and species diversity of hermatypic corals had been studied in several reef locations along the northern Gulf before the low tide (Loya 1972). Hence, the incident provided a unique opportunity to study recruitment patterns in both a chronically polluted reef (ECNR) and a clean control reef (CR).

During the low tide, the corals below the lowest water level were unharmed. Since the ECNR and the CR were similar in community structure (Loya 1975), it was reasonable to assume that a similar stock of coral propagules existed in both. Theoretically, providing human activities had not had a harmful effect on the coral communities at the ECNR, a similar recruitment rate of corals could have been expected in both localities. The parameters compared were the average number of coral species, colonies, living coverage, and diversity (Shannon and Weaver's H'n index 1948) per transect, in both reefs, during 1969 to 1986. In 1969, no significant difference was found between the coral community structure of the ECNR and that of the CR, when all four statistics were compared (Loya 1972; Fig. 1.4). A detailed account of changes in the community structure and species diversity of corals in 1969 and 1973 is given in Loya (1975, 1976a). Although both reefs suffered similar mass mortality of corals during the low tide (85% at the ECNR and 81% at the CR), a marked difference was observed in recovery 3 years later. The extent of recruitment was 23-fold greater at the CR. No significant difference was found in coral community regeneration (15% at the ECNR and 19.2% at the CR). Three years after the low tide, the number of species, number of colonies, living coverage and H'n remained very low on the reef flat of the ECNR, while the reef flat of the CR exhibited rapid recruitment and recovery (Fig. 1.4). The significantly higher coral cover at the CR was mainly due to

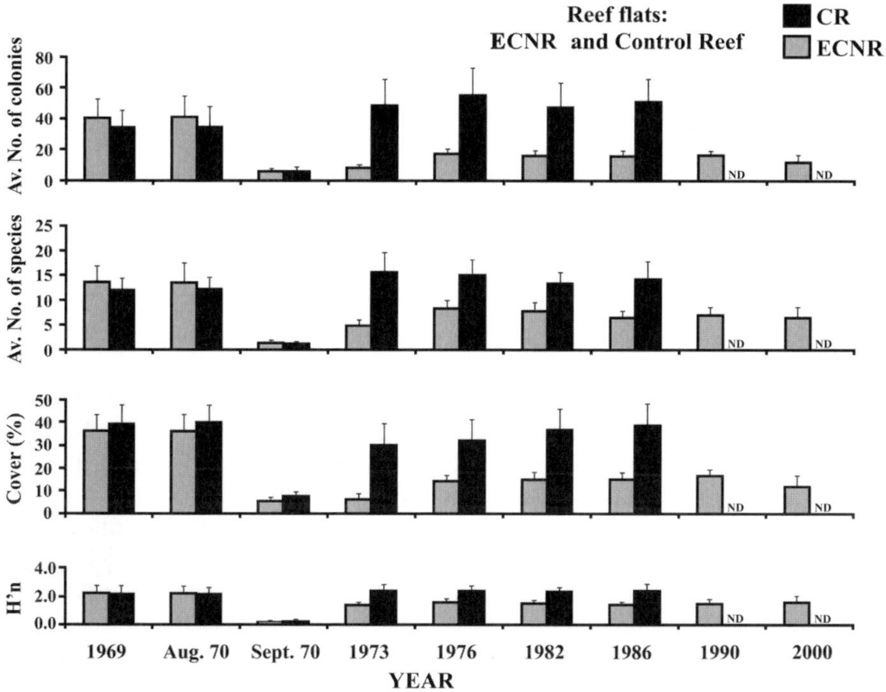

Fig. 1.4. The scleractinian coral community structure at the chronically polluted reef flat of the ECNR (Fig. 1.1) and a pristine control reef (CR), 5 km further south, were studied in detail by means of 10-m line transects in 1969 (prior to the unpredictable catastrophic low tide), during the low tide in 1970 and 3 years later in 1973 (see Loya 1976a for details). The same transects were resurveyed periodically until 1986 in both localities (in ECNR until 2000) for coral living cover, number of colonies and number of species, providing a long-term database of changes that took place in the coral community structure. No further monitoring was possible in the CR beyond1986 because of regional political reasons. (n=12 to 21 10-m transects; H'n = Shannon and Weaver's [1948] index of diversity. *ND* No data, *bars* represent ± SD

mass recruitment of the most abundant species (40% more colonies per transect in 1973 compared to 1969), in addition to the full regeneration of massive colonies that had suffered partial mortality during the low tide (Loya 1976a). Further monitoring throughout the years, in both sites, indicated that in 1976, 1982 and 1986 all monitored factors were significantly higher at the pristine CR compared to the ECNR (Loya 1990; Loya et al. 1999; Fig. 1.4).Unfortunately, further monitoring at the CR was not possible beyond 1986, due to regional political reasons. However, monitoring of the ECNR reef flat and deeper reef at 4 m depth ("Japanese Gardens") continues until today (see Sect. 1.4.1).

1.2.2 Coral Community Stability

Interpretation of stability in ecological systems has long been a matter of controversy among researchers (see selected literature: Margalef 1969; Woodwell and Smith 1969; Pimm 1991; Knowlton 1992; Holling 1996, 2001; Lewin 2000; Nystrom et al. 2000; Ives et al. 2003 and others). In this paper, I adopt, as a measure of community stability, Margalef's (1969) definition, i.e., the likelihood of a community to bounce back after external disturbance and return to its former state. Hence, there are marked differences between the ECNR and the CR. The ECNR has still not regained its former community structure and species diversity almost two decades after the low tide (Loya 1990), and unfortunately continues to remain poor in coral diversity and living cover to this day (Loya et al. 1999, this chapter). By contrast, the pristine CR showed remarkable resilience (i.e., capacity for self-repair) and only 3 years after severe perturbation, (the catastrophic low tide), it had returned to its former configuration. A significantly high correlation was found between the percentage contribution of the most common coral species to the total number of colonies and to the total living cover in the CR, before the low tide and 3 years later. Nonsignificant correlations were obtained when the same parameters were tested at the ECNR (see Loya 1976a for details). I concluded (Loya 1976a) that one of the differences between the effects of human impacts (chronic oil spills) and natural disturbances (extreme low tides) on coral reefs is the prolonged failure of a human-perturbed reef to return to its former state, whereas restoration of reef areas denuded by natural disturbances is a function of a relatively short time period. There have been, unfortunately, insufficient long-term quantitative coral reef community studies to test this hypothesis (for further discussion, see Johannes 1975; Loya 1976a; Brown and Howard 1985; Bythell et al. 1993; Dollar and Tribble 1993; Hughes 1994; Connell et al. 1997; Guzman and Cortes 2001). al. 2001).

1.2.3 Oil-Pollution Effects on Eilat Corals

Up to 1975 there was no conclusive evidence that oil floating above the reef damages stony corals (Johannes 1975). At that time, most of the studies of the effects of oil pollution on corals were either short-term incidental observations or laboratory experiments that did not reflect the effects on corals in nature. Our long-term quantitative studies in situ and in the laboratory have established the vulnerability and sensitivity of hermatypic corals to crude oil and oil components (Loya 1975, 1976a; Rinkevich and Loya 1977, 1979a; Loya and Rinkevich 1979, 1980).

As discussed before, clear differences have been shown in coral recovery between the chronically polluted ECNR (Fig. 1.2D–F) and the pristine CR. I have suggested (Loya 1975) that chronic oil pollution could damage coral communities, inhibiting coral recruitment in oil-polluted reefs, by (1) harming the reproductive system of corals, (2) decreasing the viability of coral larvae, or (3) changing some physical properties of the reef flat, thus interfering with normal settlement of coral larvae (any combination of these is also possible). We tested these hypotheses in a series of field and laboratory experiments summarized below.

We chose the coral *Stylophora pistillata* as our model experimental coral, since this is the most abundant coral in the Eilat reefs (Loya 1972), and its life history has been studied in detail (Loya 1976b–d; Rinkevich and Loya 1979b, c). Field experiments, observations, and histological studies on two populations of *S. pistillata* in a chronically oil-polluted reef near the oil terminals (ECNR, Fig. 1.1) and at the pristine CR revealed detrimental effects on the polluted corals. The coral population at the polluted reef showed a higher mortality rate of colonies, a smaller number of breeding colonies, a decrease in the number of female gonads per polyp, a smaller number of planulae produced per coral head (fecundity was four times higher in the CR than that in the ECNR), a decrease in the viability of planula larvae with increasing concentrations of oil and lower settlement success of larvae on artificial objects with the increase in oil concentrations (Rinkevich and Loya 1977).

Further studies on the effects of crude oil on the reproduction of *S. pistillata* in the laboratory, using large tanks with running seawater, supported the field results (Rinkevich and Loya 1979c). Large and mature colonies of *S. pistillata* were cut into halves at the beginning of the reproductive period; to avoid expected variation between different colonies, one was placed in a periodically oil-polluted tank, the other in a clean tank. After 2 months a significantly lower number of female gonads per polyp were recorded in 75% of the polluted halves, compared to the controls. An additional effect of low concentrations of crude oil on *S. pistillata* is the immediate indication of mouth opening followed by abortion of premature larvae. Since premature extrusion of planulae occurs during an oil spill, their chances of survival or successful settlement are very low (Loya and Rinkevich 1979).

Our field and laboratory studies on the effects of oil pollution on corals may explain the lack of recruitment of corals in the chronically polluted ECNR until 1979–1980. However, although no major oil spills have occurred in Eilat since 1980, to date the reef flats at the ECNR have not returned to their former community structure (Fig. 1.4). In the last two decades, other man-made perturbations (mainly eutrophication) have increased in severity (Sect. 1.4), preventing rehabilitation of the coral reefs of Eilat. Rather than speculating on this subject, I prefer, at this stage, to address the wider issue of the mechanisms that generate and maintain the diversity of organisms. Below, I present two hypotheses stemming from our long-term studies at the northern Gulf, to help interpret possible mechanisms that promote and maintain high coral diversity on pristine reefs of the northern Gulf.

1.3 Mechanisms That Generate and Maintain Coral Diversity in the Northern Gulf of Eilat/Aqaba

1.3.1 Hypothesis 1

Unpredictable midday low tides act as diversifying forces by preventing monopolization of the reef flat by individual or a few competitively superior species.

Because space for settlement and development is one of the most important limiting resources on coral reefs, overlap in the utilization of space may result in acute competition among coral populations (Lang 1973; Grigg and Maragos 1974; Porter 1974; Jackson and Buss 1975; Connell 1976, 1978; Loya 1976a; Jackson 1977; Buss and Jackson 1979; Lang and Chornesky 1990; Hughes 1994; Abelson and Loya 1999; Hughes and Connell 1999).

Resource monopolization by corals may take place through competitive interactions such as extracoelentric digestion, whereby the mesenterial digestive filaments of a dominant species extend onto the living tissues of an adjacent subdominant and destroy it (Lang 1973; Richardson et al. 1979), rapid growth (Connell 1973), overtopping morphology (Porter 1974), or allelopathic effects (Jackson and Buss 1975; Sammarco et al. 1983). Abilities in these competitive mechanisms are species-specific, and in areas of high densities on the reef have been shown to affect patterns of coral abundance and distribution (Lang 1973; Grigg and Maragos 1974; Porter 1974; Connell 1976, 1978; Loya 1976a; Porter et al. 1982; Sheppard 1985).

Coral mass mortality on reefs may result from a variety of causes such as (below selected references) mechanical destruction during tropical storms (Woodley et al. 1981; Rogers 1993), abnormally low or high seawater temperatures (Glynn 1981; Jokiel and Coles 1990), sedimentation (Loya 1976e; Rogers 1990) exposure to air during midday air temperature maxima (Glynn 1976; Loya 1976a), bleaching (Glynn 1993; Brown 1997; Hoegh-Guldberg 1999; Ostrander et al. 2000; Loya et al. 2001), diseases (Kushmaro et al. 1996; Harvell et al. 1999), *Acanthaster* predation (Lourey et al. 2000) and anthropogenic activities (Brown and Howard 1985; Nystrom et al. 2000). The period of time required for recruitment and recovery of reefs after such disturbances varies from very short (2 years; Shinn 1972) to prolonged (50 years or more; Grigg and Maragos 1974; Pearson 1981) and largely depends on local settings.

The role of disturbance on community structure and species diversity of natural communities has been documented and discussed by many investigators. (For selected bibliography and reviews, see Dayton 1971; Connell 1978; Paine and Levin 1981; Miller 1982; Porter et al. 1982; Sousa 1984; Pickett and White 1985; Knowlton 1992; Karlson and Hurd 1993; Hughes and Connell 1999 Nystrom et al. 2000; Holling 2001 and others.)

Grigg and Maragos (1974) suggested a model for coral community succession based on patterns of recruitment on submerged historic lava flows in Hawaii. By analyzing data sets from progressively older flows, they found that diversity at first increased and then gradually decreased as more and more species became established. They hypothesized that the decline in diversity in older flows was due to space limitation and competitive exclusion by dominant species, a process that took over 50 years. Glynn (1976) attributed the diversifying effect to recurrent and extreme tidal exposures of reef flat corals off the Pacific coast of Panama. Tide-induced mortality of pocilloporid corals, which are prime-space monopolists at that site, resulted in increased coral species diversity.

I have suggested (Loya 1976a) that the unpredictable low tides at Eilat act as a diversifying force in a way similar to storm and swell damage in Hawaii (Grigg and Maragos 1974; Grigg 1983), and the extreme tidal exposures of reef flats off the Pacific coast of Panama (Glynn 1976). Connell's (1978) 'intermediate-disturbance' hypothesis, stating that the highest diversity of natural communities will be maintained in areas suffering intermediate disturbances on the scales of frequency and intensity, fits these examples.

The higher diversity recorded at the pristine CR 3 years after the low tide, compared to the diversity at the CR prior to the devastation, may reflect a situation in which diversity continues to increase in time after a catastrophe until space for coral settlement becomes limiting and competitive interactions between species may cause a decline in diversity. The results obtained for the average number of species recorded on the reef flat of the CR between 1970 and 1986 support this pattern (Fig. 1.4). The time interval required to reach "species equilibrium" (sensu Wilson 1969) is quite short (approx. 5–7 years). The time interval for full recovery and for competitive interactions to take place is expected to be much longer and more variable between different reefs depending on local conditions (15–30 years or more; see Loya 1976a).

The frequency of extremely low tides at the Gulf of Eilat/Aqaba seems to be such that full recovery is never reached on the reef flats, and interspecific competition is not carried out to completion. Monopolization of the reef flats by competitively superior species is interrupted, providing an opportunity for more species to colonize vacant spaces. Hence, the exceptionally high diversity recorded on pristine reef flats in the northern Gulf probably reflects early successional stages.

1.3.2 Hypothesis 2

Temporal reproductive isolation (i.e., discrete breeding period) and intraspecific synchronization of spawning of northern Red Sea corals generates high diversity by preventing hybridization (i.e., minimizing gametogenic wastage) and decreasing interspecific competition.

Information on reproductive patterns in corals has progressively accumulated in the last 20 years (see Fadlallah 1983; Harrison et al. 1984; Richmond and Jokiel 1984; Shlesinger and Loya 1985; Shlesinger et al. 1998; Babcock et al. 1986; Szmant 1986; Harrison and Wallace 1990; Richmond and Hunter 1990;

Ward and Harrison 2000 for reviews and selected regional patterns). However, no study has attributed coral community structure and species diversity in relation to coral reproductive patterns. In trying to explain the high within-habitat diversity of hermatypic corals on reef flats in the northern Gulf, my second hypothesis relates to the phenomenon of temporal reproductive isolation exhibited by the most abundant coral species at Eilat (Shlesinger and Loya 1985; Shlesinger et al. 1998).

We have examined the reproductive patterns of 13 ecologically important coral species at Eilat (Fig. 1.5; Shlesinger and Loya 1985). Although these species comprise only 13% of the total known species there, they are among the most abundant, contributing approximately 60–70% of the total living cover of coral communities on the reef flats (Loya 1972). The major reproductive activities of these species (planula shedding or gamete spawning) occur in different seasons, different months, or different lunar phases within the same month (Fig. 1.5). A similar conclusion was obtained for the reproductive patterns of an additional 11 coral species studied at a later stage (Shlesinger et al. 1998). By contrast, many corals of the GBR of Australia are synchronous multispecific spawners (Harrison et al. 1984; Babcock et al. 1986, Harrison and Wallace 1990). The coral mass spawning event takes place at the GBR during a few nights in late spring, between the full and last quarter of the moon.

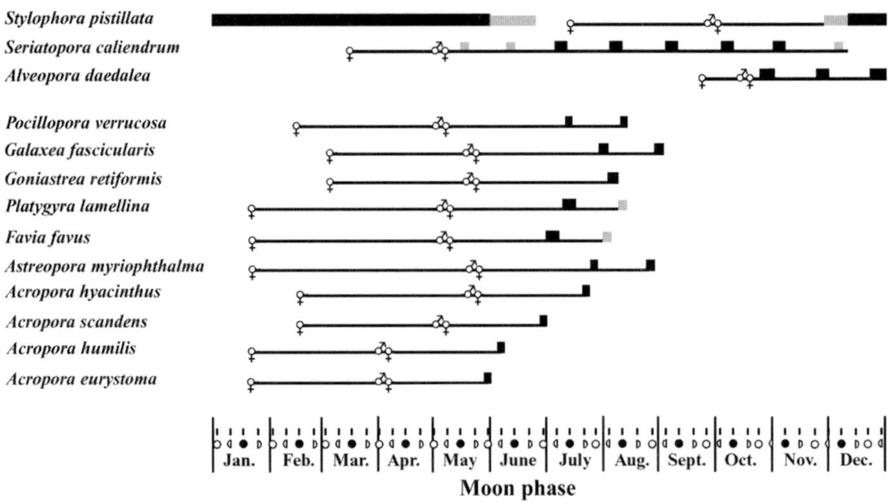

Fig. 1.5. Reproductive patterns of 13 scleractinian corals at Eilat in 1980 (after Shlesinger and Loya 1985). The first three species are brooders and the rest are broadcasting spawners. The results presented are in accordance with the lunar calendar, as follows: *black circle* new moon, *half circle to the right* first quarter, *blank circle* full moon, *half circle to the left* last quarter. The period of gonad development is represented by a *solid line* with indications of the onset of oogenesis (female sign), and spermatogenesis (male sign). Further development of both gonads is indicated by the line following the male and female signs. *Black bars* represent massive spawning (90–100% of the population), and *gray bars* represent sporadic spawning (10–20% of the population)

I have mentioned that space is a major limiting resource for settlement, growth, and development of reef corals. The reproductive activity of the vast majority of coral species at Eilat coincides with the seasonal disappearance of benthic algae, a major competitor for space on the reef flats of Eilat (i.e., mainly during June to September; Benayahu and Loya 1977a, b, 1981; Fig. 1.6C–F). Various species of algae progressively decline in abundance during the summer months, creating more space for coral settlement. During winter, algal cover on the reef flats is very high and space for settlement is scarce (Sect. 1.4; Fig. 1.6E). At any given time, space availability for settlement on the reef flats is more predictable during summer than winter. Thus, summer breeders among the corals maximize their reproductive success by concentrating their reproductive effort into a relatively short period of time (Fig. 1.5). By contrast, winter breeders like *S. pistillata* adopt an opposite strategy, spreading the risks of reproduction over a long period of time (December to May in the case of *S. pistillata*). During the winter, the availability of space in time is random, due to occasional denudation of small areas by winter storms (Fig. 1.6B) and, to a limited degree, by grazers. Therefore, a long reproductive period is advantageous for winter breeders.

Intraspecific synchronization of spawning and a discrete breeding period may be advantageous to each coral species in the northern Red Sea, not only by reducing gametogenic wastage and preventing hybridization, but also by reducing interspecific competition among corals and between corals and algae. Thus, temporal reproductive isolation may act as a mechanism that generates high coral diversity in the northern Red Sea. By contrast, the mass spawning exhibited by the GBR corals probably increases interspecific competition among them. However, whether this may be a cause for the relatively lower within-habitat diversity of corals on the reef flats of the GBR compared to that in the northern Red Sea remains to be shown.

The two hypotheses presented here are not mutually exclusive. The net result of high coral diversity on pristine reefs along the Gulf of Eilat may be a cumulative synergistic effect of external abiotic factors, such as intermediate disturbance on reef flats (sensu Connell 1978), through exposure to air during midday air temperature maxima, and biotic factors, such as temporal reproductive isolation. (reef flats and deeper reef regions).

While these hypotheses may elucidate the exceptionally high diversity of pristine reefs in the northern Gulf, the chronic disturbance of the coral reefs of Eilat, caused by a variety of human activities, are undoubtedly the major cause for their continued deterioration and critical state of health nowadays.

A
B
C
D
E
F
G
H

1.4 Man's Destructive Activities: Acute Degradation of the Coral Reefs of Eilat Due to Eutrophication

1.4.1 Acute Degradation of Coral Diversity and Living Cover at Eilat During 1986-2000 Due to Eutrophication by Urban Sewage and Net Pen Fish Farms

During the 1980s, the city of Eilat went through rapid development and population growth (20,000 people in 1980; 40,000 in 1990; 54,000 in 1995), which increased human stress on the reefs. The decline of the reefs of Eilat at that time was probably due to a combination of anthropogenic disturbances (i.e., phosphate dust, urban sewage, siltation, tourist diving activities and port ballast water). However, while most of these harmful activities have been minimized by strict environmental enforcement, the sewage of Eilat continued to flow to the sea (Fig. 1.3G–I) and stopped only after a sewage treatment plant was built north of the city in 1995. Atkinson et al. (2001) estimated that in 1980 the total annual nitrogen and phosphate release into the northern Gulf of Eilat/Aqaba originating from Eilat's urban sewage was 5×10^6 mol N and 0.24×10^6 mol P. These quantities doubled in 1990, with the doubling of the population number, and stopped in 1995 after the sewage treatment plant was built. Since coral reefs flourish in oligotrophic waters, eutrophication of reef waters is considered to be one of the main causes for their decline (see literature in Lapointe 1997; Roberts et al. 2002). Although there is no direct evidence, it is most likely that the nutrient enrichment caused by Eilat's sewage as well as the activities of Aqaba's and Eilat's phosphate ports were among the major causes for the decline of coral populations at the reefs of Eilat until 1995 (Fig. 1.3H, I).

At that time, intensive net-pen fish farms (Figs. 1.1, 1.6A), situated at the northernmost section of the Gulf of Eilat ("the North Beach") have increased their fish production exponentially. The fish farms utilize large fish cages, 12 m in diameter and 10 m in length, densely populated mainly with the gilthead sea bream *Sparus aurata* (Fig. 1.6A, insert). In 1991, the annual fish yield was

◀ **Fig. 1.6.** **A** The intensive net-pen fish farms situated at the northernmost section of the Gulf (Fig. 1.1) utilize large fish cages, mostly for the gilthead sea bream *Sparus aurata* (*insert*). **B** A southern storm detached vast amounts of macro-algae from the vicinity of the fish farm onto the northern shoreline. **C, D** The seafloor in the vicinity of the fish cages forms an anaerobic benthic habitat which supports the growth of bacterial mats and sulfide-resistant populations. **E** Vertical mixing of the water column during winter results in "algal blooms" and competitive exclusion of stony corals by algae. **F** *Stylophora pistillata*: enhanced eutrophication of the reefs encourages proliferation of filamentous algae overgrowing and smothering corals at the northern Gulf of Eilat. **G** Transplanted *Favia favus* colonies on inverted plastic boxes near the fish cages site were overgrown by algae and died within a year. **H** *Montipora lobulata*: in summers 2002 and 2003 sporadic coral bleaching was observed for the first time, in some coral species, at the northern Gulf of Eilat/Aqaba

200 tons, it grew to 700 tons in 1995 and 2300 tons in 2000 (Atkinson et al. 2001). The fish are nourished with a total of 4200 tons/year of protein-rich added food (Gordin 2000). Waste food and fish feces deposited in the vicinity of the cages, or in the direction of the prevailing currents, result in substantial accumulation of organic material (OM) on the seafloor (Fig. 1.6C, D). Such OM enrichment leads to sediment anoxia, sulfate reduction and drastic changes in the sediment pore-water chemistry (Holmer and Kristensen 1992). This anaerobic benthic habitat supports the growth of widespread bacterial mats and sulfide-resistant populations. Part of the sinking organic particles are carried a large distance southward by sea currents (Abelson et al. 1999).

After the sewage treatment plant had been built and the flow of sewage into the sea had been halted (in 1995), the fish cages remained the major source of organic enrichment (N and P) to the northern Gulf, releasing an estimated 18×10^6 mol/year N and 2×10^6 mol/year P (Atkinson et al. 2001). Gordin (2000) estimated that ca. 68% of the nitrogen and 27% of the phosphorus fed to the fish are subsequently excreted and released into the water in a dissolved form. In addition, 10% of N and 44% of P are excreted as solid excretions; hence, the annual amount of nutrients released into the surrounding water totals ca. 240 tons of N and 40 tons of P. (i.e., the input of the fish farms alone contributes 97% of N and 66% of P to the northern Gulf of Eilat. Thus, after 1995, the fish farms "replaced" Eilat's sewage, as the major source for eutrophication of the reefs continuing the mass killing of Eilat's corals (Fig. 1.7). Indeed, long term monitoring of the coral community structure at the "Japanese Gardens" site of the ECNR (has been closed to tourist diving activity in the last 8 years), at 4m depth, reveals a decrease of 76% and 73% in coral cover and abundance, re-

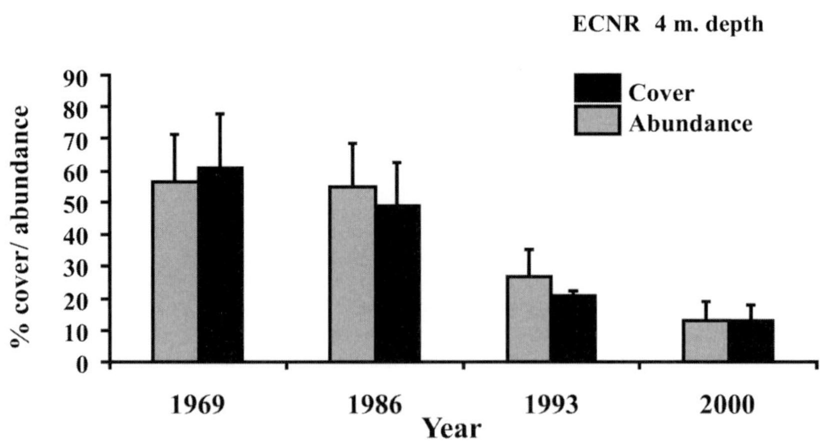

Fig. 1.7. Long-term monitoring of the coral community structure at the "Japanese Gardens" site, ECNR (4-m depth). The percent living cover of stony corals decreased in the last 15 years by 76% and the number of coral colonies decreased by 73%. Data for 1969 from Loya (1972). (n=9–21 transects, each 10 m in length, *bars* represent ± SD)

spectively. If one further examines this time period between 1986-1993 (during which Eilat's sewage was flowing to the sea but stopped in 1995)and between 1993-2000 (during which the fish cages were, and still are, the major nutrification source of the northern Gulf of Eilat/Aqaba), it is most probable that Eilat's sewage was the major cause for the decrease of 51% and 57% in coral living cover and abundance, respectively, during 1986-1993, while the fish farm industry, which grew exponentially in their fish production during 1993-2000 "replaced" Eilat's sewage contribution to nutrification of the northern Gulf, causing a further decrease of 62% and 49% in coral living cover and abundance, respectively (Fig. 1.7).

1.4.2 Eutrophication Effects on the Reef's Food Chain, Metabolism, Interspecific Interactions and Genetic Structure of Coral Populations

In 1998, the waters of the northern Gulf of Aqaba/Eilat contained at over 500-m depth a max. 4 μmol/l nitrate and 0.3 μmol/l phosphate. In 2000, they were nutrient-enriched to 7 and 0.5 μmol/l, respectively. This dramatic increase results essentially from the fish farm nutrient pollution (Lazar et al. 2000). In 1999–2000, surplus nitrate accumulating in >500-m depth equaled 200 tons; during this period the fish farms polluted the water column with 500 tons of nitrates available for primary production (i.e., N accumulating in deep water equalled ca. 40% of N originating from the fish farms; Lazar et al. 2000).

As a result of nutrient enrichment the phytoplankton, which multiplies rapidly (can double every 1–2 days), is eaten by zooplankton and the latter is preyed upon by fish. Thus, the nutrients that were passed on to higher trophic levels in the food chain, are distributed over wide areas in the northern Gulf. At each stage in this process, part of the nutrients is removed from the living system as solid excretions that drop to the depths of the sea. Finally, a surplus of nutrients is accumulating in the depths and the seafloor of the northern Gulf (Atkinson et al. 2001). During winter, cold surface waters sink to deeper layers creating vertical mixing, "pumping" nutrient-rich waters to the surface, resulting in "algal blooms" (Genin et al. 1995; Fig. 1.6B) and competitive exclusion of stony corals by algae (Fig. 1.6E, F). Genin et al. (1995) reported that the deep vertical mixing during winter 1992 resulted in excessive eutrophication of the upper illuminated reef layers, generating wide-scale algal blooms which killed 25% of the corals in Eilat.

Silverman et al. (Chap. 21, this Vol.) proposed an index of reef metabolism (RM, i.e., the proportion of reef calcification to P/R ratio) as an index for the reef's health. High RM values denote a healthy reef, while low RM values denote deteriorating reefs. The RM measured in 1991 at the ECNR was ca. 200 μM C m^{-2} day^{-1}, similar to values reported from pristine Indopacific reefs during the 1970s and 1980s. In winter 2000, RM decreased in the ECNR to a value of ca. 6 μM C m^{-2} day^{-1} (i.e., five- to sixfold lower than the values obtained in winter 1998 and over 20-fold lower than the values obtained in 1991 from the same reef). This means that the reef growth at the ECNR virtually stopped. The authors attributed these alarming results to the extreme increase in eutrophication of the Gulf in recent years.

A recent study of the genetic structure of *S. pistillata* populations in four distinct sites at the northern Gulf of Eilat/Aqaba (Zvuloni (2003) adds to our great concern for the persistent detrimental effects of the fish farms on coral populations. Zvuloni (2003) found that the genetic variation within *S. pistillata* coral populations was much greater (90.2%) than among populations (9.8%), suggesting a high level of connectivity among the coral populations in the region. Nevertheless, a significant variation was found between the adult *S. pistillata* population in the 'Peace Lagoon' (located ca. 400 m west of the fish farms) and all other adult *S. pistillata* populations in three further away southern sites. Moreover, the genetic variation within young recruits of this species in the Peace Lagoon was found to be significantly larger than within adult coral colonies at the same site. He suggested that a selection process, taking place immediately after settlement of *S. pistillata* juvenile spat, permitted only individuals from particular genotypes to survive there. Furthermore, he attributed these results to the higher pollution levels reported by other studies in the vicinity of the Peace Lagoon study site.

1.4.3 Detrimental Effects of Eutrophication on Coral Reproduction

Recently, we have assessed the impact of the fish farms on the survival of two branching corals (*Stylophora pistillata* and *Acropora variabilis*) and two massive and encrusting species (*Favia favus* and *Coscinarea monile* respectively, Fig. 1.6G) and the impact of the fish farm on the reproduction of *S. pistillata* (see details in Loya et al. 2004). For the reproduction study, we collected 40 mature *S. pistillata* colonies (larger than 20 cm in diameter) from a site mid way between the IUI and the fish farm site (at 15 m depth) and transplanted 20 colonies at each of the two sites: (1) 200 m west of the fish farm cages at a depth of 19 m, and (2) at the Interuniversity Institute (IUI) reef at the same depth. Over the next two consecutive years, histological studies of the reproductive effort of the experimental corals showed that there were more oocytes per polyp in colonies sampled from the fish farm site than at the IUI reference site (Fig. 1.8), but they remained significantly smaller than the IUI oocytes throughout the reproductive season. Monitoring changes in oocyte size frequency distribution in the colonies throughout the reproductive cycle, we found that polyps from the IUI colonies contained significantly more oocytes, which reached the mature size of over 200 μm, than those from the colonies near the cages. Moreover, the percentage of polyps with planulae was significantly higher in colonies from the IUI site than from those under the cages (Figs. 1.8, 1.9). In comparing number of planulae per polyp in the experimental colonies transplanted at both sites (Fig. 1.9), we found significantly larger numbers of polyps with planulae both in 2001 and 2002 in the IUI reference site than near the fish cages (Loya et al. 2004).

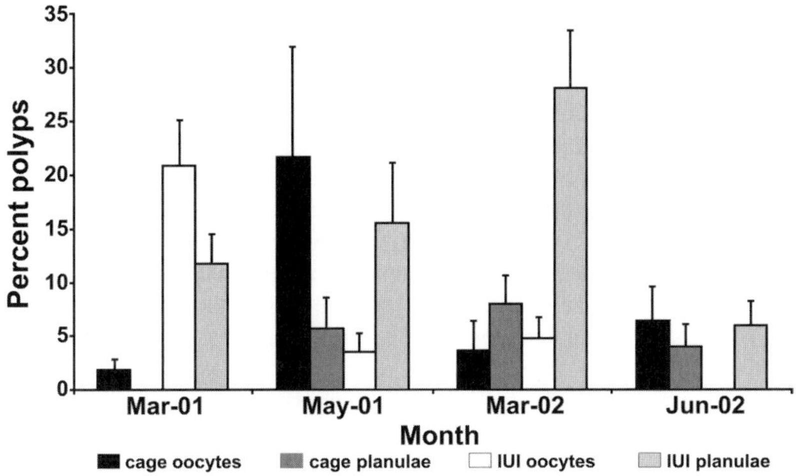

Fig. 1.8. *Stylophora pistillata*: percent polyps with oocytes and planulae (± SE) in mature experimental colonies transplanted in the vicinity of the mariculture fish farm (Fig. 1.1) and at a reference site across from the Interuniversity Institute of Eilat (IUI), during 2000–2001 (after Loya et al. 2004). (For 2001: March n=11 colonies 534 polyps, May n=11 colonies, 439 polyps, for 2002: March n=11 colonies, 840 polyps, June n=10 colonies, 577 polyps)

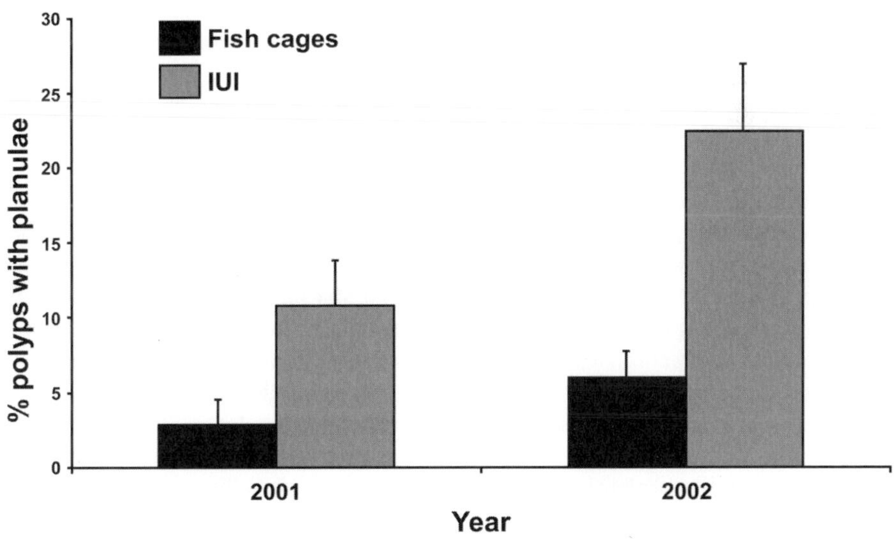

Fig. 1.9. *Stylophora pistillata*: fecundity (percent polyps with planulae, ± SE) of mature experimental colonies transplanted at the vicinity of the mariculture fish farm (Fig. 1.1) and at a reference site across from the IUI during 2000–2001 (after Loya et al. 2004)

For comparative studies on coral survival, we transplanted in November 2000, massive *Favia favus* and encrusting *Coscinarea monile* colonies (Fig. 1.6G) as well as branching mature *S. pistillata* and *A. variabilis* colonies (ten colonies of each species, at each of the two sites). The colonies were collected from the same "midway" site. Most transplanted branching colonies in both sites survived and still continue to grow well (ca. 3 years after transplantation). In contrast, almost all transplanted massive and encrusting colonies at the fish farm site were overgrown by algae and died within a year, while those at the IUI are still growing well today. These results are not surprising in view of the fact that both *Acropora variabilis* and *S. pistillata* are more abundant than any other coral species in the vicinity of the fish farms. Although these species are also abundant on reefs further south, they are among the very few coral species that may be found growing on artificial objects in the vicinity of the fish farms. Both *A. variabilis* and especially *S. pistillata* are typical "weedy" species. Among other attributes, they are the first to settle where new space becomes available, grow quickly, mature sexually at a small size (age) and have wide genetic variability. These characteristics may explain their relative survival success in the vicinity of the fish cages compared with other more specialized species, such as the *F. favus* and *C. monile*, which did not survive in this environment.

The conclusions derived from our study (Loyat et al. 2004) are in variance with the conclusions of a similar study that was carried out at the same sites (for details see Bongiorni et al. 2003; Loya and Kramarsky-Winter 2003; Rinkevich et al. 2003). In contrast to the interpretations and conclusions of Bongiorni et al. (2003), we suggest that the results presented in their paper, together with our own results, actually attest to a severe reduction in the reproductive success of coral colonies near the fish farms (Loya et al. 2004). This leads to the inference that the conditions at this site are harmful to these corals. Moreover, Abelson et al. (1999), demonstrated that particulate matter from point sources at Eilat's "North Beach" (close to the fish farms) reached coral reefs 8 km further south, both as bed-load and suspension-load particles. Indeed, much of the nutrients originating from the fish farms were found to reach the reefs further south, accumulating on the seafloor at depths >500 m (Lazar et al. 2000). Hence, these reefs are affected as well, especially during the vertical mixing events of the water column during cold winters (section 1.4.2). It is thus possible that the continuous eutrophication of the northern Gulf of Eilat/Aqaba has marked deleterious effects on coral reproductive effort even in areas distant from the fish farm site. Indeed, Zakai and Furman-Chadwick (pers. comm.) have recorded marked decreases in planulation of the coral *S. pistillata* in the ECNR and the IUI reef in the last 5 years, compared with known records from the mid-1970s (Rinkevich and Loya 1979a, b). We, therefore, concluded that nutrients released from the fish farms are the most probable and direct cause of harmful effects to *S. pistillata*'s reproduction, and most likely to the reproduction of other coral species, as reported from other reefs around the world (Tomascik and Sander 1987; Ward and Harrison 2000; Harrison and Ward 2001).

1.4.4
Coral Diseases and Syndromes

In the last 3 years (2000–2003) we have witnessed further deterioration of corals in the northern Gulf, due to coral diseases and syndromes infecting Eilat's corals (Fig. 1.10). Coral diseases are among the most recent in a series of threats affecting coral reefs (e.g., eutrophication, coral bleaching, oil and heavy metals pollution, coastal development, increased sedimentation and over-exploitation of marine species) that is challenging the resilience of coral reef communities. Since the first report of coral disease by Antonius in 1973, the rate of discovery of new diseases has increased dramatically with ca. 30 coral diseases now described (Green and Bruckner 2000; Weil 2003). Although coral disease is emerging as a serious cause of coral reef deterioration in many reefs around the world, at present we know very little about the ecology or pathology of coral diseases in Red Sea reefs in general (Antonius and Riegl 1997), and Eilat's reefs in particular (Ben-Haim et al. 2003). Figure 1.10A–D illustrates the fast spreading rate in which an unidentified disease affects a *Goniastrea* sp. colony in Eilat. Additional examples of field appearance of diseases and syndromes infecting Eilat's corals are illustrated in Fig. 1.10E–J. Increasing epizootiological knowledge of coral diseases (i.e., distribution, host range, prevalence and incidence at the species and community levels, spatial and temporal variability, epidemiology, etc.) will also help to identify the origins and reservoirs of pathogens and the vectors involved in disease transmission, all of which are important for building capacity in the management of potential disease outbreaks. At present, we are compiling a base-line data set on the presence, distribution, abundance and pathology of scleractinian coral diseases in the northern Gulf of Eilat/Aqaba (Sulam et al., Barash et al., unpubl. data).

During the summers of the last 2 years (2002/2003), we have also observed for the first time in Eilat, sporadic coral bleaching (Fig. 1.6F). The bleaching or paling of corals, through the loss of their zooxanthellae and/or their pigments, is a global phenomenon that is probably linked to global climate change and increasing ocean temperatures (Glynn 1993; Brown 1997; Hoegh-Guldberg 1999). Recent reports of wide-scale coral bleaching from all the major tropical oceans of the world during 1997–1998, and the unprecedented coral mortality that resulted (Wilkinson 1998, 2000), are a major concern among scientists and resource managers. In 1998, more than 90% of shallow corals were killed on most Indian Ocean reefs. High sea surface temperature (SST) was a primary cause, acting directly or by interacting with other factors. Even with their marked temperature acclimatization, most corals in the Arabian Gulf were killed by the 1998 peak SSTs (Sheppard 2003 and references therein). Until today, the northern Red Sea has remained relatively unaffected. However, the phenomenon of global warming, resulting in an increase in the frequency and intensity of world-wide coral bleaching events (Hoegh-Guldberg 1999), sets the ground for additional severe threats to our reefs, which is superimposed on their present fragile situation. In case of future abnormal increases in SSTs in

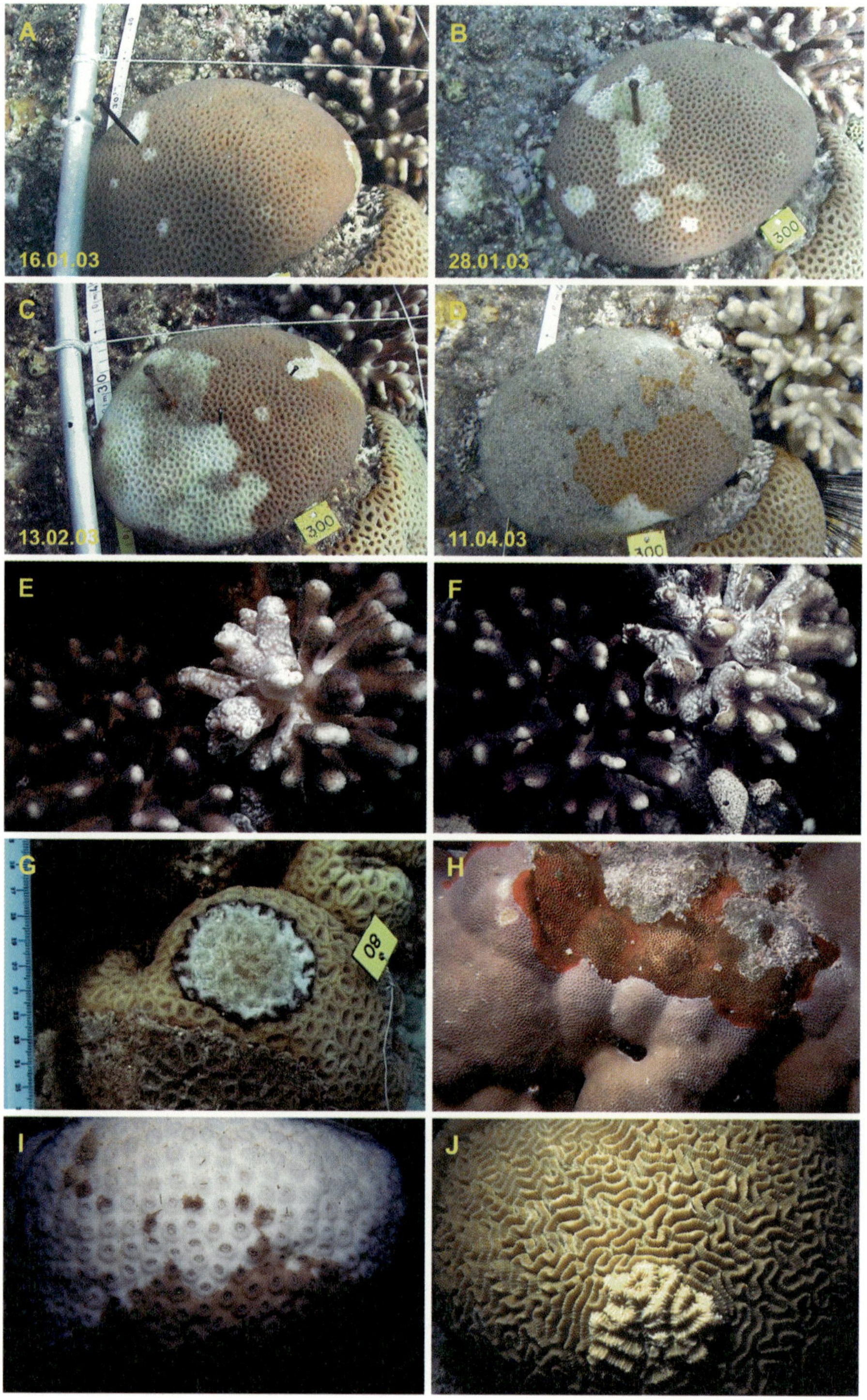
A
16.01.03
B
28.01.03
300
C
13.02.03
300
D
11.04.03
E
F
G
80
H
I
J

the northern Gulf of Eilat, which might result in mass coral bleaching, the chances of coral communities recovering are minute, in view of their critically frail state of health.

1.4.5 Local Extinction or Decrease in Abundance of Organisms Associated with Corals

The extreme deterioration in coral abundance and living cover at the northern Gulf of Eilat might have directly or indirectly caused the local extinction or severe decrease in abundance of other reef organisms, which live in close association with the corals (Fig. 1.11). Genin et al. (2001) provide a partial list of such species. It is noteworthy that echinoderms are among the most conspicuous group of such organisms (Fig. 1.11A–E). Some examples include: the sea star *Choriaster grannulatus* (Fig. 1.11A), the crinoids *Lamprometra klunzingeri* (Fig. 1.11B) and *Heterometra savignyi*, which were very abundant in the past in the ECNR (up to 50 m^2; Fishelson, pers. comm.) are virtually extinct there, the sea urchins *Asthenosoma varium*, *Heterocentrotus mammilatus*, *Lovenia elongata* (Fig. 1.11C–E, respectively), and the basket star *Astroboa nuda* (Fig. 1.11F). Distinct examples of organisms from other phyla include the hydroid coral *Millepora platyphylla* (Fig. 1.10G), the scyphozoan medusa *Cassiopea andromeda*, and the soft corals *Litophyton arboreum* and *Clavularia hamra* (Fig. 1.11H–J, respectively).

1.4.6 Recent Assessments of the Effects of the Fish Farms on the Coral Reefs of Eilat

The critical state of health of the coral reefs of Eilat (Figs. 1.5–1.11) led 14 marine scientists in Israel to express to the Israeli Minister of the Environment their growing concern with respect to the grave fate of the reefs (Genin et al. 2001). This document (known as the "Scientist's Document") lists the major detrimental effects and potential threats that the fish cage mariculture industry poses to the Gulf. There is a wide consensus that the coral reefs of Eilat are deteriorating. However, acute controversy exists between the mariculture industry vs. conser-

◄ **Fig. 1.10.** Field appearance of diseases and syndromes infecting Eilat's scleractinian corals. **A** An unidentified disease is noticeable on the surface of *Goniastrea* sp. The colony was tagged (No. 300) on 16 January2003 and a stainless steel nail marked the edge of the white area devoid of live tissue. **B, C** The disease spread very quickly and **D** within 3 months, 80% of the coral was dead and its surface covered by sediment. **E** An unidentified (fungal?) disease has recently been observed on *Stylophora pistillata* colonies. **F** The same colony 1 month later. **G** Black band disease (BBD) in *Favia favus*. A mat of filamentous cyanobacterium *Phormidium corallyticum* causes the black coloration of the black band bordering the healthy tissue. **H** *Porites lobata*. **I** Unidentified disease (patchy white spots) in *Echinopora gemmacea*. **J** Tumorous skeletal growth in the brain coral *Platygyra lamellina*

A
B
C
D
E
F
G
H
I
J

vation organizations, and a large proportion of leading Israeli marine scientists, concerning the role of the fish farms in the decline of the reefs. Hence, an International Expert Team (IET) was appointed by the Israeli Government to evaluate the pollution sources in the northern Gulf of Aqaba/Eilat and recommend steps to ameliorate the Gulf's ecological situation (Atkinson et al. 2001). Both documents evaluated published and unpublished reports on this problem, and pointed out potential harmful effects that the fish farms may be or already are inflicting on the reefs (i.e., persistent eutrophication of the reef, organic enrichment of the sea bottom in the "North Beach", a point source for the initial outburst of pathogenic diseases in wild coral reef fish and the introduction of nonindigenous fish species into the Gulf's ecosystem). It is not within the scope of this paper to go into the details of these documents. It will suffice to mention that the "Scientist's Document" recommended relocation of the mariculture operation into inland ponds. The IET document presented a list of recommendations that essentially advocated a risk management approach: recognizing that the fish farms were the unequivocal and foremost source of the persistent influx of nutrients to the Gulf, they recommended ways to substantially reduce the nitrogen loading from this source. Since then, 3 years have elapsed, but virtually non of the recommendations made to ameliorate the Gulf's ecological situation in either document have been carried out and the severe situation in the reef worsens every day.

1.5 Conclusions and a Word of Warning

At present, the coral reefs of Eilat are severely damaged and exist in a critical state. The alarming records of coral mortality in recent years, the increasing percentages of corals affected by pathogenic diseases and the first accounts of coral bleaching observed in the last 2 years, are strong indications of the fragile state of health of the coral populations in Eilat. I should perhaps emphasize the fact that the coral reefs of Eilat are no more than ca. 4 km in length. However, their uniqueness has been recognized worldwide due to their extraordinary biodiversity and beauty. There is no doubt in my mind that if a similar situation would have occurred on the Great Barrier Reef of Australia (2500 km), or for that matter, any other country that values its natural resources, there would have been immediate action to remedy the situation. The coral reefs of Eilat are a national as well as a global treasure. Therefore, even if there is the slightest doubt concerning the detrimental effect of eutrophication on reef corals, we

◀ **Fig. 1.11.** Examples of abundant species in the past at the northern Gulf of Eilat/Aqaba, which are rare or locally extinct today. **A** The sea star *Choriaster grannulatus*, **B** the feather star *Lamprometra klunzingeri*, **C** the sea urchin *Asthenosoma varium*, **D** the pencil sea urchin *Heterocentrotus mammilatus*, **E** The tiny sea urchin *Lovenia elongata* (maximum length 5 cm), **F** the basket star *Astroboa nuda*, **G** The hydroid fire coral *Millepora platyphylla*, **H** The medusa *Cassiopea andromeda*, **I** the soft coral *Litophyton arboreum*, **J** the soft coral *Clavularia hamra* cannot afford to let our reefs wait for "unequivocal scientific proof". First-rate scientific fieldwork in coral reefs requires much effort, long time and large budgets. To date, sufficient evidence has accumulated from research performed by leading marine scientists in Israel to indicate that environmental eutrophication caused by Eilat's sewage in the past (until 1995) and in recent years by the fish farms is a major cause of the deterioration of the reefs. While most other harmful man-made activities have been minimized by strict environmental enforcement (see 1.2.3., 1.4.1), the persistent influx of nutrients from the fish farms continues to pollute the Gulf without any control. The phenomenon of global warming sets the ground for additional severe threats to our reefs, which is superimposed on their critically frail state of health. In case of future abnormal increases in SSTs in the northern Gulf, which might result in mass coral bleaching, the chances of Eilat's coral communities to recover are minute, in view of their critically frail state of health. Our reefs are dying at an alarmingly fast rate. Any further bureaucratic delays imposed by the government concerning the fate of the fish farming mariculture at Eilat will condemn the reefs to their final destruction. I strongly advocate adoption of the major recommendation given in the "Scientist's Document" (Genin et al. 2001), i.e., immediate relocation of the fish farm operation to inland ponds. Continuation of the present persistent influx of nutrients originating from the fish farms constitutes a major cause for the continued deterioration of the coral reefs of Eilat and comprises a serious threat to their very existence. If eutrophication of the reefs is not stopped immediately, the total destruction and final collapse of the unique coral reefs of Eilat are a certainty. In the past, this process might have taken 15–20 years. However, in view of the ever-increasing demands we impose on our reefs, in addition to the problem of global warming, the process of the diminution of our reefs will probably be much faster. We cannot and should not let this happen. In their present fragile state of health, the only chance for restoration of the reefs is instant and extreme protection measures against all man-made perturbations.

Acknowledgments. It is a pleasure to thank my past students and present colleagues, who, throughout the years, participated and contributed to different parts of the research reviewed in this paper: A. Abelson, Y. Benayahu, I. Brickner, M. Ilan, R. Klein, E. Kramarsky-Winter, A. Kushmaro, H. Lubinevsky, O. Mokady, B. Rinkevich, M. Rosenfeld, Y. Schlesinger and M. Shpiegel. I am especially indebted to the Interuniversity Institute of Eilat (IUI) for being my "second home institute", for the continued support and hospitality during more than three decades. I am mostly indebted to the following people who kindly contributed part of the photos in this chapter: Fig. 1.2C (F. Grassle); Fig. 1.3F (R. Shem-Tov); Fig. 1.6A (I. Grinberg); A insert (D. Zakai); B (A. Diamant); C (Z. Livnat); D, G (J. Batist); H (O. Ben-Zvi); Fig. 1.10A–D (R. Sulam); Fig. 1.11H (R. Klein); Fig. 1.10E, J (A. Shoob); The research reported in Section 1.4 was partially supported by the monitoring program for the Jordanian–Israeli Red Sea Marine Peace Park Program, the Israeli Science Foundation and the Raynor Chair for Environmental Conservation Research.

References

Abelson A, Loya Y (1999) Competition in stony corals. Bull Mar Sci 65:851–860

Abelson A, Shteinman B, Fine M, Koganovsky S (1999) Mass transport from pollution sources to remote coral reefs in Eilat, Gulf of Aqaba, Red Sea. Mar Pollut Bull 38:25–29

Anthony KRN (1999) Coral suspension feeding on fine particulate matter. J. Exp Mar Biol Ecol 232:85–106

Antonius A (1973) New observations on coral destruction in reefs. Abs Assoc Isl Mar Lab Caribb 10:3

Antonius A, Riegl B (1997) A possible link between coral diseases and a corallivorous snail (*Drupella cornus*) outbreak in the Red Sea. Atoll Res Bull 47:1–9

Atkinson MJ, Carlson B, Crow GL (1995) Coral growth in high-nutrient, low pH seawater: a case study of corals cultured at the Waikiki aquarium, Honolulu, Hawaii. Coral Reefs 14:215–223

Atkinson MJ, Birk Y, Rosenthal H (2001) Evaluation of pollution in the Gulf of Eilat. Report for the Ministries of Infrastructure, Environment and Agriculture, prepared by the International Expert Team (IET), 62 pp

Babcock RC, Bull GD, Harrison PL, Heyward AJ, Oliver JK, Wallace CC, Willis BI (1986) Synchronous spawning of 105 scleractinian coral species on the Great Barrier Reef. Mar Biol 90:379–394

Benayahu Y, Loya Y (1977a) Seasonal occurrence of benthic algae communities and grazing regulation by sea urchins at the coral reefs of Eilat, Red Sea. In: Taylor DL (ed) Proceedings of the 3rd international coral reef symposium I. Biology. University of Miami, Miami, pp 383–389

Benayahu Y, Loya Y (1977b) Space partitioning by stony corals, soft corals and algae in the northern Gulf of Eilat, Red Sea. Helgolander Wiss Meeresunters 30:362–382

Benayahu Y, Loya Y (1981) Competition for space among coral reef sessile organisms at Eilat, Red Sea. Bull Mar Sci 31:514–522

Ben-Haim Y, Zicherman-Keren M, Rosenberg E (2003) Temperature-regulated bleaching and lysis of the coral *Pocillopora damicornis* by the novel pathogen *Vibrio coralliilyticus*. Appl Environ Microbiol 69:4236–4242

Bongiorni L, Shafir S, Angel D, Rinkevich B (2003) Survival, growth and reproduction of two hermatypic corals subjected to in situ fish-farm nutrient enrichment. Mar Ecol Prog Ser 253:137–144

Brown BE (1997) Coral bleaching: causes and consequences. Coral Reefs 16:129–138

Brown BE, Howard LS (1985) Assessing the effects of "stress" on reef corals. Adv Mar Biol 22:1–63

Buss LW, Jackson JBC (1979) Competitive networks: nontransitive competitive relationships in cryptic coral reef environments. Am Nat 113:223–224

Bythell JC, Bythell M, Gladfelter EH (1993) Initial results of a long-term coral reef monitoring program: impact of Hurricane Hugo at Buck Island Reef National Monument, St Croix, US Virgin Islands. J Exp Mar Biol Ecol 172:171–183

Connell JH (1973) Population ecology of reef-building corals. In: Jones OA, Endean R (eds) Biology and geology of coral reefs biology I. Academic Press, New York, pp 205–245

Connell JH (1976) Competitive interactions and the species of diversity or corals. In: Mackie GO (ed) Coelenterate ecology and behavior. Plenum Press, New York, pp 51–58

Connell JH (1978) Diversity in tropical rain forests and coral reefs. Science 199:1302–1310

Connell JH, Hughes TP, Wallace CC (1997) A 30-year study of coral abundance, recruitment and disturbance, at several scales in space and time. Ecol Monogr 67:461–488

Dayton PK (1971) Competition, disturbance, and community organization: the provision and subsequent utilization of space in a rocky intertidal community. Ecol Monogr 41:351–389

Dollar SJ, Tribble GW (1993) Recurrent storm disturbance and recovery: a long-term study of coral communities in Hawaii. Coral Reefs 12:223–233

Dubinsky Z, Stambler N (1996) Marine pollution and coral reefs. Global Change Biol 2:511–526

Fadlallah YH (1983) Sexual reproduction, development and larval biology in scleractinian corals. A review. Coral Reefs 2:129–150

Fagerstrom JA, West RR, Kershaw S, Cossey PJ (2000) Spatial competition among clonal organisms in extant and selected Paleozoic reef communities. Facies 42:1–24

Ferrier-Pagès, Gatuso C, Dallot S, Jaubert J (2000) Effect of nutrient enrichment on growth and photosynthesis of the zooxanthellate coral *Stylophora pistillata.* Coral Reefs 19:103–113

Genin A, Lazar B, Brenner S (1995)Vertical mixing and coral death in the Red Sea following the eruption of Mount Pinatubo. Nature 377:507–510

Genin A, Achituv Y, Abelson A, et al., (2001) The deterioration of the coastal ecosystem in Eilat and the fish cages: 'The scientists document' letter to Mrs. D. Itzik, Minister of the Environment, Israel April 19, 2001

Glynn PW (1968) Mass mortalities of echinoids and other reef flat organisms coincident with mid-day, low-water exposures in Puerto Rico. Mar Biol 3:226–243

Glynn PW (1976) Some physical and biological determinants of coral community structure in the Eastern Pacific. Ecol Monogr 46:431–456

Glynn PW (1981) Wide spread coral mortality and the 1982–83 El Niño warming event. Environ Conserv 11:133–146

Glynn PW (1993) Coral reef bleaching ecological perspectives. Coral Reefs 12:1–17

Gordin H (2000) Environmental effects of mariculture in the Gulf of Eilat. Ecol Environ 6:124–127 (in Hebrew)

Green E, Bruckner A (2000) The significance of coral disease epizootiology for coral reef conservation. Biol Conserv 96:347–361

Grigg RW (1983) Community structure succession and development of coral reefs in Hawaii. Mar Ecol Prog Ser 11:1–14

Grigg RW, Maragos JE (1974) Recolonization of hermatypic corals on submerged lava flows in Hawaii. Ecology 5:387–395

Grover R, Maguer JF, Reynaud-Vaganay S, Ferrier-Pagès C (2002) Uptake of ammonium by scleractinian coral *Stylophora pistillata*: effect of feeding, light, and ammonium concentration. Limnol Oceanogr 47:782–790

Guzman MG, Cortes J (2001) Changes in reef community structure after fifteen years of natural disturbances in the eastern pacific (Costa Rica). Bull Mar Sci 69:133–149

Harrison P, Ward S (2001) Elevated levels of nitrogen and phosphorus reduce fertilization success of gametes from scleractinian reef corals. Mar Biol 139:1057–1068

Harrison PL, Wallace CC (1990) Reproduction, dispersal and recruitment of scleractinian corals. In: Dubinski Z (ed) Coral reefs. Ecosystems of the world 25. Elsevier, Amsterdam, pp 133–206

Harrison PL, Babcock RC, Bull GD, Oliver JK, Wallace CC, Willis BL (1984) Mass spawning in tropical reef corals. Science 223:1186–1189

Harvell CD, Kim K, Burkholder JM, Colwell RR et al. (1999) Emerging marine diseases – climate links and anthropogenic factors. Science 285:1505–1510

Highsmith RC (1982) Reproduction by fragmentation in corals. Mar Ecol Prog Ser 7:207–226

Hoegh-Guldberg O (1999) Climate change, coral bleaching and the future of the world's coral reefs. Mar Freshwater Res 50:839–866

Holling, CS (2001) Understanding the complexity of economic, ecological, and social systems. Ecosystems 4:390-405

Holmer M, Kristensen E (1992) Impact of marine fish cage farming on metabolism and sulfate reduction of underlying sediments. Mar Ecol Prog Ser 80:191–201

Hughes TP (1994) Catastrophes, phase shifts, and large-scale degradation of a Caribbean coral reef. Science 265:1547–1551

Hughes TP, Connell JH (1999) Multiple stressors on coral reefs: a long-term perspective. Limnol Oceanogr 44:932–940

Hughes TP, Baird AH, Bellwood DR, et al. (2003) Climate change, human impacts, and the resilience of coral reefs. Science 301:929–933

Hunter CL, Evans CW (1995) Coral reefs in Kaneohe Bay, Hawaii: two centuries of western influence and two decades of data. Bull Mar Sci 57:501–515

Ives AR, Dennis B, Cottingham KL, Carpenter SR (2003) Estimating community stability and ecological interactions from time-series data. Ecol Monogr 73(2):301–330

Jackson JBC, Buss L (1975) Allelopathy and spatial competition among coral reef invertebrates. Proc Natl Acad Sci USA 72:5160–5163
Johannes RE (1975) Pollution and degradation of coral reef communities. In: Ferguson EJ, Johannes RF (eds) Tropical marine pollution. Elsevier, Amsterdam, pp 13–51
Jokiel PL, Coles SL (1990) Response of Hawaiian and other Indo-Pacific reef corals to elevated temperature. Coral Reefs 8:155–162
Karlson RH, Hurd LE (1993) Disturbance, coral reef communities, and changing ecological paradigms. Coral Reefs 12:117–125
Kinsey DW, Davies PJ (1979) Effects of elevated nitrogen and phosphorus on coral reef growth. Limnol Oceanogr 24:935–940
Knowlton N (1992) Thresholds and multiple stable states in coral reef community dynamics. Am Zool 32:674–682
Koop K, Booth D, Broadbent A, Brodie J, et al. (2001) ENCORE: the effect of nutrient enrichment on coral reefs: synthesis of results and conclusions. Mar Pollut Bull 42:91–120
Kushmaro A, Loya Y, Fine M, Rosenberg E (1996) Bacterial infection and bleaching. Nature 380:396
Lang JC (1973) Interspecific aggression by scleractinian corals: 2. Why the fact is not only to the swift. Coral Reef Project Papers in Memory of Dr TF Goreau. Bull Mar Sci 23:260–279
Lang JC, Chornesky EA (1990) Competition between scleractinian reef corals: a review of mechanisms and effects. In: Dubinsky Z (ed) Ecosystems of the world, coral reefs. Elsevier, Amsterdam, pp 133–206
Lapointe BE (1997) Nutrient thresholds for bottom-up control of macroalgal blooms on coral reefs in Jamaica and south-east Florida. Limnol Oceanogr 42:1119–1131
Lazar B, Erez J, Azulai D, et al. (2000) Eutrophication processes in the northern Gulf of Eilat and their effects on the ecological ecosystem. Ecol Environ 6:119–121 (in Hebrew)
Levin SA (1992) The problem of pattern and scale in ecology. Ecology 73:1943–1967
Levin SA (1999) Fragile dominion: complexity and the commons. Helix Books, Reading, Massachusetts
Lewin R (2000) Complexity: life at the edge of chaos. 2nd edn. University of Chicago Press, Chicago
Lourey MJ, Miller IR, Ryan DAJ (2000) Decline and recovery of Great Barrier Reef corals following damage by the crown-of-thorns Starfish *Acanthaster planci.* Mar Ecol Prog Ser 196:179–186
Loya Y (1972) Community structure and species diversity of hermatypic corals at Eilat, Red Sea. Mar Biol 13:100–123
Loya Y (1975) Possible effects of water pollution on the community structure of Red Sea corals. Mar Biol 29:177–185
Loya Y (1976a) Recolonization of Red Sea corals affected by natural catastrophes and man-made perturbations. Ecology 57:278–289
Loya Y (1976b) The Red Sea coral *Stylophora pistillata* is an r strategist. Nature 259:478–480
Loya Y (1976c) Settlement, mortality, and recruitment of a Red Sea scleractinian coral population. In: Mackie GO (ed) Coelenterate ecology and behavior. Plenum Press, New York, pp 89–100
Loya Y (1976d) Skeletal regeneration in a Red Sea scleractinian coral population. Nature 261:490–491
Loya Y (1976e) Effects of water turbidity and sedimentation on community structure of Puerto Rican corals. Bull Mar Sci 26:450–466
Loya Y (1990) Changes in a Red Sea Coral community structure: a long-term case history study. In: Woodwell GM (ed) The earth in transition: patterns and processes of biotic impoverishment. Cambridge Univ Press, Cambridge, pp 369–384
Loya Y (1995) Development and protection of the Gulf of Aqaba. In: Spiegel SL, Pervin DJ (eds) Practical peacemaking in the Middle East. Garland, New York, pp 53–63

Loya Y, Slobodkin LB (1971) The coral reefs of Eilat (Gulf of Eilat, Red Sea). In: Yonge M, Stoddart DR (eds) Regional variation of Indian Ocean coral reefs. Symp Zool Soc Lond 28. Academic Press, New York, pp 117–139

Loya Y, Rinkevich B (1979) Abortion effect in corals induced by oil pollution. Mar Ecol Prog Ser 1:77–80

Loya Y, Rinkevich B (1980) Effects of oil pollution on coral reef communities. Mar Ecol Prog Ser 3:167–180

Loya Y, Kramarsky-Winter (2003) In situ eutrophication caused by fish farms in the northern Gulf of Eilat (Aqaba) is beneficial for its coral reefs: a critique. Mar Ecol Prog Ser 261:299–303

Loya Y et al. (2004) Nutrient enrichment by in situ fish-farms is detrimental to coral reproduction. Mar Poll Bull (in press)

Loya Y, Ilan M, Al-Moghrabi SM, Crosby MP (1999) The Red Sea Marine Peace Park Coral Reef Benthic Communities: ecology and biology monitoring program. In: Maragos JE, Grober-Dunsmore R (eds) Proceedings of Hawaii's Coral Reef Monitoring Workshop, pp 239–250

Loya Y, Sakai K, Yamazato K, Nakano Y, Sembali H, van Woesik R (2001) Coral bleaching: the winners and the losers. Ecol Lett 4:122–131

MacArthur RH (1972) Geographical ecology. Harper and Row, New York

Margalef R (1969) A practical proposal and model of interdependence. In: Woodwell GM, Smith HH (eds) Diversity and stability in ecological systems. Brookhaven Symp Biol 22:25–37

Marubini F, Atkinson MJ (1999) Effects of lowered pH and elevated nitrate on coral calcification. Mar Ecol Prog Ser 188:117–121

Miller TE (1982) Community diversity and interactions between the size and frequency of disturbance. Am Nat 120:533–536

Muller-Parker G, McCloskey LR, Hoegh-Guldberg O, McAuley PJ (1994) Effect of ammonium enrichment on animal and algal biomass of the coral *Pocillopora damicornis.* Pac Sci 48:273–283

Naim O (1993) Seasonal responses of a fringing reef community to eutrophication (Reunion Island, Western Indian Ocean). Mar Ecol Prog Ser 99:137–151

Nyström M, Folke C, Moberg F (2000) Coral reef disturbance and resilience in a human dominated environment. TREE 15:413–417

Ostrander GK, Armstrong KM, Knobbe ET, Gerace D, Scully EP (2000) Rapid transition in the structure of a coral reef community: the effects of coral bleaching and physical disturbance. Proc Natl Acad Sci USA 97(10):5297–5302

Paine RT, Levin SA (1981) Intertidal landscapes: disturbance and the dynamics of pattern. Ecol Monogr 51:145–178

Pastorok RA, Bilyard GR (1985) Effects of sewage pollution on coral-reef communities. Mar Ecol Prog Ser 21:175–189

Pearson RG (1981) Recovery and recolonization of coral reefs. Mar Ecol Prog Ser 4:105–122

Peters EC (1984) A survey of cellular reactions to environmental stress and disease in Caribbean scleractinian corals. Helgol Wiss Meeresunters 37:113–137

Pichon M, Morrissey J (1981) Benthic zonation and community structure of south island reef, Lizard Island (Great Barrier Reef). Bull Mar Sci 31:581–593

Pickett STA, White PS (1985) The ecology of natural disturbance and patch dynamics. Academic Press, New York

Pimm SL (1991) The balance of nature? The University of Chicago Press, Chicago

Porter JW, Battey JF, Smith GJ (1982) Perturbation and change in coral reef communities. Proc Natl Acad Sci USA 79:1678–1681

Richardson CA, Dustan P, Lang JC (1979) Maintenance of living space by sweeper tentacles of *Montastraea cavernosa*, a Caribbean reef coral. Mar Biol 55:181–186

Richmond RH, Jokiel PL (1984) Lunar periodicity in larva release in the reef coral *Pocillopora damicornis* at Enewetak and Hawaii. Bull Mar Sci 34:280–287

Richmond RH, Hunter CL (1990) Reproduction and recruitment of corals: comparisons among the Caribbean, the tropical Pacific, and the Red Sea. Mar Ecol Prog Ser 60:185–203

Rinkevich B (1996) Do reproduction and regeneration in damage corals compete for energy allocation? Mar Ecol Prog Ser 143:297–302

Rinkevich B, Loya Y (1977) Harmful effects of chronic oil pollution on a Red Sea coral population. In: Taylor DL (ed) Proceedings of the 3rd International Coral Reef Symposium II. Geology. University of Miami, Miami, pp 585–591

Rinkevich B, Loya Y (1979a) Laboratory experiments on the effects of crude oil on the Red Sea coral *Stylophora pistillata.* Mar Pollut Bull 10:328–330

Rinkevich B, Loya Y (1979b) The reproduction of the Red Sea coral. *Stylophora pistillata*: I. Gonads and planulae. Mar Ecol Prog Ser 1:133–144

Rinkevich B, Loya Y (1979c) The reproduction of the Red Sea coral *Stylophora pistillata*: II. Synchronization in breeding and seasonality of planulae shedding. Mar Ecol Prog Ser 1:145–152

Rinkevich B, Angel D, Shafir S, Bongiorni L (2003) 'Fair is foul and foul is fair': response to a critique. Mar Ecol Prog Ser 261:305–309

Roberts CM, McLean CJ, Veron JEN, Hawkins JP, et al. (2002) Marine biodiversity hotspots and conservation priorities for tropical reefs. Science 295:1280–1284

Rogers CS (1990) Responses of coral reef and reef organisms to sedimentation. Mar Ecol Prog Ser 62:185–202

Rogers CS (1993). Hurricanes and coral reefs: the intermediate disturbance hypothesis revisited. Coral Reefs 12:127–137

Sammarco PW, Coll JC, La-Barre S, Willis B (1983) Competitive strategies of soft corals (Coelenterata: Octocorallia); allelopathic effects on selected scleractinian corals. Coral Reefs 1:173–178

Sebens KP (1982) Competition for space: growth rate, reproductive output and escape in size. Am Nat 120:189–197

Shannon CF, Weaver W (1948) The mathematical theory of communication. Univ Illinois Press, Urbana, Illinois

Sheppard C, Price A, Roberts C (1992) Marine ecology of the Arabian Region. Academic Press, London, 359 pp

Sheppard CRC (1985) Unoccupied substrate in the central Great Barrier Reef: role of coral interactions. Mar Prog Ser 25:259–268

Sheppard RC (2003) Predicted recurrences of mass coral mortality in the Indian Ocean. Nature 425:294–297

Shinn EA (1972) Coral reef recovery in Florida and the Persian Gulf. Environmental Conservation Dept, Shell Oil Co, Houston, Texas

Shlesinger Y, Loya Y (1985) Coral community reproductive patterns: Red Sea versus the Great Barrier Reef. Science 228:1333–1335

Shlesinger Y, Goulet TL, Loya Y (1998). Reproductive patterns of scleractinian corals in the northern Red Sea. Mar Biol 132:691–701

Silverman J, Lazar B, Erez J (2004) Monitoring the health of coral reef ecosystems using community metabolism. In: Rosenberg E, Loya Y (eds) Coral health and disease. Springer, Berlin Heidelberg New York (in press)

Smith SV, Kemmerer WJ, Laws EA, Brock RE, Walsh TW (1981) Kaneohe Bay sewage diversion experiment: perspectives on ecosystem responses to nutritional perturbation. Pac Sci 35:279–395

Snidvongs A, Kinzie RA III (1994) Effects of nitrogen and phosphorus enrichment on in vivo symbiotic zooxanthellae of *Pocillopora damicornis.* Mar Biol 118:705–711

Sousa WP (1984) The role of disturbance in natural communities. Annu Rev Ecol Syst 15:353–391

Stambler N, Popper N, Dubinsky Z, Stimson J (1991) Effects of nutrient enrichment and water motion on the coral *Pocillopora damicornis.* Pac Sci 45:299–307

Steven ADL, Broadbent AD (1997) Growth and metabolic responses of *Acropora palifera* to long term nutrient enrichment. Proc 8th Int Coral Reef Symp 1:867–872

Stimson J (1992) The effect of ammonium addition on coral growth rate. Proc 7th Int Coral Reef Symp Mangilao 1:383

Szmant AM (1986) Reproductive ecology of Caribbean coral reefs. Coral Reefs 5:43–53

Tanner JE, Hughes TP, Connell JH (1994) Species coexistence, keystone species, and succession: a sensitivity analysis. Ecology 75:2204–2219

Tomascik T (1990) Growth rates of two morphotypes of *Montastraea annularis* along an eutrophication gradient, Barbados WI. Mar Pollut Bull 21:376–381

Tomascik T, Sander F (1985) Effect of eutrophication on reef-building corals. I. Growth rate of the reef-building coral *Montastraea annularis*. Mar Biol 87:143–155

Veron JEN (1986) Corals of Australia and the Indo-Pacific. Angus and Robertson, London

Ward S, Harrison P (2000) Changes in gametogenesis and fecundity of acroporid corals that were exposed to elevated nitrogen and phosphor us during the ENCORE experiment. J Exp Mar Biol Ecol 246:179–221

Weil E (2003) Coral reef diseases in the wider Caribbean: status and prognosis. In: Rosenberg E, Loya Y (eds) Coral health and disease. Springer, Berlin Heidelberg New York (in press)

Wilkinson C (1998) Status of coral reefs of the world 1998. Aust Inst Mar Sci and the Global Coral Reef Monitoring Network, Townsville, Australia

Wilkinson C (2000) (ed) Status of coral reefs of the World 2000. Aust Inst Mar Sci 361 pp

Wilson EO (1969) The species equilibrium. In: Woodwell GM, Smith HH (eds) Diversity and stability in ecological systems. Brookhaven Symp Biol 22. Brookhaven Nat Lab, Upton, NY, pp 38–47

Wittenberg, M, Hunte W (1992) Effects of eutrophication and sedimentation on juvenile corals I. Abundance, mortality and community structure. Mar Biol 112:131–138

Woodley JD, Chornesky EA, Clifford PA, Jackson JBC, et al. (1981) Hurricane Allen's impact on Jamaican coral reefs. Science 213:749–755

Woodwell GM, Smith HH (eds) (1969) Diversity and stability in ecological systems. Brookhaven Symp Biol 22. Brookhaven Nat Lab, Upton, NY

Zvuloni A (2003) Intra-specific genetic variability of the corals *Stylophora pistillata* and *Heteroxenia fuscescens* from the northern tip of the Gulf of Eilat. MSc Thesis, Tel Aviv University (in Hebrew with English Abstr)

2 Coral Reef Diseases in the Wider Caribbean

Ernesto Weil

2.1 Introduction

Over the past few decades, coral reef communities around the world have been experiencing increasingly stressful conditions due to a combination of natural and detrimental anthropogenic factors. In the Caribbean, coral reefs have experienced significant losses in hard coral cover due in part to local habitat degradation, over-fishing, pollutant input, bleaching, hurricanes and more recently, diseases (Porter and Meier 1992, Ginsburg 1993; Aronson and Precht 1997; Epstein et al. 1998; Harvell et al. 1999; Wilkinson 2000; Weil 2001). These factors, acting alone or in synergy, can be highly variable on spatial and temporal scales, making it difficult to identify and characterize a single or combined cause(s) of reef deterioration. Bleaching events, for example, have increased in frequency and intensity in the last two decades and their impact has been highly variable both spatially and temporarily. In the Caribbean, bleaching has caused variable, but generally low, coral mortality, unlike the mass mortalities of the scale observed in the Indo-Pacific. In contrast, few coral diseases, with low prevalence and restricted geographic distributions, have been reported for the Indo-Pacific as compared to the Caribbean. The latter has been dubbed a "disease hot spot" because of the fast emergence, high prevalence and virulence of coral reef diseases and syndromes; their widespread geographic distribution, and the frequent epizootic events with significant coral mortalities (Epstein et al. 1998; Hayes and Goreau 1998; Green and Bruckner 2000; Weil et al. 2002). In the last few decades, for example, epizootic events have resulted in significant losses in coral cover, biodiversity and habitat (loss of spatial heterogeneity), and today, diseases have possibly become the most important factor in the deterioration dynamics of Caribbean coral reefs (Hughes 1994; Aronson and Precht 2001b; Bruckner 2002; Weil 2002).

The Caribbean Sea is a relatively small, partially enclosed and highly interconnected water body surrounded by dense human populations. The wider Caribbean includes the Gulf of Mexico, Florida, the Bahamas and Bermuda. Only about 8% of the coral reef area worldwide is found in this region (Spalding and Greenfeld 1997), yet 66% of all disease/syndrome reports up to the year 2000 came from 38 Caribbean nations (Green and Bruckner 2000). Anthropogenic impacts in combination with global warming trends and rapid currents in the basin may facilitate the emergence and spread of pathogens and/or spread of other stressful agents that could affect the natural susceptibility/resistance of coral reef organisms to pathogens (Peters 1997; Santavy and Peters 1997; Hayes and Goreau 1998; Goreau et al. 1998; Richardson 1998; Richardson and Aronson 2002; Williams and Bunkley-Williams 2000; Hayes et al. 2001; Weil 2001). Environmental changes could affect the physiological equilibrium between bacteria already present in the normal flora associated with coral reef organisms (i.e., in tissues and mucus) and their hosts, or stimulate other bacteria living in reef sediments into becoming virulent. Very little is known about the composition and dynamics of the natural microbial communities living in association with most reef organism (Ritchie et al. 2001; Rowher et al. 2001). While it has been suggested that coral reef diseases might be associated with the deterioration in reef environmental quality, few quantitative studies have been done on this topic. We are currently unable to explain the source and sudden emergence of the majority of diseases/syndromes in coral reefs.

Disease is defined as "any impairment of an organism's vital organ, system and/or body functions". This definition includes both infectious diseases (produced by biological parasites – pathogens) and noninfectious diseases (produced by genetic mutations, malnutrition and/or environmental factors; Peters 1997). Environmental bleaching, therefore, is considered a disease. The term **disease** is used here for any affliction for which the causative agent has been identified, and **syndrome** for those afflictions for which the causative agent is not known. In this context, some common names used in the literature such as yellow blotch "disease" and dark spots "disease", are referred to here as "yellow blotch syndrome (YBS)" and "dark spots syndrome (DSS)" since their causative agents have not been identified.

Even though Caribbean reef-related diseases were first reported in the early 1970s (Antonious 1973; Garret and Ducklow 1975; Dustan 1977), our knowledge of their **pathology** (isolation and identification of the pathogen), **etiology** (macroscopic signs and relationships between the host and pathogen), and **epizootiology** (local and geographic distributions, environmental correlates, host ranges, prevalence, impact, vectors and natural reservoirs, and spatial and temporal variability, etc.) of most coral reef diseases is limited.

In this chapter, the current status of diseases and syndromes and their potential impact in the wider Caribbean is discussed. A short historical perspective summarizes those epizootic events that had some significant impact at the population and community levels, followed by the actual knowledge and most recent data collected by the author on the geographic distribution, host ranges, new syndromes, prevalence and impact of the most common diseases in the region. A short discussion of the origin, causes and ecological effects is followed by some thoughts on the potential future impacts of diseases in the region and what the current research needs are.

2.2 Historical Perspective

The emergence and prevalence of diseases/syndromes in the Caribbean in the past few decades appear to be unprecedented in the geological record. Limited paleontological evidence suggests that the outbreak of white band disease (WBD) in the early 1980s, which killed acroporid corals all over the Caribbean, was unparalleled on a time scale of at least three millennia (Aronson and Precht 2001a, b). In recent years, large colonies (many over 500 years old) of the main reef-building species in the Caribbean (i.e., *Acropora* spp, *Montastraea* spp, *Colpophyllia* sp, and *Diploria* spp) have been wiped out by particular diseases, further indicating that this is a recent problem. The causes are more surely complex and possibly synergistic, involving global and local environmental changes, anthropogenic factors and old and new pathogens (Mitchell and Chet 1975; Epstein et al. 1998; Harvell et al. 1999, 2002; Hayes et al. 2001; Weil 2002).

There are few historical reports of diseases in coral reef organisms. The first documented epizootic event in the Caribbean was the mass mortality of commercial sponges in the northern Caribbean in 1938 (Gastoff et al. 1939), which killed up to 95% of the population in some localities (Table 2.1). Other minor mass mortalities (of oysters, corals, sea urchins, octocorals, seagrasses and mollusks) have occurred across the Caribbean over the years, but were more localized, short-lived, and had limited impact at the population and community levels. In many of these cases, the outbreak seems to be related to environmental changes such as unusual high temperatures.

Coral diseases were first described for the Caribbean in the early–mid-1970s (Table 2.2). Coral mortalities due to black band disease (BBD), white band-I (WBD-I) and white plague-I (WP-I) were first documented for Belize, Bermuda, Florida and the Virgin Islands (Antonius 1973; Garret and Ducklow 1975; Dustan 1977; Gladfelter et al. 1977; Peters et al. 1983). Little bacteriological or etiological work was done, however. The first report of a significant coral mortality was the WP-I epizootic event in Florida in 1975. This syndrome infected high numbers of colonies of *Mycetophyllia ferox* and predictions were made about the potential disappearance of this species from some localities in the Florida Keys (Dustan 1977). Ten years later, however, during a second epizootic event of WP-I in the same area that mostly affected massive *Montastraea* colonies, many colonies of *M. ferox* remained healthy and unaffected, suggesting either more resistant colonies or less virulent strains of the pathogen, or a combination of the two (Dustan 1987).

In the early 1980s, two widespread and virulent epizootics changed the structure and morphology of most shallow water coral reef communities in the wider Caribbean: the die-off of the black sea urchin *Diadema antillarum* (Lessios et al. 1984), and a white band disease epizootic event that significantly reduced populations of acroporid corals (*Acropora palmata* and *A. cervicornis*; Gladfelter 1982). These events had widespread ecological (and potentially

Table 2.1. Reported mass mortalities (>10%) of different marine organisms in the wider Caribbean and eastern Atlantic, their associated pathogen (if identified), estimated mortality rate and environmental correlates; *CA* Caribbean. (Modified from Harvell et al. 1999)

Date	Host organisms	Location	Pathogen	Mortality (%)	Environmental correlates	Source
1938	Sponges	North-CA	Fungus	70–95	–	Gastoff et al. (1939)
1946	Oyster	Gulf coast	Bacterium	High	High temp. – salinity	Soniat (1996)
1975	Corals	Florida	Bacterium	Med	High temp.	Dustan (1977)
1980	Urchin	NW-Atlantic	Amoeba	>50	–	Miller and Colloday (1983)
1981	Acroporid corals	Wide – CA	Bacterium	70–95	Med–High temp.	Gladfelter (1982); Aronson and Precht (2001b)
1982	Octocorals	C-America	?	High	High temp.	Guzmán and Cortés (1984)
1983	Corals	Wide – CA	Bacteria	Variable	Seasonal	Rützler et al. (1983)
1983	Black urchin	Wide – CA	?	>95	High temp.	Lessios et al. (1984)
1987	Turtle Grass	Florida	Slime mold	70–90	Seasonal	Roblee et al. (1991)
1988	Mollusk	North – CA	Protozoan	High	–	Moyer et al. (1993)
1993	Corals	Florida	Bacteria	High	–	Dustan (1999)
1993	Corals	Barbados	Bacteria?	High	Pollution	Bythell and Sheppard (1993)
1995	Octocorals	Wide – CA	Fungus	High	–	Nagelkerken et al. (1997a, b)
1995	Corals	Florida	Bacterium	High	High temp.	Richardson et al. (1998a, b)
1996	Corals	Puerto Rico	Bacterium	Medium	Seasonal	Bruckner and Bruckner (1997b)
1997	Urchin	Curacao	Bacterium	High	Pollution	Ritchie et al. (2000)
1998	Octocoral	Florida	Cyanobacteria	70	High temp.	Harvell et al. (2001)

Table 2.2. Commonly reported coral/octocoral diseases and syndromes in the Caribbean, identified pathogen(s), number of coral and/or octocoral taxa affected, and number of localities reported as of January 2003. Information compiled from the author's research, the first reports of each disease/syndrome and other sources (i.e., Green and Bruckner 2000; Bruckner 2002)

Disease (D)/ syndrome (S)	Acronym[a]	Pathogen	No. of species		Localities	Sources
			Corals	Octocorals		
Black band (D)	BBD	*P.coralliticum, Desulfovibrio, Beggiatoa* sp.	19	6	19	Antonius (1973); Rützler and Santavy (1983)
White band-I	WBD-I	Gram (–) bacterium	2	–	2	Gladfelter et al. (1977)
White band-II (D)	WBD-II	*Vibrio charcharii* ?	3	–	18	Ritchie and Smith (1998)
White plague-I	WP-I	Gram (–) bacterium	12	–	3	Dustan (1977)
White plague-II (D)[b]	WP-II	*Aurantimonas coralicida*	39	–	32	Richardson et al. (1998a); Denner et al. (2003; Weil (2002)
Aspergillosis (D)[b]	ASP	*Aspergillus sidowii*	–	10	24	Smith et al. (1996); Nagelkerken et al. (1997a. b); Geiser et al. (1998); Weil et al. (2001, 2002)
White pox (D)[b]	WPX	*Serratia marsescens*	2	–	4	Porter and Meier (1992); Patterson et al. (2002)
Tumors[c] (D)	TUM	*E.endozoica* (algae) and other causes	7	5	16	Morse et al. (1977, 1981); Goldberg et al. (1984); Loya et al. (1984)
Red band (D)	RBD	*Oscillatoria* sp. and other causes	13	1	6	Rützler et al. (1983); Santavy and Peters (1997); Richardson (1998)
Neoplasia (S)	NEP	?	4	5	3	Squires (1965); Peters (1984)
Shut-down Reaction	SDR	?	6	–	2	Antonius (1977)
Ridge mortality[d] (S)	RMS	?	4	–	6	Abbot (1979)
Yellow blotch (S)	YBS	?	11	–	22	Reeves (1994); Santavy et al. (1999); Cervino et al. (2001); Weil (2002)
Dark spots-I (S)	DSS-I	?	10	–	24	Gil-Agudelo and Garzón-Ferreira (2001); Weil (2002)
Dark spots-II (S)	DSS-II	?	8	–	10	Weil (2001, 2002);
Patchy necrosis[e] (S)	PNE	?	1	–	5	Bruckner and Bruckner (1997b)
Ring syndrome (S)	RS	?	1	–	3	Weil (2001)

[a]Commonly used and modified acronym for each disease/syndrome.
[b]Koch's postulates fulfilled.
[c]Tumors includes hyperplasia and algal tumors.
[d]Majority of ridge mortality might be caused by damselfish.
[e]Patchy necrosis and white pox are considered the same by some researchers.

evolutionary) consequences throughout the Caribbean with significant losses in coral cover, spatial heterogeneity (habitat and refuges) and biodiversity. The loss of coral cover was correlated with significant increases in algal cover (aided by lack of herbivory by *D. antillarum*), changing the community structure and dynamics of shallow coral reef habitats (Hughes 1994; Aronson and Precht 1997, 2001b). Even though these were widespread, conspicuous and detrimental events that lasted for over a year, the pathogens associated with both epizootic events were never completely identified (but see Peters et al. 1983).

Several new coral/octocoral diseases and syndromes were reported during the 1990s. These included red band disease (RBD), white band type II (WBD-II), white plague type II (WP-II), yellow blotch (YBS), dark spots (DSS), white pox disease (WPX), aspergillosis (ASP), and patchy necrosis (PNE; Table 2.2). Moreover, other important reef organisms such as sponges, hydrocorals, zoanthids and crustose algae were also affected for the first time by diseases/syndromes (Appendix 1). More epizootic events also occurred in the 1990s. A third plague event (WP-II) in the Florida Keys was more virulent, rapidly killing colonies of *Dichocoenia stockesii*. Within 2 years, it had affected 16 other coral species in the Florida Keys and other Caribbean localities (Richardson et al. 1998a). This time, however, the pathogen was successfully isolated and identified (*Aurantimonas coralicida*) and the disease was termed plague type II (WP-II; Richardson et al. 1998b; Denner et al. 2003). More recently, a fourth epizootic event has involved yet a more virulent strain (WP-III) that has produced significant coral tissue mortality in *Montastraea* spp. in Florida, the Virgin Islands, Puerto Rico and Venezuela (Richardson and Aronson 2000; Miller et al. 2001; Weil 2002). The pathogen was isolated and identified as the same one that caused the 1995 event in Florida. Aspergillosis (ASP) is a fungal disease described in the mid-1990s (Smith et al. 1996; Nagelkerken et al. 1997a, b) that affects populations of the common sea fan *Gorgonia ventalina*, and at least six other abundant octocoral species throughout the wider Caribbean, and has been responsible for widespread and rapid mortality (Weil et al. 2001; Harvell et al. 2002; Weil 2002).

Several diseases and syndromes have been persistent over the years in the wider Caribbean region, with local epizootic events in many localities. Five of them (BBD, WP-II, YBS, DSS, ASP) are widespread and may be considered chronic in many localities. Four of these have produced significant tissue and colony mortality in important reef building coral (BBD, WP-II, YBS) and common octocoral (ASP) populations (Appendix 1), endangering the future integrity of many coral reef communities (Bruckner and Bruckner 1997b; Richardson and Aronson 2000; Harvell et al. 2001; Miller et al. 2003; Bruckner 2002; Weil 2002; Weil et al. 2002).

2.3 Current Status

2.3.1 Pathology and Etiology

Recent research on coral diseases has revealed an impressive diversity of microbial pathogens, ranging from single fungal or bacterial species to loosely or tightly structured bacterial consortia that include a wide variety of phototrophic and heterotrophic bacterial species with a wide range of metabolic modes and micro-niche characteristics (Richardson et al. 2001; Ritchie et al. 2001).

For the majority of coral species (and most of the metazoans in coral reef communities), we do not know the composition of the natural microflora that inhabit their surfaces and tissues. Coral mucus is a rich protein-carbohydrate complex that harbors a diverse community of bacteria (Rowher et al. 2001). Environmental and physiological changes could alter the dynamics within epibiotic communities and/or between this community and the host. Changes in the bacterial community structure, an increase in virulence, increased overall bacterial population growth, host increased susceptibility, or any combination of these, could result in the development of an infectious disease.

Twenty two (76%) of the approximately 29 reported diseases and syndromes affecting 106 coral species and 10 octocorals worldwide are from the Caribbean (Green and Bruckner 2000, Bruckner 2002; Weil 2002). However, the associated pathogens have only been identified for seven of these diseases [BBD, WP-II, WBD-II, WPX, ASP, algal tumors (TUM) and red band disease (RBD); Table 2.2], and Koch's postulates have only been fulfilled for three of them (WP-II, ASP and WPX, but see Ritchie et al. 2001). Recently, the associated pathogens of diseases in other regions of the world have been identified. For example, pink spot is caused by a trematode *Podocatyloides stenometra* (Aeby 2000), poritid ulcerative white spot disease (PUWS) by a *Vibrio* sp. (Smith pers. comm.; Raymundo, submitted), skeletal eroding by the ciliate *Halofolliculina corallasia* (Antonius and Lipscomb 2001), bacterial bleaching (BBL) in *Oculina patagonica* by *Vibrio shiloi*, and in *Pocillopora damicorni*, by *Vibrio coralliticus* (Kushmaro et al. 1996, 1997; Ben-Haim and Rosenberg 2002). It is noteworthy that a well-defined relationship between bleaching and bacteria found in the Mediterranean and Red Sea has not been related to the mass bleaching events on Indo-Pacific and Caribbean reefs.

Only a few Caribbean coral diseases have a well-characterized etiology. Many aspects of the infection process in BBD and ASP have been studied over the years. For example, the behavior and migration patterns and some physico/chemical aspects of the bacterial components in BBD have been characterized (Carlton and Richardson 1995; Richardson 1996; Richardson et al. 1997), as well as ecological aspects of the disease (Kuta and Richardson 1996, 1997, 2002; Bruckner et al. 1997; Richardson et al. 1997; Edmunds 1991) and the ecological physiology of BBD-associated pathogens (Richardson and Kuta 2003; Chap. 18, Richardson,

Table 2.3. Ecological characteristics of common diseases/syndromes in the Caribbean. Geographic distribution (L local or few localities, *CA* Caribbean wide, *W-CA* wider Caribbean including Bermuda), host range, average prevalence at the community level (%), rates of tissue mortality (mm/day), depth distribution, identified vectors and/or reservoirs, and status of epizootic events (*L* local, *W* widespread = in more than ten localities) as of May 2003. Recent data by the author are included

Disease acronym[a]	Distribution	Host range		Prevalence (%)	Tissue mortality (mm/day)	Depth (m)	Vector/ reservoir	Epizootic	Source
		Coral	Octocoral						
BBD	W-CA	21	6	0.2–6.0	3–10	0.1–25	Sediment	No	Antonius (1981); Rützler et al. (1983); Edmunds (1991); Dustan (1993); Kuta and Richardson (1996, 2002); Bruckner et al. (1997); Richardson et al. (1997)
WBD-I	CA	3	–	0.11	?	0.1–20	?	No	Weil et al. (2002); Weil (2002) and recent observations and data; Gladfelter (1982); Smith et al. (1996); Ritchie and Smith (1998)
WBD-II	CA	2	–	0.01	?	0.1–20	?	No	Weil et al. (2002); Weil (2002) and recent observations and data; Gladfelter (1982); Smith et al. (1996); Ritchie and Smith (1998)
WP-I	Florida	12	–	3.62	3.1	10–21	?	L	Dustan (1977, 1993)
WP-II	W-CA	39	–	0.96	3–30	0.1–30	?	W	Weil et al. (2002); Weil (2002) and recent observations and data; Richardson et al. (1998a, b); Miller et al. (2001)
ASP	W-CA	–	10	1.85	?	0.1–25	African dust	W	Weil et al. (2002); Weil (2002) and recent observations and data; Geiser et al. (1998); Kim and Harvell (2001)
WPX	L	2	–	0.002	15	0.1–5.0	?	L	Weil et al. (2002); Weil (2002) and recent observations and data; Patterson et al. (2002)
DSS-I	W-CA	14	–	1.14	?	0.2–20	?	No	Weil et al. (2002); Weil (2002) and recent observations and data; Garzón-Ferreira et al. (2001)

Disease acro-nym[a]	Distribu-tion	Host range		Preva-lence (%)	Tissue mortality (mm/day)	Depth (m)	Vector/ reservoir	Epizootic	Source
		Coral	Octo-coral						
DSS-II	L	8	–	0.05	?	3.0–15	?	No	Weil et al. (2002); Weil (2002) and recent observations and data
YBS	W-CA	11	–	1.02	0.1–0.4	1.0–25	?	W	Weil et al. (2002); Weil (2002) and recent observations and data; Santavy et al. (1999); Cervino et al. (2000)
TUM	W-CA	10	6	–	–	1.0–25	?	No	Weil et al. (2002); Weil (2002) and recent observations and data; Goldberg et al. (1984); Peters (1984)
PNE	L	1	–	0.02	5–11	0.1–10	?	L	Weil et al. (2002); Weil (2002) and recent observations and data; Bruckner and Bruckner (1997b); Rodriguez-Martinez et al. (2001)
RS	L	1	–	–	0.1	2.0–10	?	No	Weil et al. (2002); Weil (2002) and recent observations and data
RBD	L	14	2	–	1	5.0–15	?	No	Weil et al. (2002); Weil (2002) and recent observations and data; Richardson and Aronson (1992)

[a]Disease/syndrome acronym as described in Table 2.2.

Fig. 2.1A–F. Different groups of reef organisms affected by diseases/syndromes in the wider Caribbean. **A** The zoanthid *Palythoa caribbaeorum* showing tissue mortality (*arrows*) by a white band syndrome in Puerto Rico. **B** A crustose algae, possibly *Paragoniolithon* sp. with a white band syndrome in Grenada. **C, D** The crustose octocoral *Erythropodium caribbaeorum* with a syndrome producing a thin white band of necrotic tissue (*arrow*) that disintegrates giving way to secondary colonization by algae. **E** A clionid sponge (*Cliona tenuis*, new sp.) with signs of a syndrome that produces fast and widespread tissue necrosis and disintegration, and **F** another sponge, possibly *Amphymedon compressa*, with tissue mortality from a fast-moving infection

this Vol.). For ASP, there is information on the cellular events occurring during the pathogenesis of ASP in *G. ventalina* (Smith et al. 1998), the mechanisms of resistance and chemical responses (Kim et al. 2000b; Kim and Harvell 2001), the potential factors (i.e., temperature requirements, susceptibility to octocoral crude extracts, and metabolic profiles) that may have facilitated the emergence of ASP (Alker et al. 2001), and the transmission mechanism in sea fan populations infected by ASP in Florida (Jolles et al. 2002; Chap. 15, Smith and Weil, this Vol.). There is an urgent need for more effort to be devoted to the pathological and etiological characterization of most syndromes, especially those that are causing the highest mortalities in populations of coral and other reef organisms.

2.3.1.1 New Syndromes

In recent years, a number of new syndromes have been observed in many localities around the Caribbean. Some of the signs include white syndromes (bands and blotches), dark syndromes (spots, bands and lines), tissue necrosis syndromes (bands or spots), and purple bands (Fig. 2.2). Some of these are affecting a single species in a few localities [i.e., ring syndrome (RS), dark band syndrome (DBS); tissue necrosis (TNE) and purple band (PBS)], while others affect several species in many localities [i.e., white syndromes (WS), and dark spots type II (DSS-II)]. Ring syndrome, for example, was first observed by the author in colonies of *D. labyrinthiformis* along the northern coast of Bermuda. This affliction begins with the fast growth of tissue and skeleton at the ridge areas, which produces high, pale and thin ridges (1.5–2.5 cm in height) over the surface of the colony, eventually encircling an area of varying diameter (3–6 cm). At this point, the tissue inside the ridges slowly starts to die (0.001 mm/day; Figs. 2.2A, B). This syndrome has also been observed in the Florida Keys and in Honduras. Dark band syndrome (DBS) was observed affecting large numbers of colonies of *M. annularis* and *M. faveolata* in shallow and intermediate depths in Puerto Rico and Mexico. It starts as medium-wide round, dark spots on healthy tissue areas and quickly produces tissue mortality, turning into a dark band that advances over the colony killing live tissue at a slow rate (Fig. 2.2B, C). This syndrome could be related, or be a different stage in the etiology of dark spots, however, some characteristics are different. Tissue necrosis (TNE) looks similar to DBS, but without the spots, while the band is wider and more irregular and the tissue looks necrotic and peels off the skeleton (Fig. 2.2D, E). It mainly affects colonies of *M. faveolata* at intermediate depths (10 m) in several reefs off the south coast of Puerto Rico. Dark spots type II usually appears as larger spots or dark, depressed areas (slower growth) and associated dead areas that appear faster than in DSS-I (Fig. 2.2G, H). This condition has been observed mostly in the south and west Caribbean, affecting several species (Appendix 1), even though these species have a wider geographic distribution. Here, again, there is the possibility that this is caused by the same unidentified pathogen as in DSS-I, and that these are signs of different stages of the same disease. Purple band was observed in colonies of *S. siderea* and *S. intersepta* in Grenada, Puerto Rico and Mexico, and is characterized by a purple band of unhealthy-looking tissue between the healthy-looking tissue and the exposed skeleton (slow advance; Fig. 2.2I, J). This coloration may be a physiological response to a pathogen; and the rate of tissue mortality appears to be faster than in DSS-I (Weil, unpubl. data). This syndrome

Fig. 2.2A–H. New, uncharacterized syndromes in Caribbean corals. **A, B** Different stages of the ring syndrome in *D. labyrinthiformis* in Bermuda. Note tissue mortality once the high ridges are developed. **C, D** Signs of a dark band type syndrome in *M. annularis* in Puerto Rico. **E, F** Tissue necrosis in *M. faveolata*; *arrows* point to necrotic areas. **G** A wide, depressed area characteristic of dark spots type II syndrome (DSS-II) in *C. natans* (note change in meandroid pattern) and **H** DSS-II in *S. intersepta* in Los Roques, Venezuela. Tissue starts to die in the center producing clear skeleton areas that are colonized by turf algae.

Fig. 2.2 I–P. I, J Purple band in colonies of *S. siderea* in Grenada (I) and Venezuela (J). Note the unhealthy looking tissue between the algae-covered skeleton and the healthy looking tissue. K–O Several coral species with different types of "white band – white spots" syndromes; *M. faveolata* showing several bands of slow dying tissue (note algae growth) in Grenada (K) and long, depressed spots devoid of tissue in Puerto Rico (L). A colony of *Undaria humilis* in Grand Cayman with ring-like areas of dead tissue that eventually lost all live tissue (M). Recent, fast tissue necrosis and mortality left these irregular white areas of clean skeleton in a colony of *S. intersepta* (N) and in a colony of *Porites astreoides* in Bermuda (O). P A large tumor growth on the side of *Diploria strigosa* in Puerto Rico

might, however, be related to DSS. White bands and blotches are common in many localities (Fig. 2.2K–P) and show different signs to those described for white band disease and white plague. Some bands, for example, start as a blotch in the middle of the colony and then spread at slow rates (Fig. 2.2K–M), or produce a skeletal depression on the surface of the coral even though the calical structure appears intact (Fig. 2.2L).

The proliferation of names for different syndromes has produced confusion when trying to evaluate host ranges and geographical distributions for both diseases and syndromes. In most instances the description of a disease/syndrome is based on limited etiological and ecological observations, often lacking pathological information and missing the temporal variability in signs and interaction between the host and pathogen. One of the most difficult issues facing coral disease researchers is in determining whether a report of a disease/syndrome in a reef/region represents a new syndrome or whether it is the same as one reported from another reef region. It is possible that in each of the white plague types I, II and III, white band types I and II, and dark spots types I and II situations, a single putative pathogen is associated with each disease/syndrome. Temporal, spatial, and/or species-specific variability in the development of a disease/syndrome (i.e., a more virulent strain that kills tissue at a higher rate, different patterns of tissue mortality, coloration, etc.) could easily lead to descriptions of them as different when limited observation are made. For example, descriptions of the two types of WBD differ in aspects that can easily be missed in short-term observations. The characteristic bleaching front in WBD-II could be an initial stage in the disease etiology, developing into a sharp edge between the live tissue and the clean skeleton (characteristic of WBD-I) after the tissue peels off (Ritchie and Smith 1998). Moreover, diseases with different names (not just types) such as WPX, PNE and shut-down reaction (SDR), which produce white areas of denuded skeleton and show similar etiologies in acroporids, might be caused by the same pathogen. At this point, it is unclear if this is the case. More resources and effort are needed to conduct extensive pathological work to identify and characterize the potentially high number of other pathogens affecting scleractinian corals and other important organisms inhabiting coral reef communities.

2.3.2 Epizootiology

2.3.2.1 Geographic Distributions

Records in the Global Coral Disease Database (GCDD) show that at least 106 species of corals from coral reefs in 54 nations worldwide have been affected by over 29 diseases and syndromes (Green and Bruckner 2000). However, confusion associated with identifying diseases in different reef organisms and regions makes it difficult at this time to establish accurate regional and/or global distribution patterns and host ranges. For example, whereas WBD had been believed to affect only two species of acroporid corals in the Caribbean, there are now recent reports of it affecting up to 15 other species in the Caribbean as well as 25 coral species in the Indo-Pacific. Without a microbiological analysis (or a bacterial probe), it is difficult to verify and validate these records since WBD signs may be confused with those of WP-II, YBS, shut-down reaction (SDR), PNE, WPX, predation from snails, and even bleaching. Since the WBD pathogen has only been tentatively identified (based on changes in the relative proportions of bacterial members of the coral community) from acroporids in the Caribbean, these records must be considered provisional at this time. Further research must be conducted in order to confirm whether the pathogen is the same in other species showing similar signs. The only infectious diseases/syndromes for which global distributions can be considered to be more or less accurate at this time are BBD and tumors.

A number of diseases/syndromes are thought to have limited geographic distribution. White pox has only been reported from Florida, the Bahamas, Puerto Rico and the Virgin Islands (and Mexico if WPX and PNE are the same). Recent data from more than 50 reef sites surveyed by the author in nine geographically distant localities throughout the wider Caribbean indicate that

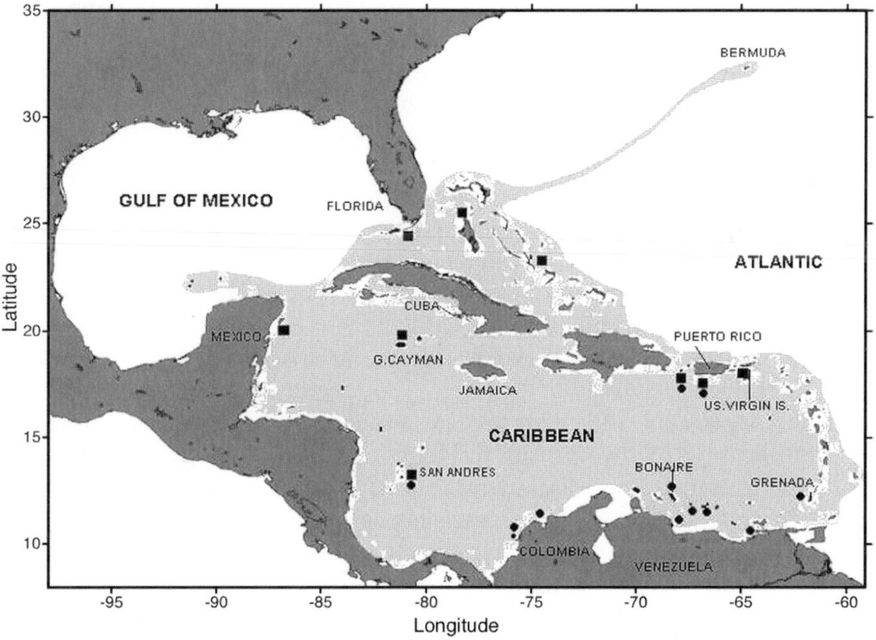

Fig. 2.3. The geographic distribution of common diseases (BBD, YBS, WP-II, DSS-I, ASP) is shown by the *gray area* covering most of the wider Caribbean. Localities where diseases/syndromes with more restricted geographic distributions are indicated by *squares* (WPX and PNE) and *circles* (DSS-II)

WP-II, YBS, BBD, ASP, and DSS-I have a much wider geographic distribution than previously believed, while other diseases/syndromes such as DSS-II, PNE and WPX have a more limited distribution. Diseases/syndromes originally restricted to Florida and a few other localities in the Caribbean, for example, have been observed in most reef sites from Bermuda to Grenada and Trinidad in the east, Venezuela and Colombia in the south, and Belize, Mexico and Grand Cayman in the west. Dark spots-II has a more restricted distribution than DSS-I even though the coral species susceptible to this syndrome (*S. intersepta*, *C. natans* and *Montastraea* spp.) have a wider Caribbean distribution. It is clear that diseases have been spreading and today, the most problematic diseases/syndromes have a wider Caribbean distribution (Fig. 2.3).

2.3.2.2 Host Ranges

Recent data and reliable sources from different localities indicate that for the Caribbean, in particular, more scleractinian taxa are now being affected by diseases than previously believed. Moreover, a suite of new syndromes are affecting different coral species and other important reef organisms (Table 2.2, Fig. 2.2) in several localities, which need to be studied and characterized. So far, 43 scleractinian coral species (close to 60% of the described zooxanthellate coral species for the Caribbean), 3 hydrocorals, 10 octocorals, 11 sponges and 3 crustose algae species are being affected by at least one disease/syndrome in the wider Caribbean (Appendix 1). Overall, WP-II is the most virulent with the widest host range of 39 species, followed by BBD with 21, DSS-I and RBD with 14, YBS, tumors and DSS-II with 11, 10 and 8, respectively. The host range of ASP has been expanded to at least eight common octocoral species in place of the original two (Weil et al. 2001, 2002; Appendix 1). The wide range of hosts for each of these diseases/syndromes is remarkable when compared with terrestrial pathogen–host interactions and when basic microbiology and coevolution principles indicate that pathogens are usually highly specific, generally infecting only one (or a very few) host(s).

Field observations clearly show that contagion can occur across colonies of the same coral species, colonies of different coral species, and even across different groups of organisms (i.e., crustose algae and corals; Fig. 2.4). This perception of wide host ranges, however, might change once the pathology of the different diseases is clarified. Some diseases, such as BBD which has a conspicuous band on the coral surface, might not change, but for other internal diseases (inside tissue or zooxanthellae) displaying physiological responses (i.e., color change) that can produce confusion, their host ranges may be reduced/expanded in the future. All the white syndromes, for example, are difficult to separate. There are a few reports of WP-II in acroporids (Green and Bruckner 2000), even though this disease is presumed to infect 39 other coral species in the Caribbean alone. Confusion is likely here because WBD, WP-II, patchy necrosis, snail predation, and sometimes bleaching, appear similar.

Fig. 2.4. A Cross-infection of white plague (WP-II) between the corals *S. siderea* and *D. labyrinthiformis*, **B** the coral *S. siderea* and a crustose coralline algae, possibly *Paragoniolithon* sp., **C** several colonies of *M. franksi*, *M. faveolata* and a single colony of *S. siderea* (*center*), and **D** a colony of *M. franksi* being infected with WP-II from the mostly dead platy coral *Mycetophyllia ferox*. *Yellow arrow* points to the small area of live tissue that is left at the top of *M. ferox*

The number of species susceptible to different diseases is highly variable. For example, 3 coral species are susceptible to 8 different diseases/syndromes, while 12 species are susceptible to 2, 6 species to 5, and 7 to just 1 disease/syndrome (Fig. 2.5A). It is highly significant that the main reef-building genera (i.e., *Montastraea*, *Colpophyllia*, *Diploria*, *Porites*, *Siderastrea*, and *Undaria*) are each affected by at least five (and up to eight) diseases/syndromes in the wider Caribbean (Fig. 2.5B). Single colonies with signs of two or more diseases at the same time are usually rare. A common occurrence, however, is the presence of several adjacent colonies of different species with each infected by what seems to be the same (i.e., showing similar signs) disease; this is a common sight in some localities (Fig. 2.4). The relevant question is whether different coral species showing similar signs in different localities are being infected by the same pathogen (or affected by the same environmental agent). To complicate matters more, afflictions showing similar signs in different species, such as bacterially induced bleaching, may be caused by different pathogens (i.e., *V. shiloi* or *V. corallilticus*). There is an urgent need for more pathological work to identify the pathogens causing these afflictions in the field.

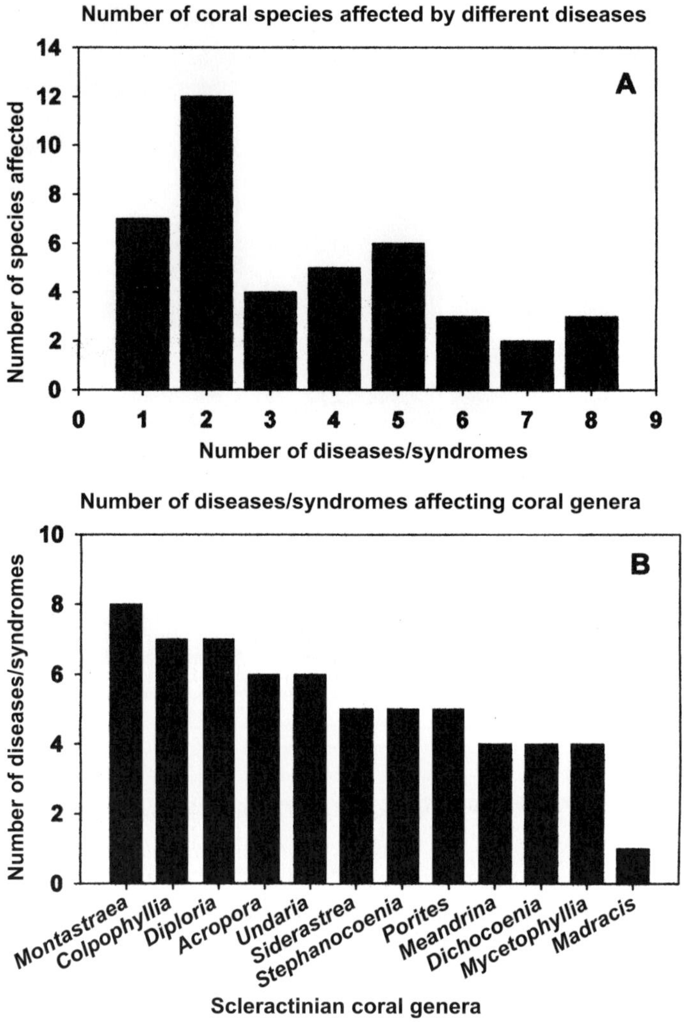

Fig. 2.5. A Number of diseases/syndromes affecting different species of scleractinian corals in the wider Caribbean. For example, there are 12 species affected (susceptible) by 2 (not necessarily the same) diseases/syndromes and 3 coral species (*Montastraea* spp.) affected by eight different diseases/syndromes. **B.** Most of the species affected by the highest number of diseases/syndrome belong in all the major reef-building genera for the Caribbean (*Montastraea, Diploria, Colpophyllia,* etc.)

2.3.2.3
Prevalence and Impact

The average prevalence of diseases at the coral community level seems to be low and highly variable in general (Weil et al. 2002). Statistically significant spatial and temporal variability in prevalence (all diseases pooled at the community level) was found within and between nine geographically distant localities in the wider Caribbean that were surveyed between 1999 and 2002 (Table 2.3, Fig. 2.6). Over this time period, average prevalence values ranged from 0.931% (±0.14) in Bermuda and 16.21% (±1.55) in Jamaica in 1999, to 4.28% (±0.46) in Bermuda and 6.60% (±0.93) in Puerto Rico in 2001, and to 6.28% (±0.64) in Puerto Rico and 10.91% (±1.57) in Akumal, Mexico, in 2002 (Fig. 2.6A). The total average prevalence of disease for 28 reef sites was 5.28% (±1.2). With the exception of Venezuela and Mexico, the average disease prevalence in octocorals was significantly higher in five of the nine localities (Fig. 2.6B). Over time, the average prevalence of diseases has increased significantly in Bermuda (from 0.931% in 1999 to 4.28% in 2001), Puerto Rico (2.13% in 1999 to 6.60 and 6.28% in 2001 and 2002 respectively) and in Venezuela (3.07% in 1999 to 5.81% in 2001).

There is geographic variability in terms of which diseases affect different species. For example, in the Virgin Islands and Culebra (eastern Puerto Rico), WP-II is more prevalent, affecting large numbers of colonies of *M. annularis* and *M. faveolata*, whereas along the southwestern coast of Puerto Rico, Mexico and Grand Cayman YBS is more prevalent. *M. faveolata* and *M. annularis* populations, for example, had between 4 and 15% (averages between 4.65 and 12.58%) of their colonies infected with at least one disease in reef sites of four geographically distant areas of the Caribbean in 1999 (Weil et al. 2002). Aspergillosis, on the other hand, affected a higher proportion of colonies (5.0–30.56% in reef sites) of the sea fan *Gorgonia ventalina* with averages between 8.80 and 18.67% for four localities (Table 2.4). In 2001, the average incidence of ASP in individual reefs varied between 2.5 and 22.5% (Grenada), 9.36 and 18.14% (Puerto Rico) and 4.17 and 21.29% in Bermuda. The average incidence of ASP in *G. ventalina* populations in Puerto Rico increased from 8.8% (±3.5) in 1999 to 13.4% (±3.53) in 2001 and 15.27% (±5.24) in 2002. It is too early to conclude that these increases in prevalence are correlated with global increases in seawater temperature, but the possibility is there and correlations with SST measured in several of the surveyed localities over the last 8 years are now being explored.

Not all diseased colonies face total mortality (as opposed to partial mortality). Some of them may respond to the infection and stop its advance, like the sea fan *G. ventalina* which reacts to the fungal disease ASP and is capable of restraining its advance by sequestering the hyphae in hard structures called galls (Smith et al. 1996). General antibacterial and antifungal activity have been documented in extracts of this gorgonian (Jensen et al. 1996; Kim et al. 2000a), and evidence of localized induction adjacent to fungal lesions has been reported (Kim et al. 2000b). The presence of chitinase and melanin, common mechanisms of fungal resistance in plants and animals, respectively, has been confirmed in diseased sea fans and work is underway to quantify the effectiveness of these mechanisms. Other affections such as DSS and YBS advance very slowly (Table 2.3), and both seem to be sensible to changes in the light and temperature regimes, usually halting their advance when water temperature drops.

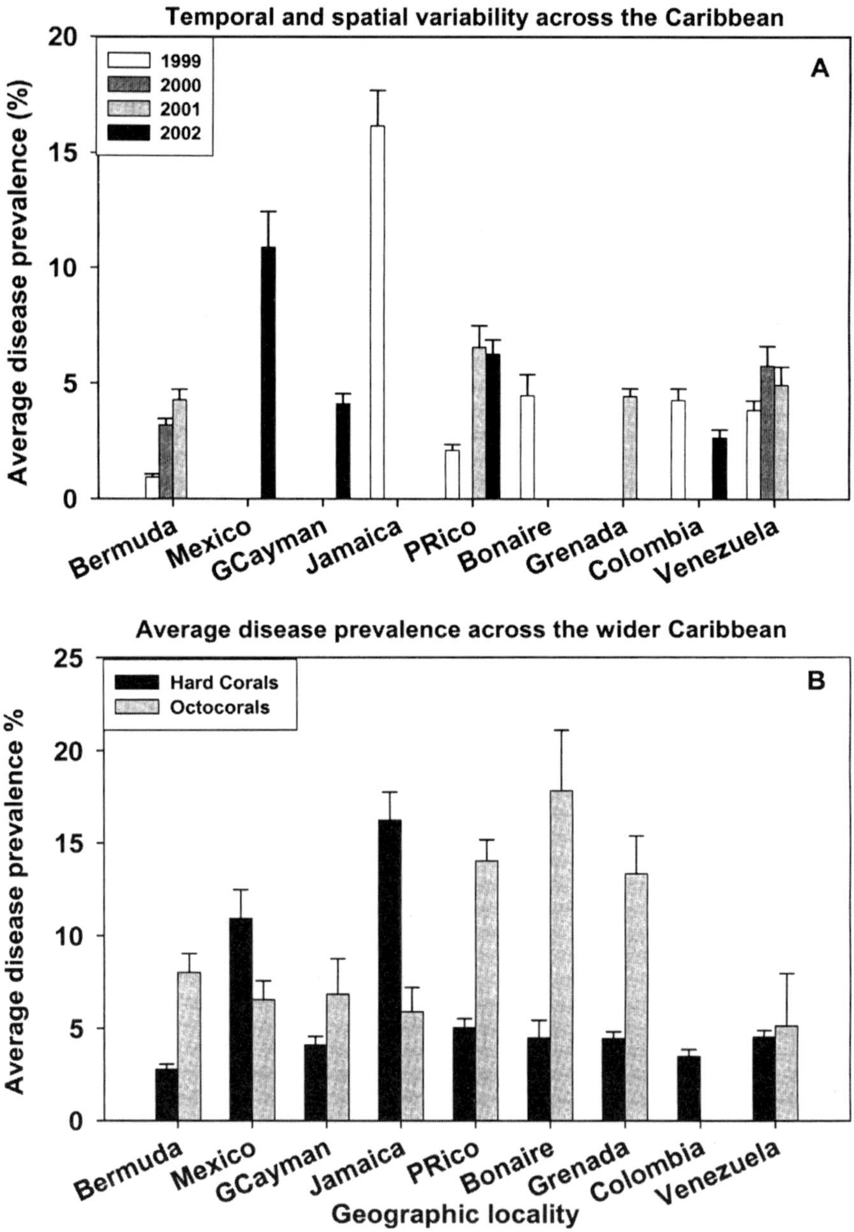

Fig. 2.6A, B. Geographic and temporal variability in disease prevalence across the wider Caribbean. A The average prevalence of all diseases (pooled at the community level) in each year (from 1999 to 2000) in which surveys were made at the particular locality. Jamaica had the highest prevalence in 1999 compared to other sites. Some localities were surveyed only once (e.g., Jamaica, Bonaire, etc.). In Bermuda, Puerto Rico and Venezuela (surveyed at least 3 years), the average disease prevalence increased significantly from 1999 to 2000. **B** In most localities, octocorals had a significantly higher prevalence of diseases compared with the coral community

Table 2.4. Average disease prevalence (%) in major reef-building coral species and *G. ventalina* across five geographic localities in the wider Caribbean in 1999

Species	Bermuda	P. Rico	Bonaire	Colombia	Venezuela
A. palmata	–	1.31	2.5	14.97	28.13
A. cervicornis	–	1.15	3.39	39.58	0.37
M. faveolata	–	4.65	9.05	10.37	8.38
M. annularis	–	5.45	12.55	12.58	5.84
M. franksi	1.1	2.29	3.32	5.60	2.02
M. cavernosa	0.16	0.53	0.43	5.99	0.83
D. strigosa	1.3	2.13	0.0	7.15	1.11
D. labyrinthiformis	0.08	0.99	0.0	7.23	0.0
S. intersepta	0.0	0.22	5.61	12.29	6.76
S. siderea	–	5.32	16.5	12.30	11.05
P. astreoides	0.16	0.79	0.0	1.54	1.3
P. porites	0.0	0.53	0.0	0.45	0.0
D. cylindrus	–	6.67	0.0	–	2.1
D. stockesii	–	4.63	0.0	16.3	0.0
C. natans	–	0.54	0.0	6.88	0.0
G. ventalina	12.55	8.08	18.67	–	10.0

In Bermuda, 35 *Diploria strigosa* colonies infected with BBD were tagged in the summer of 2000. All were alive, without traces of the disease, and regrowing lost tissue a year later. The disease disappeared during the cold winter season and colonies exhibited signs of recovery soon after. Similar recoveries have been reported for Florida and Jamaica (Bruckner and Bruckner 1997a Kuta and Richardson 1997). In contrast, tagged colonies of *M. faveolata* infected with YBS in Puerto Rico showed a reduction in the width of the band during the winter, but were never completely rid of the infection throughout 3 years of observations. Despite the potential importance of coral resistance in eventual remediation efforts, our knowledge of coral resistance to disease is still very limited.

Even though the only chronic, widespread epizootic in the wider Caribbean region at this time is ASP, local, short epizootic events, and the temporal persistence of some diseases in coral populations in some localities, may prove lethal for some reef communities. For example, YBS has been infecting large colonies of *M. faveolata* and *M. annularis* for 3–4 years in Desecheo Island, Puerto Rico, effectively reducing the live tissue area, which could affect the reproductive output in the near future. New infections of YBS started to appear in 2002 in several other reefs in the south coast of Puerto Rico where it was not present before (Weil, unpubl. data). Recent epizootic events of PNE have been reported for Mexico (Rodriguez-Martinez et al. 2001) and Puerto Rico (Weil and Ruiz 2003). In Puerto Rico, an outbreak occurred after 15 days of doldrum-like conditions (no wind, low water movement, low cloudiness and higher than normal surface water temperatures) in November 2000. Numerous fish and sea cucumber feces were observed on many colonies, and because of the calm conditions their residence time over live coral tissue was longer, possibly starting the disease. Removal of the feces showed white dead areas underneath that increased in size with time and then stopped. Between 35.29% (±10.5%) and 73.41% (±7.98%) of all colonies of *A. palmata* were affected by the initial outbreak of PNE in five reefs. The number of injuries (affected areas) per colony varied from 5 to 17 with an average of 9.57 (±2.73). Injuries varied in size from 1 to 495 cm^2. Average total loss of live tissue per colony was 16.97% (±2.71%). Average rate of tissue loss was 7.51 cm^2/day, or close to a linear advance of 2.5 cm/day during the initial phase, which lasted approximately 10 days. All but a few injuries then stopped increasing in size a few days after the environmental conditions had returned to normal. Three months after the outbreak, 6.34% of all injuries in all colonies had completely recovered (closed), 93% were recovering (tissue was actively growing back), and 0.66% have stayed the same. A few colonies were affected again late in the summer of 2001 and new injuries formed, but all had recovered by December. Results of this study and that of Rodriguez-Martinez et al. (2001) in Mexico, clearly show that PNE can produce substantial and fast tissue loss in *A. palmata*, and that colonies are capable of quick recovery when conditions return to normal.

Epizootic events such as the ones discussed above in combination with other detrimental factors (environmental and/or anthropogenic) might prove lethal for some reef communities throughout the Caribbean. However, before we speculate on the outcome of the current events, we need to gather more basic information on the spatial and temporal dynamics of the diseases (etiology and epizootiology) so that "epidemiological" models can be developed on the basis of sound quantitative data.

2.3.2.4
Disease Reservoirs and Vectors

Current research on disease reservoirs and vectors is hampered by lack of knowledge of the pathogens causing the majority of coral diseases. To date, disease reservoirs have only been identified for BBD (biofilms in reef sediments, which contain nonpathogenic versions of the BBD consortium; Richardson 1997) and ASP (atmospheric African dust, which contains spores of the fungus, *Aspergillus sydowii*; Shinn et al. 2000; Weir et al. 2000). The only disease vector that has been positively identified is the fireworm, *Hermodice carunculata* in the Mediterranean, which has been found to harbor *Vibrio shiloi* (a pathogen shown to induce bacterial bleaching in *O. patagonica*) in its gut (Sussman et al. 2003). This worm (if it is the same species) is very common in coral reef communities across the Caribbean and is one of the main predators of acroporid corals and octocorals. Therefore, it is possible that it might be a vector (reservoir?) for some disease affecting the acroporids or octocorals. Other good candidates include the snails *Coralliophylla abbreviata* and

C. caribbaea which are predators of acroporids, *Montastraea* spp., *Diploria* spp. and other coral species, parrot fishes (*Sparisoma viride*), and damselfishes that directly interact (predation, bites, etc.) with colonies of the important reef-building species that usually get infected. This is an important aspect of the dynamics of coral reef diseases and there is no information in the Caribbean. Vectors may be involved in the processes of both disease transmission and spread on both local and regional scales. More effort should be devoted to work in this research area in the near future.

2.4 Causes, Origin and Ecological Consequences

Although it is tempting to link rising disease occurrence with increasing anthropogenic impacts, or global climate change, the links are not always clear and the pool of quantitative data is at the moment too limited on a temporal scale to allow significant correlations. One hypothesis is that rising disease occurrence is related to increasing anthropogenic (direct and/or indirect) impacts (Kim and Harvell 2001; Kuta and Richardson 2002). For example, the current warming trend of the oceans, the increasing frequency of African dust storms due to desertification of the northern Sahara, and continuing sediment/organic inputs due to deforestation around the globe, are all potential causes. Warming temperatures are linked with increased disease outbreaks not only in corals, but in other marine organisms, human, agricultural and wildlife diseases too (Hayes et al. 2001; Harvell et al. 2002). However, the links are not clear. For example, the WBD outbreak in Belize occurred on reefs relatively distant from anthropogenic impacts. Data from extensive geographic surveys show that prevalence of some diseases is higher in localities distant from areas of major human development (Weil et al. 2002). The patchy necrosis outbreak that affected a high proportion of colonies of *A. palmata* in southwestern Puerto Rico (Weil et al. 2002; Weil and Ruiz 2003) occurred in reefs that are not close to any source of direct pollution although normally varying, environmental conditions were clearly different before and after the onset of the disease (see above). Increasing evidence seems to support the hypothesis that increasing ocean temperatures appear to be the common factor triggering many of the coral epizootic events (Harvell et al. 2002; Kuta and Richardson 2002; Rosenberg and Ben-Haim 2002).

At this point, it is difficult to explain the source(s) and sudden emergence of the majority of coral diseases/syndromes, except to suggest that they may be related to changing and deteriorating environmental conditions. Changes in climate, such as the current warming trend of the world's oceans, could affect basic biological and physiological properties and/or balances in parasite-host populations, making pathogens more virulent and/or hosts more susceptible to disease (Harvell et al. 2002; Rosenberg and Ben-Haim 2002). Increasing deposition of African dust and sediment from terrestrial runoff in deforested tropical areas may represent new sources for pathogens. Three different hypotheses have been put forward to explain the source of *Aspergillus sydowii*, the fungal pathogen causing aspergillosis in sea fans and other octocorals. These are terrestrial run-off (Smith et al. 1996), African dust storms (Shinn et al. 2000), and local marine sources of the fungus, which has high tolerance to salt and has been isolated from coastal systems (Roth et al. 1964). The worldwide distribution of this fungus and its ability to survive in a wide variety of habitats and environments seem to support all three hypotheses. Sources for all other coral diseases remain unknown, although suggestions have been made that some of them might be introduced through untreated sewage (Patterson et al. 2002).

The impact of widespread, intensive epizootic events may affect different ecological levels within the coral reef community, and may play important roles in the deterioration and/or significant changes in composition, structure and dynamics at local and geographic scales. This is especially true if "key" species are affected. Diseases can impact individuals, populations and communities. The WBD-II epizootic significantly impacted the populations of two of the most abundant coral species (*A. palmata* and *A. cervicornis*) in the Caribbean. Because these species actually provided most of the spatial heterogeneity (habitats and refuges for many other reef-associated species), their demise produced a cascade of effects that significantly reduced the biomass and biodiversity in many reef communities. The die-off of the black sea urchin *D. antillarum* across the wider Caribbean has significantly reduced the herbivore pressure in many reefs, allowing macroalgae to outcompete corals and changing reefs from coral-dominated to macroalgal-dominated communities in many localities of the Caribbean (Hughes 1994). In addition, because many coral reef invertebrates are modular (forming colonies of replicated modules, e.g., corals are colonies of polyps), diseases/syndromes will also affect aspects of their life history differently than in nonmodular organisms. In particular, rates of mortality are more complicated, since mortality could occur at the individual polyp level or at the colony level. Modular organisms, and especially corals, usually delay sexual reproduction until they have attained a minimum size (Szmant 1991; Szmant et al. 1997). Partial mortality could have significant consequences for reproduction by reducing fecundity and total colony reproductive output or by reducing colony size to below the reproductive threshold, effectively stopping sexual reproduction altogether (reversed puberty). Information on the minimum reproductive size is lacking for the great majority of coral species, and no information exists for octocorals. To effectively evaluate the impact of diseases on coral species, it is necessary to understand these aspects of their life cycles.

2.5 Some Unresolved Questions and Future Directions

Increasing evidence seems to point to diseases as an important factor contributing to the deterioration of coral reef communities around the Caribbean. Persistent infections and new epizootic events may result in significant deterioration over time. The combination of the impact of diseases with other detrimental factors (environmental and/or anthropogenic) might prove lethal for some reef communities around the Caribbean in the near future. However, before speculating as to the outcome of current events, it is important to gather more basic information on the pathology, etiology and epizootiology of these diseases/syndromes. Most of the limited quantitative information available on the prevalence of diseases and their impacts on coral populations and communities has been gathered in the Caribbean. However, even in the Caribbean region, there is limited information on even the most basic aspects of their pathology, etiology and epizootiology. We do not know the mechanisms by which most diseases kill their hosts, their rates of spread and host mortality, or how these parameters vary within and across host species, and at spatial and temporal scales. For most diseases and syndromes, we do not know their original sources, vectors and/or reservoirs, or how they are transmitted across colonies of the same species and of different species.

If habitat deterioration and global warming trends continue at the same rates, there is little we can do to help coral reefs communities around the world. We are still far away from any miracle "vaccine" or remedial protocol against any of the current coral reef diseases. The fact that other important members of the reef community are also being affected by new diseases and syndromes complicates the picture even more. One major concern is that marine pathogens can move faster and for longer distances than ever before due to human activities such as commercial shipping, movement of warships, and the transport of marine species for aquaculture and the aquarium trade. Another concern is how to protect coral reef communities from diseases, since marine reserves (MPA) are unlikely to protect coral reef organisms from regional-scale, waterborne pathogens such as the ones that have caused mass mortalities in the Caribbean. Coral reef managers are waiting for us to develop remedial and restoration protocols that they can use in effective management programs. For this to happen, we need to fill in the information gaps in order to understand the dynamics of disease infections, their transmission and spread, and host susceptibilities and resistance, and to develop epidemiological models for the most destructive syndromes. Furthermore, even if we can produce all the scientific information needed, and develop temporal remedial protocols, our contribution to the long-term survival of coral reef communities will be insignificant unless governments and managers increase their efforts to reduce (or stop) the input and effects of other major stressors (sediments, pesticides, nutrients, overfishing, etc.) from watersheds and coastal areas. Coral populations have the potential to recover if the environment quality of the reef is maintained. It is unrealistic to think that we can restore a 1000-year old coral reef and maintain it, without restoring the original environmental conditions. Without a concerted effort among researchers, governments and all the stakeholders, the future of tropical coral communities is in jeopardy.

Acknowledgments. I would like to thank E. Rosenberg and Y. Loya for their kind invitation to write this chapter. L. Richardson, I. Urreiztieta, G.W. Smith and Y. Loya made useful corrections and comments on the manuscript. Jaime Garzón-Ferreira, Aldo Croquer and Peter Gayle provided data for Colombia, Venezuela and Jamaica, respectively. H. Ruiz, A. Ortíz, M. Mills, G. Smith, S. Zea and I. Urreiztieta helped in the collection of data over the years. Portions of the research were supported by funds from the Total Oil Co. through the Diversitas program of UNESCO and CARICOMP. Logistical support was provided by the Department of Marine Sciences, University of Puerto Rico, Bermuda Biological Station, Fundación Científica los Roques and INTECMAR – Universidad Simón Bolívar in Venezuela, Bonaire Park Service, Grand Cayman Department of Natural Resources and the Department of Marine Biology, St. George's University, Grenada.

Appendix 1. List of coral reef organisms affected by different diseases and syndromes in Caribbean coral reefs. Disease acronym as described in Table 2.2 (*OTH* other rare syndromes).

Scleractinian corals	WP-II	BBD	YBS	DSS-I	DSS-II	WBD[a]	WPX[b]	PNE	RBD	TUM	RS	OTH	Total
A. palmata		*				*	*	*		*		*	6
A. cervicornis		*				*	*			*		*	5
A. prolifera		*				*						*	3
M. annularis	*	*	*	*	*				*	*		*	8
M. faveolata	*	*	*	*	*				*	*		*	8
M. franksi	*	*	*	*								*	5
M. cavernosa	*	*	*	*	*				*	*		*	8
D. strigosa	*	*	*						*	*		*	6
D. labyrinthiformis	*	*	*	*						*	*	*	7
D. clivosa	*	*								*		*	4
C. natans	*	*	*	*	*				*			*	7
C. amaranthus	*	*	*		*							*	5
F. fragum	*	*	*										3
M. aerolata	*	*											2
P. astreoides	*	*	*						*			*	5
P. porites	*											*	2
P."branneri"	*												1
U. agaricites	*	*	*	*					*			*	6
U. purpurea	*								*				2
U. humilis	*								*			*	3
U. tenuifolia	*												1
A. lamarcki	*								*				2

Scleractinian corals	WP-II	BBD	YBS	DSS-I	DSS-II	WBD[a]	WPX[b]	PNE	RBD	TUM	RS	OTH	Total
H. cucullata	*											*	2
D. stockesii	*	*		*								*	4
D. stellaris	*			*									2
E. fastigiata	*											*	2
M. meandrites	*	*		*								*	4
M. memorialis	*	*		*								*	4
M. danae	*												1
D. cylindrus	*											*	2
S. siderea	*	*		*	*							*	5
S. radians	*			*					*	*			4
S. intersepta	*	*			*				*			*	5
M .ferox	*								*	*			3
M. lamarckiana	*												1
M. aliciae	*								*			*	3
I rigida	*	*											2
I. sinuosa	*	*											2
M. decactis	*												1
M. mirabilis	*												1
M. pharensis	*												1
S. bournoni	*				*								2
Total species	39	21	11	14	8	3	2	1	14	10	1	25	

[a]Includes WBD-I and type II.
[b]White pox and patchy necrosis (PNE) might be the same disease. Data from the author's research and other sources (Green and Bruckner 2000; Garzón-Ferreira and Gil-Agudelo 2001).

Hydrocorals	WPlague	Fungal Disease	Total
Millepora alcicornis	*	*	2
M. complanata	*	*	2
M. squarrosa	*		1
Total	3	2	

Octocorals	ASP	BBD	RB	TUM	Other	Total
Gorgonia ventalina	*	*	*		*	4
G. flabellum	*					1
G. mariae	*					1
Pseudoyterogorgia americana	*				*	2
P. acerosa	*					1
P. keryii					*	1
Plexaura flexuosa	*	*		*	*	4
P. homomalla				*		1
Pseudoplexaura porosa	*			*	*	3
Plexaurella nutans	*			*	*	3
Briaerum asbestinum					*	1
Erythropodium caribaeorum					*	1
Total	8	2	1	4	8	

Sponges	Tissue necrosis	White syndrome	Total
Xetospongia muta	*		1
Amphimedon compressa	*	*	2
Aplysina archeri	*		1
Aplysina fistularis	*		1
Cliona tenuis (nsp.)	*	*	2
Cliona lampa	*		1
Anthosigmela varians	*		1
Neofibularia nolitangere	*		1
Callyspongia plicifera	*	*	2
Callyspongia vaginalis	*		1
Agelas confiera	*		1
Total	11	3	

Crustose algae	White band syndrome	Total
Porlython pachydermum	*	1
Paragoniolithon sp.	*	1
Peysonnelia sp.	*	1
Total	3	

References

Abbot RE (1979) Ecological processes affecting reef coral populations in the east Flower Gardens Bank, northwest Gulf of Mexico. PhD Thesis, Texas A&M University, Texas, 154 pp

Aeby GS (2000) Spatial and temporal patterns of trematode infection on the reef coral *Porites compressa* in Hawaii. Proc 9th Int Coral Reef Symp, Bali, Abstract Vol 280

Alker AP, Smith GW, Kim K (2001) Characterization of *Aspergillus sydowii* (Thom et Church), a fungal pathogen of Caribbean sea fan corals. Hydrobiologia 460:105–111

Antonious A (1973) New observations in coral destruction in reefs. Assoc Mar Lab Caribb Abstr 10:3

Antonius A (1977) Coral mortality in reefs: a problem for science and management. Proc 3rd Int Coral Reef Symp Miami 1:617–623

Antonius A (1981) The "band" diseases in coral reefs. Proc 6th Int Coral Reef Symp 2:7–14

Antonius A, Lipscomb D (2001) First Protozoan coral-killer identified in the Indo-Pacific. Atoll Res Bull 1–21

Aronson RB, Precht WF (1997) Stasis, biological disturbance, and community structure of a Holocene coral reef. Paleobiology 23:326–346

Aronson RB, Precht WF (2001a) Evolutionary palaeoecology of Caribbean coral reefs. In: Allmon WD, Bottjer DJ (eds) Evolutionary paleoecology: the ecological context of macroevolutionary change. Columbia Univ Press, New York, pp 171–233

Aronson RB, Precht WF (2001b) White-band disease and the changing face of Caribbean coral reefs. In: Porter JW (ed) The ecology and etiology of newly emerging marine diseases. Developments in hydrobiology. Kluwer, Dordrecht, pp 23–35

Ben Haim Y, Rosenberg E (2002) A novel *Vibrio* sp. pathogen of the coral *Pocillopora damicornis*. Mar Biol 141:47–55

Bruckner AW (2002) Priorities for effective management of coral diseases. NOAA technical memorandum NMFS-OPR-22. US Dept Commerce, p 54

Bruckner AW, Bruckner RJ (1997a) The persistence of black band disease in Jamaica: impact on community structure. Proc 8th Int Coral Reef Symp I:601–606

Bruckner AW, Bruckner RJ (1997b) Outbreak of coral disease in Puerto Rico. Reef Sites Coral Reef 16:260

Bruckner AW, Bruckner RJ, Williams EH (1997) Spread of a black-band disease epizootic through the coral reef system of St Ann's Bay, Jamaica. Bull Mar Sci 61:919–928

Bythell JC, Sheppard C (1993) Mass mortality of Caribbean shallow corals. Mar Pollut J 26:296–297

Carlton R, Richardson LL (1995) Oxygen and sulfide dynamics in an horizontally migrating cyanobacterial mat: black-band disease of corals. FEMS Microb Ecol 18:144–162

Cervino J, Goreau TJ, Nagelkerken I, Smith GW, Hayes R (2001) Yellow band and dark spots syndrome in Caribbean corals: distribution, rate of spread, cytology, and effects on abundance and division of zooxanthellae. In: Porter JW (ed) The ecology and etiology of newly emerging marine diseases. Developments in hydrobiology. Kluwer, Dordrecht, pp 53–63

Denner EBM, Smith GW, Busse HJ, Schumman P, Narzt T, Polson SW, Lubitz W, Richardson LL (2003) *Aurantimonas coralicida* gen. nov. sp. nov., the causative agent of white plague type II on Caribbean scleractinian corals. IJSEM 53:1115–1122

Dube D, Kiho K, Alker AP, Harvell DC (2002) Size structure and geographic variation in chemical resistance of sea fan corals *Gorgonia ventalina* to a fungal pathogen. Mar Ecol Prog Ser 231:139

Dustan P (1977) Vitality of reef coral populations off Key Largo, Florida: recruitment and mortality. Environ Geol 2:51–58

Dustan P (1987) Preliminary observations of the vitality of reef coral in San Salvador, Bahamas. In: Currant HA (ed) Proc 3rd Symp on Geol of the Bahamas. Ft Lauderdale, Florida, pp 57–65

Dustan P (1993) Developing methods for assessing coral reef vitality: a tale of two scales. In: Ginsburg B (ed) Proceedings of the Colloquium on Global aspects of coral reefs: health, hazards and history. Univ Miami, RSMAS, pp 38–44

Dustan P (1999) Coral reefs under stress: sources of mortality in the Florida Keys. Nat Res Forum 23:147–155

Dustan P, Halas JC (1987) Changes in the reef-coral community off Carisfort Reef, Key Largo, Florida: 1974–1982. Coral Reefs 6:91–106

Edmunds PJ (1991) Extent and effect of black band disease on Caribbean reefs. Coral Reefs 10:161–165

Epstein PR, Sherman K, Spanger-Siegfried E, Langston A, Prasad S, McKay B (1998) Marine ecosystems: emerging diseases as indicators of change. Health Ecological and Economic Dimensions (HEED), NOAA Global Change Program, 85 pp

Garrett P, Ducklow H (1975) Coral disease in Bermuda. Nature 253:349–350

Garzón-Ferreira J, Zea S (1992) A mass mortality of *Gorgonia ventalina* (Cnidaria: Gorgoniidae) in the Santa Marta area, Caribbean coast of Colombia. Bull Mar Sci 50:522–526

Garzón-Ferreira J, Gil-Agudelo DL, Berrios ML, Zea S (2001) Stony coral diseases observed in southwestern Caribbean reefs. In: Porter JW (ed) The ecology and etiology of newly emerging marine diseases. Developments in Hydrobiology. Kluwer, Dordrecht, pp 65–69

Gastoff P, Brown H, Smith C, Smith F (1939) Sponge mortalities in the Caribbean. Nature, p 807, Smith F ibid, p 785

Geiser DM, Taylor JW, Ritchie KB, Smith GW (1998) Cause of sea fan death in the West Indies. Nature 394:137–138

Gil-Agudelo DL, Garzón-Ferreira J (2001) Spatial and seasonal variation of dark spots disease in coral communities of the Santa Marta area (Colombian Caribbean). Proc Int Conf on Scientific Aspects of Coral Reef Assessment, Monitoring and Restoration. Bull Mar Sci 69:619–629

Ginsburg R (ed) (1993) Global aspects of coral reefs: health, hazards and history. RSMAS, Univ Miami

Gladfelter WB (1982) White Band disease in *Acropora palmata*: implications for the structure and growth of shallow reefs. Bull Mar Sci 32:639–643

Gladfelter WB, Gladfelter EH, Monaham RK, Ogden JC, Diil RF (1977) Coral destruction, environmental studies of Buck Island Reef National Monument, St Croix US Virgin Islands. National Park Service, US Dept of the Interior, Report 6

Goldberg WM, Makemson JC, Colley SB (1984) *Entocladia endozoica* sp. nov., a pathogenic Chlorophyte: structure, life history, physiology, and effect on its coral host. Biol Bull 166:368–383

Goreau TJ, Cervino J, Goreau M, Hayes R, Hayes M, Richardson L, Smith GW, DeMeyer G, Nagelkerken I, Garzon-Ferreira J, Gill D, Peters EC, Garrison G, Williams EH, Bunkley-Williams L, Quirolo C, Patterson K (1998) Rapid spread of diseases in Caribbean coral reefs. Rev Bio Trop 46:157–171

Green EP, Bruckner AW (2000) The significance of coral disease epizootiology for coral reef conservation. Biol Cons 96:347–461

Guzmán HM, Cortés J (1984) Mass death of *Gorgonia flabellum* L. (Octocorallia: Gorgonidae) in the Caribbean coast of Costa Rica. Rev Biol Trop 32:305–308

Harvell CD, Kim K, Burkholder JM, Colwell RR, Epstein PR, Grimes DJ, Hofmann EE, Lipp EK, Osterhaus ADME, Overstreet AM, Porter JW, Smith GW, Vasta GR (1999) Emerging marine diseases – climate links and anthropogenic factors. Science 285:1505–1510

Harvell CD, Kim K, Quirolo C, Weir J, Smith GW (2001) Coral bleaching and disease: contribution to 1998 mass mortality of *Briaeorum asbestinum* (Octocorallia, Gorgonacea). In: Porter JW (ed) The ecology and etiology of newly emerging marine diseases. Developments in hydrobiology. Kluwer, Dordrecht, pp 97–104

Harvell CD, Mitchell CE, Ward JR, Altizer S, Dobson AP, Ostfeld RS, Samuel MD (2002) Climate warming and disease risk for terrestrial and marine biota. Science 296:2158–2162

Hayes RL, Goreau NI (1998) The significance of emerging diseases in the tropical coral reef ecosystems. Rev Bio Trop 46:172–185

Hayes ML, Bonaventura J, Mitchel TP, Prospero M, Sheen EA, Van Dolah F, Barber R (2001) How are climate and marine biological outbreaks linked? In: Porter JW (ed) The ecology and etiology of newly emerging marine diseases. Developments in hydrobiology. Kluwer, Dordrecht, pp 213–222

Hughes TP (1994) Catastrophes, phase shifts and large scale degradation of a Caribbean coral reef. Science 265:1547–1549

Jensen PR, Harvell CD, Wirtz K, Fenical W (1996) Antimicrobial activity of extracts of Caribbean gorgonian corals. Mar Biol 125:411–419

Jolles AE, Sullivan P, Alker AP, Harvell DC (2002) Disease transmission of aspergillosis in sea fans: Inferring process from spatial pattern. Ecology 83(9):2373–2378

Kim K, Harvell CD (2001) Aspergillosis of sea fan corals: disease dynamics in the Florida Keys, USA. In: Porter KG, Porter J (eds) Linkages among ecosystems in the Florida hydroscape. CRC Press, Boca Raton, pp 813–823

Kim K, Harvell CD, Kim PD, Smith GW, Markel SW (2000a) Fungal disease resistance of Caribbean sea fan corals (*Gorgonia* spp). Mar Biol 136:259–267

Kim K, Kim PD, Alker AP, Harvell CD (2000b) Chemical resistance of gorgonian corals against fungal infections. Mar Biol 137:393–401

Kushmaro A, Loya Y, Fine M, Rosenberg E (1996) Bacterial infection and coral bleaching. Science 380:396

Kushmaro A, Rosenberg E, Fine M, Loya Y (1997) Bleaching in the coral *Oculina patagononica* by *Vibrio* AK-1. Mar Ecol Prog Ser 147:159–165

Kuta KG, Richardson L (1996) Abundance and distribution of black-band disease in the northern Florida Keys. Coral Reefs 15:219–223

Kuta KG, Richardson L (1997) Black band disease and the fate of diseased coral colonies in the Florida Keys. Proc 8th Int Coral Reef Symp 1:575–578

Kuta KG, Richardson LL (2002) Ecological aspects of black band disease of corals: relationships between disease incidence and environmental factors. Coral Reefs 21:393–398

Lessios HA (1988) Population dynamics of *Diadema antillarum* (Echinodermata; Echinodea) following mass mortality in Panama. Mar Biol 95:515–526

Lessios HA, Robertson DR, Cubit JD (1984) Spread of *Diadema* mass mortality throughout the Caribbean. Science 226:335–337

Loya Y, Bull G, Pichon M (1984) Tumor formations in scleractinian corals. Helgol Wiss Meeresunters 37:99–112

Miller J, Rogers C, Waara R (2003) Monitoring the coral disease plague type II on coral reefs of St. St John, US Virgin Islands. Proceedings of the 30th scientific meeting of the Association of Marine Laboratories of the Caribbean, Puerto Rico. Rev Biol Trop 51(4):47–55

Mitchell R, Chet I (1975) Bacterial attack of corals in polluted sea-water. Microb Ecol 2:227–233

Morse DE, Morse ANC, Duncan H (1977) Algal tumors in the Caribbean sea fan *Gorgonia ventalina*. Proc 3rd Int Coral Ref Symp Miami 1:623–629

Morse DE, Morse ANC, Duncan H, Trench RC (1981) Algal tumors in the Caribbean octocorallian *Gorgonia ventalina* II. Biochemical characterization of the algae, and first epidemiological observations. Bull Mar Sci 31:399–409

Moyer, MA, Blake NJ, Arnold WS (1993). An ascetosporan disease causing mass mortality in the Atlantic calico scallop *Argopecten gibbus* (Linnaeus, 1758), J Shellfish Res (12):305–310

Nagelkerken I, Buchan K, Smith GW, Bonair K, Bush P, Garzon-Ferreira J, Botero L, Gayle P,Harvell CD, Heberer C, Kim K, Petrovic C, Pots L, Yoshioka P (1997a) Widespread disease in Caribbean sea fans I. Spreading and general characteristics. Proc 8th Int Coral Reef Symp I:679–682

Nagelkerken I, Buchan K, Smith GW, Bonair K, Bush P, Garzon-Ferreira J, Botero L, Gayle P, Harvell CD, Heberer C, Kim K, Petrovic C, Pots L, Yoshioka P (1997b) Widespread disease in Caribbean sea fans II. Pattern of infection, and tissue loss. Mar Ecol Prog Ser 160:255–263

Patterson KL, Porter JW, Ritchie KB, Polson SW, Mueller E, Peters EC, Santavy DL, Smith GW (2002) The etiology of white pox, a lethal disease of the Caribbean elkhorn coral, *Acropora palmata*. Proc Natl Acad Sci USA 99:8725–8730

Peters EC (1984) A survey of cellular reactions to environmental stress and disease in Caribbean scleractinian corals. Helgol Wiss Meeresunters 37:113–137

Peters EC (1997) Diseases of coral-reef organisms. In: Birkeland C (ed) Life and death of coral reefs. Chapman and Hall, London, pp 114–136

Peters EC, Oprandy JJ, Yevich PP (1983) Possible cause of "white band disease" in Caribbean corals. J Invert Pathol 41:394–396

Porter JW, Meier O (1992) Quantification of loss and change in Floridian reef coral populations. Am Zool 23:625–640

Richardson LL (1996) Motility patterns of *Phormidium corallyticum* and *Beggiatoa* sp. associated with black-band disease of corals. Bio Bull 165:429–435

Richardson LL (1997) Occurrence of the black band disease cyanobacterium on healthy corals of the Florida Keys. Bull Mar Sci 61:485–490

Richardson LL (1998) Coral diseases: what is really known? Trends Ecol Evol 13:438–443

Richardson LL, Aronson RR (2002) Infectious diseases of reef corals. Proc 9th Int Coral Reef Symp Bali Indonesia 2:1225–1230

Richardson LL, Kuta KG (2003) Ecological physiology of the black band disease cyanobacterium *Phormidium corallyticum*. FEMS Microbiol Ecol 43:287–296

Richardson LL, Kuta K, Schnell S, Carlton R (1997) Ecology of the black-band disease microbial consortium. Proc 8th Int Coral Reef Symp Panama 1:597–600

Richardson LL, Goldberg WM, Carlton RG, Halas JC (1998a) Coral disease outbreak in the Florida Keys: plague type II. Rev Biol Trop 46:187–198

Richardson LL, Goldberg WM, Kuta KG, Aronson RB, Smith GW, Ritchie KB, Halas JC, Feingold JS, Miller M (1998b) Florida's mystery coral killer identified. Nature 392:557–558

Richardson LL, Smith GW, Ritchie KB, Carlton RG (2001) Integrating microbiological, microsensor, molecular, and physiologic techniques in the study of coral disease pathogenesis. In: Porter JW (ed) The ecology and etiology of newly emerging marine diseases. Developments in hydrobiology. Kluwer, Dordrecht, pp 71–89

Ritchie KB, Smith GW (1998) Description of type II white band disease in acroporid corals. Rev Biol Trop 46:199–203

Ritchie KB, Nagelkerken I, James S, Smith GW (2000) A tetrodotoxin-producing marine pathogen. Nature 404:354

Ritchie KB, Polson SW, Smith GW (2001) Microbial disease causation in marine invertebrates: problems, practices and future prospects. In: Porter JW (ed) The ecology and etiology of newly emerging marine diseases. Developments in hydrobiology. Kluwer, Dordrecht, pp 131–139

Roblee M, Barber T, Carlson P, Durako M, Fourqurean J, Muehkstein L, Porter D, Yarbro L, Zieman R, Zieman J (1991) Mass mortality of the tropical seagrass *Thalassia testudinum* in Florida Bay (USA). Mar Ecol Prog Ser 71:297–299

Rodriguez-Martinez RE, Benaszak AT, Jordan-Dahlgren W (2001) Necrotic patches affect *Acropora palmata* (Scleractinia:Acroporidae) in the Mexican Caribbean. Dis Aquat Org 47:229–234

Rosenberg E, Ben-Haim Y (2002) Microbial diseases of corals and global warming. Environ Microbiol 4(6):318–326

Roth FJ, Orpurt PA, Ahearn DG (1964) Occurrence and distribution of fungi in a subtropical marine environment. Can J Bot 42:375–383

Rowher, F Breitbar M, Jara J, Azam F, Knowlton N (2001) Diversity of bacteria associated with the Caribbean coral *Montastraea franksi*. Coral Reefs 120(1):85–91

Rützler K, Santavy DL (1983) The black band disease of Atlantic reef corals. I. Description of the cyanophyte pathogen. PSZNI Mar Ecol 4:301–319

Rützler K, Santavy DL, Antonious A (1983) The black band disease of Atlantic reef corals III. Distribution, ecology and development. PSZNI Mar Ecol 4:329–335

Santavy DL, Peters EC (1997) Microbial pests: coral disease in the Western Atlantic. Proc 8th Int Coral Reef Symp 1:607–612

Santavy DL, Peters EC, Quirolo C, Porter JW, Bianchi CN (1999) Yellow-blotch disease outbreak on reefs of the San Blas Islands, Panamá. Reef sites, Coral Reefs 18:97

Shinn EA, Smith GW, Prospero JM, Betzer P, Hayes ML, Garrison V, Barber RT (2000) African Dust and the demise of Caribbean coral reefs. Geo Res Let 27:3029–3032

Smith GW, Ives ID, Nagelkerken IA, Ritchie KB (1996) Aspergillosis associated with Caribbean sea fan mortalities. Nature 382:487
Smith GW, Harvell CD, Kim K (1998) Observations on the pathogenesis of sea fans infected with *Aspergillus* sp. Rev Biol Trop 46:205–208
Spalding DL, Greenfeld A (1997) New estimates of global and regional coral reef areas. Coral Reefs 16:225–230
Soniat TJ (1996). Epizootiology of *Perkinsus marinus* disease of eastern oysters in the Gulf of Mexico. J Shellfish Res 15:35–43
Squires DF (1965) Neoplasia in a coral? Science 148:503–505
Sussman M, Loya Y, Fine M, Rosenberg E (2003) The marine fireworm *Hermodice carunculata* is a winter reservoir and spring-summer vector for the coral-bleaching pathogen *Vibrio shiloi*. Environ Microbiol 5:250–255
Szmant AM (1991) Sexual reproduction by the Caribbean reef corals *Montastrea annularis* and *M. cavernosa*. Mar Ecol Prog Ser 74:13–25
Szmant AM, Weil E, Miller MW, Colon DE (1997) Hybridization within the species complex of the scleractinian coral *Montastraea annularis*. Mar Biol 129:561–572
Taylor D (1983) The black band disease of Atlantic reef corals. II. Isolation, cultivation, and growth of *Phormidium corallyticum*. PSZNI Mar Ecol 4:320–328
Weil E (2001) Caribbean coral reef diseases. Status and research needs. In: McManus J (ed) Priorities for Caribbean Coral Reef Research. National Center for Caribbean Coral Reef Research. RSMAS, University of Miami, Florida, 10 pp
Weil E (2002) Coral disease epizootiology: status and research needs. Coral health and disease: developing a national research plan. Coral Health and Disease Consortium, Charleston, South Carolina, 2002, 14 pp
Weil E, Ruiz H (2003) Tissue mortality and recovery in *Acropora palmata* (Scleractinia, Acroporidae) after a patchy necrosis outbreak in southwest Puerto Rico. Abstract, 31st scientific meeting of the Association of Marine Laboratories of the Caribbean, Port of Spain, Trinidad
Weil E, Smith GW, Mills M (2001) Spatial and temporal variability in coral and octocoral diseases in Bermuda. Abstract book, 30th scientific meeting of the Association of Marine Laboratories of the Caribbean (AMLC), La Parguera, Puerto Rico, p 20
Weil E, Urreiztieta I, Garzón-Ferreira J (2002) Geographic variability in the incidence of coral and octocoral diseases in the wider Caribbean. Proc 9th Int Coral Reef Symp Bali Indonesia 2:1231–1238
Weir JR, Garrison V, Smith GW, Shinn EW (2000) The relationship between gorgonian coral (Cnidaria: Gorgonacea) disease and African dust storms. 9th Int Coral Reef Symp, Abstract
Wilkinson C (ed) (2000) Status of coral reefs of the world 2000. Aust Inst Mar Sci 361 pp
Williams EH Jr, Bunkley-Williams L (2000). Major marine ecological disturbances of the Caribbean. Infect Dis Rev 2(3):110–127

3 Coral Disease on the Great Barrier Reef

Bette L. Willis, Cathie A. Page, and Elizabeth A. Dinsdale

3.1 Introduction

Coral disease is one of the most recent in a series of threats that is challenging the resilience of coral reef communities and is of particular concern because it may interact with and augment the impacts of other commonly recognised threats to coral health (e.g. bleaching, over-exploitation of fish stocks, destructive fishing practices and coastal developments). Since the first report of coral disease by Antonius in 1973, the rate of discovery of new diseases has increased dramatically with more than 29 coral diseases now described (Green and Bruckner 2000, Weil, this Vol.). Although coral disease is emerging as one of the major causes of coral reef deterioration in the Caribbean (Hayes and Goreau 1998; Harvell et al. 2002; Weil et al. 2002), at present we know very little about the ecology or pathology of coral disease on Indo-Pacific reefs. The comparatively few reports of coral disease from Indo-Pacific reefs, despite the region encompassing more than 80% of reefs worldwide (Bryant et al. 1998) is in contrast to the high proportion (>65%) of records in the Global Disease Database from the Caribbean reef region, now widely considered to be a coral disease hotspot (Green and Bruckner 2000; Weil, this Vol.). Such comparisons suggest that either disease is genuinely more prevalent in the Caribbean or lack of studies in other reef regions is underestimating its distribution and abundance. Distinguishing between these two alternatives represents an important step in advancing global epizootiological studies.

The rising incidence of marine diseases worldwide in the past few decades (Harvell et al. 1999), and particularly of coral diseases in the Caribbean, underscores the need for assessment of the status of disease on a region-by-region basis. Such assessments will help to identify the origins and reservoirs of pathogens and vectors involved in disease transmission. The Great Barrier Reef (GBR) stretches over 2000 km along the eastern coastline of Australia, representing the largest reef tract under management worldwide. Its unique status as one of the few reef systems under government jurisdiction for timescales that have preceded recent increases in the prevalence of coral disease has the potential to provide important insights into factors influencing disease occurrence and the underlying causes of escalating disease incidence. In this chapter, we summarise the current state of knowledge of coral disease on the Great Barrier Reef by (1) describing syndromes and diseases observed in our studies on GBR reefs and interpreted in the light of published literature and (2) presenting the results of a 5-year, large-scale study in conjunction with a regional disease prevalence study that together provide an overview of the current status of disease occurring on reefs extending over 1200 km of the Great Barrier Reef.

3.2 Overview of Diseases Infecting Great Barrier Reef and Indo-Pacific Corals

Until recently, it has been tacitly assumed that disease has had little impact on the population dynamics or community structure of coral assemblages on the Great Barrier Reef (GBR). However, there have been only two detailed studies of coral diseases on the GBR, both at Lizard Island in the northern sector: one on black band disease (BBD; Dinsdale 2002) and the other on skeletal eroding band (SEB; Antonius 1999; Antonius and Lipscomb 2001). A few additional sightings of coral diseases have been reported in anecdotal notes, i.e. BBD (Miller 1996) and white band disease (WBD; Baird 2000), although the report of WBD must be viewed with caution since a number of diseases are now known to produce white band-like symptoms (e.g. WBDI, WBDII, white plague I and II). Fungal pathogens have also been reported in gorgonians (Morrison-Gardiner 2001) and tumours in scleractinian corals (Loya et al. 1984). However, in general, there have been few studies specifically targeting coral disease, a factor likely to have contributed to the current paradigm of apparently low occurrence of coral disease on the GBR.

Elsewhere in the Indo-Pacific, in addition to BBD, SEB and WBD (Antonius 1985), there are isolated reports of diseases generally not yet described from the Caribbean. For example, yellow band disease (YBD) affected ten species primarily from the families Acroporidae and Poritidae in the Arabian Gulf (Korrubel and Riegl 1998); the encysting stage of a trematode has infected *Porites compressa* in Hawaii causing enlarged pink polyps (Aeby 1991); and *Porites* ulcerative white spot disease (PUWSD) infected more than 20% of *Porites* colonies on 8 out of 10 reefs surveyed in the Philippines (Raymundo et al. 2003). In addition, fungal-algal associations have affected *Porites lobata* in French Polynesia (Le Champion-Alsumard et al. 1995), cyanobacteria have affected *Porites lutea* in the Indian Ocean (Ravindran and Raghukumar 2002), and a bacterial pathogen has infected coralline algae [coralline lethal orange disease (CLOD)] throughout a large part of the South Pacific (Cook Islands, Fiji, Solomon Islands and Papua New Guinea, GBR; Littler and Littler 1995; C. Page, pers. observ.). Thus, despite the paucity of studies of coral disease in the Indo-Pacific region, the occurrence of the more common and infectious Caribbean diseases, in combination with reports of diseases unique to the region, suggest that infectious pathogens are a common component of Indo-Pacific reef communities and that disease may have a greater role in structuring coral communities in the region than previously thought.

3.2.1
Black Band Disease on the Great Barrier Reef

Black-band disease was first observed on GBR reefs in 1994 (Dinsdale 1994), more than two decades after the first Caribbean record (Antonius 1973) and a decade after the first Indo-Pacific record (Antonius 1985). Initial Indo-Pacific records were for two massive faviid species, *Goniastrea pectinata* and *Platygyra lamellina*, from the Philippines and a further seven massive faviids from the Red Sea (ibid), but BBD has subsequently been recorded on 21 species in five families from Lizard Island, GBR (Dinsdale 2002). Unlike in the Caribbean, where BBD primarily infects massive species (Kuta and Richardson 1996), branching pocilloporid and acroporid corals are important host species on the GBR (6.0 and 3.6% of corals in the families Pocilloporidae and Acroporidae, compared to 3.0 and 2.1% in the primarily massive families Faviidae and Poritidae; Dinsdale 2002; Fig. 3.1a, b). Dinsdale (2002) found a mean prevalence of 2.8% (and range of 1.3–4.9%) on Lizard Island reefs in the summer of 1994, which is comparable to the prevalence of BBD on most Caribbean reefs (Green and Bruckner 2000, Weil 2003). Seasonal increases in the prevalence of BBD on reefs in the Caribbean region are related to high summer seawater temperatures, but may also be related to water depth, coral diversity, population density and elevated nutrients (Kuta and Richardson 1996, 2002; Bruckner and Bruckner 1997; Bruckner et al. 1997). However, there are no studies of factors associated with the prevalence of BBD on GBR reefs, so the global generality of these patterns, particularly the associations between high temperatures and nutrients and increased abundance of BBD, remains unclear.

BBD isolated from Caribbean corals was originally described as a consortium of microorganisms dominated by *Phormidium corallyticum*, a gliding filamentous cyanobacteria, but including heterotrophic bacteria, marine fungus, sulphide-oxidising bacteria (*Beggiatoa*) and sulphate-reducing bacteria (*Desulfovibrio*; Ducklow and Mitchell 1979; Richardson 1996). However, recent molecular studies have identified anomalies in the identification of the cyanobacteria suspected to be the causative agent and a range of additional microorganisms associated with BBD mats of corals from St Croix, US Virgin Islands, Curacao, Netherlands Antilles and New Britain, Papua New Guinea (Cooney et al. 2002; Frias-Lopez et al. 2002, 2003). rDNA sequence analysis of microorganisms isolated from BBD mats have revealed the presence of up to three unidentified taxa of cyanobacteria, whereas *P. corallyticum*, the previously identified causative agent, was not detected (Cooney et al. 2002; Frias-Lopez et al. 2002, 2003). The lack of concordance in the cyanobacterial species associated with BBD mats between these and earlier studies and the differences in cyanobacterial taxa between Caribbean and Indo-Pacific (PNG) corals (Frias-Lopez et al. 2003) raise questions about the causative agent. They also highlight the need for further microbial and molecular studies of BBD from different coral species and from different reef regions. There have been no studies of microorganisms associated with BBD mats on GBR corals and it is possible that cyano-

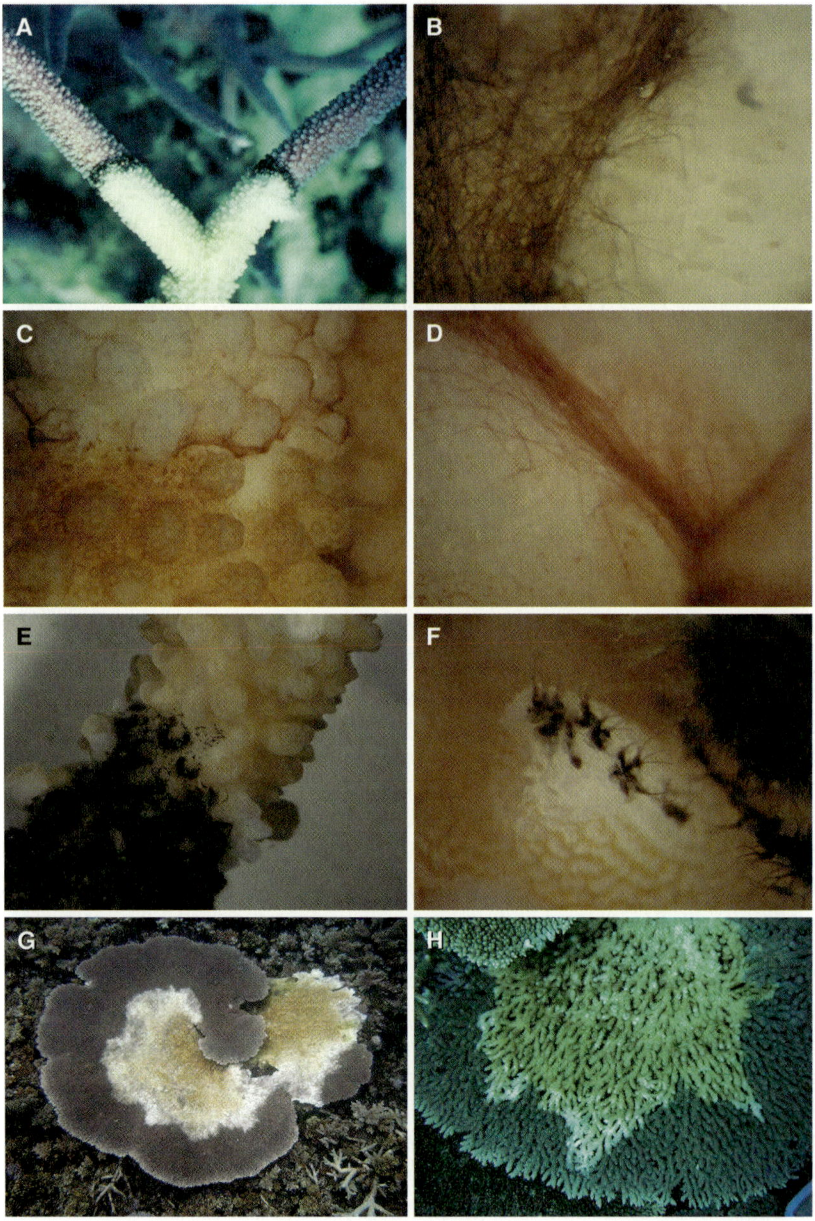

Fig. 3.1A–H. Field appearance of diseases and syndromes infecting corals and gorgonians on the Great Barrier Reef: **a** black band disease (BBD) on *Acropora intermedia*, **b** cyanobacterial mat, responsible for the black colouration of BBD, **c** unknown cyanobacteria forming a mat at tissue-skeleton interface, **d** unknown red-brown cyanobacteria forming BBD-like mat, **e** skeletal eroding band (SEB) on *A. intermedia* showing speckled appearance of band near tissue interface, **f** clusters of the ciliate, *Halofolliculina corallasia*, on live coral tissue ahead of the main SEB front, **g** white syndrome (WS) on *Acropora hyacinthus* in the Capricorn Bunker sector (photo AIMS LTMP), **h** WS on *Acropora clathrata* in the Lizard Is./Cooktown sector (photo AIMS LTMP)

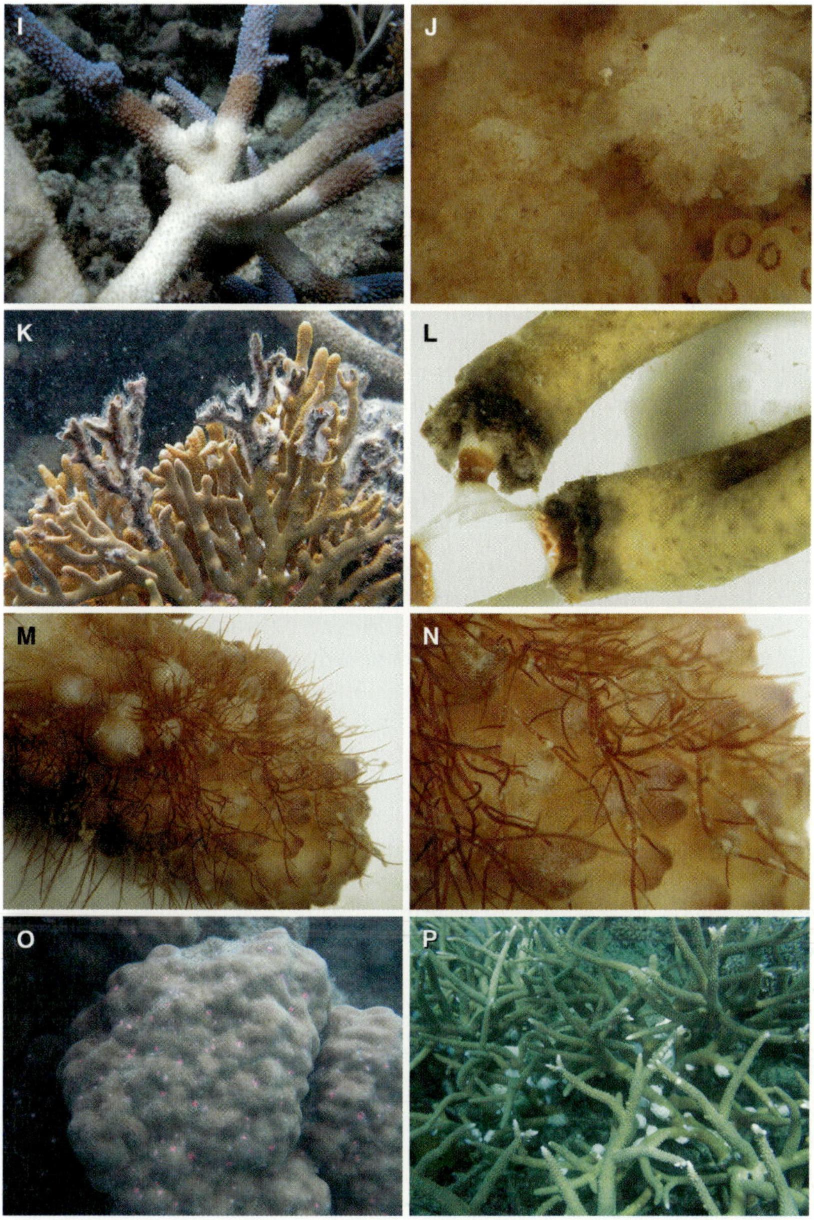

Fig. 3.1I–P. **i** brown band (BrB) on *A. muricata*, **j** clusters of the ciliates that cause the distinctive colouration of BrB, **k** black necrosing syndrome (BNS) on the gorgonian, *Isis* sp., **l** skeletal axis of *Isis* sp. exposed by BNS, **m** coral-algal interactions, **n** detail of filamentous algae overgrowing live coral tissue, **o** pink pigmented spots (PS) on massive *Porites* sp., **p** coral tumours on *Acropora* (photo L. Vail). All photographs were taken on Lizard Is. reefs by authors unless otherwise indicated

bacteria associated with GBR infections may differ from those isolated from Caribbean and even PNG corals. During our regional disease prevalence surveys (see Sects. 3.3.1.2, 3.3.3), we identified more than one type of cyanobacteria associated with coral disease states that resembled BBD (Fig. 3.1c, d). Therefore, in our analysis we have included unidentified cyanobacterial syndromes in the BBD category.

3.2.2 Skeletal Eroding Band : an Indo-Pacific Coral Disease?

Skeletal eroding band (SEB) is the only disease condition other than BBD for which there are more than anecdotal reports on the Great Barrier Reef. SEB is caused by the protozoan, *Halofolliculina corallasia* (Fig. 3.1e, f), which erodes the tissue and skeleton of corals as it produces a black lorica or test (Antonius 1999). Tissue damage occurs when the ciliates mechanically disrupt and lyse coral tissues through spinning and secretion of chemicals in the process of embedding their loricae within the coral skeletal matrix. Clusters of ciliates along the tissue-skeleton interface produce a black band (Fig. 3.1e) similar in appearance to black band disease, but the skeleton behind the advancing SEB is speckled with the remains of empty black loricae (Antonius and Lipscomb 2001), unlike the uniformly white skeleton exposed as BBD advances. Antonius and Libscomb (2001) report that the progression of SEB can be relatively slow, approximately 1 mm per week, further distinguishing it from BBD, but that it may also advance at rates up to 1 mm per day, comparable to BBD.

SEB affects at least 24 species of corals on reefs throughout the Indo-Pacific, but despite searching, there are no records from the Caribbean or the Atlantic Ocean (Antonius and Lipscomb 2001). A qualitative, 6-point scale was used to measure the prevalence of SEB on Indo-Pacific reefs, scoring the abundance of disease from rare (1–3 cases of SEB/30-min swim) to catastrophic (>100 cases per 30-min swim) (ibid). Prevalence of SEB increased in all reef regions revisited; from rare to moderate (4–12 cases/30-min swim) in the 10 years between visits to Lizard Island, GBR (1988–1998), and from rare to frequent (13–25 cases) in the 8 years between visits to Mauritius (1990–1998) and in the 3 years between visits to the Sinai (1994–1997) (ibid). Apart from these records at Lizard Island in the northern sector, the geographic extent of SEB on the GBR is currently unknown (but see Sect. 3.3.3).

3.2.3 White Syndrome – a Collective Term for Conditions Producing White Symptoms on the Great Barrier Reef

A proliferation of names for coral diseases that produce white symptoms in Caribbean corals presents challenges for relating Indo-Pacific white syndromes to the Caribbean white diseases based on macroscopic field characters. Rather than attempt to identify features such as the variable zone of bleached tissue that distinguishes white band II (WBII) from white band I (WBI), or differences in the rates of movement that distinguish the faster moving white plague II (WPII) from white plague I (WPI; reviewed in Richardson 1998), we have chosen to use the collective term white syndrome (WS) to describe conditions resulting in white bands of tissue and/or skeleton on GBR corals (Fig. 3.1g, h). In addition to WBI/II and WPI/II, white syndrome could potentially encompass white pox (Patterson et al. 2002), patchy necrosis (Bruckner and Bruckner 1997; Rodriguez-Martinez et al. 2001), and even shut down reaction (Antonius 1977). However, WS is distinguished from feeding scars by the narrow width of the zone of recently exposed, white skeleton and the relatively regular appearance of the tissue front. These features are in contrast to the wide zone of white skeleton commonly exposed following *Acanthaster planci* predation and the scalloped or irregular tissue front produced by *Drupella* spp.

Determining the relationship(s) between the Caribbean white diseases and WS and applying the appropriate name(s) will not be possible until pathogens infecting GBR corals are isolated and compared to those producing white symptoms in Caribbean corals. It is thus difficult to determine the accuracy of records of white band disease on the GBR (Baird 2000; Antonius and Lipscomb 2001) and of records of WBD infecting 20 coral species in the Philippines (Antonius 1985). However, since white band disease and white plague have caused major changes to coral communities in the Caribbean region (Aronson et al. 1998; Green and Bruckner 2000; Aronson and Precht 2001), the potential for their presence and impact on coral communities on the GBR should be viewed with concern (see Sect. 3.3.2 for current distribution and abundance of WS on the GBR).

3.2.4
Brown Band: a New Syndrome on the Great Barrier Reef

Brown band (BrB) is a new syndrome that we have recorded for the first time infecting corals on surveys in the northern and southern sectors of the GBR (see Sect. 3.3.3). The distinctive macroscopic field symptom of corals infected with BrB is a brown zone of variable width, flanked by healthy tissue at the advancing front and exposed white skeleton at the trailing edge as the band progresses over the surface of the colony (Fig. 3.1i). There is often a white zone between the healthy tissue and brown band, which may comprise bleached tissue and/or denuded skeleton. Dense populations of ciliates, packed with zooxanthellae from engulfed coral tissue, cause the brown coloration of the band (Fig. 3.1j). As densities of ciliates decrease, the zone becomes lighter and may appear white at very low ciliate densities. In these latter cases, the condition would be assigned to the WS category based solely on field observations. It is possible that BrB is caused by the ciliate, *Helicostoma nonatum*, which is thought to produce a brown jelly-like condition on corals grown in aquaria (Borneman 2001), but to our knowledge, this ciliate infestation has not been reported previously from in situ corals. Note that an earlier report of a brown band on a colony of *Acropora formosa* (Dinsdale 1994) referred to a different, but unknown syndrome, and has subsequently been mistakenly quoted as affecting 20 coral species on the GBR (Santavy and Peters 1997; Borneman 2001). While it is possible that the unknown syndrome was caused by a cyanobacterium similar to the one causing red-band disease in the Caribbean, as suggested by Santavy and Peters (1997), in the absence of the specimen it is not useful to speculate further about this isolated observation; it is not to be considered a record of BrB as described here.

3.2.5 Gorgonian Infections on the Great Barrier Reef: Black Necrosing Syndrome

Gorgonians are highly susceptible to disease in the Caribbean, where the fungal disease Aspergillosis has infected 12–90% of gorgonians on reefs in 13 countries (Nagelkerken et al. 1997a, b; Smith 2003) and black band disease has infected 13.8% of some species in the Florida Keys (Fengold 1988). However, little is known about gorgonian diseases on the GBR. The only study of GBR gorgonians to date reports that 10% of populations of *Isis hippuris* on Davies Reef were infected with a fungal disease that manifested as black necrotic areas and led to loss of both tissues and skeleton (Morrison-Gardiner 2001). Although two species of *Penicillium* isolated from infected gorgonians were able to infect healthy colonies of *I. hippuris* and *Pinnigorgia* sp., and could be re-isolated, they did not produce the typical symptoms of the disease (Morrison-Gardiner 2001). We have also observed black necrotic patches on many gorgonians at Lizard Island during our regional disease prevalence surveys (see Sect. 3.3.3) and will refer to the disease state as black necrosing syndrome (BNS; Fig. 3.1k, l). Whether gorgonian species on the GBR produce antifungal compounds similar to those produced by Caribbean gorgonians (Kim et al. 2000a, b), or vary in their susceptibility to fungal infections (Nagelkerken et al. 1997a) is unknown, but merits further study.

3.2.6 Coral-Algal Interactions: Algal Infections?

The impacts of coral-algal interactions may be positive, neutral or negative for the coral (reviewed in McCook et al. 2001), with negative interactions generally being discussed in the context of competition. However, when interactions that negatively affect corals (1) result in net positive benefits for algae and (2) impede the functioning and growth of coral polyps (e.g. through direct overgrowth and/or invasion of coral tissue), they take on the character of a disease. On reefs in the central GBR, examples that appear to cross the boundary between a competitive interaction and disease include overgrowth of coral by (1) the filamentous algae, *Coralliophila hurysmansii* causing tissue swelling, and (2) by *Anotrichium tenue*, which traps mucus, sediments and possibly microbes damaging the underlying tissues (McCook et al. 2001). We also found filamentous algae overgrowing live coral tissue in both the southern and northern GBR (Fig. 3.1m, n). What is unclear at this stage is whether some other stress or pathogen had previously weakened the corals' resistance allowing algae to invade their tissues. Therefore, rather than attribute coral mortality solely to algal overgrowth in our disease prevalence surveys (Sect. 3.3.3), we assigned such cases to an unidentified syndrome category. However, reports of a coralline red alga, *Pneophyllum conicum*, overgrowing and killing up to 100% of colonies of nearly all coral species present on a patch of reef in Mauritius (Antonius and Afonso-Carillo 2001) suggest that algal overgrowth can reach epizootic status. Controlled experimental studies on the ability of algal species to infect healthy coral tissues will clarify the pathogenic nature these coral-algae interactions.

3.2.7 Pigmentation Response in *Porites:* A symptom with a variety of causes?

The reef coral, *Porites,* appears to respond to a variety of competitive, invasive and parasitic challenges by producing pink or purple pigmentation in polyps adjacent to interaction sites (Fig. 3.1o). Hence pink lines, rings or spots are often visible in coral tissue bordering the margins of competing or boring organisms. The pigmentation appears to be a symptom of a response mounted by the coral to contain invading or competing organisms such as cyanobacteria (Ravindran and Raghukumar 2002), polychaetes, molluscs, and the intermediate metacercariae stage of the digenetic trematode, *Podocotyloides stenometra* (Aeby 1991, 1998). The trematode has been reported to encyst in tissues of the massive coral, *Porites compressa,* on Hawaiian reefs causing coral polyps to appear swollen and pink in colour (Aeby 1998). Infected polyps are unable to retract, reducing their function and increasing their vulnerability to predation by butterflyfish, the final host for the trematode. On Hawaiian reefs, the pink spots represent a parasitic infection, which reduces growth of heavily infected colonies by up to 50% (Aeby 1991). When the cysts were removed (through fish predation), healthy coral polyps were regenerated. We recorded the presence of pigmented spots (PS) on *Porites* colonies as a potential indicator of a parasitic infection in our GBR disease prevalence studies (see section 3.3.3). The pigmented spots appeared as small raised pink areas surrounded by healthy tissue, however the presence of trematodes has not been confirmed. Their location in the midst of healthy tissue is more consistent with a parasitic infection than a competitive interaction, unlike a variety of pink lines or rings that were commonly seen bordering dead patches and could generally be interpreted as a response to competitive interactions.

3.2.8 Coral Tumours

Coral tumours, manifesting as raised roughly spherical masses projecting about 4.5 cm above the surface of the colony, were reported to affect 18–24% of populations of *Platygyra pini* and *P. sinensis* on Magnetic Island, central GBR (Loya et al. 1984). Tumours were associated with increased growth rates of polyps and a general proliferation of all cell types, some atrophied and others normal, but in all cases macroscopic polyp structures were discernible and tissues remained pigmented (Loya et al. 1984). This type of abnormal growth has been termed a hyperplasia, in contrast to the bleached neoplasms that have been classified as calicoblastic epitheliomas. The latter appear as white, globular masses of skeleton raised above the surface of the colony and have few discernible polyp structures (reviewed in Peters et al. 1986). Tumours identified in our disease prevalence surveys were similar to the latter bleached neoplasms (Fig. 3.1p; see Sect. 3.3.3). Such tumours tend to be largest and most concentrated in the centre of colonies of table acroporids in the Gulf of Oman, whereas they tend to be similar in size along the length of branches in arborescent species (Coles and Seapy 1998). In high densities, tumours may reduce UV absorption rates (Coles and Seapy 1998), lipid storage capacity (Yamashiro et al. 2001) and linear growth rates of colonies (Bak 1983). Bleached neoplasms occur mainly on corals in the family Acroporidae and have been reported from throughout the Indo-Pacific, i.e. from Guam and Enewetak (Cheney 1975), French Polynesia (Le Champion-Alsumard et al. 1995), Japan (Yamashiro et al. 2001) and the Gulf of Oman (Coles and Seapy 1998).

3.3 Coral Disease Surveys on the Great Barrier Reef

The diversity of diseases and syndromes infecting GBR corals as described above highlights the need for targeted surveys of coral disease in the region. Here, we present the results of two types of studies designed to redress this need: (1) a large-scale study comprising rapid annual surveys of coral disease abundance (# cases per site) on 48 reefs as part of the Australian Institute of Marine Science (AIMS) long-term monitoring program (LTMP; Sweatman et al. 2001), and (2) a regional study comprising belt transect surveys to estimate disease prevalence (i.e. the total number of cases of disease expressed as a proportion of the total number of colonies examined per reef, site, family/order or disease category as appropriate) at selected sites in the northern and southern GBR. The large-scale AIMS LTMP surveys provide a broad overview of the abundance of two coral diseases (WS and BBD) on reefs throughout the Great Barrier Reef and follow changes in the number of cases of each disease over the last 5 years. The regional disease prevalence surveys are designed to detect all diseases and syndromes present at selected GBR sites, to determine their prevalence with respect to species and family groups, and to determine changes in prevalence associated with season, coral cover and wave exposure.

3.3.1 Survey Protocols

3.3.1.1 Large-scale Australian Institute of Marine Science Long-Term Monitoring Program Surveys

Forty-eight reefs spanning 1200 km of the Great Barrier Reef were surveyed for coral disease annually between 1998 and 2003 (Sweatman et al. 2001). Reefs were partitioned into six latitudinal sectors (i.e. Cooktown/Lizard Is., Cairns, Townsville, Whitsundays, Swains and Capricorn Bunkers sectors; Fig. 3.2) and three cross-shelf locations (inner, mid and outer-shelf). Within each sector, generally three reefs were surveyed in each of the three cross-shelf locations

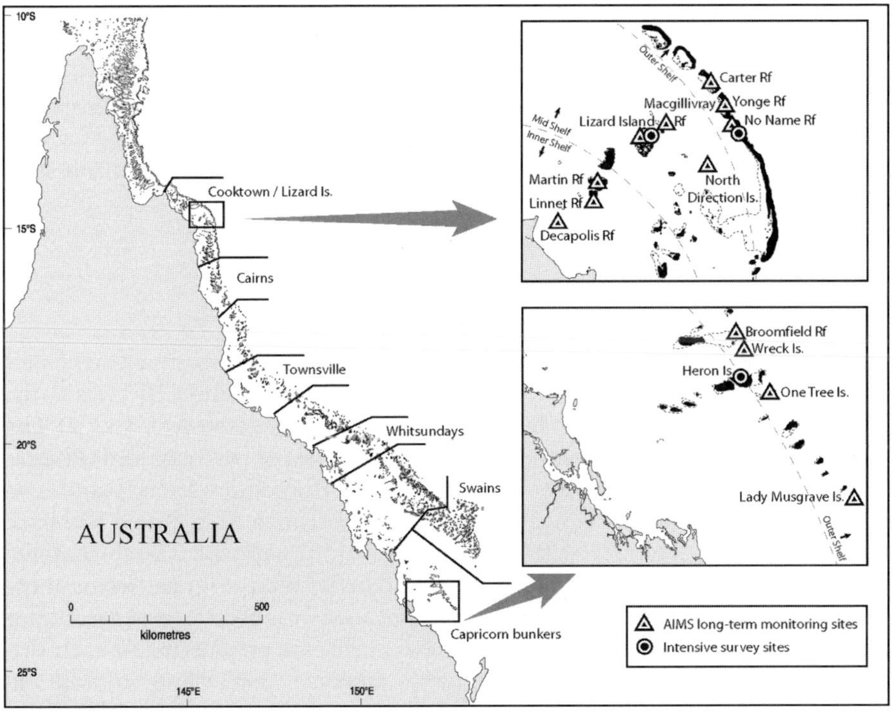

Fig. 3.2. Map of the Great Barrier Reef showing (1) the six sectors, and (2) the inner-, mid- and outer-shelf reefs in the Cooktown/Lizard Island sector and the outer-shelf reefs in the Capricorn Bunker sector that were surveyed as part of the Australian Institute of Marine Science Long Term Monitoring Program (AIMS LTMP), and (3) sites for the detailed surveys at Lizard Island and No Name Reef in the Cooktown/Lizard Is. sector and Heron Island in the Capricorn Bunker sector

(full methods in Sweatman et al. 2001). In total, there were 15 cross-shelf/sector combinations, which we will refer to as regions. Five 50-m transects were surveyed at each of three sites on the northeast flank of each reef. Transects were permanently marked and followed depth contours on the reef slope at 6–9 m. Surveys in the first 2 years (1998/1999, 1999/2000) were spread over the warmer months (September–May), whereas in the last 3 years, surveys in some sectors included the austral winter months of July and August. Changes in the timing of the surveys are discussed further in the context of their impact on disease prevalence in Section 3.3.2.2.

Coral mortality attributable to disease (BBD, WS), predation (*Acanthaster planci*, *Drupella*) and unknown sources was recorded in visual censuses (as per Bass and Miller 1996) of 2-m belts along each 50-m transect; thus an area of 1500 m^2 was surveyed on each reef. Diseases were identified from macroscopic field symptoms as outlined in Sections 3.2.1 for BBD and 3.2.3 for WS. Counts of the number of coral colonies manifesting symptoms of the two disease states on each transect are hereafter referred to as the number of cases of BBD or WS. It is likely that some cases of skeletal eroding band (SEB) and brown band (BrB) are included in the WS category because both can appear as white zones when ciliate densities are low (discussed in Sect. 3.2.4). Mortality was attributed to *A. planci* or *Drupella* when white zones were consistent with the appearance of feeding scars (see Sect. 3.2.3) and/or these predators were visible in the vicinity of white zones adjacent to healthy coral tissue. If coral mortality could not be clearly attributable to disease or predation, it was recorded in the unknown category. Percent cover estimates of benthic groups were determined from video transects (further details in Page et al. 2001).

3.3.1.2
Regional Disease Prevalence Surveys

To determine the prevalence of coral disease in summer, we surveyed eight sites in January 2003 in the northern and southern sectors of the GBR, where the AIMS LTMP found the highest number of cases of disease (see Sect. 3.3.2.2). The eight sites comprised: four mid-shelf sites at Lizard Island [two exposed (Bird Is., Lizard Head) and two sheltered (Vicki's and Horseshoe Reefs)] and two outer-shelf sites at No Name Reef (the exposed NE front and sheltered NW back reef) in the northernmost sector; and two sites [one exposed (Coral Gardens) and one sheltered (Little Bay)] at Heron Island in the southernmost sector of the GBR (Fig. 3.2). The two sheltered Lizard Island sites were also surveyed in winter (July 2002) to initiate seasonal comparisons of disease prevalence. At each site, three random 20×2 m belt transects were surveyed along depth contours at 3–6 m and all hard corals, soft corals and gorgonians were identified to the lowest taxonomic level recognised or morphological group as appropriate. Each colony was then categorised as healthy, bleached, or assigned to one of eight disease categories: BBD (including BBD-like mats associated with a number of different cyanobacteria), SEB, WS, BrB, tumour, BNS, PS (pigmented spots on *Porites*), or to an unidentified syndrome category. The unidentified syndrome category included filamentous algae overgrowing live coral tissue and unidentified syndromes causing deterioration in soft corals. Samples of diseased colonies were collected and examined microscopically to identify associated microorganisms and verify field identifications of disease states. To enable comparisons of disease prevalence with coral cover, we used line intercept surveys to record percent cover of the major benthic categories along the first 10 m of each transect.

3.3.1.3
Statistical Analysis

Differences in the abundance of WS detected in the AIMS LTMP surveys among shelf positions, sectors and years were tested using split-plot ANOVA. The total number of diseased colonies were summed over transects on each reef. Data were log transformed [log (X+0.1)] to satisfy assumptions of normality and homogeneity of variances. Where significant changes in disease abundance over time among sectors and shelf positions were identified, available degrees of freedom were partitioned into single degree of freedom contrasts to determine the specific years in which changes occurred within each sector by shelf combination. The abundances of BBD were too low to allow formal analysis of change.

Differences in distribution of WS among shelf positions, sectors and years were also examined by comparing changes in the proportion of transects on which WS was recorded using split-plot ANOVA. The number of transects with disease present was summed on each reef and divided by the number of transects sampled. The data were square root transformed to satisfy assumptions of normality and homogeneity of variances. As for WS abundance above, when significant changes over time in the proportion of transects with disease were identified among sectors and shelf positions, contrasts were used to determine the specific years in which changes occurred.

The relationships between WS abundance and (1) hard coral cover and (2) *Drupella* spp. abundance were examined by including hard coral cover and abundance of *Drupella* as covariates in a split-plot ANOVA model. Interaction terms in the model were used to estimate how consistent differences in relationships with WS abundance were among sectors and shelf positions. The abundances of WS and *Drupella* were log (X+0.1) transformed for analysis as described above. Similarly, single degree of freedom contrasts were used to determine when the relationship between disease abundance and coral cover or *Drupella* abundances differed among sectors and shelf positions.

The relationship between change in percent hard coral cover and change in WS abundance was also examined by including the change in cover of hard corals between years as a covariate in an additional split-plot ANOVA model.

Variations in disease prevalence detected in the regional disease prevalence surveys were compared among reefs (Lizard Is., No Name, Heron Is.) and among seasons (winter vs. summer) and exposures (sheltered vs. exposed) on Lizard Is. reefs using separate 1-way ANOVAs. When Levene's test determined that variances were heterogeneous, data were arcsine transformed. Differences in the distribution of the number of diseased vs. healthy colonies, pooled for the two sheltered and two exposed sites at Lizard Is., among the five scleractinian families in summer 2003 were tested using a χ^2 homogeneity test.

3.3.2
Results of Large-Scale AIMS LTMP Surveys

3.3.2.1
Patterns in the Distribution and Abundance of Black Band Disease

BBD is widespread throughout the GBR, occurring in all six sectors and all three cross-shelf locations. There were only three regions (mid-shelf Cooktown/Lizard Is., inner-shelf Cairns, and outer-shelf Townsville), of the 15 surveyed, in which BBD was not detected in any of the surveys. However, in any one year, BBD was recorded on a maximum of 2.5% of transects (n=720) from a maximum of 47% of regions (n=15).

The abundances of BBD were too low to allow formal analysis of change, however, the number of colonies infected by BBD did not appear to change markedly between 1998 and 2003 (Fig. 3.3), infections occurring on 0.04–0.47 colo-

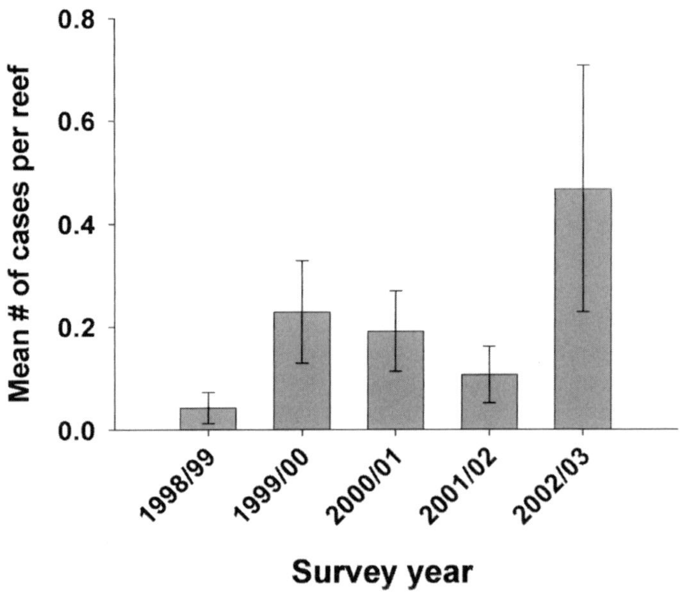

Fig. 3.3. Mean abundance (±SE) of black band disease (BBD) in survey years between 1998 and 2003. Histograms represent the mean of the total number of cases of BBD (±SE) in the 1500-m^2 area surveyed on each of the n=48 reefs per survey season

nies per reef in any given year. The highest occurrence of BBD was a total of 22 cases across all reefs in 2002/2003. Thus, despite its widespread distribution, the general abundance of BBD has been very low and stable, for the last 5 years.

3.3.2.2
Patterns in the Distribution and Abundance of White Syndrome

Abundance of WS

In contrast to the stable abundance of BBD over the past 5 years, white syndrome has increased 20-fold (F_{year}=52.12, *df*=4, *P*<0.001), from a mean of 1.7±0.58 cases in 1998 to 47.7±12.60 cases in 2002/2003 (Fig. 3.4). Mean occurrence of WS has increased at all three cross-shelf locations (Fig. 3.5), with significantly greater increases occurring on outer-shelf reefs, where there was a mean of 85±29.5 cases per reef in 2002/2003 (F_{shelf}=13.28, *df*=2, *P*<0.001). Overall, there is a pattern of increasing occurrence of WS with increasing distance from the coast over the 5 years ($F_{year \times shelf}$=1.36, *df*= 8, *P*=0.221), a pattern that is particularly pronounced in the last two survey years (Fig. 3.5). However, the pattern breaks down when variances due to the sector level are factored in ($F_{year \times shelf \times sector}$=1.91, *df*=28, *P*=0.008) because of the comparatively constant abundance of WS on all cross-shelf transects (within each year) in the Townsville, Whitsundays and Swains sectors and the higher abundance of WS on transects on the mid-shelf reefs in the Cairns sector in 2002/2003.

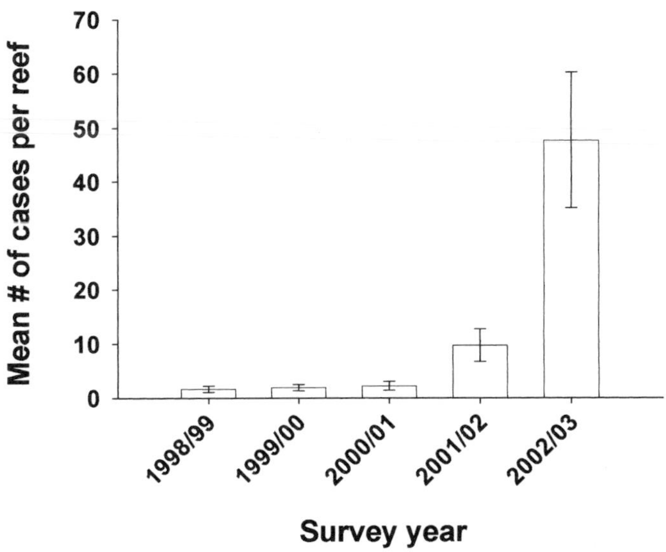

Fig. 3.4. Mean abundance (±SE) of white syndrome (WS) in survey years between 1998 and 2003. Histograms represent the mean (±SE) of the total number of cases of WS in the 1500-m^2 area surveyed on each of the reefs (n=48) per survey year

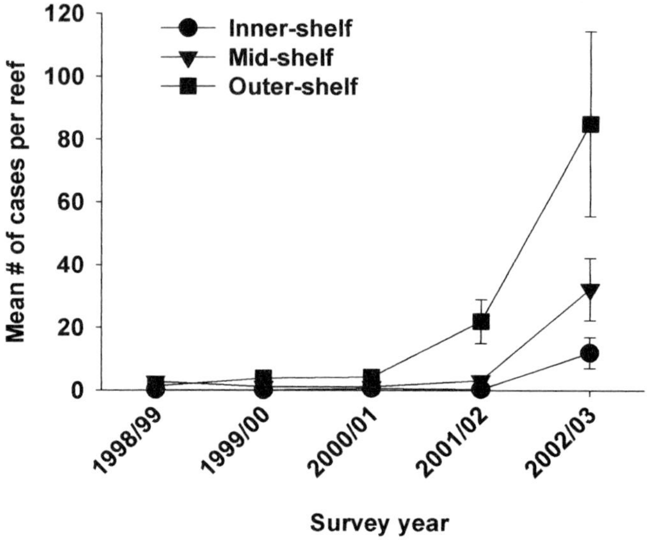

Fig. 3.5. Comparison of patterns in the mean (±SE) number of white syndrome (WS) cases per reef throughout the 5-year study period among reefs in three cross-shelf locations [i.e. inner-shelf (n=12 reefs), mid-shelf (n=18 reefs), and outer-shelf (n=18 reefs)]

The increase in WS abundance through time occurred in every sector except Townsville, where it has remained at low levels (≤11 cases per reef) since 1998/1999 (Fig. 3.6). Increases were first detected on outer-shelf reefs in the northernmost Cooktown/Lizard Is. sector in 2001/2002, when a more than 20-fold increase in the number of cases was recorded (an increase from a mean of 3 to 67 cases per reef). Smaller increases were also detected on both mid- and outer-shelf reefs from the central Whitsundays sector south to the southernmost sector (i.e. on outer-shelf reefs in the Whitsundays, mid-shelf reefs in the Swains and outer-shelf reefs in the Capricorn Bunker sectors). A dramatic, 30-fold increase in WS to a mean of 304 cases per reef occurred in the following year (2002/2003) on outer-shelf reefs in the southernmost Capricorn Bunker sector. The greatest increases in WS also occurred this year in regions representing all cross-shelf locations and all sectors except Townsville (i.e. on inner-shelf reefs of Cooktown/Lizard Is., mid-shelf reefs of Cairns, inner- and mid-shelf reefs of the Whitsundays, and mid- and outer-shelf reefs of the Swains sectors; Fig. 3.6). In summary, mean occurrence of WS has either increased (9 regions) or remained constant (6 regions) in all regions surveyed (n=15) on the GBR between 1998 and 2003 ($F_{\text{year} \times \text{sector} \times \text{shelf}}$=1.92, df=28, P=0.008).

Given that prevalence of coral diseases like BBD and white pox increase with high summer temperatures (Rodriguez-Martinez et al. 2001; Kuta and Richardson 2002), changes in the timing of survey seasons from warmer to cooler months in the Cooktown/Lizard Is and Capricorn Bunker sectors in 2000/2001 would be predicted to have underestimated the potential magnitude of changes

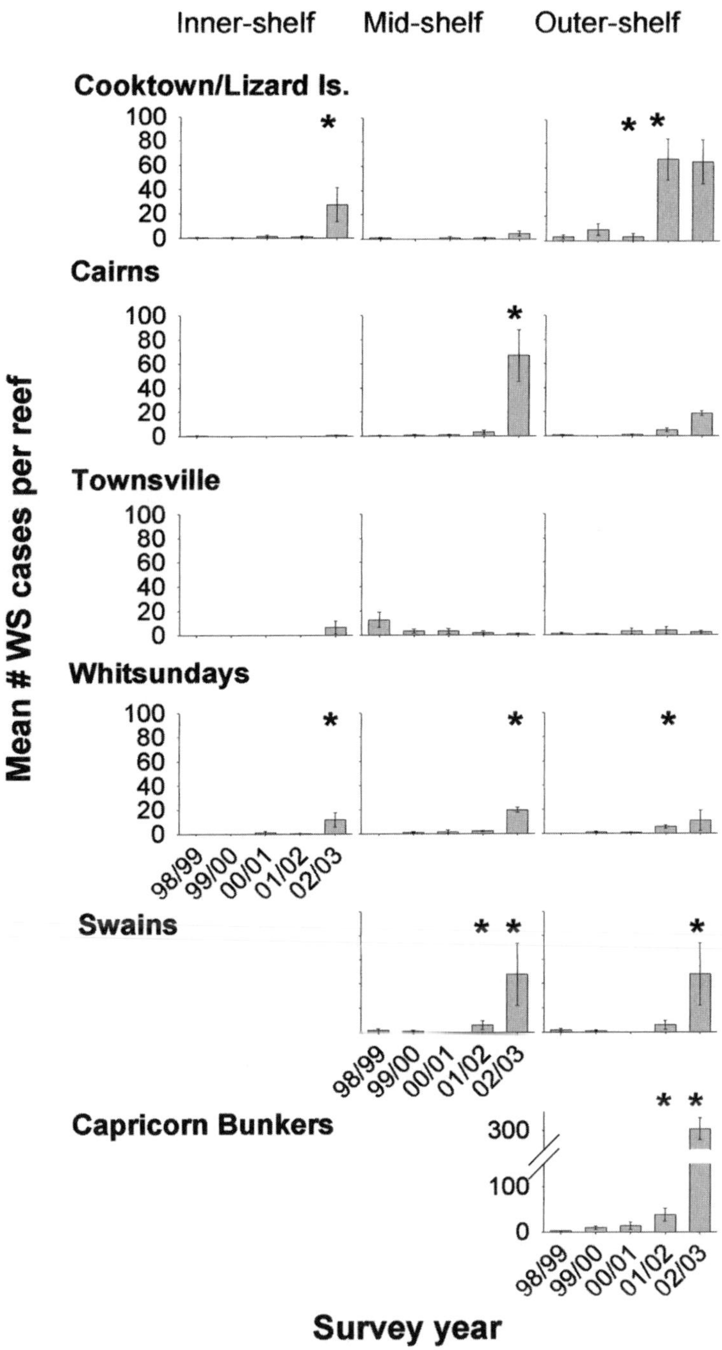

Fig. 3.6. Mean (±SE) of the total number of white syndrome (WS) cases per 1500-m^2 area surveyed on each reef compared among the 15 regions (i.e. combinations of the 6 sectors and 3 cross-shelf locations). Significant change from previous year is denoted by *

in the distribution and abundance patterns for WS. Thus, despite the striking 22- and 150-fold increases in the abundance of WS in outer-shelf reefs in these two sectors over the five years (Fig. 3.6), their magnitude might have been even greater if reefs in these sectors had been surveyed during summer in the last 3 years. In contrast, despite surveying the Cairns and Townsville sectors in the warmer months from 2000/2001 onwards, there was no increase in the abundance of WS. In addition, since WS was erected as a category representing a new source of mortality 6 years after the AIMS LTMP began, it is conceivable that researchers were changing the categorisation of colonies from unknown to WS. However, despite the continuously increasing abundance of WS, records in the unknown category remained relatively constant, suggesting that the rise in WS abundance is not accounted for by a decrease in the unknown category (data not shown).

Distribution of WS

In addition to WS becoming more abundant, infections have increased in distribution over the 5 years. In 1998, WS was distributed across approximately 75% of regions (11 of 15 regions) and 45% of reefs (22 of 48 reefs). However, by 2002/2003, WS had spread to all regions and 89% of reefs. Furthermore, the number of transects with WS increased more than ten-fold, from 3% of transects in 1998/1999 to 39% of transects in 2002/2003 (F_{year}=57.05, df=4, P<0.001). Patterns of increasing distribution of WS across sectors and regions are similar to those described above for abundance. In particular, we found the same pattern of consistently increasing occurrence of WS through time on transects in all cross-shelf locations (F_{shelf}=12.07, df=2, P<0.001; $F_{shelf \times year}$=1.83, df=8, P=0.077), but patterns of occurrence through time differed among sectors ($F_{sector \times year}$=3.57, df=20, P<0.001) and regions ($F_{sector \times shelf \times year}$=2.47, df=28, P<0.001). In summary, the number of transects with WS increased with increasing distance of cross-shelf location from the coast (when transects at each shelf location were combined across sectors), from lows of <1–5% of transects in 1998/1999 in all cross-shelf locations to maxima which differed with cross-shelf location in 2002/2003 (i.e. from 17% of transects on inner-shelf reefs to 45% of transects on mid- and 51% of transects on outer-shelf reefs). However, this pattern broke down because there were six regions spread across all shelf locations in which the percent of transects with WS did not increase in at least 1 year.

3.3.2.3 Relationship Between Percent Coral Cover and Abundance of White Syndrome

Given the profound increases of WS on reefs in some sectors of the GBR (up to 150-fold on outer-shelf reefs in the Capricorn Bunker sector, Fig. 3.6), we analysed WS abundance in relation to both percent coral cover and abundance of *Drupella* spp. to (1) provide insights into factors promoting the spread of WS and (2) evaluate the possible effects of WS on coral abundance. We found a significant relationship between mean percent cover of scleractinian corals and abundance of white syndrome (F_{cover}=14.55, *df*=1, *P*<0.001), which was consistent among shelf locations ($F_{cover \times shelf}$=2.11, *df*=2, *P*=0.127) and regions ($F_{cover \times sector \times shelf}$=1.70, *df*=7, *P*=0.117), and only marginally inconsistent between sectors ($F_{cover \times sector}$=2.38, *df*=5, *P*=0.044). However, single degree of freedom contrasts revealed that, although there were positive trends between percent coral cover and WS abundance in all sectors but Cairns, the relationship was only significant on reefs within the Capricorn Bunkers sector (*P*=0.012). After accounting for the association between percent cover of hard corals and WS abundance, we found no relationship between the abundance of *Drupella* spp. and WS (F=2.45, *df*=1, *P*=0.121). In summary, there is a general trend for abundance of WS to be greatest on reefs with the highest percent hard coral cover that was most pronounced in the Capricorn Bunkers sector.

Mean percent cover of scleractinian corals has fluctuated between about 20–40% in the past decade in all, except four of the regions surveyed (see Sweatman et al. 2001). On outer-shelf reefs in the Cooktown/Lizard Is. and Capricorn Bunker sectors, percent hard coral cover has increased continuously between 1995 and 1998 to approximately 60%, but this pattern altered in 1998, with percent coral cover remaining about the same or declining slightly during the last 5 years (Fig. 3.7). This decline in the rate of change of coral cover coincides roughly with the rising incidence of WS (Fig. 3.7a, b), which reached its highest abundances on outer-shelf reefs in these two sectors (Fig. 3.6). Thus, in the northern Cooktown/Lizard Is. sector, the mean number of WS cases has been highest for the last 2 years on Carter and No Name Reefs, where there is a declining, but non-significant, trend in coral cover (Fig. 3.7a). In the southern Capricorn Bunker Sector, there has been no change in coral cover in the past 5 years since percent cover has stabilised, but WS infections have only risen dramatically in the last year (Fig. 3.7b).

Given that large increases in WS appeared to have occurred concurrently with increases in hard coral cover on some reefs, we examined the relationship between changes in hard coral cover and changes in WS abundance between survey years. Thus we asked: "Does an increase or decrease in hard coral cover correlate with a corresponding increase or decrease in WS?". We found that change in hard coral cover did not always coincide with a similar change in WS abundance ($F_{\Delta cover \times \Delta WS}$=5.50, *df*=1, *P*=0.022); in particular, it was variable across cross-shelf positions ($F_{\Delta cover \times \Delta WS \times shelf}$=3.49, *df*=2, *P*=0.035). Single degree of freedom contrasts indicated that there was a significant association between changes in WS abundance and changes in coral cover only on outer-shelf reefs (*P*=0.025).

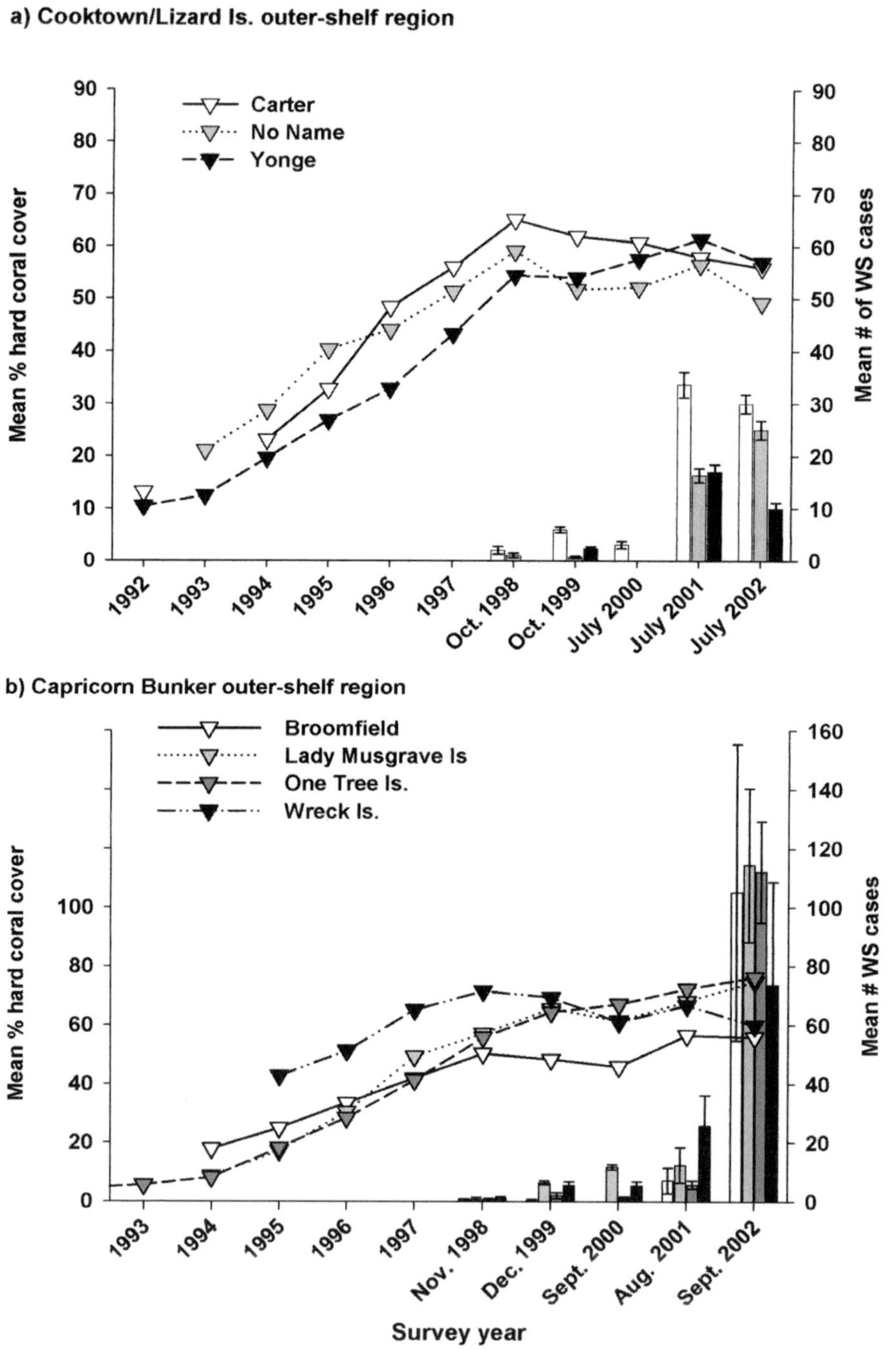

Fig. 3.7. Comparison of the mean percent hard coral cover (n=15 transects per reef) compared to mean (±SE) number of WS cases (n=15 transects per reef) in **a** Cooktown/Lizard Is. outer-shelf region, and **b** Capricorn Bunker outer-shelf region

3.3.3
Results of Regional Disease Prevalence Surveys

3.3.3.1
Comparison of Disease Prevalence Between the Northern and Southern Sectors of the Great Barrier Reef

Overall, symptoms of disease were detected in 8.97±0.79% of colonies (n=27,842) examined in the northern Cooktown/Lizard Is. and southern Capricorn Bunker sectors in summer 2003. Combining records for all scleractinians, alcyonaceans and gorgonians, mean disease prevalence was similar on the northern Lizard Is. (9.4±0.53%) and southern Heron Is. (10.4±2.07%) reefs, but marginally lower on the northern outer-shelf No Name reef (7.2±1.13%; Fig. 3.8). Patterns in overall mean prevalence of disease were

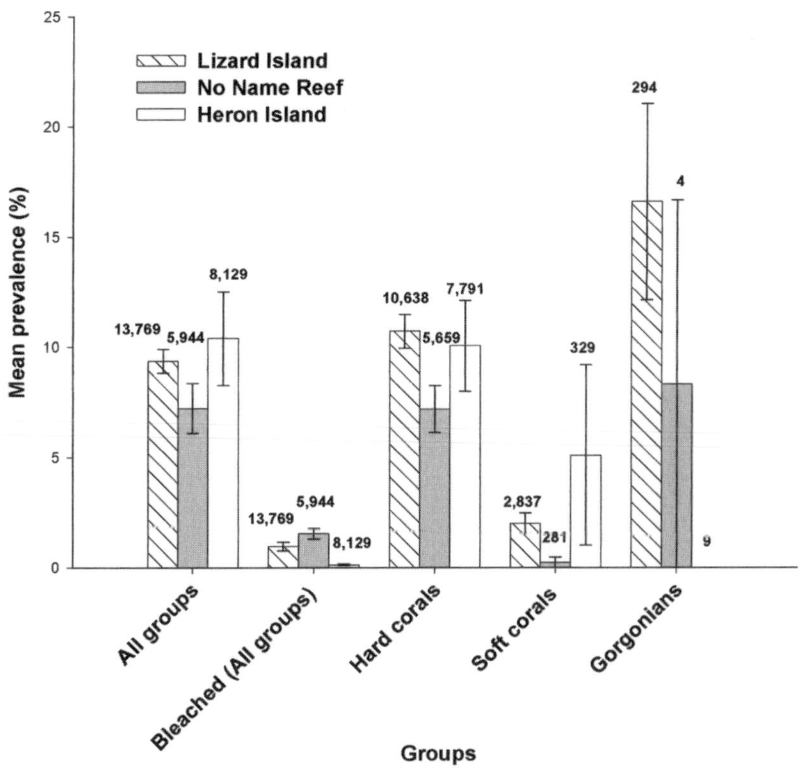

Fig. 3.8. Mean (±SE) prevalence of disease per reef for all taxonomic groups (hard and soft corals and gorgonians) combined compared to mean (±SE) prevalence of disease in hard corals, soft corals, gorgonians and bleached colonies (from all groups) at Lizard Is. (n=4 sites × 3 transects), No Name Reef (n=2 sites × 3 transects), and Heron Is. (n=2 sites × 3 transects). Disease prevalence (per taxonomic group) is calculated relative to the total number of colonies examined in each group, at each reef, as shown above the appropriate histogram

driven by patterns of disease prevalence in hard corals, which dominate cnidarian communities on these reefs (Fig. 3.8). Disease prevalence in hard corals ranged from a minimum of 7.2±1.06% at No Name reef to a maximum of 10.7±0.76% on Lizard Is. reefs, both in the northern sector. Gorgonian assemblages on Lizard Is. reefs were most affected by disease, with a mean of 16.6±4.50% of colonies infected on these reefs (Fig. 3.8). Disease was least prevalent amongst soft coral assemblages. Bleaching affected less than 1.7% of colonies from all three cnidarian groups.

3.3.3.2
Patterns in Prevalence of Disease Categories Among Sectors and Coral Families

The prevalence of four (SEB, BBD, BNS and PS) of the seven major disease categories was greatest at Lizard Is. in the northern sector, whereas WS was most prevalent at Heron Is. in the southern sector (Fig. 3.9). BrB and tumours were

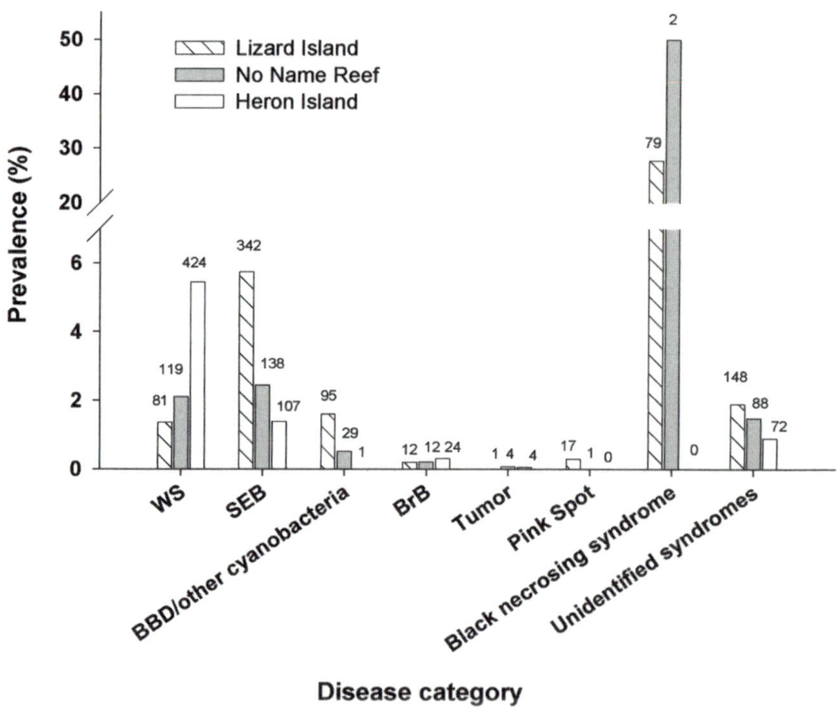

Fig. 3.9. Prevalence of each disease category at Lizard Is., No Name Reef and Heron Island in summer 2003, based on surveys of two sites per reef. Prevalence (per disease category) is calculated relative to the number of hard coral colonies examined at each reef for all disease categories except BNS (calculated relative to total number of gorgonian colonies) and unidentified syndromes (calculated relative to number of scleractinian and alcyonacean colonies combined). Total number of cases of each disease category recorded at each reef is shown above the appropriate histogram

uncommon on all reefs, although twice as many cases of BrB were recorded at Heron Is. in the southern sector compared to the two reefs in the northern sector. Among disease states, black necrosing syndrome (BNS) had the highest prevalence, but the syndrome was restricted to gorgonians, which represent only a minor component of the community. Among hard corals, white syndrome (WS) attained the highest prevalence of any of the disease categories at Heron Is., where 424 cases were recorded (survey area = 240 m^2; Fig. 3.9). Although WS was the most prevalent hard coral disease on Heron Is. reefs, skeletal eroding band (SEB) was the most prevalent on Lizard Is. reefs. Overall, WS and SEB were the two most common diseases on all reefs. Black band disease

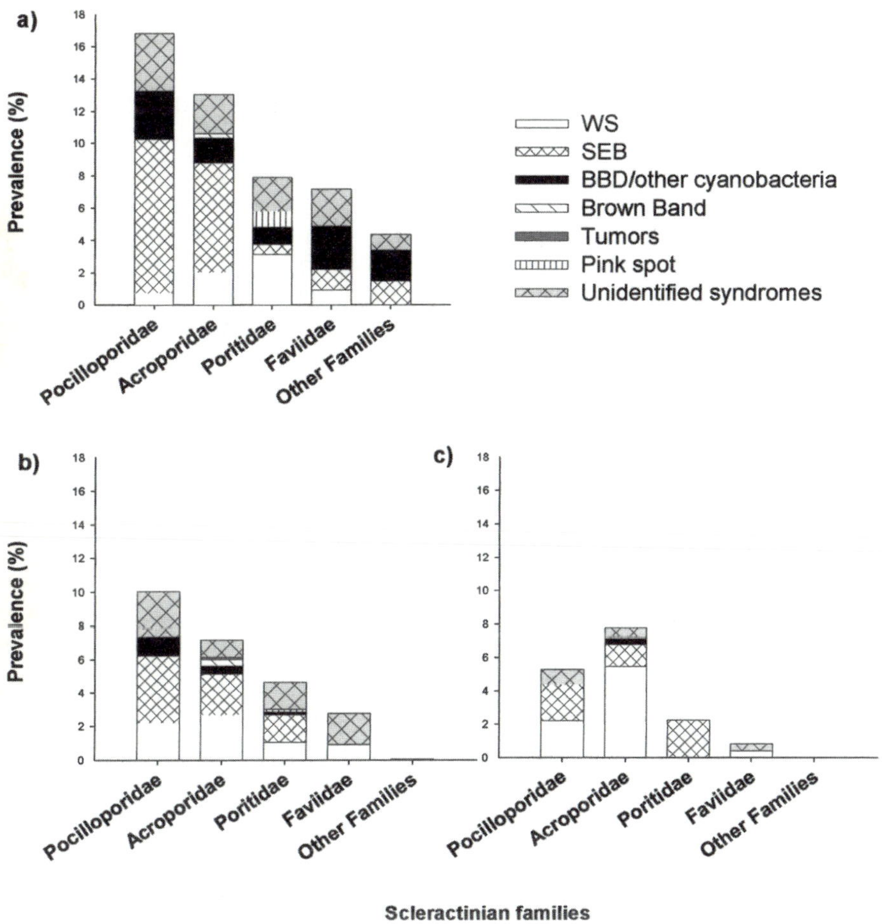

Fig. 3.10. Prevalence of seven disease categories in scleractinian families at **a** Lizard Is. (n=4 sites × 3 transects), **b** No Name reef (n=2 sites × 3 transects) and **c** Heron Is. (n=2 sites × 3 transects) in summer 2003. Prevalence (per family) is calculated relative to the total number of colonies examined in the respective family at each site

(BBD), which was grouped with unidentified cyanobacterial syndromes, tended to be the third most common disease and was present on all three reefs, although it affected only a very low proportion of colonies on Heron Is. reefs. Brown band (BrB) was also present on all three reefs, but generally with lower prevalence than BBD, although the pattern was reversed on Heron Is. reefs. In summary, five of the seven disease categories were present on reefs in all three locations (i.e. WS, SEB, BBD, BrB, and tumours), the exceptions being pigmented spots on *Porites* (PS) and black necrosing syndrome (BNS), which were not recorded from southern sector reefs (Fig. 3.9).

Disease prevalence varied among scleractinian families (χ^2=130.460, *df*=4, *P*<0.001), being greatest in the Pocilloporidae and Acroporidae at all three reefs (Fig. 3.10). When all disease categories were combined, disease prevalence in the northern sector was greatest (16.8%) in the family Pocilloporidae, but in the southern sector, it was greatest (7.8%) in the family Acroporidae. Otherwise, patterns of disease prevalence were consistent at all three reefs, decreasing in the Poritidae and further still in the Faviidae to a minimum prevalence of 0.8% in

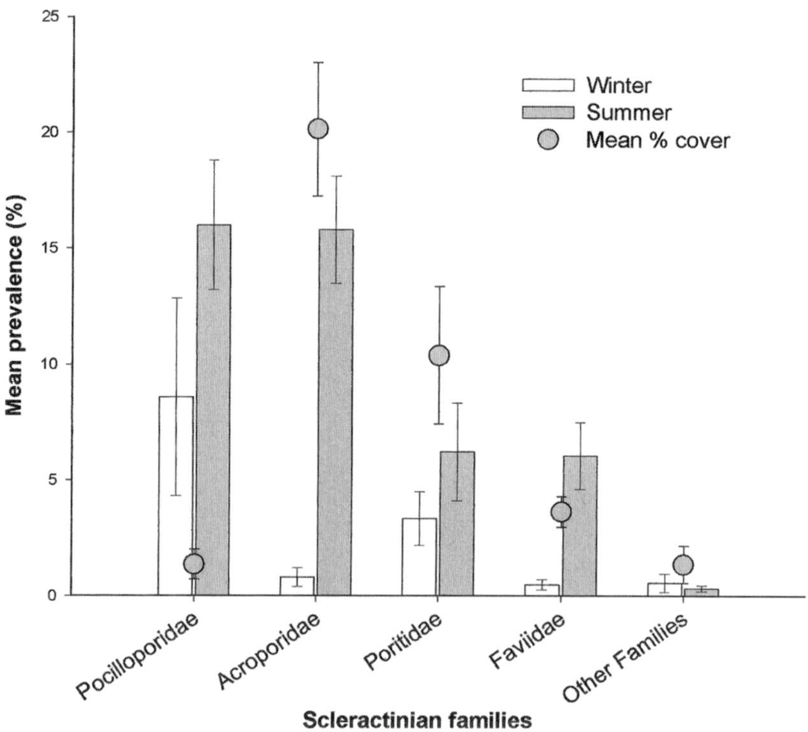

Fig. 3.11. Mean prevalence (±SE) of all diseases in scleractinian families at Lizard Island in winter 2002 vs. summer 2003 (*n*=2 sites × 3 transects) compared to mean (±SE) percent cover of each family in summer 2003 (*n*=2 sites × 3 transects). Prevalence (per family) is calculated relative to the total number of colonies examined in the respective family in each season

faviids on Heron Is. reefs. Interestingly, the high prevalence of disease in the pocilloporids on Lizard Is. reefs was despite the mean percent cover of this family being the lowest of the five family groups on these reefs (Fig. 3.11).

The major families (Pocilloporidae, Acroporidae, Poritidae and Faviidae) were each host to four to five diseases, with WS, SEB and BBD infecting corals in all four families (Fig. 3.10). SEB and WS were generally the most common diseases affecting pocilloporids and acroporids in both sectors, followed by BBD and BrB. The high prevalence of WS on Heron Is. sites (Fig. 3.9) is mostly explained by its high prevalence in acroporids; the proportion of colonies affected by WS being two times greater in acroporids than in pocilloporids at these sites. SEB was the dominant disease affecting pocilloporids and acroporids on Lizard Is. and No Name reefs. BBD showed highest prevalence at Lizard Is., where it disproportionately affected faviid corals. BrB affected a low proportion of corals (≤0.31%) on all three reefs, targeting particularly acroporids, but also pocilloporids and faviids. Poritids were host to WS, SEB, BBD and PS on both Lizard Is. and No Name reefs in the northern sector, but were only affected by SEB on Heron Is. in the southern sector. In the northern sector, pink pigmented spots (PS) were found on 0.97% of poritids on Lizard Is. reefs and 0.17% on No Name Reef, the only syndrome other than tumours (in this study) that was restricted to one family. Tumours were found only on acroporids, and only on a low proportion (≤0.13%) of colonies on each reef.

Within the Acroporidae and Pocilloporidae, WS affected at least 9 and 4 species respectively, and SEB at least 18 and 5 species respectively (Table 3.1). Colonies of other scleractinian families were also observed to be diseased, notably SEB affected fungid and merulinid colonies and BBD and unidentified cyanobacterial syndromes affected pectinid, mussid, dendrophylliid and siderastreid colonies (Table 3.1).

Table 3.1. Species and growth forms or taxonomic groups (if species not identified) of cnidarians displaying symptoms of seven potential disease states during regional prevalence surveys on Lizard Is. (L), No Name (N) and Heron Is. (H) reefs in January 2003. Total number of 1) scleractinian families or alcyonarian orders and 2) minimum number of species affected by each of the potential disease states are shown

		Disease state						
Family or order	**Species/ growth form**	**WS**	**SEB**	**BrB**	**BNS**	**BBD**	**Tumour**	**Unidentified**
Acroporidae	*Acropora hyacinthus*	N L	N H	H N L	–	L	–	H L
	A. cytherea	–	H L	L	–	–	–	H
	A. nasuta	–	L	–	–	L	–	–
	A. millepora	–	L	–	–	L	–	L
	A. subulata	–	–	H	–	–	–	–
	A. tenuis	–	–	–	–	–	–	H
	A. latistella	L	–	–	–	L	–	L
	A. cerealis	–	–	N	–	–	–	–

Table 3.1 (continued)

Family or order	Species/ growth form	Disease state WS	SEB	BrB	BNS	BBD	Tumour	Unidentified
	A. valida	–	H	H	–	H	–	H
	A. secale	–	–	N	–	–	–	–
	A. humilis	–	N	–	–	N	–	–
	A. gemmifera	–	N L	–	–	L	–	–
	A. monticulosa	–	L	N	–	–	–	L
	A. elseyi	–	–	–	–	L	–	–
	A. microphthalma	–	H L	–	–	L	–	–
	A. loripes	–	L	–	–	L	–	–
	A. austera	–	H	–	–	–	–	H
	A. yongei	–	L	L	–	L	–	–
	A. muricata	–	H N L	H N L	–	H L	–	H L
	A. grandis	–	L	L	–	–	–	–
	A. florida	–	–	–	–	L	–	–
	A. intermedia	–	H L	H L	–	–	–	H
	A. palifera	–	H	L	–	–	–	H
	A. cuneata	–	–	–	–	N	–	–
	A. cuneata/palifera	–	–	–	–	–	–	N
	A. brueggemanni	–	–	–	–	–	–	L
	Tabular *Acropora*	N L	N L	N L	–	N L	N	N
	Staghorn *Acropora*	H N L	H N L	N L	–	L	–	H L
	Corymbose Acropora	H N L	H N L	H N	–	L	–	H N L
	Digitate *Acropora*	H N L	H N L	N	–	N L	–	N L
	Bottlebrush *Acropora*	H L	H N L	–	–	H L	–	N L
	Bushy *Acropora*	N L	N L	–	–	L	L	L
	Isoporan *Acropora*	H N	H N L	H	–	N L	H	H N L
	Astreopora spp.	L	H L	–	–	L	–	L
	Montipora spp.	L	H NL	H	–	H N L	L	H L
Pocilloporidae	*Pocillopora meandrina/verrucosa*	–	N	N	–	–	–	–
	P. verrucosa	–	L	–	–	L	–	–
	P. eydouxi	–	N L	–	–	L	–	N
	P. damicornis	H N L	H N L	L	–	N L	–	H N L
	Seriatopora hystrix	–	L	–	–	L	–	–
	Seriatopora spp.	H N	H L	–	–	N L	–	L
	Stylophora pistillata	HN	H N L	–	–	N L	–	N L
	Other pocilloporids	N L	H N L	N	–	N L	–	H N L
Poritidae	*Porites* spp.	N L	N L	–	–	N L	–	N L
Fungiidae		–	H	–	–	–	–	–
Pectiniidae		–	–	–	–	L	–	L
Mussidae		–	–	–	–	L	–	N L

Table 3.1 (continued)

Family or order	Species/ growth form	Disease state WS	SEB	BrB	BNS	BBD	Tumour	Unidentified
Merulinidae	*Hydnophora rigida*	–	L	–	–	–	–	L
	H. microconos	–	–	–	–	–	–	L
Faviidae	*Favia, Favites* or *montastrea* spp.	H N L	L	–	–	L	–	H N L
	Favia stelligera	N	–	–	–	–	–	–
	Goniastrea or *Platygyra* spp.	N L	L	–	–	L	–	H N L
	Echinopora sp.	–	L	L	–	L	–	–
	E. horrida	–	L	–	–	–	–	–
	E. mammiformis	L	–	–	–	–	–	–
	E. lamellosa	–	L	–	–	–	–	L
	Other faviids	N L	L	L	–	L	–	N L
Dendrophylliidae	*Turbinaria* sp.	–	–	–	–	L	–	L
Siderastreidae	*Psammocora digitata*	–	–	–	–	L	–	–
O: Alcyonacea	*Sinularia* sp.	–	–	–	–	H	–	–
	Lobophytum sp.	–	–	–	–	L	–	–
	Other alcyonaceans	–	–	–	–	L	–	N H L
O: Gorgonacea	*Isis* sp.	–	–	–	N L	–	–	L
O: Hydrocorallina	*Millepora* sp.	–	–	–	–	L	–	L
	Number of families/ orders affected	4	6	3	1	10	1	11
	Minimum number of species affected	17	31	16	1	32	4	30

3.3.3.3
Seasonal and Habitat (Wave Exposure) Patterns in Disease Prevalence

Disease prevalence was higher in summer than in winter on sheltered Lizard Is. reefs ($F=78.13$, $df=1$, $P<0.001$; Fig. 3.11). In particular, mean disease prevalence in summer (January 2003) was more than 15-fold greater in acroporids, more than 12-fold greater in faviids and approximately doubled in pocilloporids compared to the preceding austral winter (July 2002). Seasonal patterns of increased disease prevalence in summer in most coral families corresponded to striking increases in the number of cases of disease in all categories except tumours and pigmented spots on *Porites* (Fig. 3.12). In particular, disease incidence was high for WS (increased from 0 to 81 cases), SEB (~20-fold increase to 342 cases), BBD and unidentified cyanobacterial syndromes (~8-fold increase to 95 cases), but moderate for BrB (increase from 1 to 12 cases) (Fig. 3.12). In addition, the number of cases of BNS on gorgonians and unidentified syndromes increased in summer (the latter 14-fold).

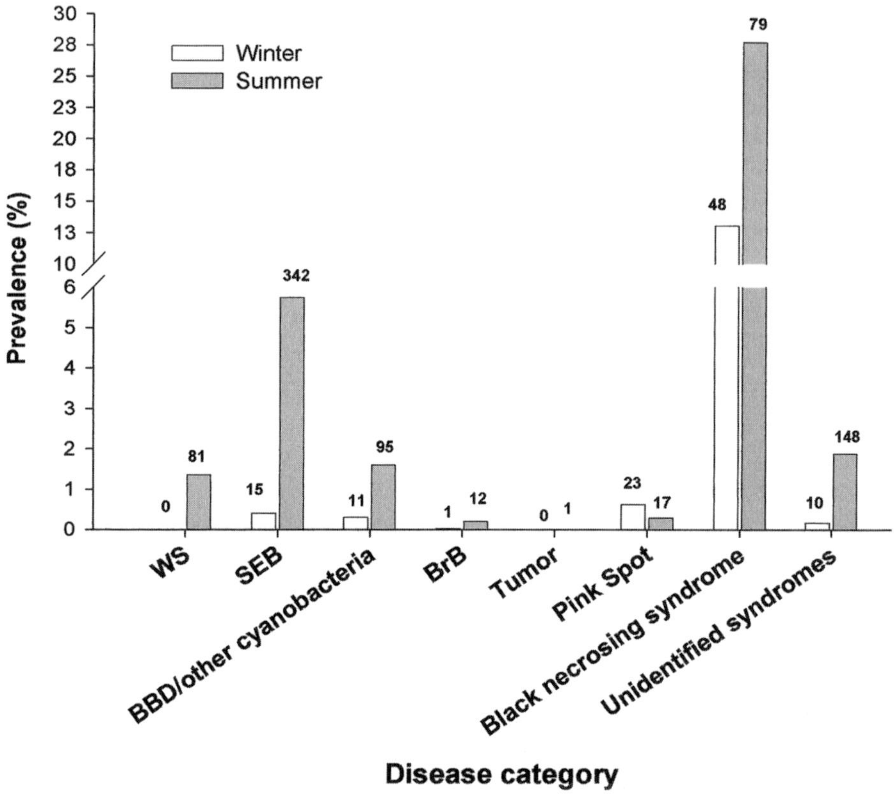

Fig. 3.12. Prevalence of seven disease categories of hard corals at Lizard Island in winter 2002 compared to summer 2003, based on surveys of two sites. Prevalence (per disease category) calculated as in Fig. 3.9. Total number of cases of each disease category is shown above the appropriate histogram

Among hard corals, disease prevalence was greater on sheltered than exposed sites at Lizard Is. ($F=298.23$, $df=1$, $P=0.003$). This pattern was largely the result of the more than two-fold greater disease prevalence on acroporids at sheltered compared to exposed sites ($F=31.22$, $df=1$, $P=0.031$). Disease prevalence did not differ significantly between the two exposure habitats for the three other major families (Pocilloporidae, Poritidae, Faviidae).

3.3.3.4
Relationship Between Percent Coral Cover and Disease Prevalence

There was no correlation between mean disease prevalence and mean percent cover of all groups (scleractinian, alcyonaceans and gorgonians combined) at the eight sites surveyed (Fig. 3.13). There was more than a two-fold range in both mean disease prevalence (varying between 5.2±0.98% on the reef front at No Name Reef in the northern sector to 12.8±3.02% at Little Bay, Heron Is. in

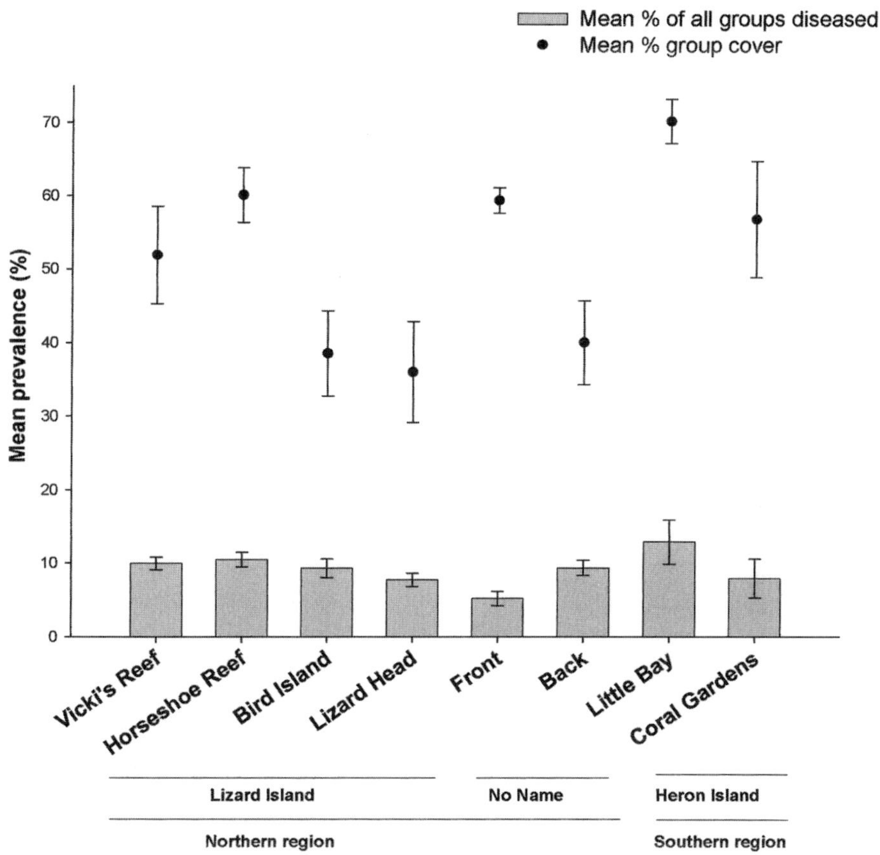

Fig. 3.13. Geographic patterns in mean disease prevalence (±SE) in all groups (hard corals, soft corals and gorgonians combined) compared to mean percent cover (±SE) of all groups at each reef in summer (January) 2003 (n=3 transects per site). Prevalence (per site) is calculated relative to the total number of scleractinian, gorgonian and alcyonacean colonies examined at each site

the southern sector) and mean percent cover (varying between 36.0±6.84% at Lizard Head to 70.2±3.05% at Little Bay) over the eight sites. However, despite mean disease prevalence being highest where mean percent cover was highest, it was also lowest where mean percent cover was second highest.

3.3.4 Discussion and Conclusions

Our review of disease records to date reveals the presence of at least eight disease states on the Great Barrier Reef (GBR) and at least another four elsewhere in the Indo-Pacific. Although this approximates to only half of the 22 diseases so far recorded from the Caribbean, the global coral disease hot spot (Green and Bruckner 2000; Weil, this Vol.), most of these records are from the last

10 years and as a result of comparatively minimal research effort. Given that the first record of coral disease in the Caribbean was 30 years ago (Antonius 1973) and that observations have been escalating ever since (Harvell et al. 1999), the rate of discovery of new syndromes and diseases may accelerate in the Indo-Pacific as research becomes more focused and reefs come under increasing pressure from a plethora of environmental issues. In particular, with more than 80% of reefs in SE Asia under medium to high threat from activities like over-exploitation of resources and coastal developments (Bryant et al. 1998) and with predicted increased sea temperatures associated with global climate change (IPCC 2002) likely to augment pathogen virulence (Harvell et al. 2002; Rosenberg and Ben-Haim 2002), environmental conditions in the next few decades are poised to foster increasing incidence and spread of disease on Indo-Pacific reefs. The presence of a number of pathogens on the Great Barrier Reef that have had major impacts on the structure of Caribbean coral communities, such as black band disease and potentially one or more of the white band or plague diseases within the white syndrome category (Gladfelter 1982; Bruckner and Bruckner 1997; Richardson 1998; Richardson and Aronson 2002), emphasises the gravity of the threat posed by predicted environmental changes for coral reefs in this region.

One of the greatest causes for concern is the 22- to 150-fold increases in the abundance of white syndrome on outer-shelf reefs in the northern and southern sectors of the GBR that have been detected over the last 5 years by the Australian Institute of Marine Science Long Term Monitoring Program (AIMS LTMP). The occurrence of these striking increases on reefs separated by 1200 km indicates that conditions promoting the spread of the syndrome are widespread and extend from the northern to southern sectors of the GBR. The lack of association between the abundance of *Drupella* spp. and the abundance of WS suggests that GBR *Drupella* species are not major vectors in the transmission of WS, in contrast to previous positive correlations between *D. cornus* and abundance of white diseases found in the Red Sea (Antonius and Riegl 1997, 1998). Curiously, patterns of increasing abundance of WS are correlated with increasing distance of reefs from the coast and are associated with high mean percent cover of hard corals. Furthermore, the significant and positive relationship between changes in percent coral cover and changes in WS abundance on outer-shelf reefs, where the greatest increases in WS abundance have been recorded, further implicates increases in coral cover as playing a role in the spread of WS. The pattern of greatest occurrence of WS on outer-shelf reefs, where anthropogenic impacts are least, indicates that WS abundance is unlikely to be directly caused by human activities or terrestrial sources of pollution. The increasing abundance of WS with increasing percent coral cover could reflect either increased pathogen transmission or host vulnerability as coral assemblages become more crowded and approach carrying capacity, or it could reflect increased pathogen susceptibility as colonies age.

In the absence of other identifiable disturbances, the association between rising WS abundance and declining rates of increase or negative trends in percent coral cover on some reefs, suggests high WS abundance may be contributing to the lower than predicted (from rates of change prior to the rise of WS) percent covers attained on these reefs. Given that we found the prevalence of WS to be higher on two of the three reefs surveyed in the regional disease prevalence surveys (5.44% on Heron Is. Reef and 2.11% on No Name Reef) than prevalences recorded for most of the white band (WB) and white plague (WP) diseases that have been so destructive in the Caribbean [range: 0.01–1.85%, except for 3.62% prevalence of WPI in Florida (Dustan 1977); reviewed in Weil, this Vol.], establishing the causative agent(s) of white syndrome on the GBR must be considered an urgent priority.

The regional disease prevalence surveys revealed equally disturbing patterns in disease occurrence on the GBR. The overall mean disease prevalence of 8.97 ±0.79% at eight sites in the northern and southern sectors of the GBR in summer 2003 is higher than the 5.38±1.2% mean disease prevalence that has been recorded in comparable surveys of 28 Caribbean sites in the past 4 years (Weil, this Vol.). Although we surveyed sites in regions identified as having the highest abundance of WS in the large-scale surveys, nevertheless the Caribbean surveys included some of the most impacted sites within the biogeographic reef region, for example Jamaica, which had a mean disease prevalence of 16.21±1.55% in 1999, and Mexico, which had mean prevalence of 10.91±1.57% in 2002 (Weil, this Vol.). The lowest mean prevalence on GBR sites (5.16±0.98% at the reef front site at No Name Reef, an outer-shelf reef in the northern sector) was similar to the overall mean prevalence of the Caribbean sites, although the highest GBR disease prevalence (12.89±3.02% at the Little Bay site on Heron Is. in the southern sector) was somewhat less than the highest Caribbean prevalence (16.21±1.55% in Jamaica; Weil, this Vol.). The lower disease prevalence at No Name than at Lizard Is. sites is contrary to the patterns of WS abundance recorded on the same reefs in the large-scale AIMS LTMP surveys, but may reflect the shallower depth of transects and, hence, the higher wave energy habitats that were surveyed in the disease prevalence study. The lower disease prevalence found on the shallower transects accords with the pattern of lower disease prevalence on exposed sites found for hard corals on Lizard Is. reefs, particularly for the family Acroporidae. Overall, the increasing abundance of WS recorded by the AIMS LTMP surveys in all GBR sectors but one (Townsville sector) and in all cross-shelf locations (inner-, mid- and outer-shelf) in the past 5 years highlights the need for a co-ordinated, large-scale program to establish baseline levels of disease prevalence at key sites throughout the GBR, against which to judge whether disease incidence is increasing.

Until more is known about the etiology of GBR and Indo-Pacific coral diseases, it is difficult to compare prevalence of specific diseases reported here with those in the Caribbean. Black band disease (BBD) is the only disease that is common to the two reef regions, although it appears that there are further cyanobacterial species associated with BBD-type infections in both the GBR and the Caribbean (see Sect. 3.2.1). Assuming that BBD records in both regions may encompass a variety of cyanobacterial agents and are thus comparable, the BBD prevalence we found on Lizard Is. reefs (1.7%) is similar to BBD prevalence on Caribbean reefs (0.2–6.0%; reviewed in Weil, this Vol.), however, BBD prevalence was lower on the two outer-shelf GBR reefs (0.01% at Heron Is., 0.51% at No Name Reef) than was generally found on Caribbean reefs. The very low and stable abundance of BBD throughout the past 5 years in the AIMS large-scale survey program, in combination with the higher prevalence found a decade ago at the Lizard Is. study sites (Dinsdale 2002), suggests that BBD is a common component of pathogenic assemblages on GBR corals but, as in the Caribbean, it rarely reaches outbreak proportions.

Prevalences of other diseases on the GBR were generally low (<1%), with the exception of skeletal eroding band (SEB) on hard corals and black necrosing syndrome (BNS) on gorgonians. The occurrence of SEB at all sites in the southern and northern sectors (1.4–5.7% prevalence) and the range of hosts (at least 32 species in 6 scleractinian families) suggest that it is a widespread, generalist pathogen. Although direct comparisons of the prevalence of SEB found in our surveys at Lizard Is. (5.7%) with previous records of SEB at the same reef in 1998 (season unknown) are not possible given the semi-quantitative nature of the latter surveys (Antonius and Lipscomb 2001), it is likely that the 344 SEB cases in 240 m^2 represents, if anything, an increase in abundance over the 13–25 cases/30-min swim recorded in the previous study. However, the nearly eight-fold increase in prevalence of SEB that we found between winter and summer on Lizard Is. reefs suggests that comparisons are only valid if reefs are surveyed in the same season. Prevalence of BNS also increased in summer on Lizard Is. reefs, infecting more than 25% of gorgonian populations compared to 13% in winter. The year-round high prevalence of BNS suggests that it may have a major impact on gorgonian populations on Lizard Is. reefs.

Patterns in disease prevalence among families suggest the faster growing corals in the families Acroporidae and Pocilloporidae are disproportionately targeted by pathogenic microorganisms, including cyanobacteria and protozoans. Although acroporids dominated coral benthic cover at Lizard Is. sites (and at all other sites), the pattern of very low percent cover of pocilloporids despite high disease prevalence in both winter and summer, indicates that pathogens are not necessarily keying into the most abundant or spatially dominant corals. In contrast, disease prevalence in the two slow-growing massive families, Poritidae and Faviidae, was less than half that of pocilloporids, despite their 2.5–7.5 times greater percent cover. It is possible that corals with fast growth have less well-developed disease resistance strategies as a consequence of life histories that channel resources into growth for space monopolisation rather than into maintenance activities, whereas massive corals that tend to be more committed to confrontational strategies (Jackson 1979), may have evolved greater disease resistance. The tendency for WBD epizootics in the Caribbean to disproportionately affect acroporids (e.g. Gladfelter 1982) supports the hypothesis that faster growing corals may have decreased disease resistance. However, more extensive testing of patterns in host susceptibility among coral families is required before life history patterns in disease resistance can be identified.

3.3.5 Some Unresolved Questions and Future Research

There are a myriad of unresolved questions that should be tackled with some urgency to begin to address questions concerning the impact of coral disease on the Great Barrier Reef and the wider Indo-Pacific. Foremost, surveys of disease prevalence on reefs representative of the major habitats and community types throughout the region are required to better document the full range of pathogens and establish a baseline against which to judge whether disease incidence is increasing. Although rapid surveys of the number of cases of disease identify general trends, more detailed disease prevalence surveys are preferred to accurately estimate the impact of disease on coral populations and assemblages. A key objective will be to determine rates of mortality caused by disease and to put them into context with mortality caused by other disturbance agents on Indo-Pacific reefs, such as bleaching events, cyclones and *Acanthaster planci* outbreaks.

The modular nature of corals raises another set of issues regarding the impact of disease on coral populations that should be addressed concurrently. Whereas disease impacts the whole animal in unitary organisms, modular organisms may suffer partial mortality, which compromises colony fecundity, but does not reduce population size. Thus, in addition to establishing disease prevalence and rates of mortality attributable to disease within populations, it will be equally important to determine rates of disease spread and tissue loss within colonies and their associated impacts on colony fecundity and growth, to fully understand the impact of diseases on coral populations. The limited nature of current knowledge of the etiology of Indo-Pacific coral diseases is a major impediment to determining reservoirs and vectors involved in disease transmission, both of which are keys to the management of potential epizootics. Therefore, another critical focus for future research is molecular and microbiological studies to characterise and identify pathogens associated with currently uncharacterised disease states, particularly white syndrome. There is some urgency to initiate research in each of these areas given that impacts of global climate change are likely to include decreased resistance and increased susceptibility of coral hosts, potentially in combination with increased virulence of pathogens. In conclusion, without a concerted effort to characterise the impacts of coral disease on GBR coral communities as well as the pathogens associated with coral diseases, including their patterns of spread, origins, reservoirs and vectors, our ability to develop effective strategies to manage disease on the Great Barrier Reef is limited.

Acknowledgements. We would like to thank past and present members of the AIMS LTMP for their assistance with data collection in the large-scale surveys. We are particularly grateful to Hugh Sweatman for management of the AIMS LTMP and for permission to publish these data, Greg Coleman for assistance with data management, Steve Delean for assistance with statistical analysis, Lyle Vail and Anne Hoggett for logistical support at Lizard Is. Research Station, and Heron Is. Research Station staff for logistical support at Heron Is. We thank E. Rosenberg and Y. Loya for their invitation to write this chapter and for their comments, and also S. Connolly, S. Neale, A. Thompson and R. Willis for comments that improved it. This research was supported by the Australian Institute of Marine Science and by grants from the World Bank and James Cook University. This is publication number 210 from the Coral Group at James Cook University.

References

Aeby GS (1991) Behavioural and ecological relationship of a parasite and its hosts within a coral reef system. Pac Sci 45:263–269

Aeby GS (1998) A digenean metacercaria from the reef coral, *Porites compressa*, experimentally identified as *Podocotyloides stenometra*. J Parasitol 84:1259–1261

Antonius A (1973) New observations on coral destruction in reefs. Abs Ass Isl Mar Lab Carib 10:3

Antonius A (1977) Coral mortality in reefs: a problem for science and management. Proc 3rd Int Coral Reef Symp 2:3–6

Antonius A (1985) Coral disease in the Indo-Pacific: a first recording. PSZNI Mar Ecol 6:197–218

Antonius A (1999) *Halofolliculina corallasia*, a new coral-killing cilicate on Indo-Pacific reefs. Coral Reefs 18:300

Antonius A, Riegl B (1997) A possible link between coral diseases and a corallivorous snail (*Drupella cornus*) outbreak in the Red Sea. Atoll Res Bull 47:1–9

Antonius A, Riegl B (1998) Coral diseases and *Drupella cornus* invasion in the Red Sea. Coral Reefs 17:48

Antonius A, Afonso-Carillo J (2001) *Pneophyllum conicum* killing reef-corals in Mauritius: a new Indo-Pacific syndrome? Bull Mar Sci 69:613–618

Antonius A, Lipscomb D (2001) First protozoan coral-killer identified in the Indo-Pacific. Atoll Res Bull 481–493:1–21

Aronson RB, Precht WF (2001) White-band disease and the changing face of Caribbean reefs. In: Porter JW (ed) The ecology and etiology of newly emerging marine diseases. Kluwer, London, pp 25–38

Aronson RB, Precht WF, Macintyre I (1998) Extrinsic control of species replacement on a Holocene reef in Belize: the role of coral disease. Coral Reefs 17:223–230

Baird A (2000) Microbial menace. CRC Reef Research note, 2 pp

Bak R (1983) Neoplasia, regeneration and growth in the reef-building coral *Acropora palmata*. Mar Biol 77:221–227

Bass DK, Miller IR (1996) Crown-of-thorns starfish and coral surveys using manta tow and SCUBA search techniques. Long-term monitoring of the GBR, Standard Operational Procedure no 1. Aust Inst Mar Sci, Townsville, 42 pp

Borneman EH (2001) Aquarium corals: selection, husbandry, and natural history. TFH Publishing, Neptune City, NJ, USA

Bruckner A, Bruckner R (1997) The persistence of black-band disease in Jamaica: impact on community structure. Proc 8th Int Coral Reef Symp 1:601–606

Bruckner A, Bruckner J, Wiliams E Jr (1997) Spread of a black-band disease epizootic though the coral reef system in St Ann's Bay, Jamaica. Bull Mar Sci 61:919–928

Bryant D, Burke L, McManus J, Spalding M (1998) Reefs at risk: a map-based indicator of threats to the World's coral reefs. World Resource Institute, Washington, DC, 56 pp

Cheney D (1975) Hard tissue tumors of scleractinian corals. In: Hildemann W, Benedict A (eds) Immunologic phylogeny. Plenum Press, New York, pp 77–87

Coles S, Seapy D (1998) Ultra-violet absorbing compounds and tumorous growths on acroporid corals from Bandar Khayran, Gulf of Oman, Indian Ocean. Coral Reefs 17:195–198

Cooney RP, Pantos O, Tissier MDAL, Barer MR, O'Donnell AG, Bythell JC (2002) Characterization of the bacterial consortium associated with black band disease in coral using molecular microbiological techniques. Environ Microbiol 4:401–413

Dinsdale EA (1994) Coral disease on the Great Barrier Reef. Joint scientific conference on science, management and sustainability of marine habitats in the 21st century. Abstract

Dinsdale EA (2002) Abundance of black-band disease on corals from one location on the Great Barrier Reef: a comparison with abundance in the Caribbean region. Proc 9th Int Coral Reef Symp 2:1239–1243

Ducklow HW, Mitchell R (1979) Observations on naturally and artificially diseased tropical corals: a scanning electron microscope study. Microbial Ecol 5:215–223

Dustan P (1977) Vitality of reef coral populations of Key Largo, Florida: recruitment and mortality. Environ Geol 2:51–58

Fengold JS (1988) Ecological studies of a cyanobacterial infection on the Caribbean sea plum *Pseudopterogorgia acerosa* (Coelenterata: Octocorallia). Proc 6th Int Coral Reef Symp 3:157–162

Frias-Lopez J, Zerkle AL, Bonheyo GT, Fouke BW (2002) Partitioning of bacterial communities between seawater and healthy, black-band diseased, and dead coral surfaces. Appl Environ Microbiol 68:2214–2228

Frias-Lopez J, Bonheyo GT, Jin Q, Fouke BW (2003) Cyanobacteria associated with coral black band disease in Caribbean and Indo-Pacific reefs. Appl Environ Microbiol 69:2409–2413

Gladfelter W (1982) White-band disease in *Acropora palmata*, implications for structure and growth of shallow reefs. Bull Mar Sci 32:639–643

Green E, Bruckner A (2000) The significance of coral disease epizootiology for coral reef conservation. Biol Conser 96:347–361

Harvell CD, Kim K, Burkholder J, Colwell RR, Epstein PR, Grimes J, Hofmann EE, Lipp EK, Osterhaus ADME, Overstreet R, Porter JW, Smith GW, Vasta GR (1999) Emerging marine diseases – climate links and anthropogenic factors. Science 285:1505–1510

Harvell CD, Mitchell CE, Ward JR, Altizer S, Dobson AP, Ostfeld RS, Samuel MD (2002) Climate warming and disease risk for terrestrial and marine biota. Science 296:2158–2162

Hayes R, Goreau T (1998) The significance of emerging diseases in the tropical reef ecosystem. Rev Biol Trop 46 [Suppl 5]:173–185

IPCC (2002) Climate change and biodiversity. In: Gitay H, Suarez A, Watson R, Dokden D (eds) Intergovernmental panel on climate change. Technical Paper V, 86 pp

Jackson JBC (1979) Morphological strategies of sessile animals. In: Larwood G, Rosen BR (eds) Biology and systematics of colonial organisms. Academic Press, New York, pp 499–555

Kim K, Harvell CD, Kim PD, Smith GW, Merkel SM (2000a) Fungal disease resistance of Caribbean sea fan corals. Mar Biol 136:259–267

Kim K, Kim PD, Alker AP, Harvell CD (2000b) Chemical resistance of gorgonian corals against fungal infections. Mar Biol 137:393–401

Korrubel J, Riegl B (1998) A new coral disease from the southern Arabian Gulf. Coral Reefs 17:22

Kuta KG, Richardson LL (1996) Abundance and distribution of black band disease on coral reefs in the northern Florida Keys. Coral Reefs 15:219–223

Kuta KG, Richardson LL (2002) Ecological aspects of black band disease of corals: relationships between disease incidence and environmental factors. Coral Reefs 21:393–398

Le Champion-Alsumard T, Golubic T, Priess K (1995) Fungi in corals: symbiosis or disease? Interactions between polyps and fungi causes pearl-like skeleton biomineralization. Mar Ecol Progr Ser 117:137–147

Littler M, Littler D (1995) Impact of CLOD pathogen on Pacific coral reefs. Science 267:1356–1360

Loya Y, Bull G, Pichon M (1984) Tumor formations in scleractinian corals. Helgol Meer 37:99–112

McCook LJ, Jompa J, Diaz-Pulido G (2001) Competition between corals and algae on coral reefs: a review of evidence and mechanisms. Coral Reefs 19:400–417

Miller I (1996) Black-band disease on the Great Barrier Reef. Coral Reefs 15:58
Morrison-Gardiner S (2001) Studies on the morphology and ecology of fungi associated with the Australian marine environment. PhD Thesis, James Cook University, Townsville, 246 pp
Nagelkerken I, Buchan K, Smith G, Boniar K, Bush P, Garzon-Ferreira J, Botero L, Gayle P, Harvell CD, Heberer C, Kim K, Petrovic C, Pors L, Yoshioka P (1997a) Widespread disease in Caribbean Sea Fans. II. Patterns of infection and tissue loss. Mar Ecol Progr Ser 160:255–263
Nagelkerken I, Buchan K, Smith G, Boniar K, Bush P, Garzon-Ferreira J, Botero L, Gayle P, Heberer C, Petrovic C, Pors L, Yoshioka P (1997b) Widespread disease in Caribbean Sea fans. I. Spreading and general characteristics. Proc 8th Int Coral Reef Symp 1:679–682
Page C, Coleman G, Ninio R, Osborne K (2001) Surveys of sessile benthic communities using underwater video. Standard operational procedure no 7. Aust Inst Mar Sci, Townsville, 49 pp
Patterson K, Porter JW, Ritchie KB, Polson SW, Mueller E, Peters E, Santavy D, Smith GW (2002) Etiology of white pox, a lethal disease of the Caribbean elkhorn coral, *Acropora palmata*. Proc Natl Acad Sci USA 99:8725–8730
Peters E, Halas J, McCarty H (1986) Calicoblastic neoplasms in *Acropora palmata* with a review of reports on anomalies of growth and form in corals. J Natl Cancer Inst 76:895–912
Ravindran J, Raghukumar C (2002) Pink line syndrome (PLS) in the scleractinian coral *Porites lutea*. Coral Reefs 21:252
Raymundo L, Harvell CD, Reynolds T (2003) Porites ulcerative white spot disease: description, prevalence, and host range: a new disease impacting Indo-pacific reefs. Dis Aquat Org 56(2):95–104
Richardson LL (1996) Horizontal and vertical migration patterns of *Phormidium corallyticum* and *Beggiatoa* spp. associated with black-band disease. Microbial Ecol 32:323–335
Richardson LL (1998) Coral diseases: what is really known. Trends Evol Ecol 13:438–443
Richardson LL, Aronson RB (2002) Infectious diseases of reef corals. Proc 9th Int Coral Reef Symp 2:1225–1230
Rodriguez-Martinez RE, Banaszak A, Jordan-Dahlgren E (2001) Necrotic patches affect *Acropora palmata* (Scleractinia: Acroporidae) in the Mexican Caribbean. Dis Aquat Org 47:229–234
Rosenberg E, Ben-Haim Y (2002) Microbial diseases of corals and global warming. Environ Microbiol 4:318–326
Santavy D, Peters E (1997) Microbial pests: coral disease in the Western Atlantic. Proc 8th Int Coral Reef Symp 1:607–612
Sweatman H, Cheal A, Coleman G, Delean S, Fitzpatrick B, Miller I, Ninio R, Osborne K, Page C, Thompson A (2001) Long-term monitoring of the Great Barrier Reef. Status Report no 5, 106 pp
Weil E, Urreiztieta I, Garzón-Ferreira J (2002) Geographic variability in the incidence of coral and octocoral disease in the wider Caribbean. Proc 9th Int Coral Reef Symp 2:1231–1238
Yamashiro H, Oku H, Onaga K, Iwasaki H, Takara K (2001) Coral tumors store reduced levels of lipids. J Exp Mar Biol Ecol 265:171–179

4 Coral Diseases in Gulf of México Reefs

Eric Jordán-Dahlgren and Rosa E. Rodríguez-Martínez

4.1 Introduction

The potentially catastrophic effects of hard coral diseases and syndromes on the population and community scale have been clearly demonstrated by the widespread demise of key species in the Caribbean region, such as *Acropora palmata* and *A. cervicornis*, presumably by white-band disease (Gladfelter 1982; Aronson and Pretch 2001). Other important coral species in the Caribbean, such as those of the *Montastraea annularis* species complex, have also recently been threatened by a number of diseases and syndromes (Richardson et al. 1998; Santavy et al. 1999; Cervino et al. 2001; Porter et al. 2001). Even the mass mortality of other non-coral species, such as the sea-urchin *Diadema antillarum*, can also impede the survival and recovery of coral populations affected by other natural catastrophes or human impacts (Hughes 1994; Lessios et al. 1984). Therefore, the ecological importance of diseases is not exclusively dependent on their capability to wipe out entire populations, but to the extent of affecting the community and population dynamics of their hosts, and that of related species. Furthermore, some potentially lethal coral diseases, such as white plague, black band, and yellow band, do not always cause mass mortalities, and some authors consider that environmental deterioration may be responsible for the spread and morbidity of diseases, both on global (Daszak et al. 2001; Kim and Harvell 2002; Kuta and Richardson 2002) and local scales (Green and Bruckner 2000; Cervino et al. 2001; Porter et al. 2001).

At present, our understanding of the potential ecological effect of coral diseases is confounded by contradictory reports on disease incidence and severity. This results partly from the difficulties associated with identifying poorly characterized diseases or syndromes in their different stages, but also because some reports are based on data from a single reef site and focus their effort on single species that are usually already severely affected by diseases or syndromes. In addition, several reports fail to follow up disease evolution in appropriate ecological/epizootic time frames.

The ample surveys required to allow appropriate spatial and temporal representation of the coral community are expensive and time consuming. However, the few studies that have been performed (Ginsburg et al. 2001; Porter et al. 2001) show that coral diseases and partial mortality patterns (not necessarily related to diseases) may vary within a given reef system and that a scale issue is present in those patterns. Moreover, Green and Bruckner (2000) have shown that disease effects can be quite variable among reefs. As the host–pathogen relationship may be influenced by local conditions (Harvell et al. 1999), a first step in understanding such relationships is to assess disease incidence and virulence patterns in the natural variability range of the chosen system. By doing this, a comprehensive ecological baseline could be established to allow for the analysis of disease effect in time and space at meaningful population and community levels.

In this report, we analyze the incidence of different diseases and syndromes at the coral community level. We take into account the natural variability found in the shallow environment of two widely separated oceanic reefs that lie far from terrestrial influence, and outside the main Caribbean superficial circulation patterns. The purpose is twofold: (1) to assess if there is a relationship between community structure variability and disease incidence, or if it is species-specific, and (2) to compile a baseline data set on the presence, distribution and abundance of hard coral diseases for reefs lying far from terrestrial influence that may function as a standard for future comparisons.

4.2 Study Sites

The study was conducted on two reefs located on the western edge of the Campeche bank in the southeastern Gulf of México: Triángulos Este and Cayos Arcas (Fig. 4.1). Both are small (±10 km^2) reef groups, more than 150 km off the coast, outside the main surface circulation patterns of the Gulf of México

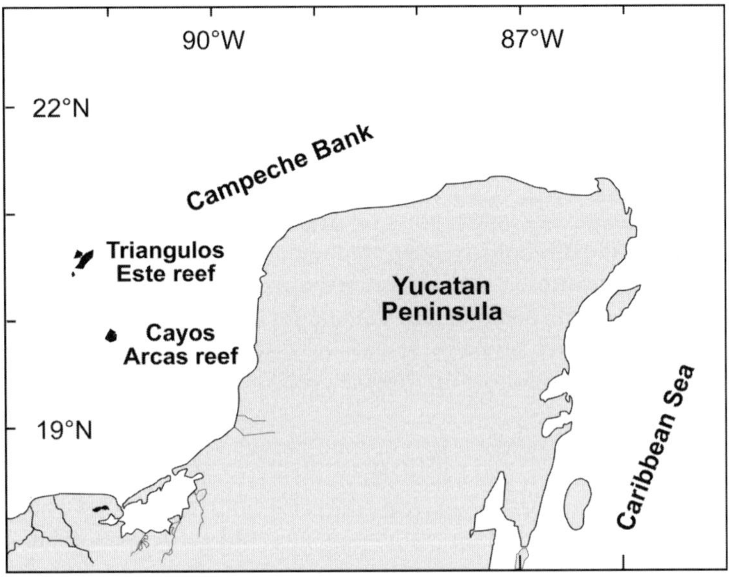

Fig. 4.1. Location of the coral reefs surveyed in this study. Reef sizes are exaggerated

(Vukovich and Crissman 1986; Vidal et al. 1999; Jordán-Dahlgren 2002). They lie far from any population center and terrestrial influence, but while Triángulos Este is quite isolated, Cayos Arcas is about 4 km upstream from a large filling platform for crude oil tankers. Fishing activities are relatively intense in Triángulos Este, but not so on Cayos Arcas because access to the area is controlled. The climate in the area is mostly defined by changes in the wind regimes: trade winds and hurricanes in summer, and cold polar fronts in winter, eventually of hurricane force (Vidal et al. 1990). In 1995, both reefs were affected by two category I hurricanes (Saffir-Simpson scale): Opal and Roxanne. Acroporids have been decimated in both reefs, probably by the white-band disease. *Diadema antillarum* die-off was severe in these reefs, but recovery is underway. It is unknown if severe bleaching events have occurred on these reefs.

4.3 Methods

Surveys for benthic community structure analyses were carried out in 1996 and 2001 on both reefs by means of high-resolution video transects. For this paper, we use data on prevalence of diseases and syndromes affecting hard corals, as well as the degree of colony partial mortality. In order to compare the same habitat types, we use data from sampling sites on shallow and protected areas (depth range between 1.5–5.0 m).

For the 1996 survey, three separate sites on each reef were chosen from the video-library, corresponding to the same areas surveyed in 2001. In the 1996 surveys, each site was composed of two parallel 20-m-long video transects (0.6 m wide). In 2001, a more ample sampling with high replication was carried out. Six sites were sampled in Triángulos Este and 17 in Cayos Arcas, randomly dispersed over the reefs. Each sampling site was composed of five to ten randomly parallel (1–10 m apart) 10-m-long video transects (0.4 m wide). All samplings were carried out in the summer when diseases are prone to flourishing (Bruckner and Bruckner 1997; Bruckner et al. 1997; Kuta and Richardson 2002). In all video-surveys, the camera to surface distance was controlled by means of a projecting aluminum rod, ending in a horizontal scale; image resolution with wide-angle lenses at this distance was in the order of 1 mm. Videorecordings were digitized and the still images were analyzed using the software program SigmaScan Pro Versión 4.0 (SPSS, Chicago, IL). The area of diseased and healthy tissue, as well as the area of recent and old mortality, was measured for every coral colony registered in the video transects. Colonies with recent mortality areas, but without evident signs of diseases/syndromes, were classified as necrotic colonies.

4.4 Results

4.4.1 Community Structure of the Reefs

The shallow water hard coral community of both reefs is relatively similar. A total of 24 species were registered (including the hydrocoral *Millepora complanata*), 18 of which were found in at least three surveys. Six other species were registered in the more intense 2001 surveys; however, these are rare species for the habitat (a single or very few colonies per species; Table 4.1). Similarity in

Table 4.1. Number of colonies registered per coral species on each of the four surveys. *CA* Cayos Arcas reef; *TE* Triángulos Este reef

	CA 96	CA 01	TE 96	TE 01
Porites astreoides	149	2702	102	404
Favia fragum	74	1629	184	1278
Millepora complanata	49	467	144	531
Montastraea annularis[a]	153	253	86	533
Acropora palmata	127	137	171	359
Diploria clivosa	39	188	11	294
Diploria strigosa	24	189	4	268
Porites porites	5	274	24	35
Acropora prolifera	11	83	187	2
Montastraea cavernosa	1	56	2	114
Agaricia agaricites	1	45	25	35
Siderastrea siderea	2	38	6	43
Mycetophyllia lamarckiana	3	7	3	11
Acropora cervicornis	1	25	–	1
Madracis decactis	–	23	–	3
Colpophyllia natans	–	12	2	10
Diploria labyrinthiformis	3	13	5	3
Siderastrea radians	6	11	–	4
Meandrina meandrites	–	3	–	1
Agaricia fragilis	–	1	1	1
Dichochoenia stokesii	–	–	–	3
Manicina areolata	–	2	–	–
Mycetophyllia aliciae	–	–	–	1
Stephanocoenia michelini	–	–	–	1
n	648	6158	957	3935
Species	16	21	16	23

[a]Refers to the species complex, comprised of *Montastraea annularis* and *M. faveolata* colonies in the shallow habitat sampled.

species richness was therefore greater among 2001 surveys than between consecutive surveys on the same reef (Table 4.2). If rare species are not included, all surveys have very similar species composition and richness, indicating that all surveys were carried out in a similar habitat type.

Of the 14 common species, *Porites astreoides*, *Favia fragum*, *Millepora complanata*, *Montastraea annularis* species complex (composed of *M. annularis* and *M. faveolata*) and *Acropora palmata* had the highest abundances in most surveys, *Diploria clivosa*, *D. strigosa* and *P. porites* had intermediate abundances, and *M. cavernosa*, *Agaricia agaricites*, *Siderastrea siderea*, *Mycetophyllia lamarckiana* and *D. labyrinthiformis* had low abundances (Table 4.1). However, large differences in dominant species abundances were found among the four surveys. In 1996, the dominant species in Cayos Arcas were those of the *M. annularis* species complex (23.6% of the colonies), *P. astreoides* (23.0%) and *A. palmata* (19.6%) while in Triángulos Este *A. prolifera* (19.5%), *Favia fragum* (19.2%) and *A. palmata* (17.9%) were dominant. By 2001, massive recruitment of *P. astreoides* and *F. fragum* occurred, as indicated by the small-sized colonies, resulting in single species dominance in both reefs: *P. astreoides* (43.9%) in Cayos Arcas and *F. fragum* (32.5%) in Triángulos Este.

Even though species richness was similar among Cayos Arcas and Triángulos Este reefs, both in the 1996 and 2001surveys (Table 4.1), species diversity and dominance indices were different, as indicated by the 95% confidence intervals obtained by bootstrapping (Table 4.3). Comparison of the Shannon diversity indices by means of a t-test, also showed significant differences among both reefs in the 1996 (t=9.24; v=32; P<0.05) and 2001 (t=19.11; v=44; P<0.05) surveys. However, rarefaction curves obtained from the largest sample (Cayos

Table 4.2. Species presence-absence Jaccard's matrix

	CA 96	CA 01	TE 96	TE 01
CA 96	1	–	–	–
CA 01	0.762	1	–	–
TE 96	0.778	0.762	1	–
TE 01	0.696	0.833	0.696	1

Table 4.3. Species diversity and dominance indices based on standardized number of colonies per survey. Lower (*L*) and upper (*U*) values correspond to 95% confidence intervals obtained by bootstrapping re-sampling

	CA 96	L	U	CA 01	L	U	TE 96	L	U	TE 01	L	U
Species	16	–	–	21	–	–	16	–	–	23	–	–
n	648	–	–	6158	–	–	957	–	–	3935	–	–
H'n	1.98	1.96	1.99	1.74	1.71	1.76	2.06	2.04	2.07	2.04	2.02	2.06
Equitability	0.72	0.71	0.72	0.57	0.56	0.59	0.74	0.74	0.75	0.65	0.65	0.67
Dominance	0.17	0.17	0.18	0.28	0.27	0.28	0.15	0.15	0.15	0.17	0.17	0.18

Arcas 2001) indicate that a sample of 900 colonies is necessary to include the 18 commonest species, and thus the sample size in Cayos Arcas 1996 (n=648) may be insufficient to compare diversity. For that reason, we compared the square-root transformed data of species proportions between both reefs, which also yielded significant differences for the 1996 (χ^2=69.9; df=23; P<0.05) and 2001 (χ^2=257.6; df=23; P<0.05) surveys. We conclude that all four surveys correspond to the same habitat type as indicated by their species richness and species composition, allowing us to make direct comparisons, but also that all four surveys show relatively different community structures, thus the data sets also reflect the natural variability in this habitat.

4.4.2 Diseases and Syndromes

4.4.2.1 2001 Surveys

Of the 24 registered species, only 10 had colonies with some affection. However, more than 97% of all affected colonies in both reefs belonged to six species: *Montastraea annularis* species complex (considering the *M. annularis* species complex as one), *M. cavernosa*, *D. clivosa*, *D. strigosa*, *P. astreoides*, and *S. siderea* (Table 4.4). Of the 10,093 colonies registered in both reefs, only 5.6% had affections and only 3.0% of these corresponded to characterized diseases and syndromes. Approximately twice as many diseased colonies were found in Triángulos Este reef as in Cayos Arcas reef, despite having an almost 35% lower sample size. Three diseases (white plague, white band and black band), one syndrome (yellow band) and one new symptom, which we named the Thin Dark-line, were registered. This symptom is characterized by a thin dark-line (1–5 mm) at the boundary of the living tissue, but unlike black band, it seems stationary or progression occurs at a very slow rate (<1 cm/year; preliminary observations at western Caribbean reefs; unpubl. data). Microscope observations of the thin dark line-affected tissue revealed the presence of cyanobacteria, and presently it is not known if this affection is related to black band disease or what the causative agent is. It is also not known if this symptom is causing tissue mortality, but it is a distinct one and was relatively important in the surveyed reefs, so we consider it as an affection.

Overall, larger incidence values of characterized diseases and/or syndromes per species were apparently associated with greater number of colonies (Table 4.4). Analyzing the relationship between colony abundance and number of affected colonies by regression analysis on square-root transformed data for both reefs, it was found to be nonsignificant (r^2=0.07, P=0.41). Excluding *P. astreoides* from the data set, as its much larger sample size and low affection levels render it as an outlier in residual analysis, a significant linear regression (r^2=0.61, P<0.05) was obtained, even though the explained change in number of affected colonies by colony abundance was still relatively low. We conclude

Table 4.4. Species affected by diseases, syndromes and symptoms on each of the four surveys. *CA* Cayos Arcas reefs; *TE* Triángulos Este reefs. *Montastraea annularis* refers to the species complex. *BB* Black band, WB white band, WP white plague, YB yellow band, TDL thin dark line. Note that one colony could be affected by more than one disease

	Colonies		Colonies affected by disease/syndrome/symptom					Colonies		Colonies affected by disease/syndrome/symptom				
	n	Affected	BB	WB	WP	YP	TDL	*n*	Affected	BB	WB	WP	YB	TDL
Cayos Arcas 1996								Cayos Arcas 2001						
Aaga	1	–	–	–	–	–	–	45	–	–	–	–	–	–
Apal	127	–	–	–	–	–	–	138	–	–	–	–	–	–
Dcli	39	10	2	–	–	–	8	188	7	1	–	–	–	6
Dlab	3	–	–	–	–	–	–	13	5	–	–	–	–	5
Dstr	24	10	2	–	–	–	8	189	14	4	–	–	–	10
Mann	153	50	1	–	–	5	44	253	120	2	–	–	84	64
Mcav	1	–	–	–	–	–	–	56	6	–	–	–	–	6
Mcom	49	–	–	–	–	–	–	467	1	1	–	–	–	–
Past	149	5	–	–	–	–	5	2702	31	–	–	–	–	31
Ssid	2	1	–	–	–	–	1	38	2	–	–	–	–	2
Total		76	5	0	0	5	66		186	8	0	0	84	124
Triángulos Este 1996								Triángulos Este 2001						
Aaga	25	–	–	–	–	–	–	35	1	–	–	–	–	1
Apal	171	1	–	1	–	–	–	359	1	–	1	–	–	–
Dcli	11	–	–	–	–	–	–	294	20	3	–	–	–	17
Dlab	5	3	–	–	–	–	3	3	–	–	–	–	–	–
Dstr	4	1	–	–	–	–	1	268	26	1	–	1	–	24
Mann	86	27	–	–	2	–	25	533	298	5	–	8	186	154
Mcav	2	1	–	–	–	–	1	114	19	–	–	2	–	17
Mcom	144	–	–	–	–	–	–	531	–	–	–	–	–	–
Past	102	–	–	–	–	–	–	404	1	–	–	–	–	1
Ssid	6	1	–	–	–	–	1	43	10	–	–	–	–	10
Total		34	0	1	2	0	31		376	9	1	11	186	224

that although there is an effect of sample size, levels of affection incidence for these species under study in both reefs were not linearly related to the local abundance of that species in the 2001 surveys.

The *Montastraea annularis* species complex had a higher number of affections than any other species in our surveys (Fig. 4.2). This species complex also had by far the highest incidence of affections of any species in the 2001 surveys, with 55.9% of the colonies being affected in Triángulos Este and 47.4% in Cayos Arcas (Fig. 4.2, Table 4.4). *S. siderea* and *M. cavernosa* also had relatively high incidence values, but just in Triángulos Este reef (23.3 and 16.7%, respectively). *Diploria strigosa* and *D. clivosa* consistently showed diseased colonies in both reefs, but the degree of incidence was rather low (Fig. 4.2.). The other species had less than 10% of colonies affected, and particularly *P. astreoides* had less than 1.2% of affected colonies despite the large number of colonies that were sampled. Acroporids also had a low incidence of affections at both

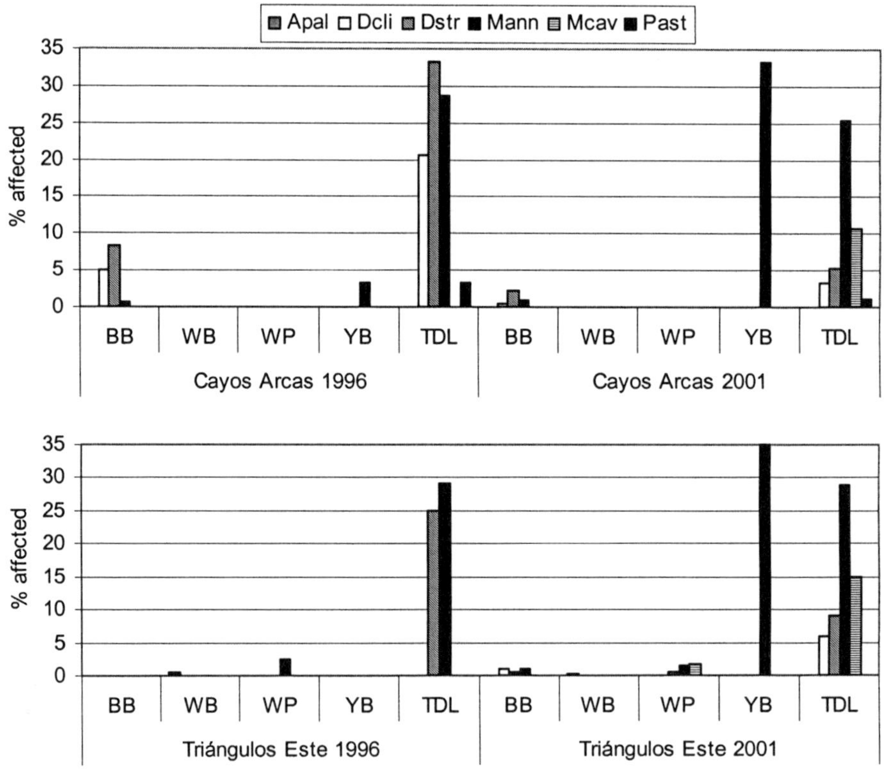

Fig. 4.2. Percentage of coral colonies with different type of affections for diverse species. Sample sizes as in Table 4.4. *BB* Black band, *WB* white band, *WP* white plague, *YB* yellow band, *TDL* thin dark line. Data for *Montastraea cavernosa* were not included for both reefs in 1996 due to the low sample sizes (one and two colonies registered, respectively, see Table 4.1)

sampling periods on both reefs. Although the incidence of affections was rather low in all species, except those of the *M. annularis* complex, data suggest that sensitivity to affections in all of the above mentioned species is species-dependent, rather than reef-dependent.

In terms of diseases and syndromes, the higher incidence in both reefs corresponded to the yellow-band syndrome which affected only colonies of the *M. annularis* species complex: 34.9% of the colonies in Triángulos Este and 33.2% in Cayos Arcas in 2001. Other potentially lethal diseases such as black band, white band and white plague were rare in both reefs (Table 4.4). The thin dark-line symptom (TDL; Table 4.4), was registered in eight species and 3.4% of all colonies (Table 4.4). The highest incidence of TDL was registered in the *M. annularis* species complex. In many colonies (25% in Cayos Arcas and 33% in Triángulos Este), the TDL symptom was associated with the yellow-band syndrome (Table 4.4). The incidence of this symptom was low in other species, although it was higher than that of characterized diseases and syndromes (Table 4.4). All 24 species, except *Favia fragum* and *Diploria labyrinthiformis* had colonies with necrotic injuries of unknown origin, but in small proportions: 0.4% of all registered colonies in Cayos Arcas, and 1.2% in Triángulos Este.

4.4.2.2 Change in the Incidence of Affections over Time

The percentage of change in the incidence of characterized diseases, syndromes and TDL between 1996 and 2001 was estimated for the species that had more than 20 colonies registered in both sampling periods for both reefs (Fig. 4.3). With the exception of the *Montastraea annularis* species complex, no important changes in the incidence of diseases, syndromes and symptoms per species, within or between reefs was found between the 1996 and 2001 surveys. Furthermore, several of the affected species had only one or two diseased colonies (Table 4.4).

The change observed in the *M. annularis* species complex was due to the proliferation of the yellow-band syndrome in 2001. In the 1996 survey, yellow band was not registered in Triángulos Este reef and only five colonies were affected by this syndrome in a single site in Cayos Arcas reef. The overall incidence of the yellow band in colonies of the *M. annularis* species complex increased in a 5-year period from 0 to 35% in Triángulos Este and from 3 to 33% in Cayos Arcas (Fig. 4.2). In contrast, there was an apparent decrease in the incidence of affections in *Diploria strigosa* and *D. clivosa* in Cayos Arcas. In Triángulos Este, a similar pattern was observed, although *D. strigosa* had an apparent slight increase in diseases while *D. clivosa* showed a slight increase in TDL between the 1996 and 2001 surveys. For other species the incidence of TDL appears to decrease in 2001, even though the number of species affected by TDL increased from 5 to 7 both in Cayos Arcas and Triángulos Este reefs between sampling periods (Table 4.4).

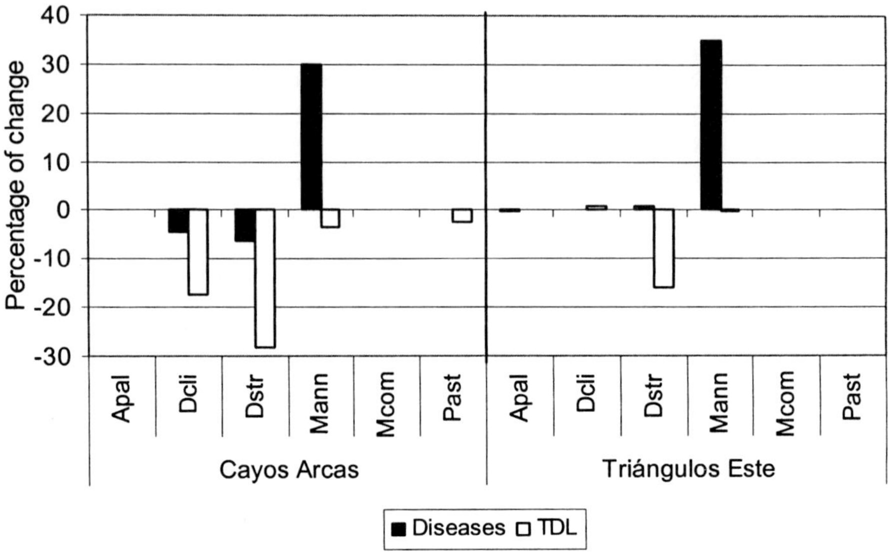

Fig. 4.3. Percentage of change in the incidence of diseases and the thin dark line symptom between 1996 and 2001 in Cayos Arcas and Triángulos Este reefs. Only species with more than 20 colonies registered in both sampling periods for each reef are included

4.5 Discussion

Our data show that relative differences in hard coral community structure are not reflected in the type and incidence of affections, and suggest that the sensitivity of coral species to affections in the shallow reef habitat was primarily species-related rather than site- or density-dependent. Despite Cayos Arcas and Triángulos Este reefs being separated by more than 90 km and having geomorphologic differences, they both showed similar sets of disease-affected species and incidence patterns in the 2001 surveys. Moreover, both reefs show similar patterns of proportion of change in disease incidence from 1996 to 2001, both in severely affected species, such as those of the *M. annularis* species complex, and in lightly affected ones.

Although the surveyed reefs are outside the main superficial water currents of the Caribbean, five out of six of the most common coral diseases reported for Atlantic reefs (Harvell et al. 1999; Green and Bruckner 2000; Porter et al. 2001; García et al. 2002) were registered in the Campeche Bank reefs: black band, white plague, white band, white pox (or patchy necrosis) and the yellow-band syndrome. Only the dark-spot syndrome, which is common in Caribbean reefs affecting *Siderastrea siderea* and *Stephanochoenia michelini* (Gil-Agudelo and Garzón-Ferreira 2001) was not observed in our surveys, but the number of registered colonies of these species was low. On the other hand, a *Millepora complanata* colony affected by black band was found in Cayos Arcas reef (Fig. 4.4).

The incidence of diseases in both reefs surveyed, however, was remarkably low, with the exception of the yellow-band syndrome in the *M. annularis* species complex. Incidence values were much lower than those reported for instance in the Florida Keys reefs (Kuta and Richardson 1996; Porter et al. 2001); although García et al. (2002) report low values of black-band incidence for some Venezuelan reefs. Moreover, white-plague disease incidence was considerably lower than that reported for reefs in Florida and Puerto Rico (Table 4.5). The same low disease incidence trend was found for the white-band and white-pox diseases in Acroporids, as compared to what has been reported for other reefs (Table 4.5). This is interesting because a massive mortality event, presumably due to white-and disease (Gladfelter 1982), occurred in these reefs prior to May 1995, when our first exploratory survey was carried out in these reefs. A qualitative estimation rendered more than 95% of the acroporid colonies dead, and while *A. palmata* skeletons remained mostly in standing position, *A. cervicornis* beds were mostly gravel deposits at that time (Jordán-Dahlgren 2003). Later in 1995, hurricanes Roxane and Opal severely hit these reefs, further increasing the mortality of surviving acroporids and destroying many of the dead *Acropora* stands. By the summer of 2001, acroporid recovery was still low in most of the shallow reef areas of both reefs (unpubl. data).

Fig. 4.4. *Millepora complanata* colony affected by black-band disease

Only the yellow-band syndrome incidence values in *Montastraea annularis* species complex in the 2001 surveys are comparable to that reported for several Caribbean reefs for the 1997–1998 period (Table 4.5). Our data coincide with other reports, indicating that species of the *Montastraea annularis* complex are affected by more diseases and syndromes than other Caribbean scleractinian corals (Rützler and Santavy 1983; Santavy and Peters 1997; Santavy et al. 1999; Cervino et al. 2001; Gil-Agudelo and Garzón-Ferreira 2001; Porter et al. 2001).

Our results suggest that, for these oceanic reefs, assumed differences in local environmental degradation (i.e., the potential influence of the oil rig on the health of Cayos Arcas reef corals) have little bearing on disease incidence. It also seems that species-specific poor resistance to pathogens is the main cause

Table 4.5. Incidence of diseases and syndromes in several Atlantic reefs. N sample size; nd no data. *M. annularis** species complex

Disease	Site	Sampling year	Incidence (%)	Species
Black band	USA, Florida[a]	1992–93	0–12.2	Several
	USA, Florida[b]	1998	0–16	Several
	Venezuela[c]	1999–2000	0.03–0.4	Several
	México, Campeche Bank[d]	2001	0.13–0.23	Several
White band	St. Croix[e], British Virgin Islands[f], US Virgin Islands[f], Jamaica[f]	Several	5–81	Acroporids
	México, Campeche Bank[d]	2001	0–0.03	*Acroporids*
White plague	USA, Florida[g]	1995–1997	0–33	*D. stokesii*
	Puerto Rico[h]	1996	47	Several
	USA, Florida[i]	nd	7–24	*S. siderea*
	USA, Florida[i]	nd	54–73	*M. ferox*
	México, Campeche Bank[d]	2001	0–0.28	Several
Yellow band	Panamá[j], Venezuela[c]	1996 and 2000	<2–10	*M. annularis**
	Curacao[f], Grenada[k], St. John[l], Turks and Caicos[k], Bonaire[l]	1997	18–91	*M. annularis**
	México, Campeche Bank[d]	2001	33–35	*M. annularis**
Dark spot	Bonaire[k], Turks and Caicos[k], Grenada[k]	1997	42–56	*S. siderea*
	Colombia[l], Venezuela[c]	1997	0.7–25.8	*S. siderea*
	México, Campeche Bank[d]	2001	0	*S. siderea*

[a]Kuta and Richardson (1996).
[b]Porter et al. (2001).
[c]García et al. (2002).
[d]This work.
[e]Gladfelter (1982).
[f]Green and Bruckner (2000).
[g]Richardson et al. (1998).
[h]Bruckner and Bruckner (1997).
[i]Dustan (1977).
[j]Santavy et al. (1999).
[k]Cervino et al. (2001).
[l]Gil-Agudelo and Garzón-Ferreira (2001).

of widespread mass mortalities, as evidenced by the case of the acroporids and the species of the *Montastraea annularis* complex. These results suggest that coral diseases may be difficult to use as an indicator of reef health. However, it must be remembered that although our surveys are quite ample and were carried out in late summer when diseases are expected to proliferate (Bruckner and Bruckner 1997), our data record a snapshot of a time-dependent process.

This work points out that in order to better understand the ecological effects of diseases, and also the nature of epizootiologies at the reef scale, it is necessary to include the natural variability in a given system, but also to design a sampling method capable of assess change in a time frame relevant to the etiology of the disease, as well as to the dynamics of the coral community or population affected. And to include in the design all possible controls to isolate local anthropogenic effects from the potential global and regional ones

Acknowledgments. The following people are warmly thanked for assisting in this work: M.A. Maldonado, G. Jordán and H. Solís for their continuous collaboration in different stages of this work; E. Martín, F. Negrete, C. Barradas, C. Meiner, L. Filip, F. Cat and E. Torrens, for their assistance with the field work and F. Escobar who contributed in the field work logistics. This study was partially funded by PEM-PEMEX and by ICMyL-UNAM, both through direct funding and by providing us with ship time on the B/O Justo Sierra.

References

Aronson RB, Precht WF (2001) White-band disease and the changing face of Caribbean coral reefs. Hydrobiologia 460:25–38

Bruckner AW, Bruckner RJ (1997) Outbreak of coral disease in Puerto Rico. Coral Reefs 16:260

Bruckner AW, Bruckner RJ, Williams EH Jr (1997) Spread of a black-band disease epizootic through the coral reef system in St. Ann's Bay, Jamaica. Bull Mar Sci 61:919–928

Cervino J, Goreau TJ, Nagelkerken I, Smith GW, Hayes R (2001) Yellow band and dark spot syndromes in Caribbean corals: distribution, rate of spread, cytology and effects on abundance and division rate of zooxanthellae. Hydrobiologia 460:53–63

Daszak P, Cunningham AA, Hyatt AD (2001) Anthropogenic environmental change and the emergence of infectious diseases in wildlife. Acta Tropica 78:103–116

Dustan P (1977) Vitality of reef coral populations off Key Largo, Florida: recruitment and mortality. Environ Geol 2:51–58

García A, Cróquer A, Paul SM (2002) Relación entre la incidencia de enfermedades y la estructura de tallas y especies en corales del Parque Nacional Archipiélago de los Roques, Venezuela. Interciencia 27(9):448–453

Gil-Agudelo DL, Garzón-Ferreira J (2001) Spatial and seasonal variation of dark spots disease in coral communities of the Santa Marta area (Colombian Caribbean). Bull Mar Sci 69(2):619–629

Ginsburg RN, Gischler E, Kiene WE (2001) Partial mortality of massive reef-building corals: an index of patch reef condition, Florida Reef Tract. Bull Mar Sci 69(3):1149–1173

Gladfelter WB (1982) White-band disease in *Acropora palmata*: implications for the structure and growth of shallow reefs. Bull Mar Sci 32(3):639–643

Green EP, Bruckner AW (2000) The significance of coral disease epizootiology for coral reef conservation. Biol Cons 96:347–361

Harvell CD, Kim K, Burkholder JM, Colwell RR, Epstein PR, Grimes DJ, Hofmann EE, Lipp EK, Osterhaus ADME, Overstreet RM, Porter JW, Smith GW, Vasta GR (1999) Emerging marine diseases – climate links and anthropogenic factors. Science 285:150

Hughes TP (1994) Catastrophes, phase shifts, and large-scale degradation of a Caribbean coral reef. Science 265:1548–1551

Jordán-Dahlgren E (2002) Gorgonian distribution patterns in coral reef environments of the Gulf of Mexico: evidence of sporadic ecological connectivity? Coral Reefs 21:205–215

Jordán-Dahlgren E (2003) Status of Acroporids in the Mexican Atlantic. In: Bruckner AW (ed) Proceedings of the Caribbean Acropora Workshop: Potential application of the US endangered species act as a conservation strategy, Miami, Florida, pp 156–159

Kim K, Harvell CD (2002) Aspergillosis of sea fan corals: disease dynamics in the Florida Keys. In: Porter JW, KG. Porter (eds) The everglades, Florida Bay, and coral reefs of the Florida Keys. CRC Press, Boca Raton, pp 813–824

Kuta KG, Richardson LL (1996) Abundance and distribution of black band disease on coral reefs in the Northern Florida. Coral Reefs 15(4):219–223

Kuta KG, Richardson LL (2002) Ecological aspects of black band disease of corals: relationships between disease incidence and environmental factors. Coral Reefs 21:393–398

Lessios HA, Robertson DR, Cubit JD (1984) Spread of *Diadema* mass mortality through the Caribbean. Science 226:335–337

Porter JW, Dustan P, Jaap WC, Patterson KL, Kosmynin V, Meier OW, Patterson ME, Parsons M (2001) Patterns of spread of coral disease in the Florida Keys. Hydrobiologia 460:1–24

Richardson LL, Goldberg WM, Carlton RG, Halas JC (1998) Coral disease outbreak in the Florida Keys: Plague Type II. Rev Biol Trop 46:187–198

Rützler K, Santavy DL (1983) The black band disease of Atlantic reef corals. Mar Ecol 4(4):301–319

Santavy DL, Peters EC (1997) Microbial Pests: coral disease in the western Atlantic. Proc 8th Int Coral Reef Symp, Panama 1:607–612

Santavy DL, Peters EC, Quirolo C, Porter JW, Bianchi CN (1999) Yellow-blotch disease outbreak on reefs of the San Blas Islands, Panama. Coral Reefs 18:97

Vidal VMV, Vidal FV, Hernández A (1990) Atlas oceanográfico del Golfo de México. Inst de Invest Electr, Cuernavaca, Morelos, México, vol 2, 707 pp

Vidal VMV, Vidal FV, Mesa E, Portilla J, Zambrano L, Jaimes B (1999) Ring-slope interactions and the formation of the western boundary current in the Gulf of Mexico. J Geophys Res 104:20523–20550

Vukovich FM, Crissman BW (1986) Aspects of warm rings in the Gulf of Mexico. J Geophys Res 91:2645–2660

5 Coral Bleaching: Signs of Change in Southern Japan

Robert van Woesik, Akiyuki Irikawa, and Yossi Loya

5.1 Introduction

The coral reefs of Japan extend over 8° in latitude (25°N to 33°N); there is a distinct transformation from well-developed reefs in the south to coral communities growing on rocky shores north of 29°N. Coral diversity in the Ryukyu Islands of southern Japan is as high as anywhere in the tropical Pacific, with over 340 coral species (van Woesik, unpubl. data), but biodiversity rapidly declines towards the north (Veron and Minchin 1992). Solar insolation and sea surface temperatures (SSTs) play a role in differential selection against low temperature-intolerant species on more northern reefs, just as warm waters have facilitated high diversity and reef development at southern latitudes over the millennia. Ironically, recent high SST events have led to declines in coral species richness and shifts in relative colony abundance in the low latitude reefs (Loya et al. 2001). This species loss is primarily a result of coral bleaching and subsequent coral mortality. Other disturbances pale in comparison with recent coral bleaching and subsequent mortality events around the world; however, coral bleaching and other impacts are not mutually exclusive and research directions are heading toward understanding these interrelated phenomena.

Coral bleaching is generally defined as the loss of symbiotic algae (i.e., zooxanthellae) and/or pigment because of stress at and above the corals' acclimation capacity (Hoegh-Guldberg and Smith 1989; Glynn 1993; Fitt and Warner 1995; Brown 1997; Kuroki and van Woesik 1999). Some authors argue that bleaching is an adaptive process (Buddemeier and Fautin 1993), while others argue that coral morbidity and the adverse effects on coral reproduction are counterintuitive to any adaptive process (van Woesik 2001; Hoegh-Guldberg et al. 2002). Clearly, the coral bleaching condition is attributable to an environmental change and affects the vital functions of coral colonies (Loya et al. 2001); the global coral-bleaching event witnessed in 1997–1998 coincided with the highest amplitude and duration of SSTs on record in the Indian and Pacific Oceans (Berkelmans and Oliver 1999; Marshall and Baird 2000; McClanahan 2000; Strong et al. 2000; Edwards et al. 2001; McClanahan et al. 2001). If bleaching is a sign of change, what is the cause of morbidity and mortality?

Under high temperature and high irradiance, carbon fixation may be reduced significantly, either at the dark reaction of the Calvin cycle (see Jones et al. 1998), leading to stress at photosystem II (PS II), and consequential photoinhibition of photosynthesis and bleaching, or the D1 protein in the thylakoid membranes may become dysfunctional, leading to photoinhibition of photosynthesis at PS II directly (see Warner et al. 1999). Bleaching is also thought to be a consequence of opportunistic infection secondary to environmental stress which produces an immuno-compromised host, and even a primary infectious process (i.e., *Vibrio shiloi*, Kushmaro et al. 1996, and *Vibrio coralliitycus*, Ben-Haim and Rosenberg 2002), where the high temperatures initiate the virulence in coral-associated bacteria that reduce endosymbiotic zooxanthellae populations.

Other considerations toward understanding the bleaching mechanism and coral morbidity involve research on mass-transfer processes. Previously, Nakamura and van Woesik (2001) showed that coral bleaching occurred in low-flow treatments, but not in moderate-flow treatments when corals were subjected to high temperature and high irradiance. Under high temperature and high irradiance, low-water flow rates cause slow mass transfer and carbon fixation may be reduced significantly. More recently, Nakamura et al. (2003) has shown that recovery from bleaching is facilitated by flow treatment. In combination, the prevention of and rapid recovery from bleaching by high water flow rates suggest that the mechanisms that lead to dysfunction of the coral–algal symbiosis are driven, in part, by mass transfer-limited processes. Therefore, understanding coral bleaching and recovery extends beyond mere temperature–light effects, since the rate of water flow alters the rate of mass transfer which, in turn, modifies any adverse effects of temperature and light.

Bleaching mechanisms and ecological outcomes are far from resolved, some critical questions remain: (1) what is the level of morbidity associated with bleaching? (2) What relative differences in morbidity and mortality are there among coral species? (3) How does the environmental setting affect the bleaching response? (4) What past circumstances influence coral morbidity? (5) Which coral species, with which zooxanthellae, will acclimate? (6) Which species will adapt? (7) How is adaptation influenced by geographical circumstances and reef proximity? Here, we address some of these ecological issues on coral mortality, resilience, resistance and recovery of 12 coral populations in southern Japan that were subjected to two thermal stress events, one in 1998 and the other in 2001.

5.1.1 Environmental Setting

In 1998, coral bleaching was evident from 25°N to 33°N in Japan (Fig. 5.1). Nearshore SSTs in Okinawa were 2.5 °C above the 10-year average (Loya et al. 2001; Nadaoka et al. 2001). An absence of typhoons and the presence of low cloud cover developed a strong thermocline and high SSTs that caused the bleaching and subsequent mass mortality of many of the stony and soft coral populations (Loya et al. 2001). Anomalous SSTs were evident again in summer 2001 (Fig. 5.2), although during 2001 coral bleaching was restricted to the southern Japanese islands. The onset, progress, and magnitude of the warming

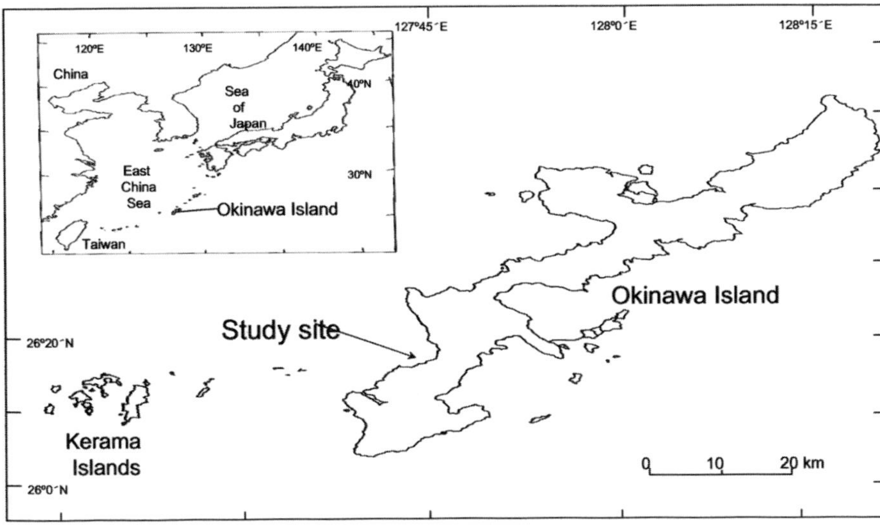

Fig. 5.1. Study site at Sunabe, Okinawa

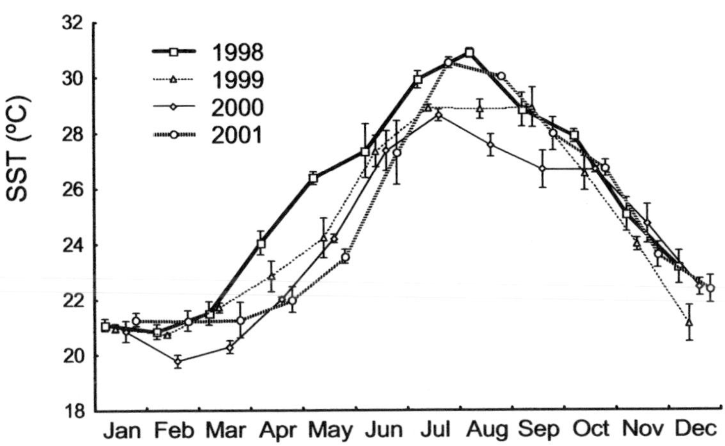

Fig. 5.2. Seawater thermal history of study site (1998–2001)

events were clearly detected on the US National Oceanic and Atmospheric Administration coral reef bleaching hotspot and degree heating weeks charts derived from satellite remotely sensed SSTs (Strong et al. 2002). The warming event in 1998 began at the end of June and thermal stress continued to accumulate through to mid-August 1998, while the 2001 event began in mid-July and thermal stress continued in the region through to early September 2001 (Fig. 5.2).

5.1.2 Objectives

If we are to examine these bleaching events in a historical context, to define baselines and differentiate responses, then the temporal response of coral tolerance, or a shift in the tolerance due to global climate change, is best resolved through inference at the population level (Williams and Bunkley-Williams 1990; Buddemeier and Fautin 1993; Hoegh-Guldberg 1999; van Woesik 2001). However, field reports on bleaching have generated a plethora of 'apparent' coral responses from slight paling and partial bleaching to absolute bleaching followed by mortality (Fitt and Warner 1995; Marshall and Baird 2000; Fitt et al. 2001). Here, we restrict our colorimetric field interpretations of coral bleaching to either (1) mortality, (2) resilience (which is defined as bleaching and recovery), and (3) resistance (i.e., where no discernible changes to the colonies were evident). Our main objectives were to assess whether the six coral species pairs (n=12) varied in mortality, resilience and resistance, and whether these parameters varied spatially and temporally. The corals examined were *Pocillopora damicornis*, *Pocillopora verrucosa*, *Acropora gemmifera*, *Acropora digitifera*, *Porites lutea*, *Porites cylindrica*, *Pavona minuta*, *Pavona varians*, *Pachyseris speciosa*, *Pachyseris gemmae*, *Favia pallida* and *Favia favus*. We clearly show intra- and interspecific differences in mortality, resilience, and resistance that varied in accordance with depth. While the environmental conditions were similarly extreme in 1998 and in 2001, 33% of the coral populations showed an increase in tolerance, suggesting the populations had adjusted their constitution to the thermal stresses.

5.2 Materials and Methods

5.2.1 Field Study

Permanent study sites were established in mid-July 1998 on the fringing reef (26°20′N, 127°45′E) at Sunabe, Okinawa, southern Japan. Monitoring was conducted at two stations, a few hundred meters apart, and at two depths, 3 and 10 m below the low water datum. At each depth we established 20×1 m^2 permanent quadrats, clearly marked with plastic hoses and nails at each corner. Each of the 80 quadrats was photographed in four subsections. In 1998–2000, slides were taken using an AF-NIKKOR 28-mm (F-3.5~4.5) SLR camera in a SEA & SEA steel housing NX90 with electronic flash YS50 (GN22), while in 2001 and 2002 we used a SONY cyber-shot digital camera with a Carl Zeiss lens. Photographs used for analysis were taken six times: August 1998, May 1999, May 2000, May 2001, September 2001, May 2002. Two thermal stress events occurred during the study period, one began at the end of June 1998 and continued to August 1998, the other began in July 2001 and continued until September 2001 (Fig. 5.2).

5.2.2
Data Acquisition and Processing

From the dataset we selected six coral species pairs (12 coral species) for detailed analysis – *Pocillopora damicornis*, *P. verrucosa*, *Acropora gemmifera*, *Acropora digitifera*, *Porites lutea*, *Porites cylindrica*, *Pavona minuta*, *Pavona varians*, *Pachyseris speciosa*, *Pachyseris gemmae*, *Favia pallida* and *Favia favus*. Each coral colony within the permanent quadrats was identified to species level in the field and their response was determined from the time-series photographs (in 1998, n=1629; in 2001, n=720). Colonies were categorized into one of two size classes (4 or >4 cm) and classified as either dead (mortality), or resilient, if they bleached and recovered, or resistant, if they showed no discernible (visual) change. The frequencies of the three response variables were used to test a series of null hypotheses regarding effects of size, depth, and time using the McNemar's Chi-square test for *dependent* samples. We compared the relative frequencies of resistance and mortality of the 1998 and 2001 data sets for the different species, at different depths for different colony sizes.

5.3
Results

5.3.1
Community Responses

Branched and corymbose *Acropora* (mainly *A. nobilis*, *A. nasuta*, *A. gemmifera*, and *A. digitifera*) and encrusting *Montipora* dominated the shallow Okinawan reef in early 1998 and accounted for 80% of the coral cover at 3 m, while faviids and *Porites* dominated the reef at 10 m. Visible coral bleaching commenced in mid-July 1998. Coral mortality was more immediate and significantly higher on the shallow 3-m reef than on the deeper 10-m slope, producing a more marked and eventual structural shift to the shallow community. Although most colonies of *Acropora* died in August or September 1998, 2 weeks after loss of color, several *Acropora* colonies retained some pigmentation on the lower or shaded parts of branches; the upper branches died, however, and were overgrown by micro-algae, which led to colony death.

5.3.2
Species-Specific Responses

Ninety percent of the *Pocillopora damicornis* colonies died during the high SST conditions in August 1998, while many *P. verrucosa* colonies were resilient or resistant, and recovered (Figs. 5.3, 5.4). Notably *Pocillopora verrucosa* showed higher tolerance in 2001 than in 1998; where tolerance is the reduction in relative colony mortality (McNemar Chi-square, χ^2=27.4, p<0.001; χ^2=12.9, p<0.003; χ^2=19.3, p<0.001; χ^2=57.1, p<0.001, for 4 and >4 cm at 3 and 10 m re-

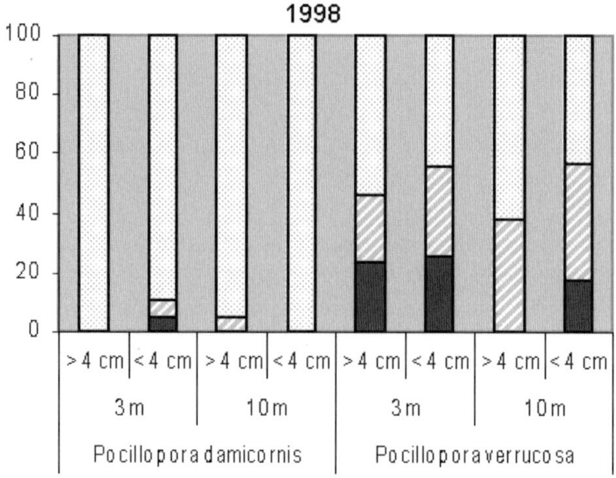

Fig. 5.3. Responses of *Pocillopora damicornis* and *Pocillopora verrucosa* to the 1998 thermal stress event, assessed at two depths (3 and 10 m) and categorized into two size classes (4 and >4 cm), where mortality is indicated with *light dots*, resilience suggests that the colonies bleached and recovered and is indicated by *stripes*, and resistance suggests that no discernible (visual) change occurred to the colonies, and is indicated by *solid segments*

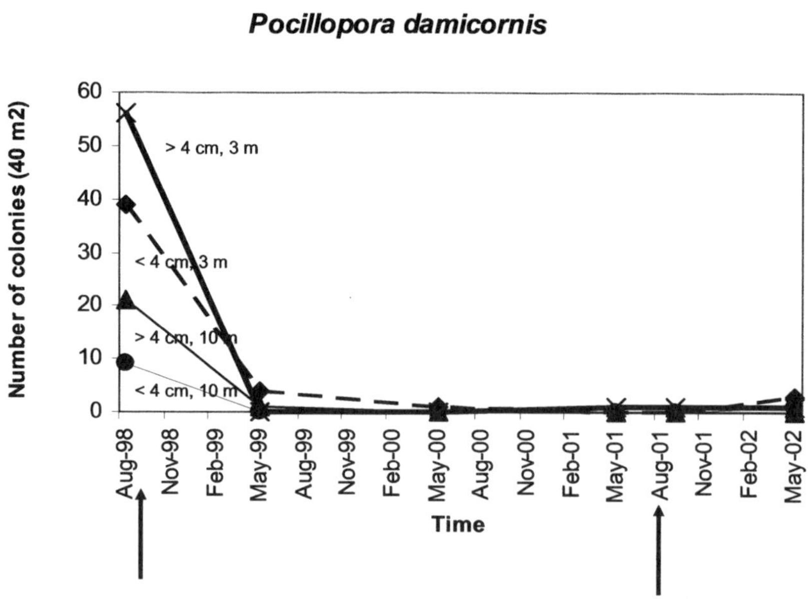

Fig. 5.4.a. Comparative responses of *Pocillopora damicornis* between 1998 and 2002. *Arrows* indicate bleaching events

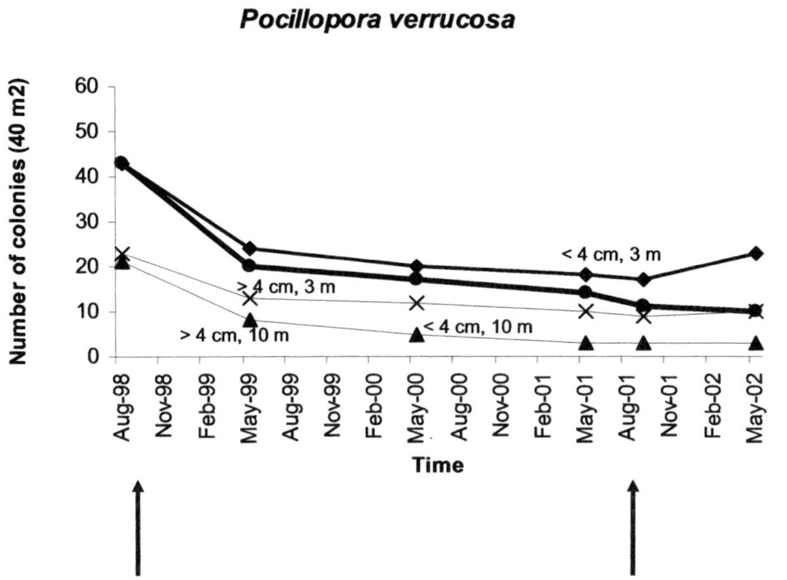

Fig. 5.4.b. Comparative responses of *Pocillopora verrucosa* between 1998 and 2002. *Arrows* indicate bleaching events

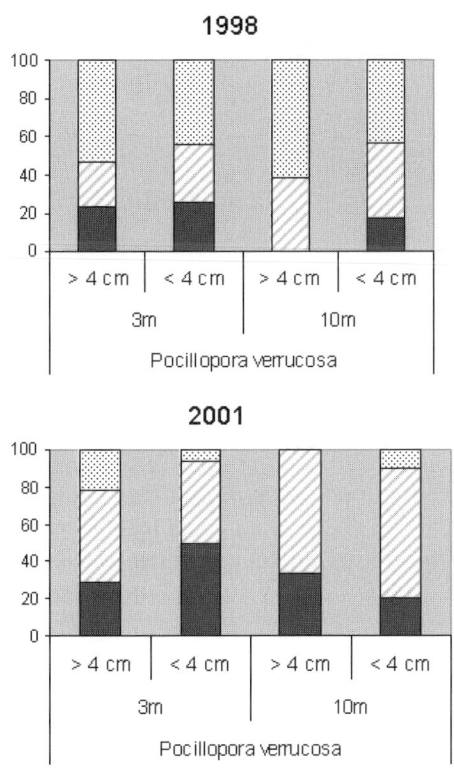

Fig. 5.5. Comparative responses of *Pocillopora verrucosa* colonies to the 1998 and 2001 thermal stress events (densities per m^2 at 4 and >4 cm, at 3 and 10 m, respectively) that survived the 1998 thermal anomaly were 0.6, 0.5, 0.32, and 0.2, while in May 2002 the densities were 0.58, 0.25, 0.25, and 0.08. Criteria specified in Fig. 5.3

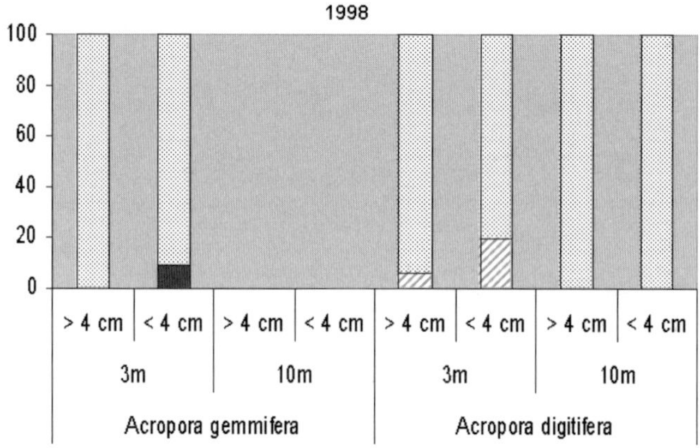

Fig. 5.6. Responses of *Acropora gemmifera* and *Acropora digitifera* to the 1998 thermal stress event. Criteria specified in Fig. 5.3

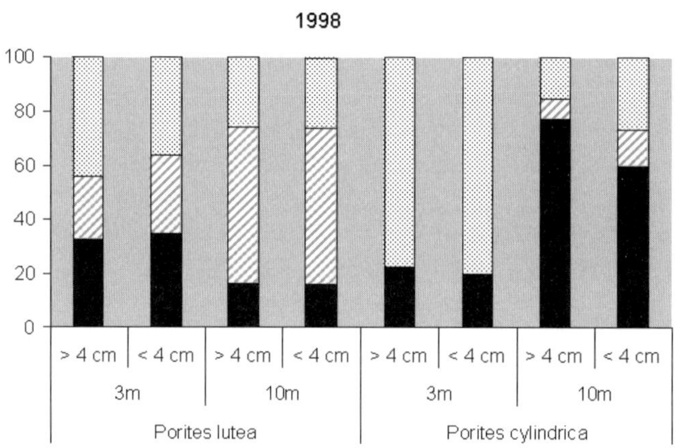

Fig. 5.7. Responses of *Porites lutea* and *Porites cylindrica* to the 1998 thermal stress event. Criteria specified in Fig. 5.3

spectively; Fig. 5.5). Nearly all of the *Montipora* (data not shown) and *Acropora* colonies died in 1998, except for some small (<4 cm) colonies (Fig. 5.6), but there were considerable *Acropora* recruits on the shallow reefs by 2001 (data not shown). *Porites lutea* showed a higher frequency of resilience and lower frequency of mortality at 10 than at 3 m (Figs. 5.7, 5.8); *Porites cylindrica* was more tolerant at 10 than at 3 m depth (Fig. 5.7). *Porites lutea* showed a similar response in both years (Fig. 5.9), but *Porites cylindrica* showed a higher tolerance in 2001 than in 1998 at both depths (Fig. 5.10) (for 3 m McNemar Chi-square was χ^2=48.6, p<0.001, for >4 cm and χ^2=52.6, p<0.001 for 4 cm (Fig. 5.9). Note that we do not display the 10-m depth results for *Porites cylin-*

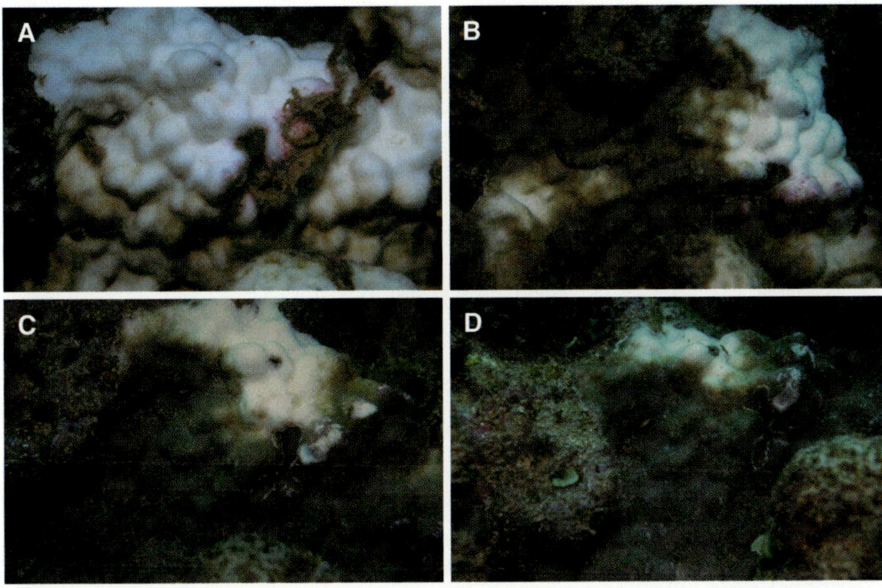

Fig. 5.8. An example of bleaching and recovery (i.e., resilience) of a *Porites lutea* colony found at 10 m in 1998. **A** 2 Months, **B** 3 months, **C** 4 months, **D** 5 months. (Photos by A. Irikawa)

drica because no colonies died in 2001, therefore, our frequencies were zero and we cannot perform McNemar's test.

Colonies within the genus *Pavona* (family Agariciidae) were relatively resistant to bleaching (Fig. 5.11), but interestingly *Pavona varians* showed a significant *reduction* in resistance in 2001 (for colonies at 3 m, McNemar Chi-square was χ^2=17.8, p<0.001 for 4 cm, and at 10 m for colonies >4 cm, χ^2=40.5, p<0.001) (Fig. 5.12). *Pachyseris* species suffered high mortality at 3 m in 1998 (Figs. 5.13, 5.14). The surviving colonies of *Pachyseris speciosa* (1.35 and 0.68 m^{-2}, for 4- and >4-cm-sized colonies, respectively) and *Pachyseris gemmae* at 10 m in 1999 (0.9 and 0.53 m^{-2}, for 4- and >4-cm sized colonies, respectively) declined to approximately half their densities between the thermal stress events (between 1998 and 2001) because of *Acanthaster planci* predation. The remaining colonies of *Pachyseris speciosa* at 10 m and the *Pachyseris gemmae* at both 3 and 10 m showed little change in tolerance between 1998 and 2001 (Figs. 5.15, 5.16). Clearly, small (4 cm) *Favia pallida* and *Favia favus* colonies were more resistant than large colonies (>4 cm) in 1998 (Figs. 5.17, 5.18), and both *Favia pallida* and *Favia favus* showed a significant increase in tolerance in 2001 (*F. pallida*, McNemar Chi-square, χ^2=5.3, p<0.02; χ^2=20.6, p<0.001; χ^2=4.1, p<0.04, χ^2=15.7, p<0.001, for 4 and >4 cm at 3 and 10 m, respectively) (*F. favus*, McNemar Chi-square, χ^2=14.9, p<0.01; χ^2=46.8, p<0.001; χ^2=29.5, p<0.001, for 4 and >4 cm at 3 m and for >4 cm at 10 m, respectively; Figs. 5.19, 5.20).

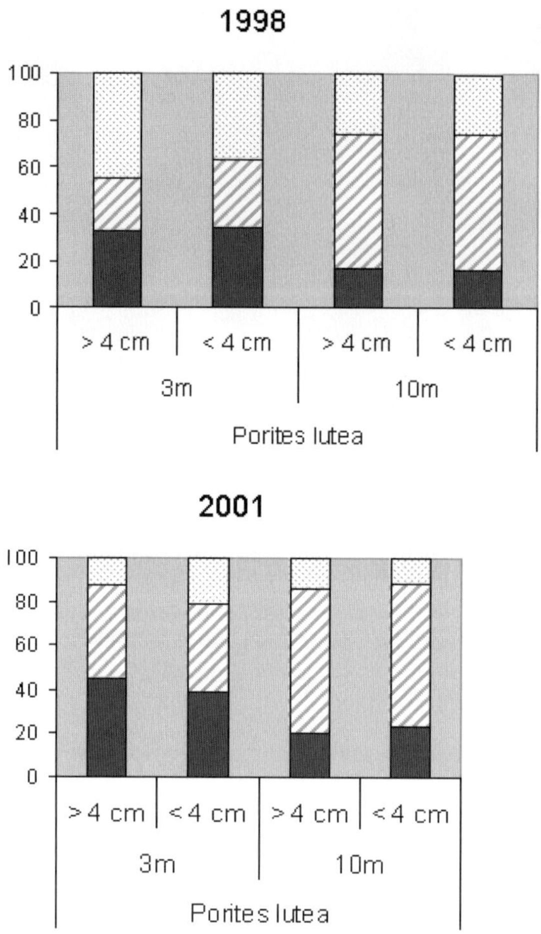

Fig. 5.9. Comparative responses of *Porites lutea* colonies to the 1998 and 2001 thermal stress events (densities per m^2 at 4 and >4 cm, at 3 and 10 m, respectively) that survived the 1998 thermal anomaly were 2.68, 1.05, 5.15, and 2.93, while in May 2002 the densities were 1.73, 0.6, 3.28, and 1.93. Criteria specified in Fig. 5.3

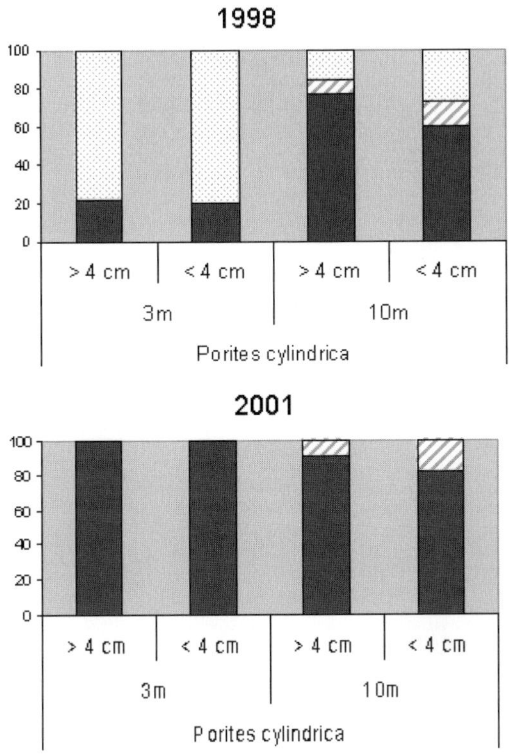

Fig. 5.10. Comparative responses of *Porites cylindrica* colonies to the 1998 and 2001 thermal stress events (densities per m^2 at 4 and >4 cm, at 3 and 10 m, respectively) that survived the 1998 thermal anomaly were 0.05, 0.05, 0.28, and 0.33, while in May 2002 the densities were 0.03, 0.03, 0.1, and 0.08. Criteria specified in Fig. 5.3

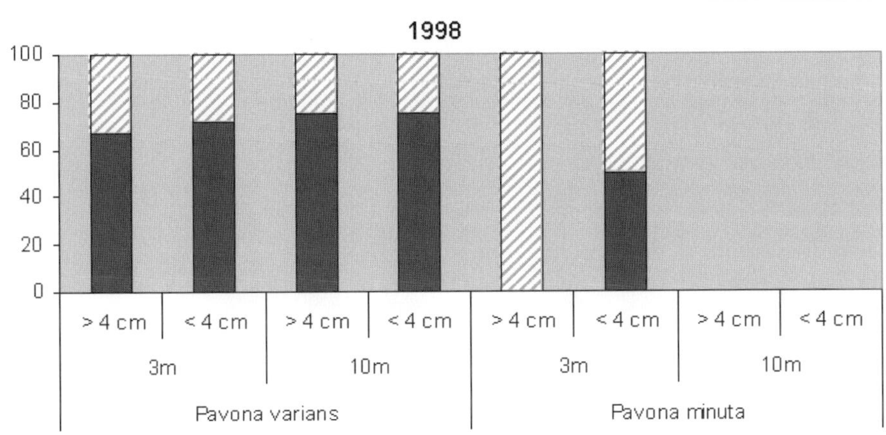

Fig. 5.11. Responses of *Pavona varians* and *Pavona minuta* to the 1998 thermal stress event. Criteria specified in Fig. 5.3

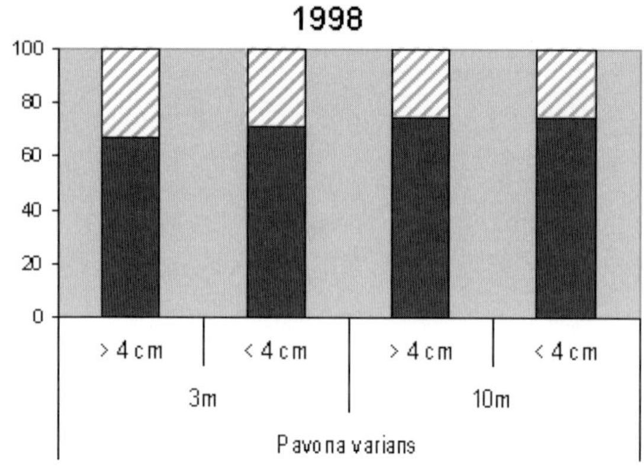

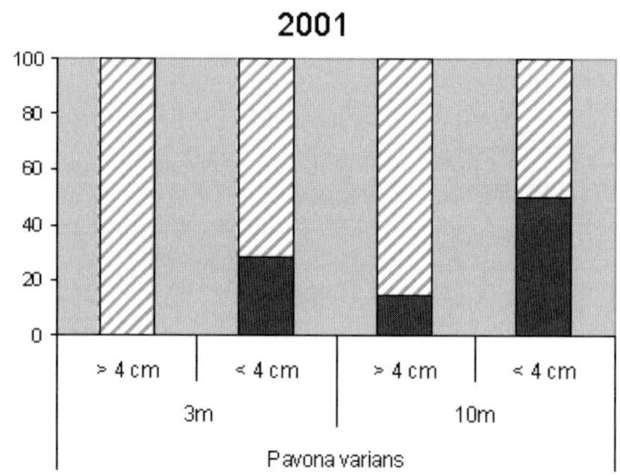

Fig. 5.12. Comparative responses of *Pavona varians* colonies to the 1998 and 2001 thermal stress events (densities per m^2 at 4 and >4 cm, at 3 and 10 m, respectively) that survived the 1998 thermal anomaly were 0.18, 0.08, 0.2, and 0.2, while in May 2002 the densities were 0.18, 0.08, 0.15, and 0.18). Criteria specified in Fig. 5.3

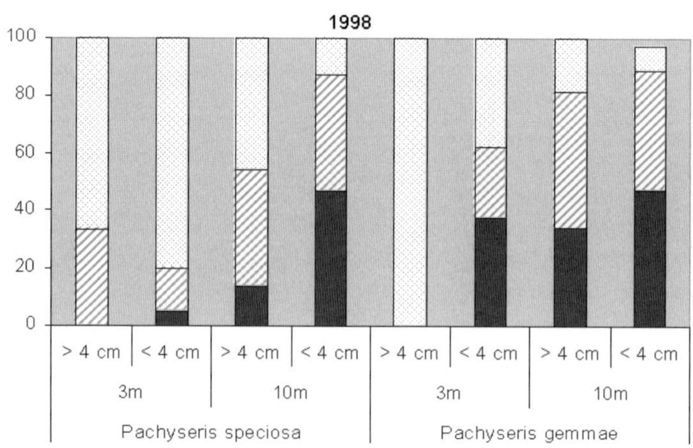

Fig. 5.13. Responses of *Pachyseris speciosa* and *Pachyseris gemmae* to the 1998 thermal stress event. Criteria specified in Fig. 5.3

Fig. 5.14. An example of bleaching and recovery of a *Pachyseris gemmae* colony in 1998. (Photos by A. Irikawa)

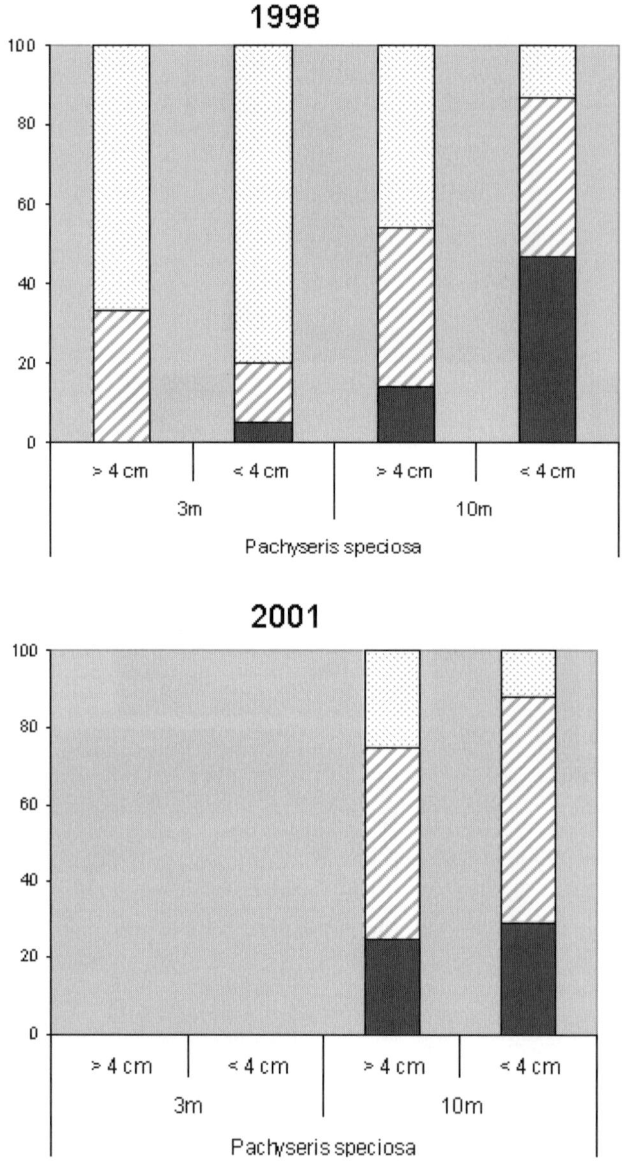

Fig. 5.15. Comparative responses of *Pachyseris speciosa* colonies to the 1998 and 2001 thermal stress events (densities per m^2 at 4 and >4 cm, at 3 and 10 m, respectively) that survived the 1998 thermal anomaly were 0.1, 0.05, 1.32, and 0.68, while in May 2002 the densities were 0, 0, 0.38, and 0.3. Criteria specified in Fig. 5.3

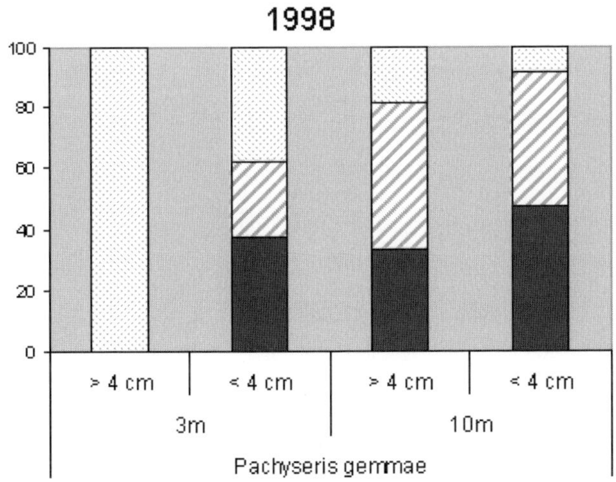

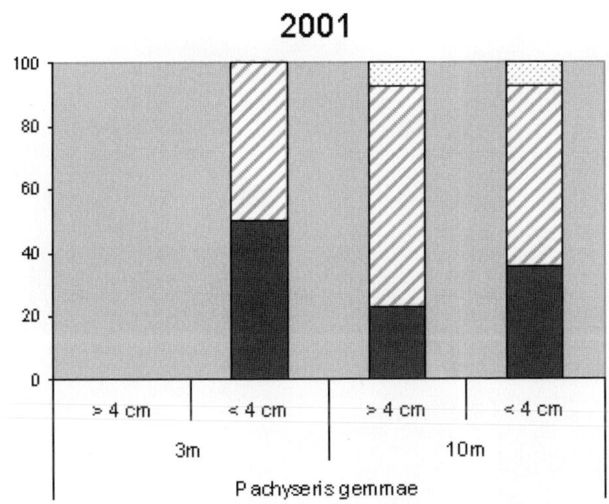

Fig. 5.16. Comparative responses of *Pachyseris gemmae* colonies to the 1998 and 2001 thermal stress events (densities per m^2 at 4 and >4 cm, at 3 and 10 m, respectively) that survived the 1998 thermal anomaly were 0.13, 0, 0.8, and 0.43, while in May 2002 the densities were 0.05, 0, 0.33, and 0.3. Criteria specified in Fig. 5.3

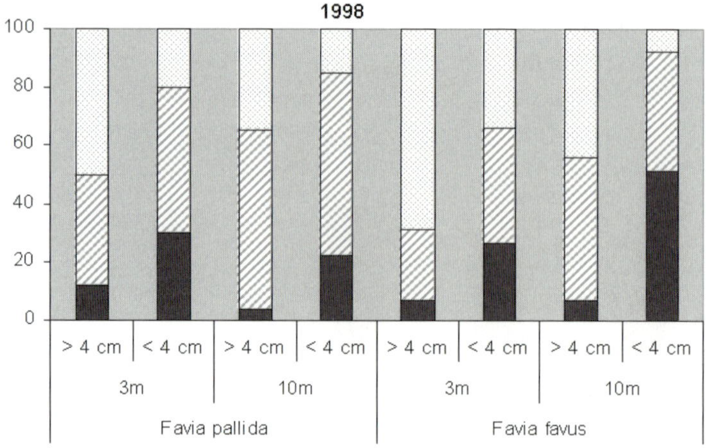

Fig. 5.17. Responses of *Favia pallida* and *Favia favus* to the 1998 thermal stress event. Criteria specified in Fig. 5.3

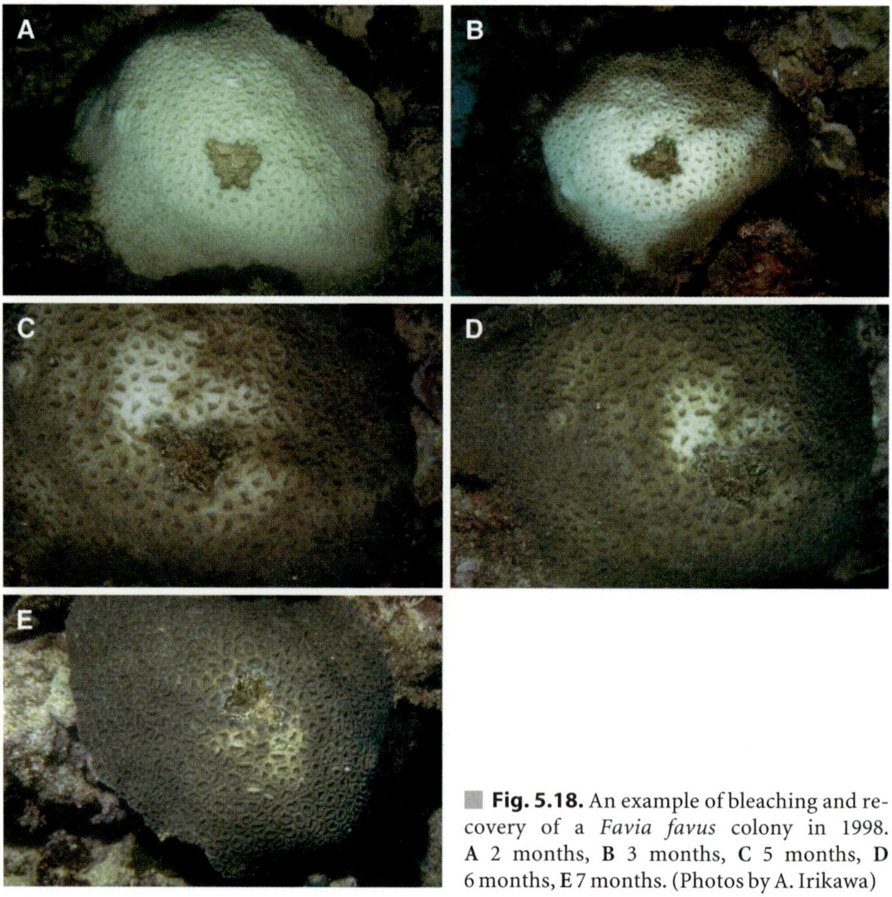

Fig. 5.18. An example of bleaching and recovery of a *Favia favus* colony in 1998. **A** 2 months, **B** 3 months, **C** 5 months, **D** 6 months, **E** 7 months. (Photos by A. Irikawa)

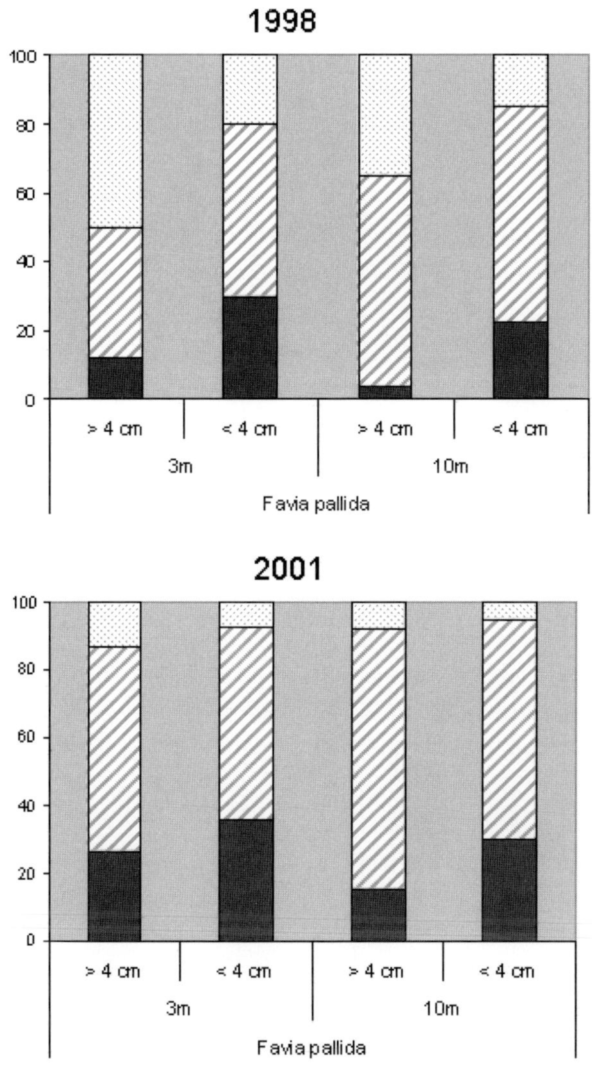

Fig. 5.19. Comparative responses of *Favia pallida* colonies to the 1998 and 2001 thermal stress events (densities per m^2 at 4 and >4 cm, at 3 and 10 m) that survived the 1998 thermal anomaly were 0.6, 0.52, 0.58, and 0.43, while in May 2002 the densities were 0.83, 0.4, 0.65, and 0.35. Criteria specified in Fig. 5.3

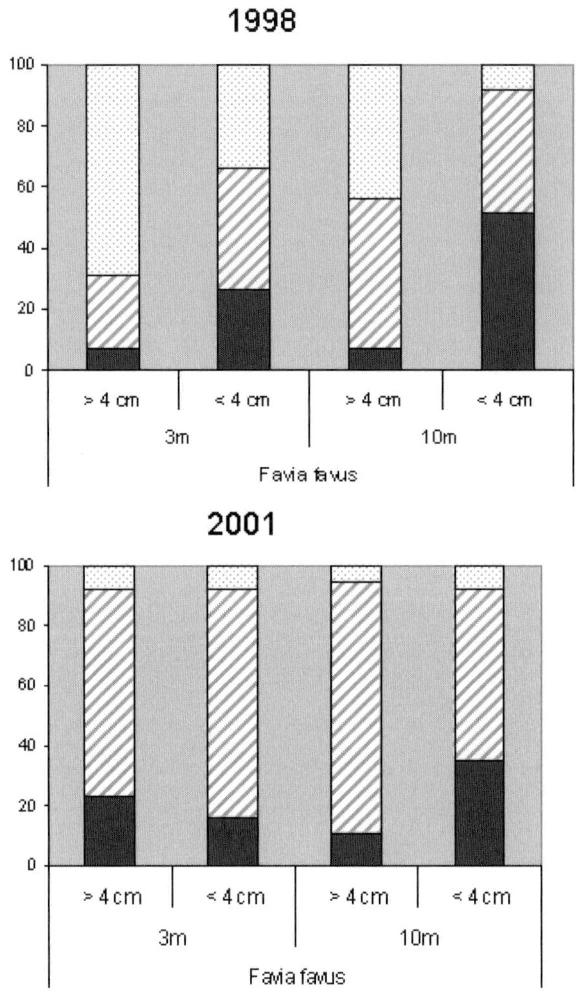

Fig. 5.20. Comparative responses of *Favia favus* colonies to the 1998 and 2001 thermal stress events (densities per m^2 at 4 and >4 cm, at 3 and 10 m, respectively) that survived the 1998 thermal anomaly were 0.88, 0.45, 1.75, and 1.05; while in May 2002 the densities were 1.02, 0.38, 17, and 0.98. Criteria specified in Fig. 5.3

5.4 Discussion

The two bleaching events, in 1998 and 2001, in southern Japan varied considerably in their visual impact. Mortality in 1998 was most evident for *Pocillopora* and *Acropora* species at 3 m, which agrees with Loya et al. (2001) who reported similar effects and dramatic reductions in coral cover and diversity. While the temperatures in 2001 were as extreme and as long-lasting as temperatures in 1998, relative shifts to coral communities in 2001 were only slight compared with shifts in 1998. Such differences were a consequence of five possibilities, which are not mutually exclusive: (1) there were fewer coral colonies in 2001; (2) the two events were only 3 years apart and the reefs of Okinawa had little time to recover; (3) a structural shift in community composition had occurred in 1998, and selected for tolerant colonies and tolerant zooxanthellae; (4) these same coral colonies and their zooxanthellae may have acclimatized to the thermal stresses; and (5) adaptation had occurred, the corals that survived the 1998 bleaching reproduced and the recruits were descendants of a more tolerant parent stock.

While the first three statements are certain, there is some evidence of acclimatization (the fourth scenario) in the present study. The fifth scenario is also probable, but needs further testing. Some coral populations appeared hardly affected by the bleaching events, for example, we recorded no mortality for *Pavona varians* and *Pavona minuta* in 1998; but *Pavona varians* was less resistant in 2001, while *Porites cylindrica, Favia pallida* and *Favia favus* were clearly more tolerant in 2001. Glynn et al. (2001), however, reported *Pavona* spp. in the east Pacific as susceptible to high SSTs, showing high mortality. Although Hueerkamp et al. (2001) reported that *Pavona gigantea* was very tolerant, while *Pavona clavus* was more susceptible. Indeed, there is considerable variation within each genus that may mask, or average out, any genera-specific effects. Here, we show that although pocilloporids suffered considerable mortality the reefs continued to support a moderate number of *Pocillopora verrucosa* colonies; only 50% died.

The present study also shows that small colonies were more tolerant to the anomalous SSTs than large colonies; such small size facilitates mass transfer which in turn aids survival (Loya et al. 2001; Nakamura and van Woesik 2001). This further suggests an increasing trend toward smaller colonies if bleaching events become more frequent. Smaller colonies may not be reproductively competent, if they are remnants of much larger colonies, and certainly when they are new recruits and thus immature. Since the foundation of adaptation is based upon the notion that differential reproductive rates are facilitated by environmental influences on different individuals within populations, then the shift in size-frequency distributions toward smaller corals may be a sign that adaptation is less likely to occur if bleaching events are frequent – because of repeated setbacks of the coral colonies toward immaturity. It is also important to understand where the new recruits came from. If the new recruits came from reefs offshore, characterized by high coral survival because the 1998 maximum sea surface temperatures were not as anomalously high as reefs nearshore (Nadaoka et al. 2001), then recruits from these reefs are less likely to be of more heat-tolerant offspring. In contrast, local recruitment may indicate a trend toward adaptation because selection against less tolerant corals in 1998 infers that more tolerant offspring survived and reproduced.

Genetic similarity over large geographic areas where corals are found (Ayre and Hughes 2000; Ridgway et al. 2001) may restrict the degree of adaptation of some populations. There is also some evidence in the plant literature that the predicted rates of evolutionary response of prairie plants are much slower than the predicted rate of climate change (Etterson and Shaw 2001). Nevertheless, there is palaeontological evidence for scleractinian coral adaptation and survival through warmer tropical sea surface temperatures in the late Cretaceous (145–65 Ma) and Eocene (54–38 Ma) epochs, when the sea surface temperatures were at least 28–32 °C (Pearson et al. 2001). What we need to compare are the rates of change that occurred in these warmer periods in the past compared with contemporary rates of environmental change.

The period between bleaching events, 1999–2000, was not without change. We show clear reductions in colony numbers between bleaching events, particularly for *Pachyseris speciosa* and *Pachyseris gemmae*. These species-specific changes were a result of selective predation by resident *Acanthaster planci*. Chronic *A. planci* populations have survived in the Ryukyu Islands for more than 20 years (Yamaguchi 1986). Because *A. planci* populations were not affected by the thermal stress event in 1998, a reduction in coral abundance in 1998 increased the relative predation pressure on the colonies that survived. Therefore, in order to survive, coral populations need to tolerate not only high and increasingly frequent thermal stress events, but they also need to be relatively unpalatable to *A. planci* – a coevolutionary process that has received little attention. Selective pressures of this magnitude may see shifts to the coral communities unprecedented since the Triassic when scleractinian corals first evolved.

5.5 Unresolved Questions and Future Research

The present study provides a comparative baseline to assess the important, but yet unanswerable, questions: (1) is bleaching or paling of corals, through the loss of their symbiotic algae (zooxanthellae) and/or their pigments, linked to global climate change and increasing ocean temperatures? and (2) will the communities tolerate and to some degree acclimatize and even adapt to these conditions? The 3-year interval between thermal stress events in Okinawa contrasts markedly with Glynn et al.'s (2001) study in the eastern Pacific that recorded bleaching events in 1982–1983 and in 1997–1998, which were both related to El-Niño Southern Oscillations. Guzman and Cortes (2001) reported that populations of massive and branched corals in 1997–1998 were more tolerant to elevated thermal stress than populations in 1982–1983 in the eastern Pacific (Costa Rica), where both events recorded similar warming trends and temperature maxima. Guzman and Cotes (2001) suggested that some of the coral populations had adapted to these warmer conditions.

Points of future research that this study raises include the following:

1. There is a need to test the degree of adaptation in different geographical locations and assess which coral species are adapting most readily and which are rapidly losing individuals and becoming less viable.
2. How long is the period of reproductive incompetence after a bleaching event?
3. What is the thermal stress threshold before gametogenesis is terminated?
4. There is a need to consider whether there is a shift in the size of maturity of these colonies undergoing frequent thermal stress events.
5. Do repeated bleaching events interfere with other life-history processes, and even lead to senescence?
6. More information is needed on whether the gametes from different depths show a similar community restructuring capacity, or are the gametes that stem from deep-water coral less competent in shallow high-light and high-temperature habitats.
7. A critical point is, how does the survival of recruits differ in accordance with micro-habitat, depth, and locality.
8. Are local rivers affecting recruit survival and the onset of coral bleaching? Clearly, rivers have the capacity to alter coral communities through selective processes at and above the recruitment processes, and there are analytical techniques to assess the degree of these impacts (West and van Woesik 2001).

The challenge remains in understanding the links among physico-chemical influences when large-scale regional events, which include thermal stresses, differentially impact local coral communities. Survival of thermal stress-tolerant corals is certain, but adaptation may also occur for coral populations that have high genotypic diversity and specific zooxanthellae types. Here, we show a high interspecific response during thermal stress events, some acclimation of the populations over time, and even a hint of adaptation, and a co-evolutionary response between thermal stress and increased predation pressure. Yet, an assessment of 12 populations is only the beginning in a coral reef region supporting over 340 coral species.

Acknowledgments. Our thanks are extended to Sandra van Woesik for editorial comments on the manuscript.

References

Ayre DJ, Hughes TP (2000) Genotypic diversity and gene flow in brooding and spawning corals along the Great Barrier Reef, Australia. Evolution 54:1590–1605

Ben-Haim Y, Rosenberg E (2002) A novel *Vibrio* sp. pathogen of the coral *Pocillopora damicornis*. Mar Biol 141:47–55

Berkelmans R, Oliver JK (1999) Large-scale bleaching of corals on the Great Barrier Reef. Coral Reefs 18:55–60

Brown BE (1997) Coral bleaching: causes and consequences. Coral Reefs 16:S129–S138

Buddemeier RW, Fautin DG (1993) Coral bleaching as an adaptive mechanism: a testable hypothesis. BioScience 43(5):320–325

Edwards AJ, Clark S, Zahir H, Rajasuriya A, Naseer A, Rubens J (2001) Coral bleaching and mortality on artificial and natural reefs in Maldives in 1998, sea surface temperature anomalies and initial recovery. Mar Pollut Bull 42:7–15

Etterson JR, Shaw RG (2001) Constraint to adaptive evolution in response to global warming. Science 294:151–154

Fitt WK, Warner ME (1995) Bleaching patterns of four species of Caribbean reef corals. Biol Bull 189:298–307

Fitt WK, Brown BE, Warner ME, Dunne RP (2001) Coral bleaching: interpretation of thermal tolerance limits and thermal thresholds in tropical corals. Coral Reefs 20:51–65

Glynn PW (1993) Coral reef bleaching: ecological perspectives. Coral Reefs 12:1–17

Glynn PW, Mate JL, Baker AC, Calderon MO (2001) Coral bleaching and mortality in Panama and Ecuador during the 1997–1998 El Niño-Southern Oscillation event: spatial/temporal patterns and comparisons with the 1982–1983 event. Bull Mar Sci 69(1):79–109

Guzman HM, Cortes J (2001) Changes in reef community structure after fifteen years of natural disturbances in the eastern Pacific (Costa Rica). Bull Mar Sci 69(1):133–149

Hoegh-Gulberg O (1999) Climate change, coral bleaching and the future of the world's coral reefs. Mar Freshwater Res 50(8):839–866

Hoegh-Guldberg O, Smith GJ (1989) The effect of sudden changes in temperature, light and salinity on the population density and export of zooxanthellae from the reef corals *Stylophora pistillata* Esper and *Seriatopora hystrix* Dana. J Exp Mar Biol Ecol 129:279–304

Hoegh-Guldberg O, Jones RJ, Ward S, Loh WK (2002) Is coral bleaching really adaptive? Nature 415:601–602

Hueerkamp C, Glynn PW, D'Croz L, Mate JL, Colley SB (2001) Bleaching and recovery of five eastern Pacific corals in an El Niño-related temperature experiment. Bull Mar Sci 69(1):215–236

Jones RJ, Hoegh-Guldberg O, Larkum AWD, Schreiber U (1998) Temperature-induced bleaching of corals begins with impairment of the CO_2 fixation mechanism in zooxanthellae. Plant Cell Environ 21:1219–1230

Kuroki T, VanWoesik R (1999) Changes in zooxanthellae characteristics in the coral *Stylophora pistillata* during the 1998 bleaching event. Galaxea 1:97–101

Kushmaro A, Loya Y, Fine M, Rosenberg E (1996) Bacterial infection and coral bleaching. Nature 380:396

Loya Y, Sakai K, Yamazato K, Nakano H, Sambali H, van Woesik R (2001) Coral bleaching: the winners and the losers. Ecol Lett 4:122–131

Marshall PA, Baird AH (2000) Bleaching of corals on the Great Barrier Reef: differential susceptibilities among taxa. Coral Reefs 19:155–163

McClanahan TR (2000) Bleaching damage and recovery potential of Maldivian coral reefs. Mar Pollut Bull 40:587–597

McClanahan TR, Muthiga NA, Mangi S (2001) Coral and algal changes after the 1998 coral bleaching: interaction with reef management and herbivores on Kenyan reefs. Coral Reefs 19:380–391

Nadaoka K, Nihei Y, Wakaki K, Kumano R, Kakuma S, Moromizato S, Omija T, Iwao K, Shimoike K, Taniguchi H, Nakano Y, Ikema T (2001) Spatial variation of water temperature around Okinawa coasts and its relationship to offshore thermal environments and coral bleaching. Coral Reefs 20:373–384

Nakamura T, van Woesik R (2001) Water-flow rates and passive diffusion partially explain differential survival of corals during 1998 bleaching event. Mar Ecol Prog Ser 212:301–304

Nakamura T, Yamasaki H, van Woesik R (2003) Water flow facilitates recovery from bleaching in the coral *Stylophora pistillata*. Mar Ecol Prog Ser 256:287–291

Pearson PN, Ditchfield PW, Singano J, Harcourt-Brown KG, Nicholas CJ, Olsson RK, Shackleton NJ, Hall MA (2001) Warm tropical sea surface temperatures in the Late Cretaceous and Eocene epochs. Nature 413:481–487

Ridgway T, Hoegh-Guldberg O, Ayre DJ (2001) Panmixia in *Pocillopora verrucosa* from South Africa. Mar Biol 139:175–181

Strong AE, Kearns E, Gjovig KK (2000) Sea surface temperature signals from satellites - an update. Geophys Res Lett 27(11):1667–1670

Strong AE, Liu G, Kimura T, Yamano H, Tsuchiya M, Kakuma S, van Woesik R (2002) Detecting and monitoring 2001 coral reef bleaching events in Ryukyu Islands, Japan using satellite bleaching hotspot remote sensing technique. Proc 2002 IEEE Int Geosci Remote Sensing Symp and 24th Can Symp Remote Sensing, Toronto, Canada, CD-ROM

Van Woesik R (2001) Coral bleaching: transcending spatial and temporal scales. Trends Ecol Evol 16:119–121

Veron J, Minchin P (1992) Correlations between sea surface temperature, circulation patterns and the distribution of hermatypic corals of Japan. Cont Shelf Res 12:835–857

Warner ME, Fitt WK, Schmidt GW (1999) Damage to Photosystem II in symbiotic dinoflagellates: a determinant of coral bleaching. Proc Natl Acad Sci USA 96:8007–8012

West K, van Woesik R (2001) Spatial and temporal variance of river discharge on Okinawa (Japan): inferring the temporal impact on adjacent coral reefs. Mar Pollut Bull 42(10):864–872

Williams EH, Bunkley-Williams L (1990) The world-wide coral reef bleaching cycle and related sources of coral mortality. Atoll Res Bull 355:1–72

Yamaguchi M (1986) *Acanthaster planci* infestations of reefs and coral assemblages in Japan: a retrospective analysis of control efforts. Coral Reefs 5:23–30

6 Coral Bleaching in a Temperate Sea: From Colony Physiology to Population Ecology

Maoz Fine and Yossi Loya

6.1 Introduction

While shifts in coral reef species composition following bleaching events have been reported (Glynn 1988; Brown and Suharsono 1990; Gleason 1993; Guzman and Cortes 2001; Loya et al. 2001) and widely discussed (Glynn 1993; Brown 1997; Done 1999; Hoegh-Guldberg 1999; Wilkinson 2000), little is known about the effect of bleaching on coral populations in temperate zones or nonreef environments.

A major factor influencing shifts in coral community structure is differential susceptibility of species to bleaching and their recovery capabilities following a bleaching event. Reef-building corals are not all equally susceptible to the influence of increased temperature. Some coral species have been shown to survive severe bleaching events (Glynn 1993; Brown 1997; Hoegh-Guldberg 1999; Wilkinson 2000; Loya et al. 2001), leading to major structural shifts in coral communities (Glynn 1993; Brown 1997; Ostrander et al. 2000; Loya et al. 2001). Once these communities have shifted, it would probably require a lengthy period of time to return to their original states (Hoegh-Guldberg 1999).

The population of *Oculina patagonica*, an invading species to the Mediterranean Sea from the southwestern Atlantic Ocean, has been studied extensively during the last decade (Fine et al. 2001). *O. patagonica* was first recorded in the Mediterranean Sea in 1966 near Savona harbor (Gulf of Genova; Zibrowius 1974). In 1973, it was first noticed in southeastern Spain (Zibrowius and Ramos 1983; Zibrowius 1992). At present, it is known to inhabit many natural sites and harbors along hundreds of kilometers of coastline in southeastern Spain. Along the Mediterranean coast of Israel, *O. patagonica* was first recorded in 1993 (Fine et al. 2001).

The intrinsic and extrinsic factors shaping the current population of *O. patagonica* were studied, including the way that bleaching events affect its population dynamics through reduction of gametogenesis, regeneration, and competitive abilities. Bleaching in *O. patagonica* was first recorded in 1993 and explained by a bacterial infection, to which high water temperature was a contributing factor (Kushmaro et al. 1996, 1998). Since then, 70–90% of coral colonies in the study population have exhibited bleaching during the summer months, when water temperature reaches a maximum of 29–30 °C, and recovered in the following winter (Figs. 6.1, 6.2; Kushmaro et al. 1998; Fine et al. 2001). The studies investigating this phenomenon revealed that the causative

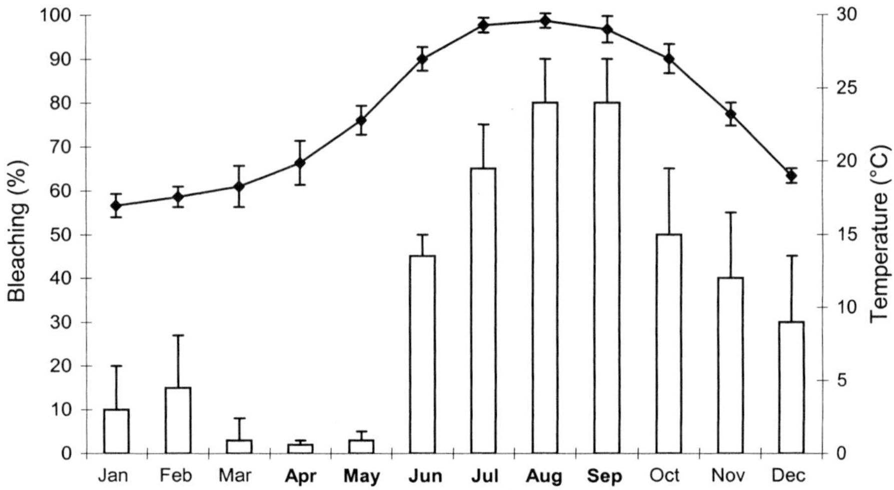

Fig. 6.1. *Oculina patagonica.* Monthly bleaching percentages (*bars*, ±SD visual estimates) of 200 tagged colonies, and monthly surface seawater temperatures (SST, *line*, ±SD) recorded from 1993 to 2002. During the summer, with an increase in SST, a peak in the number of bleached colonies is evident, while during the winter, with a decrease in SST, coral recovery takes place (less bleached colonies). Months in *bold letters* represent gametogenesis. In bleached corals gametogenesis is terminated with the onset of a bleaching event

agent for bleaching is the bacterium *Vibrio shiloi* (Kushmaro et al. 1996, 1997), whereas the environmental cause of the disease is increased seawater temperature (Kushmaro et al. 1998). In contrast to reports from other regions of mass mortality following bleaching events, over 90% of the bleached *O. patagonica* colonies fully recovered each winter through recruitment and/or cell division of their symbiotic zooxanthellae (Kushmaro et al. 1998). Since summer 2000, however, following the eighth consecutive bleaching event recorded, higher percentages of mortality (up to 25%) were recorded in the major study site at Sdot-Yam, Israel (Fine, Levin and Loya, unpubl. data).

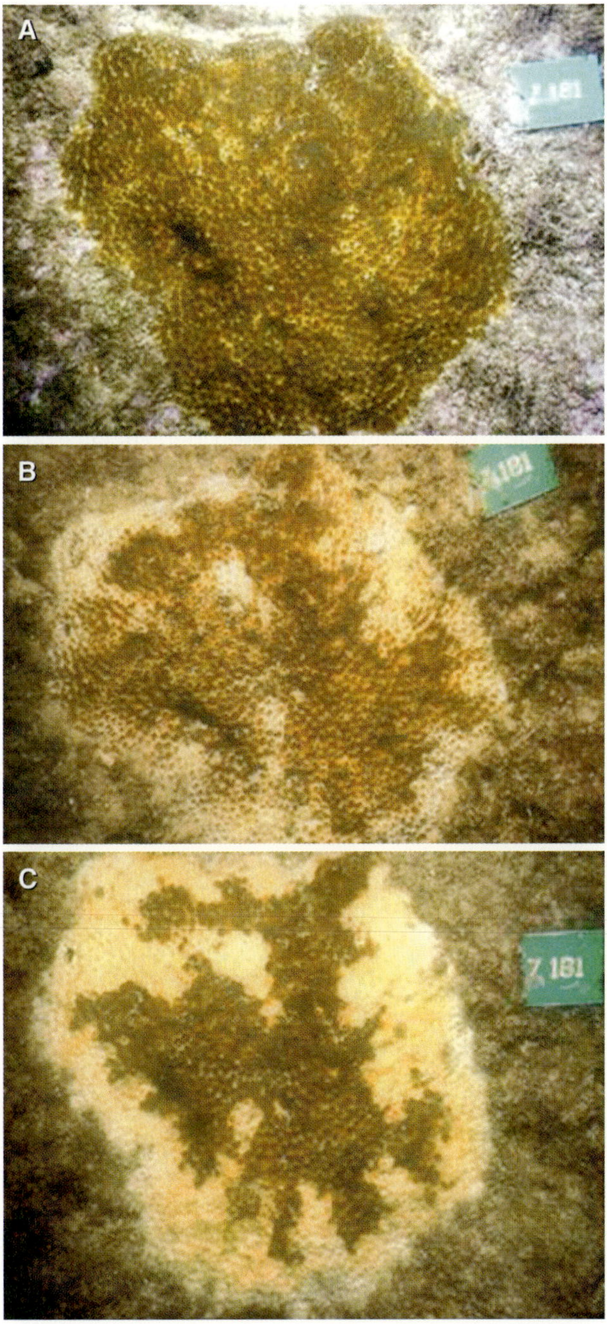

Fig. 6.2. *Oculina patagonica*. Stages of bleaching of tagged colony Z181 beginning in early May (**A**) when most of the population is healthy, through late June (**B**) when colonies begin to bleach in their marginal areas. With increasing seawater temperature, the bleached surface area of the colony increases gradually (**C**) and reaches more central areas

6.2
Spatial Distribution of Bleaching

The bleaching percentage of *O. patagonica* varies considerably in the Mediterranean Sea. No bleaching of *O. patagonica* was recorded along the coasts of the western Mediterranean Sea, whereas along the coasts of the southeastern Mediterranean, a high bleaching percentage (70–90%) has been recorded annually since 1993 (Fine et al. 2001). This may be explained by the colder seawater temperatures in the western Mediterranean (12–24 °C), while in the south-eastern Mediterranean seawater temperature is much higher (16–30 °C).

Along the Israeli coast, an increasing bleaching percentage was recorded from south to north although the distance from the southernmost study site (Ashqelon) to the northernmost site (Rosh-Hanikra) is only 170 km. In northern sites of Achziv and Rosh-Hanikra, a significantly lower percentage of the population bleached throughout the study (10–50%) and for considerably shorter periods of time (1–3 months). Interestingly, when the distribution of bleached colonies was surveyed along a depth gradient in the study site in Sdot-Yam, colonies in shallow water (0–80 cm depth) showed little (2–15%) bleaching, despite being exposed to higher temperatures (by 2–3 °C) than the deeper water corals (Fig. 6.3). Colonies inhabiting shallow tidal pools (where seawater temperature may exceed 35 °C) showed a low bleaching percentage as well. Fragments transplanted from 4-m depth to a shallow reef flat in May showed no bleaching during the summer months, whereas intact colonies at 4 m and fragments transplanted to that depth from the reef flat underwent ex-

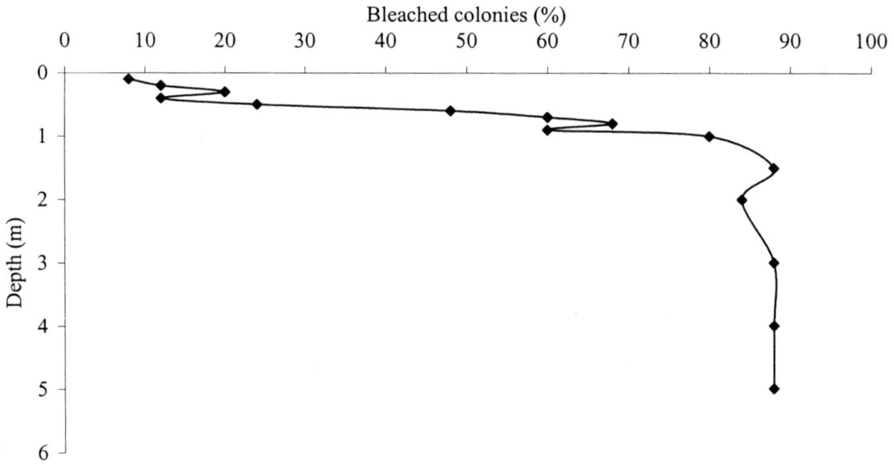

Fig. 6.3. *Oculina patagonica.* Bleaching along a depth gradient, from the reef flat (<30-cm depth) down to tide pools (30–80 cm) and at 1–6 m, demonstrating the low bleaching percentage in populations in very sallow waters as compared with the deeper water population

tensive bleaching. These findings contradicted previous studies which suggested that ultraviolet radiation (UVR) is one of the major causes for bleaching. In the case of *O. patagonica*, corals that experience higher degrees of UVR are in fact less susceptible to bleaching due to the damage caused by UVR to the causative agent of bleaching (Fine et al. 2002a). Brown et al. (2002) reported a similar observation for the coral *Goniastrea aspera*, but concluded that the susceptibility to bleaching is shaped by experience, and that the algae may acquire cross-protection against bleaching from solar to temperature stresses.

6.3 Physiological Effects of Bleaching on *Oculina patagonica*

Bleaching leads to a decline in the supply of photoassimilates provided by the zooxanthellae to the host and reduces the coral's ability to activate costly (energetic) processes such as maintenance, growth, and reproduction.

6.3.1 Reproduction

Increased temperature and bleaching affect coral populations by reducing reproductive capacity (Szmant and Gassman 1990). In a comparison of the fecundity of 200 bleached and unbleached colonies of reef-flat corals at Heron Island (Great Barrier Reef, Australia) after the 1998 bleaching event, bleached colonies of many reef-flat species contained no eggs at all (Ward et al. 1998). These bleached corals, despite having recovered their zooxanthellae, did not spawn during the normal spawning period in November. Furthermore, in other prolific reef-flat species (*A. aspera*, *A. palifera*, *A. pulchra* and *M. digitata*), there were significantly fewer eggs in bleached than in unbleached corals (Ward et al. 1998).

O. patagonica is a gonochoric coral. Onset of gametogenesis was found to be temperature dependent. Female gonads were first observed in May with a rise in surface seawater temperature (SST) and male gonads in July, both reaching maturity in late August and early September (Fig. 6.1: Fine et al. 2001). Spawning occurred on two consecutive nights during full moon of September, after a drop in SST. Colonies sampled in the western Mediterranean (Italy, France, and Spain) showed a similar pattern of gametogenesis to colonies sampled in the southeastern Mediterranean (Egypt, Israel, and Lebanon). Interestingly, naturally occurring azooxanthellate colonies of *O. patagonica* inhabiting small dark caves developed gonads and spawned in parallel to zooxanthellate colonies exposed to light, indicating that a healthy energetic state (which can also be achieved through heterotrophic nourishment) of the coral, rather than the existence of the symbiotic dinoflagellates, is essential for normal gametogenesis. Zooxanthellate colonies that underwent bleaching during the reproductive season did not develop gonads. The high incidence of bleaching events along the Israeli coast observed throughout the years of this study may explain the low recruitment of new colonies during the same period. In the western Mediterranean, however, where no bleaching was recorded, recruitment was very high following the September spawning on any natural or artificial submerged object.

6.3.2 Regeneration and Growth

The physiological constraints on *O. patagonica* colonies were studied by examining translocation capabilities of photosynthetic products during and between bleaching events. We associated translocation and integration abilities with health of the colony. *O. patagonica*'s ability to orient its resources to reproduction (Fine et al. 2001), to recuperating lesions (Fine et al. 2002b), and to zones of competitive interaction (Fine and Loya 2003) is an indication of the coral's integration capacity.

Fig. 6.4. *Oculina patagonica*. Round tissue lesions on the surface of a colony (**A**) following predation by polychaetes of the species *Hermodice carunculata* (**B**). Recovery of lesions inflicted on the unbleached *O. patagonica* colonies (**C**) was significantly higher compared to percentage recovery of the lesions inflicted on the partly bleached colonies (**D**). Whereas most lesions in the unbleached colonies healed completely over the 45-day period, the bleached colonies (n=10) showed no recovery from the inflicted lesions even after 65 days

Using ^{14}C point labeling technique on coral tissue, we examined possible intracolonial translocation of photosynthetic products from the point of tissue labeling, towards recuperating lesions in partly bleached vs. nonbleached intact colonies. Colonies of *O. patagonica* are often preyed upon by polychaetes of the species *Hermodice carunculata*. These nocturnal predatory worms use their jaws to scrape the coral tissue off its skeleton, causing round tissue lesions (Fig. 6.4A, B). Predation lesions can also be observed during the bleaching event. Lesion regeneration in unbleached *O. patagonica* resulted in oriented intracolonial translocation of ^{14}C products towards recuperating edges of the lesions. Translocation was evident in nonbleached and partly bleached colonies, in contrast to colonies that underwent severe bleaching (40–80%), where such translocation did not occur. This probably explains the reduced recovery rates of lesions in these colonies (Fig. 6.4C, D). We also examined intracolonial translocation of photosynthetic products from the point of tissue labeling towards zones of competitive interaction (Fig. 6.5A). In nonbleached *O. patagonica*, competition with the bryozoan *Watersipora* sp. resulted in preferentially oriented translocation of ^{14}C products to the interaction zone, at distances of up to 4–8 cm away from the tissue-labeling site. Sites opposite the interaction zone received significantly less labeled photoassimilates compared to the interaction zone. This allowed competitive superiority of *O. patagonica* over the bryozoan (Fig. 6.5C). In bleached colonies (40–85% bleached surface area), translocation did not occur at the sites of interaction, explaining the failure to compete with the encrusting neighboring *Watersipora* sp. (Fig. 6.5D). These findings demonstrate the importance of colonial integration and resource orientation for lesion recovery and competitive superiority in *O. patagonica.*

This led us to conclude that there is a bleaching threshold of ca. 30% in *O. patagonica* colonies, which determines the levels of colony integration and intracolonial translocation of resources towards regions of maximal demand. Termination of intracolonial resource translocation allows the colony to keep its life-preserving resources within the compartments of the colony that have a greater chance of survival through the bleaching event. We suggest that this threshold is one of the vital mechanisms for the survival of *O. patagonica* through a bleaching event and its recovery during the winter months. The ability to terminate resource translocation to one physiological activity while allowing another is of great importance to the survival of the species. For example, investment in somatic tissues, which can expand or fight along the margins, is presumably more important than investment in reproductive tissues, which may ultimately be lost (Hughes and Jackson 1985).

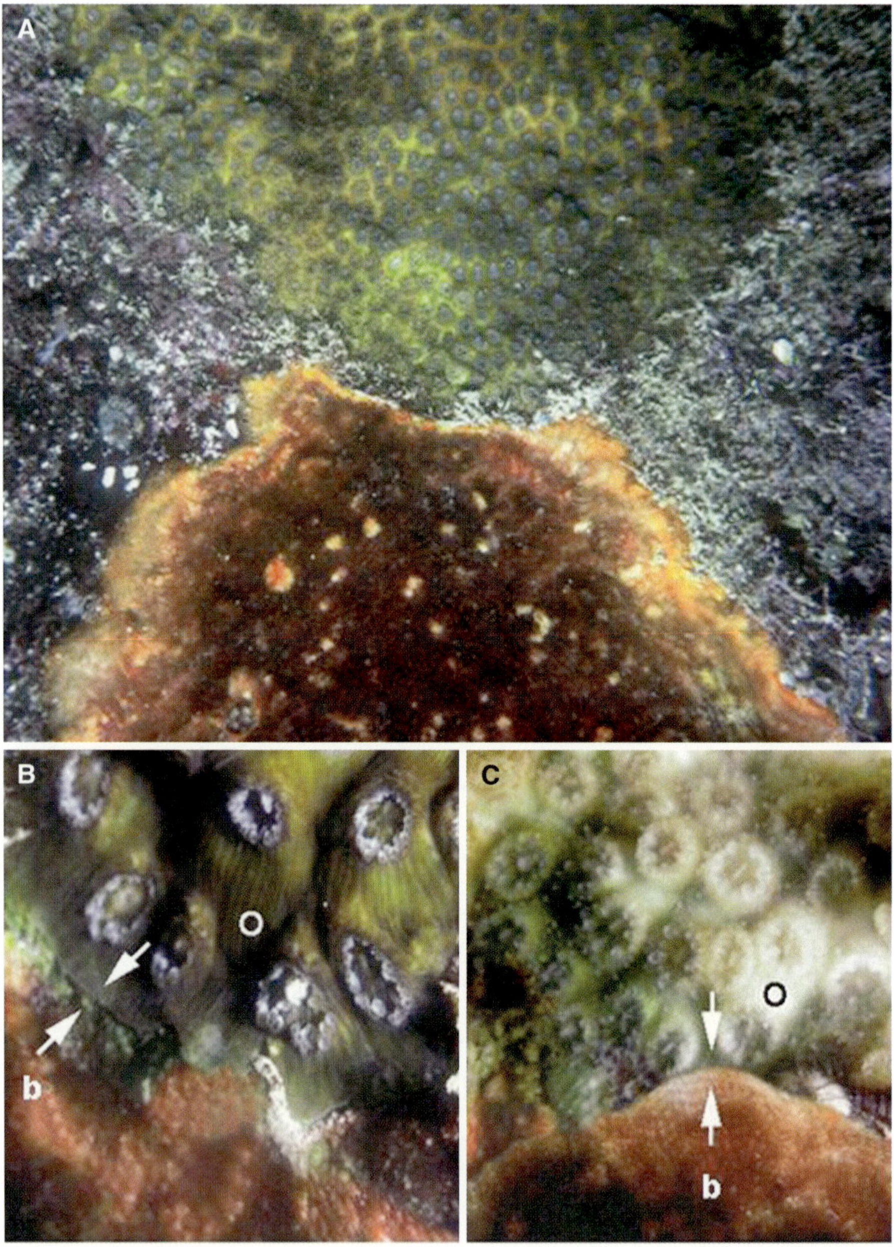

Fig. 6.5. *Oculina patagonica.* **A** Competitive interactions between *O. patagonica* and the bryozoan *Watersipora* sp. A seasonal close-up observation on these competitive interactions shows **B** a healthy nonbleached *O. patagonica* colony overgrowing the bryozoan *Watersipora* sp. and **C** *Watersipora* sp. overgrowing a bleached *O. patagonica* colony (photographs are 200% actual size)

6.4 Alternative Source of Photosynthetic Products During Coral Bleaching

One factor possibly contributing to the survival of coral species during bleaching events is the presence of the endolithic algae harbored in the coral's skeleton, and which provide an alternative source of energy. We studied the dynamics of photosynthetic pigment concentrations and biomass of endolithic algae in the skeleton of *O. patagonica* throughout a bleaching event (Fig. 6.6). During repeated summer bleaching events, these endolithic algae receive increased photosynthetic active radiation (PAR), markedly increase in chlorophyll and biomass, and produce increasing amounts of photoassimilates, which are translocated to the coral (Fine and Loya 2002). Chlorophyll concentrations and biomass of endoliths were up to fivefold higher in skeletons of bleached colonies compared with skeletons of relatively healthy colonies. The translocation dynamics of ^{14}C-labeled photoassimilates from the endoliths to bleached coral tissue showed significantly higher ^{14}C activity of the endoliths harbored within the skeletons of bleached corals than that of nonbleached corals.

Schlichter et al. (1995) demonstrated that photoassimilates released from the phototrophic endoliths reach the coral tissue; following uptake of dissolved organic substances by the coral tissue, these photosynthesis products can be used. Our study, however (Fine and Loya 2002), is the first to show the significant contribution of the endolithic algae to the coral during bleaching. While the energetic contribution of zooxanthellae to coral tissue decreases with increased bleaching, the contribution of the phototrophic endoliths increases.

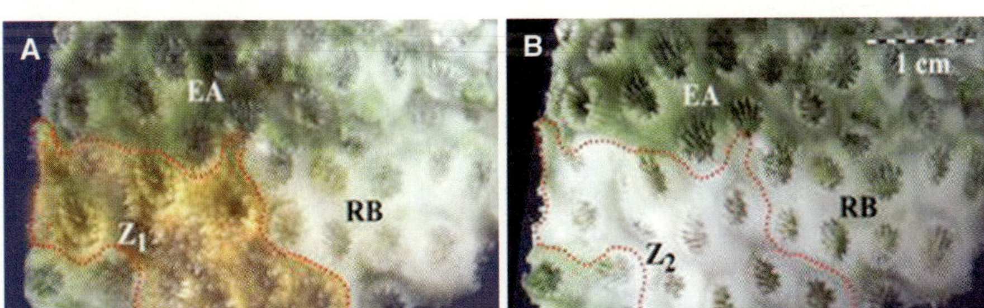

Fig. 6.6. Distribution of endolithic algae during bleaching of the coral *Oculina patagonica* demonstrated on the same colony before and after removal of the tissue. **A** Prior to removal of the tissue, distinct stages of bleaching and tissue condition can be observed. Z_1 Area of healthy coral tissue, *RB* an area of recently bleached and transparent coral tissue revealing skeleton free of pigmented endoliths, *EA* an area of coral that was bleached for 3 weeks, showing the heavily pigmented endoliths beneath the transparent coral tissue. **B** After removal of the tissue, the endolith's condition can be detected and related to the previous tissue condition. Z_2 After removal of the living coral tissue from area Z_1, no pigments were observed in the skeleton beneath the previously healthy tissue

Although this source of energy is nutritionally less important than the zooxanthellae (Schlichter et al. 1997) and is insufficient for sexual reproduction (Fine et al. 2001), it may be vital for the survival of bleached colonies until the recruitment of zooxanthellae enables recovery from bleaching during the following winter.

6.5 Prospects for *Oculina patagonica* Population

A question that remains to be answered relates to the chances of *O. patagonica* populations persisting or further establishing along the eastern Mediterranean coast.

To answer this question, we should first reconsider the factors shaping the current population dynamics (Fig. 6.7). Spatially, bleaching events of *O. patagonica* were not recorded in the western Mediterranean. In the southeastern Mediterranean, mass bleaching events were recorded except for reef flats and tidal pools where the bleaching percentage is very low. Colonies inhabiting a depth of 1–6 m are most likely to experience bleaching annually as does 70–90% of the population (Fig. 6.1).

In terms of reproduction, since bleached colonies cease gametogenesis, the only major source of gametes is likely to be from colonies from the western Mediterranean, but these are expected to mostly supply the western basin. In the eastern Mediterranean, possible sources of gametes are colonies that inhabit reef flats and tidal pools (ca. 10% of the species population), colonies inhabiting dark caves and overhangs (ca. 5% of the species population), and unbleached colonies at a depth of 1–6 m. The latter are, on average, 10–30% of the population. However, since the bleaching percentage significantly decreases from south to north, it is possible that there is a potentially higher supply of gametes in the northern sites.

Colonies found at a depth of 1–6 m differ in their physiological and ecological performances in accordance with the bleaching percentage or bleaching threshold. Colonies with less than 40% bleached surface will show high regenerative and competitive capabilities, whereas colonies with over 40% bleached surface area lack regenerative and competitive capabilities and are expected to lose a large portion of the colony surface area. Once these colonies regain their zooxanthellae, they may possibly regain their lost tissues during the next winter through a higher growth rate. In both cases (either more or less than 40% bleached surface area), the colony is expected to survive through the bleaching event, especially if endolithic algae within the coral skeleton are blooming and supplying the coral with increasing amounts of photoassimilates. Thus, in *O. patagonica*'s population, colony size can be a misleading measure of colony age in view of the fact that small colonies are the result of both young colonies and large older colonies that underwent bleaching, lost in competition with neighboring organisms and lost up to 50% of their surface area (Fine and Loya 2003).

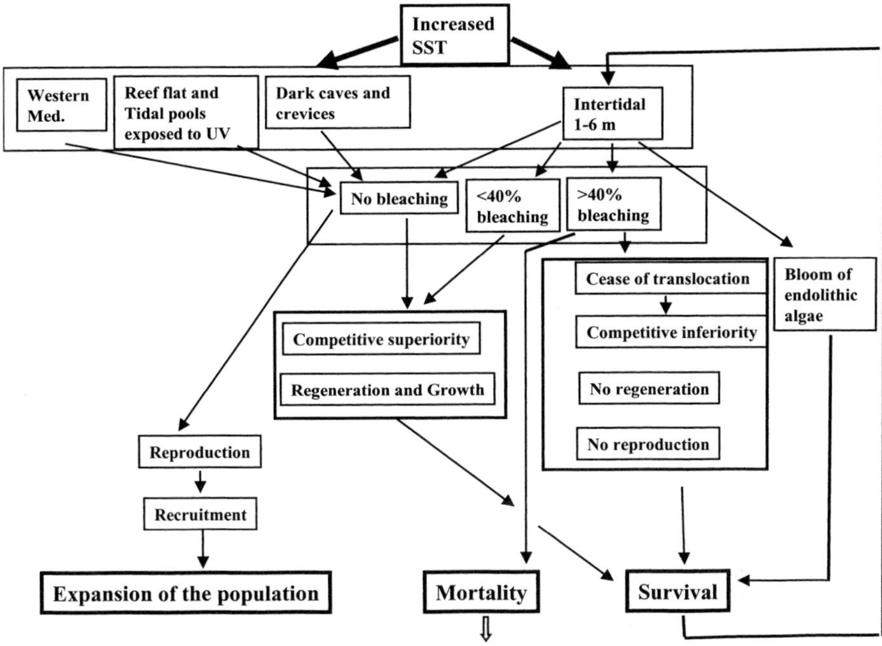

Fig. 6.7. A diagram presenting ecological and physiological pathways shaping the population of *O. patagonica*. The outcome of a bleaching event to populations of *O. patagonica* depends on various levels of circumstances such as the habitat and its environmental conditions, severity of bleaching and a range of physiological processes depending on the energetic constrains. Thus, a coral inhabiting bleaching-free habitats is likely to fulfill all physiological functions and contribute to population expansion. Colonies inhabiting bleaching-affected habitats restrain basic physiological functions such as reproduction or even regeneration and growth with an increase in bleached surface area. Resources are then conserved for life preservation to facilitate survival of the colony. Nevertheless, colonies that experience repetitive bleaching with a high percentage of bleached surface area may in some cases die

Since summer 2000, we have observed an increasing percentage in mortality of labeled colonies that bleached repetitively during the last decade. We suggest that repetitive bleaching of colonies and a very short period of well-being (1–2 months) had adverse affects on the colonies, resulting in increased mortality. Thus, if a high seawater temperature persists in the eastern Mediterranean, we expect certain phase shifts in this species population to take place. We expect the *O. patagonica*'s population in the southern and central part of the Israeli coast to decrease in coral cover (colony size and number of colonies) due to the high occurrence of bleaching, lack of gametogenesis, very low growth rate, and increased mortality. In the northern part of the coast, we expect an increase in coral cover (colony size and number of colonies) as the bleaching percentage in that region is lower and short termed, resulting in longer periods of growth and potentially higher number of propagules. We also expect a further spread of

O. patagonica northward to sites previously not inhabited by that species as recent observations indicate (Fine, unpubl.; Zibrowius, pers. comm.).

Hermatypic corals may sometimes suffer damage from a multitude of agents such as storms (Scoffin 1993), sedimentation (Loya 1976a; Rogers 1990), temperature fluctuations (Jokiel and Coles 1990), emersion at low tide (Loya 1972, 1976b), diseases (Peters 1984; Brown 1997), competitive interactions (Chornesky 1989), anthropogenic stresses (Brown and Howard 1985) and predation (Cameron et al. 1991). Any changes in the above-mentioned factors may modify the current population dynamics recorded and our prospects accordingly. During a bleaching event, any such change in *O. patagonica*'s competitors and/or predators population may lead to an array of events that will divert the establishment of that invading species.

6.6 Unresolved Questions

Oculina patagonica presents the unique case of a coral population undergoing repetitive bleaching events for a decade. It is also a unique case of a coral invading a new region.

Considering the species population in the western (no bleaching recorded) and eastern (high bleaching recorded) Mediterranean, we expect a different fate for each of the populations. *O. patagonica* is an invading species, hence, we expect low genetic divergence of the population. A low percentage of sexual reproduction in the eastern Mediterranean might lead to an even lower genetic divergence. However, to date, no data are available concerning the population genetics of that species. How do the populations (west and east) differ genetically due to bleaching?

A second intriguing unresolved question relates to the ecological fitness of *O. patagonica* being an invasive species. Bleaching seems to decrease the ecological fitness of that species, yet it is, after a decade of bleaching events, the most abundant scleractinian coral along the Israeli coast and is taking over the habitat of indigenous Mediterranean coral species which had no recorded bleaching, such as *Cladocora ceaspitosa.* Are nonbleaching species population structures indirectly affected by bleaching events?

Finally, we would like to address the generality of bleaching thresholds and the way it relates to susceptibility/resilience to bleaching, and mortality/survival following these events. Is there a species-specific bleaching threshold and can that explain shifts in coral community structure in coral reefs? To answer this question, we should now use the techniques successfully used for *O. patagonica* in reef environments and on reef coral species. This may allow us to better predict future changes following bleaching events in the reefs of the world.

References

Brown BE (1997) Coral bleaching: causes and consequences. Coral Reefs 16:129–138
Brown BE, Howard LS (1985) Assessing the effects of stress on reef corals. Adv Mar Biol 22:1–63
Brown BE, Suharsono (1990) Damage and recovery of coral reefs affected by El Niño related seawater warming in the thousand islands, Indonesia. Coral Reefs 8:163–170
Brown BE, Dunne RP, Goodson MS Douglas AE (2002) Experience shapes the susceptibility of a reef coral to bleaching. Coral Reefs 21:119–126
Cameron AM, Endean R, DeVantier LM (1991) The effect of *Acanthaster planci* predation on populations of two species of massive corals. Hydrobiologia 216/217:257–262
Chornesky EA (1989) Repeated reversals during spatial competition between corals. Ecology 70:843–855
Done TJ (1999) Coral community adaptability to environmental change at the scales of regions, reefs and reef zones. Am Zool 39:66–79
Fine M, Loya Y (2002) Endolithic algae – an alternative source of energy during coral bleaching. Proc R Soc Lon B 269:1205–1210
Fine M, Loya Y (2003) Alternate coral-bryozoan competitive superiority during coral bleaching. Mar Biol 142:989–996
Fine M, Zibrowius H, Loya Y (2001) *Oculina patagonica*: a non-lessepsian scleractinian coral invading the Mediterranean Sea. Mar Biol 138:1195–1203
Fine M, Banin E, Israely T, Rosenberg E, Loya Y (2002a) Ultraviolet (UV) radiation prevents bleaching in the Mediterranean coral *Oculina patagonica*. Mar Ecol Prog Ser 226:249–254
Fine M, Oren U, Loya Y (2002b) Bleaching effect on regeneration and resource translocation in the coral *Oculina patagonica*. Mar Ecol Prog Ser 234:119–125
Gleason MG (1993) Effects of disturbance on coral communities: bleaching in Moorea, French Polynesia. Coral Reefs 12:193–201
Glynn PW (1988) El Niño Southern Oscillation 1982–1983: nearshore population, community, and ecosystem responses. Annu Rev Ecol Sys 19:309–345
Glynn PW (1993) Coral-reef bleaching – ecological perspectives. Coral Reefs 12:1–17
Guzman MG, Cortes J (2001) Changes in reef community structure after fifteen years of natural disturbances in the eastern pacific (Costa Rica). Bull Mar Sci 69:133–149
Hoegh-Guldberg O (1999) Climate change, coral bleaching and the future of the world's coral reefs. Mar Freshwater Res 50:839–866
Hughes TP, Jackson JBC (1985) Population dynamics and life histories of foliaceous corals. Ecol Monogr 55:141–166
Jokiel PL, Coles SL (1990) Response of Hawaiian and other Indo-Pacific reef corals to elevated temperatures. Coral Reefs 8:155–162
Kushmaro A, Loya Y, Fine M, Rosenberg E (1996) Bacterial infection and coral bleaching. Nature 380:396
Kushmaro A, Rosenberg E, Fine M, Loya Y (1997) Bleaching of the coral *Oculina patagonica* by Vibrio AK-1. Mar Ecol Prog Ser 147:159–165
Kushmaro A, Rosenberg E, Fine M, Ben Haim Y, Loya Y (1998) Effect of temperature on bleaching of the coral *Oculina patagonica* by *Vibrio* AK-1. Mar Ecol Prog Ser 171:131–137
Loya Y (1972) Community structure and species diversity of hermatypic corals at Eilat, Red Sea. Mar Biol 13:100–123
Loya Y (1976a) Effects of water turbidity and sedimentation on community structure of Puerto Rican corals. Bull Mar Sci 26:450–466
Loya Y (1976b) Recolonization of Red Sea corals affected by natural catastrophes and man-made perturbations. Ecology 57:278–289
Loya Y, Sakai K, Yamazato K, Nakano Y, Sambali H, van Woesik R (2001) Coral bleaching, the winners and the losers. Ecol Lett 4:122–131

Ostrander GK, Armstrong KM, Knobbe ET, Gerace D, Scully EP (2000) Rapid transition in the structure of a coral reef community: the effects of coral bleaching and physical disturbance. Proc Natl Acad Sci USA 97:5297–5302

Peters EC (1984) A survey of cellular reactions to environmental stress and disease in Caribbean scleractinian corals. Helg Meeresunters 37:113–137

Rogers CS (1990) Responses of coral reef and reef organisms to sedimentation. Mar Ecol Prog Ser 62:185–202

Schlichter D, Zscharnack B, Kerisch H (1995) Transfer of photoassimilates from endolithic algae to coral tissue. Naturwissenschaften 82:561–564

Schlichter D, Kampmann H, Conrady S (1997) Trophic potential and photoecology of endolithic algae living within coral skeletons. PSZN Mar Ecol 18:299–317

Scoffin TP (1993) The geological effects of hurricanes on coral reefs and the interpretation of storm deposits. Coral Reefs 12:203–221

Szmant AM, Gassman NJ (1990) The effects of prolonged bleaching on the tissue biomass and reproduction of the reef coral *Montastrea annularis.* Coral Reefs 8:217–224

Ward S, Jones R, Harrison P, Hoegh-Guldberg O (1998) Changes in the reproduction, lipids and MAAs of corals following the GBR mass bleaching event. In: Ward S (ed) Abstract, Australian Coral Reef Society annual meeting in Port Douglas. Univ Queensland Press, Queensland, p 10

Wilkinson CR (ed) (2000) Status of coral reefs of the world 2000. Australian Institute of Marine Science, Townsville, Australia, pp 21–34

Zibrowius H (1974) *Oculina patagonica*, scleractiniaire hermatypique introduit en Méditerranée. Helg Wiss Meeresunters 26:153–173

Zibrowius H (1992) Ongoing modification of the Mediterranean marine fauna and flora by the establishment of exotic species. Mesogee 51:83–107

Zibrowius H, Ramos A (1983) *Oculina patagonica*, scleractiniare exotique en Méditerranée-nouvelles observations dans le sud-est de l'Espagne. Rapp Comm Int Mer Medit 28:297–301

7 Coral Bleaching, Diseases and Mortality in the Western Indian Ocean

Tim McClanahan

7.1 Introduction

The western Indian Ocean generally receives less scientific attention regarding corals and coral reefs than other regions, but it has not been spared many of the deteriorating effects of either bleaching or diseases (McClanahan et al. 2000). Both factors, but particularly bleaching, are among the leading causes of reef degradation in the region and this was made abundantly clear in 1998 when most of the region was exposed to one of the warmest El Niño Southern Oscillation (ENSO) events in at least the past 200 years (McPhaden 1999; Cole et al. 2000). This ENSO caused 50–60% mortality and reduced coral cover to <10% of the substratum in many of the region's reefs (Goreau et al. 2000). It was followed by an additional ENSO and bleaching event in 2003 and somewhat mysterious and large-scale mortality of corals by unknown disease in 2002. These events have elevated these past curiosities to the top of the coral reef scientific and conservation agenda. Coral reef investigators are now attempting to understand these events, and their full consequences to reef and fisheries ecology and possible mitigation by management. Below, I describe a few of the recent notable events and discuss some questions and topics for future research.

7.2 Bleaching

7.2.1 The Environmental Background

Coral bleaching and large-scale mortality of corals have been reported from the Indian Ocean, like much of the world. Since the first bleaching reports, it is being recognized that most large-scale bleaching has followed the El Niño Southern Oscillation (ENSO) with observations of coral bleaching following the warm phases, which recently include the years 1987/1988, 1990/1992, 1994/1995, 2002/2003 and the largest event in 1997/1998 (Huppert and Stone 1998). ENSO warm events from the 1950s to 1983 differed from the post-1983 events in that they peaked early. After 1983, the warming peaked later, has taken longer and appears to be reaching greater intensity with the 1997/1998 event being the strongest recorded since 1877/1878 (McPhaden 1999), but based on a longer record from corals to 1800, it was the warmest year in the

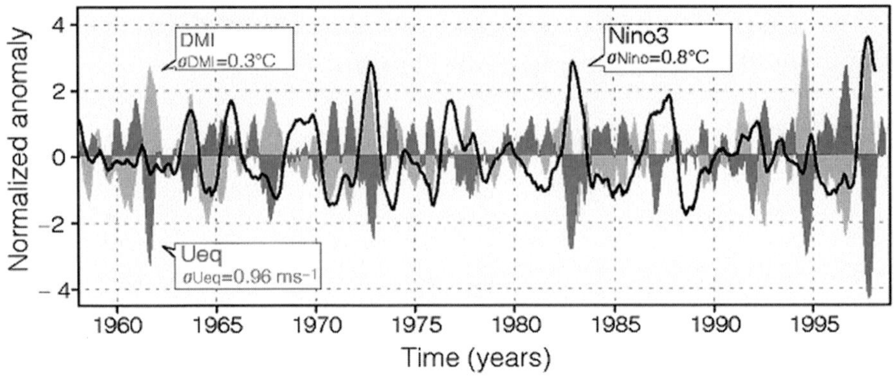

Fig. 7.1. Time series of the ENSO and the Indian Ocean dipole in the western Indian Ocean. The *solid line* is the ENSO, *dark gray* is the equatorial wind and *light gray* is the Indian Ocean dipole intensity over the time series. Slide prepared by H. Kayanne and permission to reprint granted by N.H. Saji

past two centuries (Cole et al. 2000), and probably the last 400 years (Anderson et al. 2002). ENSO in the Indian Ocean appears to oscillate between 3.5- and 5.3-year periods, with the shorter cycle dominant before the 1920s and after the 1960s and the longer cycle dominant in the interim, but both cycles are contained within a stronger decadal cycle of 11.8–12.3 years driven by the annual Asian monsoon system known as the Indian Ocean dipole (Fig. 7.1; Charles et al. 1997; Saji et al. 1999). Therefore, the strongest warm events may occur when decadal strong dipole and ENSO oscillations are temporarily in phase, which was the case in 1998.

The apparent predictability of ENSO-associated patterns is weakened by evidence that both ENSO events and bleaching are spatially patchy and sometimes show surprising relationships. A recent modeling study suggests bleaching is chaotic, but bounded with oscillations every 3–4 years (Huppert and Stone 1998). Some of this patchiness may be due to differences in the monsoon patterns in northern and southern hemispheres or the existence of cyclone systems and cycles, but this is poorly understood and studied. A number of recent 150–200 year temperature record studies that rely on the isotopic ratio of oxygen left in coral cores have found strong connections between Indian Ocean and Pacific oscillations, but with occasional diversions from this association (Charles et al. 1997; Kuhnert et al. 1999; Cole et al. 2000). Curiously, the East African temperature pattern is more strongly related to ENSO events than are western Australia and the Seychelles patterns (Kuhnert et al. 1999; Cole et al. 2000), and it may be the latitudinal position, local currents, monsoonal or local landmass effects that periodically override the ENSO and oceanic forces, even for fairly oceanic environments like the oceanic islands of the northern Indian Ocean (Charles et al. 1997) and the Houtman Abrolhos Island off western Australia (Kuhnert et al. 1999).

Despite this spatial and temporal patchiness, the long-term temperature records from the west, central and eastern Indian Ocean are showing the ominous and predicted rise in water temperatures, rising 0.8 °C in the Seychelles core and ~1.2–1.4 °C in both the western Australia and East African coral cores (Charles et al. 1997; Kuhnert et al. 1999; Cole et al. 2000). This rising pattern is also occurring in other ocean basins (Parker et al. 1995). Most of the Indian Ocean temperature rise has occurred in the last half century, which is the prediction for global climatic models with greenhouse gas forcing (Houghton et al. 1990). This warming may result in more uniform bleaching throughout the region as more and more areas reach the temperature and light levels that cause bleaching (Goreau and Hayes 1994). The 1997/1998 bleaching was an example of this, as the largest number of reports of bleaching were made in that year, with only a few areas, such as South Africa (M. Samways, pers. comm.) and the northern Red Sea, not reporting extensive bleaching (Wilkinson et al. 1999; Goreau et al. 2000; Pillay et al. 2002).

7.2.2 Bleaching in 1998

Reports of bleaching generally followed the monsoon cycle, bleaching was associated with the calm and warm conditions of the Inter-Tropical Convergence Zone (ITCZ), with the first bleaching reports starting in the southern Indian Ocean in late 1997, following the warm low-pressure monsoon moving north until bleaching was observed in the northern Indian Ocean in April and May of 1998. This year was also associated with a weak northeast monsoon. In some years, this monsoon can drive water temperatures down to 24.5 °C, but it was notably weak in 1998 (Fig. 7.2). Water temperatures during the peak warming periods in 1998 were frequently 2 °C above long-term means and exceeding 30 °C in areas with extensive bleaching. Comparing the water temperature patterns in bleaching and nonbleaching years indicates that it is probably an interaction between water temperature in the months before peak temperatures and the amount of time above the bleaching threshold that causes bleaching rather than just having temperatures above the 30.25 °C threshold (Fig. 7.2). A number of years without bleaching had temperatures above the threshold, but they were generally not persistent and often preceded by cooler water earlier in the season. One observation of the bleaching event in 1998 was that even after the ITCZ moved on, and temperatures cooled, many of the corals remained bleached. Species with branching morphologies are known to be most sensitive to warm water, and these were the first to die, often within a month of the peak temperatures, but it was some months later when the more resistant massive and encrusting forms died in water as cool as 26 °C that the severity of this event was fully recognized (McClanahan et al. 2001).

The bleaching event in 1998 was generally preceded by an early warning from the NOAA satellite data and this led to the many, albeit casual, observations of this event. The one exception was a detailed study of the response of the major

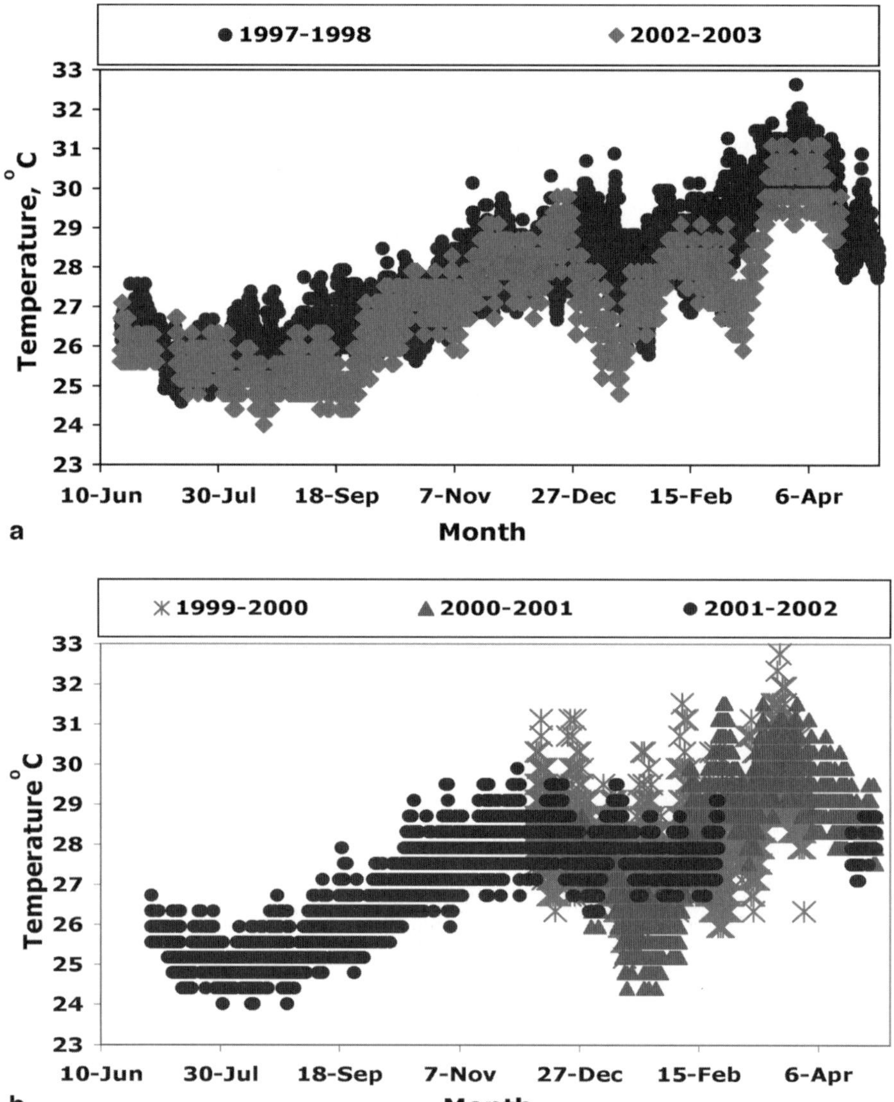

Fig. 7.2. Water temperatures in the shallow waters of the Mombasa MNP during two recent bleaching events, 1998 and 2003 (**a**) and nonbleaching years (**b**). The fungal syndrome was observed from February to March 2002. Missing data occur for various logistic reasons when the gauge was not deployed or working.

taxa to this event (McClanahan et al. 2001), which would not have been possible without this early warning system. Observations on coral mortality throughout the region suggest average mortalities of over 50% of the live coral cover, but with some restricted areas, often noncontinental reef areas, like Reunion Island, western Madagascar, Zanzibar and an island off Sri Lanka, having much lower mortality (Wilkinson et al. 1999; Goreau et al. 2000; Pillay et al. 2002). A continental area with relatively low water visibility in the Gulf of Kutch, India, was also reported to have experienced low levels of bleaching and mortality (R. Arthur, unpubl. data). The largest and longest term field study in the western Indian Ocean, which has a number of study sites in and out of marine protected areas in Kenya, reported that nearly all reefs were reduced to around 10% coral cover, the parks having 45% and the unprotected areas having 25% cover before the bleaching (McClanahan et al. 2001). The parks contained many of the ornamental, sensitive and more delicate species with branching morphologies and so protection from resource extraction, that has helped to protect these species from resource extraction, changes in the food web, and destructive gear (McClanahan and Mutere 1994) was ineffective in preventing bleaching-related mortality. This suggests that already degraded reefs may be the least likely to show the full effects of bleaching, as populations of sensitive species may already be greatly reduced.

Where post-bleaching field surveys were performed, it was found that reefs throughout the region presently have around 10% live coral cover, and include areas of Western Australia (A. Halford, unpubl. data) and the whole chain of oceanic islands from the Lakshwadweep (R. Arthur, unpubl. data), Maldives (McClanahan 2000) and Chagos (Sheppard 2000a). These studies also indicate the loss in coral cover in some of the remote atolls was not just attributable to the 1997/1998 event, but coral mortality in excess of growth has probably been ongoing since the 1987/1988 event (Allison 1995; Sheppard 2000a); however, the 1997/1998 event was particularly devastating to many species (McClanahan 2000). The ironic and most troubling aspect of mortality observations is that these islands are the center of coral diversity in the Indian Ocean and also have the region's warmest water temperatures, which was originally hypothesized to be the cause for this high diversity (Rosen 1971; Sheppard 2000b). A fairly extensive post-bleaching survey of Maldivian corals in three atolls found conservatively, however that more than a dozen of the branching species went locally extinct since surveys in the early 1960s (McClanahan 2000). Perhaps water temperatures and environmental disturbances do promote coral diversity, but beyond some threshold, probably just above 30 °C, the numbers of species may drop as high-temperature-sensitive species are lost.

7.2.3 Bleaching in 2003

In contrast to 1998, when there was good early warning, the 2003 event was not predicted well in advance and the reports of bleaching were also more variable, with Kenya reporting the highest incidence of bleaching and other regions experiencing some bleaching, but mostly milder paling of corals. Satellite records indicated small and ephemeral hotspots along the Kenya and Tanzanian coasts throughout March and April 2003, but these were not persistent and considerably less extreme than 1998. Water temperatures in 2003 started cooler in August and were cooler during the northeast monsoon from January to February, peak temperatures were about 1 °C less, and the warm water did not persist as long as the 1998 event (Fig. 7.2). Whereas bleaching in Kenya began at the end of February in 1998, the first paling corals were not observed until the end of March in 2003. Other reports in the region came from northern Tanzania (Tanga and Zanzibar Island), Reunion, Mauritius and the Seychelles, but all reported mild to moderate bleaching. The most affected species in Kenya were *Pocillopora verrucosa* and *Porites lutea*. *Pocillopora verrucosa* was quite extensively bleached, but its cogener *P. damicornis* was considerably less affected. Other less affected genera included *Acropora*, *Echinopora*, *Galaxea*, *Montipora*, *Pavona*, *Platygyra*, and some faviids and fungiids. Many of the species that are commonly reported as being affected, such as *Acropora*, *Seriatopora* and *Stylophora* (McClanahan et al. 2001), had not recovered well from 1998 and there may have been insufficient observations to determine their response. Coral mortality occurred, but was probably low and cannot be determined at this time. As described below, it is often difficult to determine the amount of coral mortality from direct observations.

7.2.4 Coral Bleaching and Mortality

The relationship between coral bleaching and mortality is not obvious and it cannot be assumed that bleached corals die. In fact, some investigators view coral bleaching as a healthy adaptation to reduce coral mortality (Buddemeier and Fautin 1993; Ware et al. 1995; Kinzie 1999). From this perspective, bleaching is an interim period where switching of the *Symbiodinium* towards taxa that have different thermal optima occurs and this may lead to improved host survival (Jokiel and Coles 1977; Rowan et al. 1997; Gates and Edmunds 1999; Baker 2001). The 1998 event provided an opportunity to determine the relationship between bleaching and mortality as there was sufficient warning to undertake observations on the full course of the bleaching event and a coral benthic monitoring program was in place such that estimates of mortality could be determined from both observation and line transects.

Based on observations of 6800 colonies and 180 m of line transect just before and just after the bleaching event, I was able to determine the response of the18 common coral taxa over this exceptionally warm period, where the average mortality was 40% (McClanahan 2004). This study indicated that there was a great deal of scatter and there was no tight relationship between the loss of color or bleaching and either direct observation or transect-based estimates of mortality (Fig. 7.3). The morphology of the coral did not influence color loss,

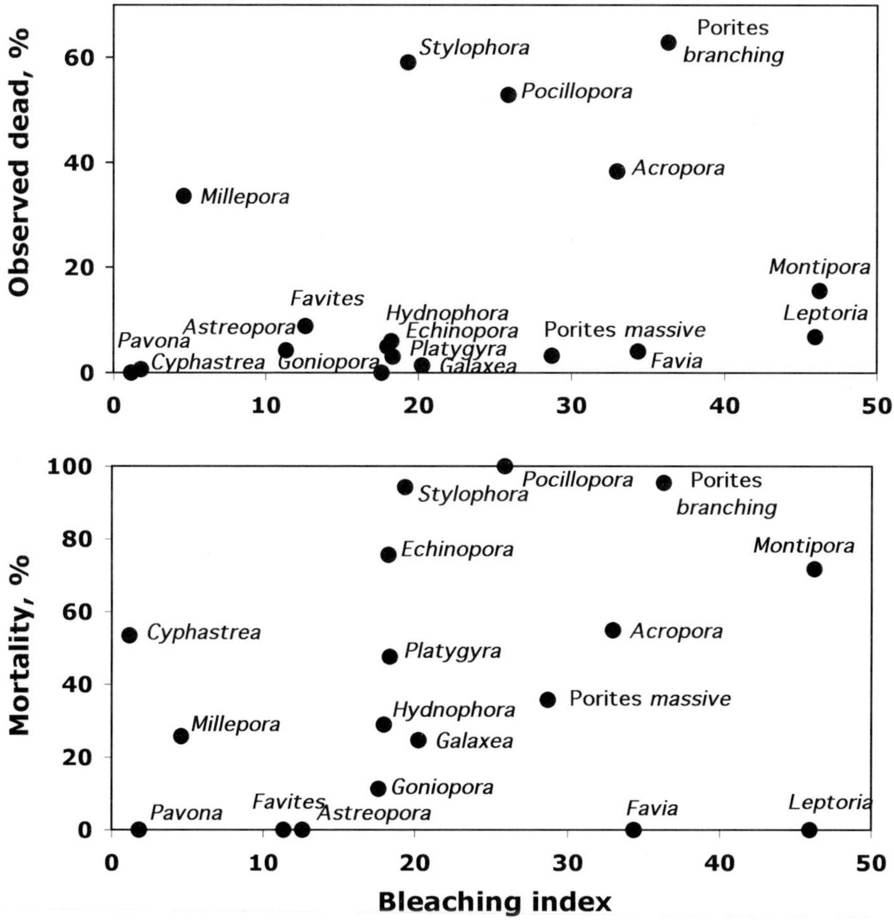

Fig. 7.3. The relationship between coral bleaching and mortality in the Mombasa MNP based on observations of 6800 coral colonies and visual estimates of mortality and line transect estimates from just before and after the bleaching event

but branching and encrusting taxa had higher mortality than massive and submassive taxa. Nonetheless, the loss of color and mortality were the most common responses to this anomalous warm water. Only *Pavona* did not lose color or die and only two taxa, *Cyphastrea* and *Millepora*, did not significantly lose color but died. *Astreopora*, *Favia*, *Favites*, *Goniopora* and *Leptoria* lost color, but did not die and these taxa may have reduced mortality by losing pigments and their algal symbionts. The death of the branching taxa was detected reasonably well by direct field observation, but some taxa were underestimated compared to mortality estimates from line transects. Death of encrusting and massive taxa including *Echinopora*, *Galaxea*, *Hydnophora*, *Montipora*, *Platygyra*, and massive *Porites* were not well detected from direct observations, but these genera proved to have modest to high mortality based on line transects, which suggests that casual field observations may be poor at detecting the full effects of bleaching. There was no single response to warm water, although bleaching was most frequently a sign of morbidity, particularly for branching and encrusting taxa.

7.2.5 Acclimation and Adaptation to Bleaching

For many regions the observed temperature threshold for bleaching lies close to 31 °C (Glynn 1984; Goreau and Hayes 1994; Goreau et al. 2000). It is recognized, however, that this threshold varies in different regions and the thermal histories of these regions (Berkelmans and Willis 1999; Dunne and Brown 2001; Coles and Brown 2003). The threshold can vary among coral species, the clade of the *Symbiodinium*, their position in the coral colony, habitats, light, water depth, water flow, regions, interannual variation and seasons (Coles et al. 1976; Marcus and Thorhaug 1981; Gleason and Wellington 1993; Hoegh-Guldberg and Salvat 1995; Brown 1997; Coles 1997; Rowan et al. 1997; Berkelmans and Willis 1999; Baker 2001; D'Croz et al. 2001; Dunne and Brown 2001; Nakamura and van Woesik 2001).

An opportunity to study the effects of a site's thermal history on the outcome of anomalous warm water was found in Kenyan reef lagoons, which often differ in their depth, distance from shore and isolation from open water (McClanahan and Maina 2003). In this study, there was a 120–275% range in between-site warm season temperature standard deviations, but only minor differences in their mean temperatures (Fig. 7.4). Given this naturally high temperature variation, what would be the consequences for the corals living in these sites across the warm 1998 ENSO? One finding was that in the high-temperature variation reefs the 1998 event produced fewer changes in coral cover and community similarity than the low-temperature variation reefs. However, there were consequences for the diversity of corals at these eurythermal sites. The eurythermal sites had lower taxonomic richness at the beginning of the study and there was an additional loss of diversity in the high and none in the stenothermal reefs. This loss of diversity was largely among species that were not dominant in the community and, therefore, not an important part of the measure of community structure. In all sites there was a loss of many of the branching taxa such as branching *Porites*, *Pocillopora* and *Stylophora* and a relative increase in massive *Porites* and *Favia*. Consequently, a site having high background temperature variation may be more likely to maintain coral cover and the dominant species, but these sites are those with the lowest diversity and probably not those reefs of major concern for conservation. The site with the highest diversity lost most of its coral cover, but not diversity. Therefore, if climate change does not change the background temperature variation, there is some hope for the persistence of species across the rare or pulsed warm events. This study is intriguing, but given the discussion above concerning

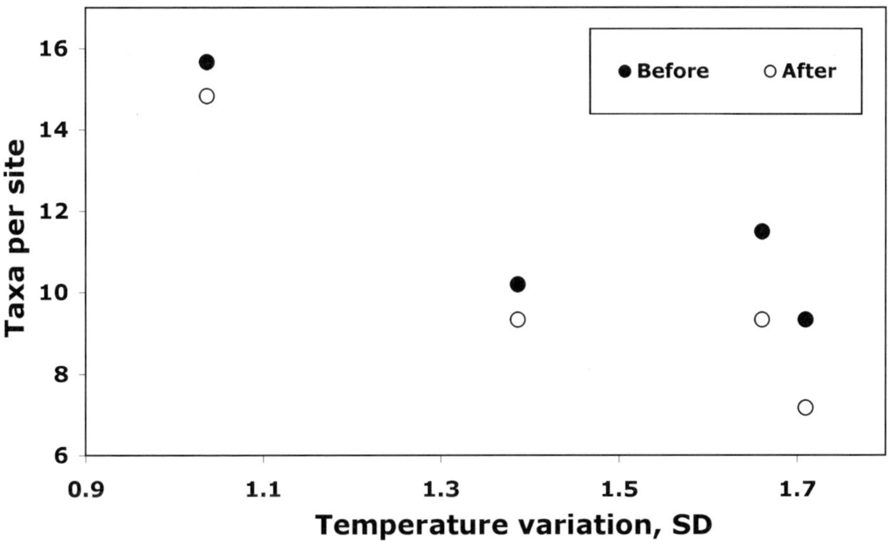

Fig. 7.4. The relationship between seawater temperature variation and coral diversity in four Kenyan reef lagoons before and after 1998

species losses in the Maldives, it really needs to be repeated in many other habitats and regions to determine the validity of conclusions to other regions and reef habitats.

7.2.6 The Effects of Coral Bleaching and Mortality on Fish and Fisheries

There are a number of competing hypotheses concerning the effects of coral bleaching and mortality on the coral reef benthos and fisheries. One hypothesis preferred by those with a focus on reef energetics and production is that the loss of coral cover will increase space for fast-growing turf algae and this will increase organic production resulting in an elevated abundance of herbivores and perhaps other trophic groups influenced by this increased primary production (Hart et al. 1996). A second hypothesis favored by those who view the habitat provided by corals as the controlling aspect of coral reef ecology is that the loss of coral will lead to a loss of reef complexity, refuge for fishes and a consequent loss of fish production (Sano 2000). A third alternative that recognizes the successional stages of algae as important to the ecology of reef fishes is that the loss of coral will open up space, reduce the intensity of herbivory and, therefore, lead to colonization of the reef benthos by late-successional frondose algae (Williams and Polunin 2001; McClanahan et al. 2002a). Dominance by frondose algae is predicted to lower benthic production and increase cover of less palatable algae and result in the reduced abundance and diversity of fish (McClanahan et al. 2002a). It is likely that all of these processes occur simultaneously or at different time scales, but the factors that will dominate are important to predict the consequences of coral mortality on fish and fisheries production.

A study to test these alternative outcomes and possible influence was undertaken in Kenya by studying the changes in the reef benthos and fisheries 3 years before and after the 1998 coral mortality event (McClanahan et al. 2002b). Data were collected in southern Kenya on benthic cover and fish abundance in unfished marine parks and fished reefs and the reef-associated fisheries 3 years before and after 1998. Across the 1998 ENSO, turf algae increased in marine parks and sponge and fleshy algae increased in the fished reefs while the fish community changed less than benthic cover. Butterflyfish, damselfish and wrasse numbers were, however, reduced while surgeonfish and a few uncommon families were positively influenced by the substratum change. Fishing effort increased by 17%, while the total demersal catch declined 8% and the catch per man by 21%. The increase in effort and decline in total catch indicate an overexploited fishery and this makes it difficult to distinguish changes caused by coral mortality or fishing effort. The price of fish increased over this period and this caused an 18% increase in the total value of the fishery, but no difference in the net income of individual fishermen.

The results can be interpreted to support the various hypotheses. The loss of damselfish, butterflyfish and wrasse would support the hypothesis that the loss of coral and reef complexity results in a loss of fish. These taxa are among those least fished, with the exception of a few of the larger wrasses (i.e., *Coris* and *Cheilinus*). The increase in surgeonfish would support the increased organic production hypothesis (McClanahan et al. 2002b). Evidence that increased frondose algae will reduce fish numbers is restricted to the observed decrease in damselfish numbers in fished reefs. Experimental removal of fleshy algae in Kenya indicates that damselfish exhibit both positive and negative responses to increased algae (McClanahan et al. 2002a). Frondose algae can inhibit fish populations including some herbivorous and detritus-feeding fish and the rates of surgeonfish grazing (McClanahan et al. 2002b), but this was not found in the fished study sites reported here. This was probably due to the observation that frondose algae in our fished study sites never exceeded 17% compared to >40% of the benthic cover in sites used for the above experimental manipulation. It may require further increases in frondose algae to see the full effects of the late-successional algae on fish. Overall, it may be ornamental species that are most affected by coral mortality while commercial herbivorous food fish may have some positive influence from increased organic production. In this fishery, however, the high level of fishing effort is likely to remain as the primary force on fish populations and reef degradation.

7.2.7 Recovery from Coral Mortality

The potential for recovery of these reefs and corals may depend on the future temperature regimes and the ability of species, taxonomic groups and ecological communities to adapt to these external environmental and internal ecological changes. The Kenyan monitoring study has shown that the ecological outcome of the coral mortality depends on the background temperature regimes, the amount of coral mortality and the interaction with herbivory (McClanahan et al. 2001; McClanahan and Maina 2003). Reefs with low herbivory and high coral mortality experienced the greatest ecological changes, with those reefs becoming dominated with large frondose algae and having little indication of recovery (McClanahan et al., in review). Large erect frondose algae are expected to suppress coral growth and recovery (Tanner 1995). Surprisingly, increased erect algal and sponge cover have occurred in reefs with high levels of herbivory (Sheppard 2000a; McClanahan et al. 2002b), but this increase was largely temporary in areas with high fish grazing (McClanahan, unpubl. data) and perhaps the levels were sufficiently low to not retard coral recruitment (McClanahan et al., in review).

Since large-scale bleaching disturbances are a rather new phenomenon to coral reef science, the ability to predict the recovery of reefs in the western Indian Ocean may be difficult due to the ecological surprises that have already affected the Caribbean (Hughes 1994; Aronson et al. 1998; McClanahan and Muthiga 1998) and Eastern Pacific (Glynn 1988; Reaka-Kudla et al. 1996). The recovery of reefs may be retarded by the spatial extent of this disturbance, but also by local-scale ecological forces that may inhibit coral recruitment and delay reef recovery, such as coral predators or competition with frondose algae (Knowlton et al. 1988; Hughes and Tanner 2000). Recovery may also interact with other factors such as the large-scale loss of a grazer (Lessios 1988), or other local factors such as fishing and pollution.

As described above, we found that one shallow reef lagoon (Kanamai) that naturally experienced both high seawater temperature and variation (23–36 °C) had low coral mortality of the dominant species (McClanahan and Maina 2003). It is possible that reefs such as these may form a refuge for coral and their symbiotic *Symbiodinium* taxa or genotypes that will form source populations for the colonization of disturbed reefs. An alternative is that these eurythermal taxa are not tolerant of conditions outside their stressful environments and will not be able to colonize stenothermal environments. The near future of coral reefs could depend on the importance of these two alternatives and the metapopulation dynamics of corals.

These alternative hypotheses were tested by examining the early recovery process of Kenyan coral reefs in unfished and heavily fished coral reefs after 1998 by counting recruits, measuring coral cover, observing predation on corals and experimentally transplanting common coral taxa from the Kanamai to three unfished and heavily fished reefs (McClanahan et al., in review). Natural coral

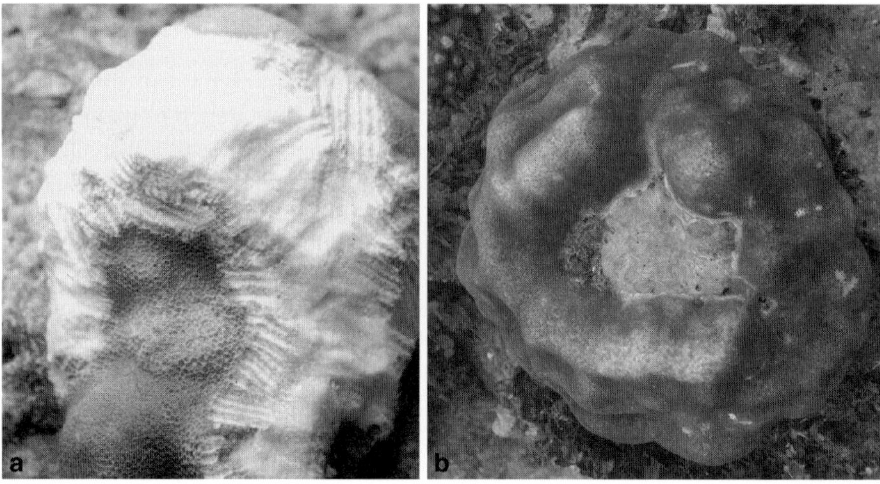

Fig. 7.5. *Porites lutea* transplanted from Kanamai (a high-temperature variation reef) to Malindi (a low-temperature variation reef) after **a** 1 month and **b** 3 years after the transplant. Note that the coral has recovered from the edges leaving a dead spot in the middle

recruitment was moderate (~40–60 per m^2) and not different in fished and unfished reefs. Coral recovery between 1998 and 2002 was variable, low (6.5% in 4 years) and also did not differ between fished and unfished reefs. There was high variability in the mortality of coral transplanted from Kanamai to other reefs and particularly among the three unfished reefs despite low variation in estimates of predator biomass. The study found that the substratum was more important than predator abundance and that coralline alga was positively, and turf alga was negatively associated with coral mortality. This suggests that the prey choices of coral predators may be more important than their abundance. High coralline alga may indicate high herbivory relative to production and the persistence of the most predator-resistant algal forms. Under these conditions, some of the more predator-susceptible corals may be vulnerable to predators and direct observations suggest that large parrotfish and triggerfish caused the most damage to these transplanted corals. This may retard the most predator-susceptible taxa, notably *Pavona decussata*. We found, however, that some taxa, such as *Porites lutea*, which were heavily damaged in the initial stages of transplantation, would often recover and the damage became less obvious with time (Fig. 7.5). Therefore, some eurythermal, but predator-susceptible taxa may recover poorly in stenothermal reefs, but some taxa, such as *Porites lutea*, may have the ability to adapt to the higher predation in stenothermal reefs and high predation will not necessarily retard total hard coral recovery.

7.3 Coral Diseases

Coral diseases were first discovered in the Caribbean in the 1970s, reported from the Indo-Pacific in the 1980s and shortly thereafter in the western Indian Ocean (Antonius 1988; Coles 1994). Black, white, and yellow band diseases have been reported from this region, but have not received extensive attention (Korrubel 2000). Recently, *Vibrio coralyticus* has been reported from Zanzibar (Ben-Haim and Rosenberg 2002). Little is known of these coral diseases other than their presence in the region and that they are pathogenic. The lack of reports may indicate that they are not the serious problem they are elsewhere, but this could also simply result from the lack of trained observers or an understanding of what are the recognizable symptoms. Below, I describe one source of a largely unexplained coral mortality that caused extensive damage in Kenya in early 2002 and this example of a possible disease or symbiotic imbalance reflects the difficulty in identifying and studying coral health and diseases.

7.3.1 The "Fungal Syndrome"

This source of coral mortality was difficult to identify in the field and in the laboratory (McClanahan, in press). Problems in the field occurred because the death was fast and there were no clear bands or clear infestations on the edge between living and dead tissue making it difficult to identify and sample (Fig. 7.6a, b). The syndrome largely affected *Astreopora*, *Montipora* and *Echinopora*, but also *Acropora*, *Goniopora*, *Platygyra* and massive *Porites*. The syndrome was not localized, but occurred on a scale of several hundred kilometers along the East African coast from at least northern Kenya to Tanzania. The syndrome along with the coral bleaching in 1998 has greatly reduced *Montipora* from Kenyan reefs (Fig. 7.7). The mortality was discovered while doing extensive field work in which many sites were visited daily and this allowed repeated observations on both the same coral and also the same taxa in different reefs. Discussions with other field biologists (J. Church) determined that this mortality was occurring in both southern and northern Kenya and exhibited the same symptoms in both regions.

In the early stages, affected corals develop an ashy dull coloration with a brittle or weak skeleton while in the intermediate stages they become covered with mucus and this collects debris. Once the mucus and debris clears, a white calcareous dust is left on the surface and at times a black layer is formed underneath, probably due to anaerobic microbial decay of the tissue under the mucus. Death is very quick in less than 2 weeks. This description is particularly consistent for *Montipora* and *Echinopora*. In the case of *Astreopora*, the tissue develops a dull pale color and then leaves a bare white skeleton, seldom producing mucus. *Echinopora* is more variable, with patches of dead skeleton

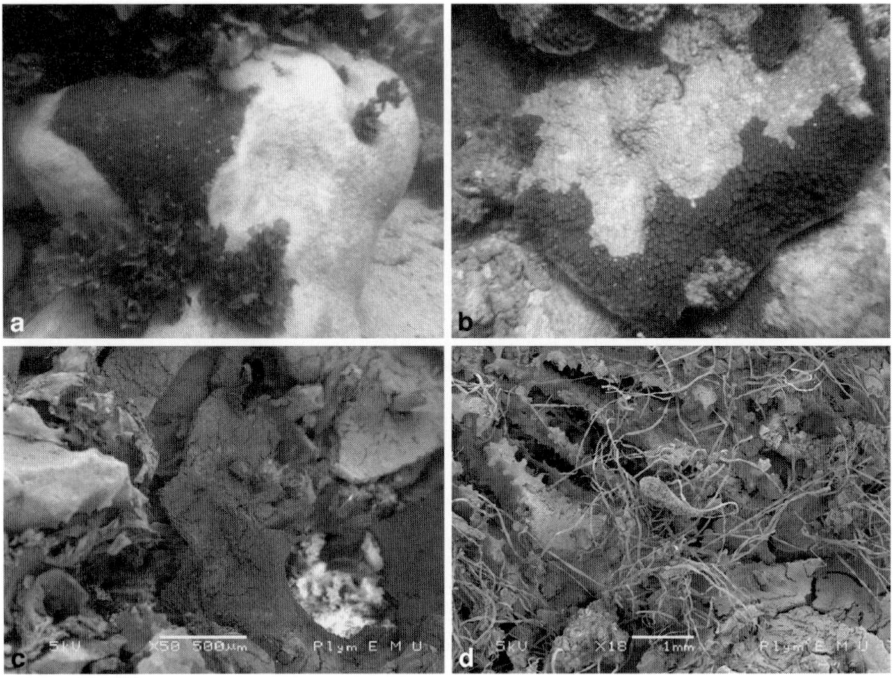

Fig. 7.6. Examples of corals affected by the "fungal syndrome". Examples of the two species, **a** *Astreopora myriophthalma* and **b** *Echinopora gemmacea*, affected by the syndrome (T. McClanahan) and a scanning electron microscope image of **c** unaffected and **d** affected tissue. (Courtesy of W. Wilson)

among living patches. *Porites lutea* turns from brown to ashy gray, becomes covered with mucus, but does not appear to die.

The reason the syndrome cannot be confused with thermal-induced coral bleaching is that from visual inspection it only affects a few taxa, other taxa that might pale or bleach during thermal anomalies are not affected by any change in color. Bleaching of some massive and submassive species is often protracted taking weeks to months to die and the color changes are more vibrant. In this case, death is rapid and the colors are dull and do not attract attention. Further, *Astreopora* and *Echinopora* are fairly resistant to thermal bleaching (McClanahan et al. 2001), but both seem greatly affected by this syndrome. Unlike commonly reported microbial diseases that form bands, there is no band formation, just an ashy or loss of color, mucus and death beneath the mucus. Tissue samples from the dying edges and from nonaffected portions of the coral viewed under a scanning electron microscope (SEM; Fig. 7.6c, d) differ in the extensive fungal hyphae seen in the affected tissue and their complete absence in the unaffected tissue.

Observations of this death in northern Kenya corresponded to a red tide during the warm part of the year and this was, therefore, a possible cause of the

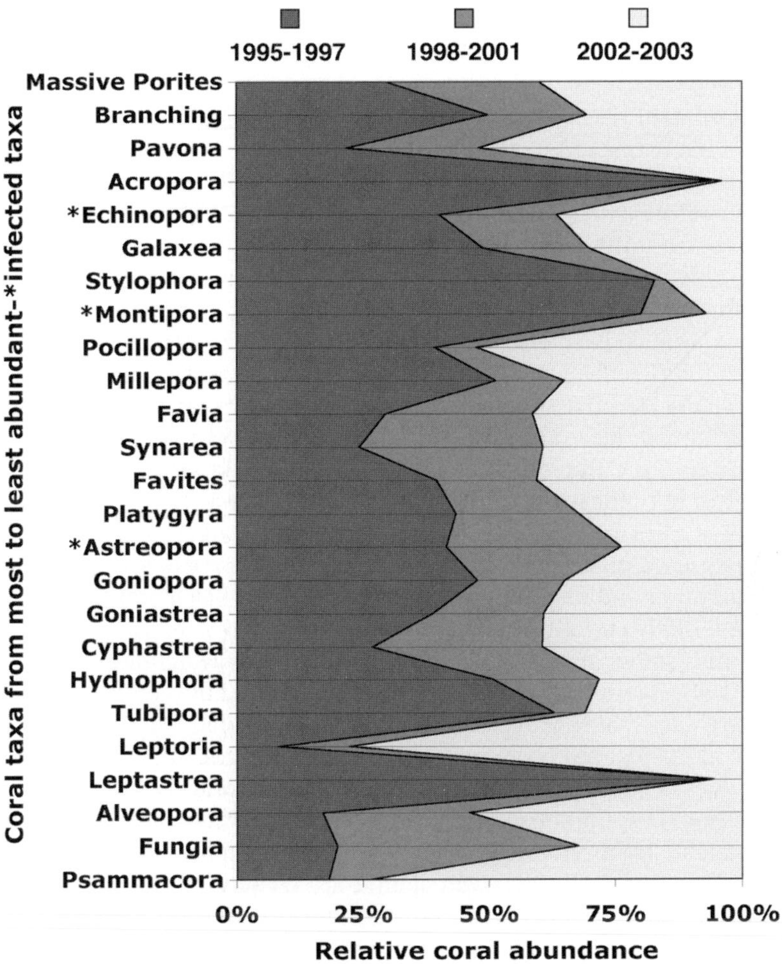

Fig. 7.7. Changes in the relative dominance of Kenyan corals over the period of 1995–2003 when bleaching occurred in 1998 and the fungal syndrome in 2002. * indicates genera observed to be affected by the fungal syndrome

death. There was, however, no red tide in southern Kenya where, instead, a short 3–4 day current-associated increase in gelatinous zooplankton was observed, but it was not associated with rich phytoplankton. The water temperature during this year started cool in August, did not experience a strong northeast monsoon and the associated cold water in January and February, but the water after that period was not particularly warm for this season (Fig. 7.2). The weather during the period was windy and the warm season water was about 28 °C in southern Kenya and 26 °C in northern Kenya, not unusually warm compared to recent measurements (Fig. 7.2). Once the seasonal change in the weather occurred in April 2002, the syndrome disappeared and has not been seen to date.

7.3.2 The Working Hypotheses

There are a number of reasonable explanations for the observed mortality, but the three most likely ones from field and SEM observations are (1) the tissue was invaded by a foreign fungus possibly associated with the current water mass and is, therefore a new and infectious fungal disease, (2) the fungus is either a foreign or endolithic taxa and feeds on the coral tissue that died from other means, either another disease or unfavorable physicochemical conditions during that season, and (3) the observed fungal hyphae are extensions of endolithic fungi that coexist with corals, but the usual coexistence was tipped in favor of the fungi such that the fungi were able to invade the coral tissue without the usually effective defense response from the coral. At present, it is difficult to distinguish these alternatives, but recent observations on coral and fungi associations (Le Campion-Alsumard et al. 1995; Bentis et al. 2000) suggest the later two hypotheses as the most likely contenders.

Many recent studies indicate that endolithic fungi live in close association with coral. They are a mix of species found on land while some are unique to the sea (Raghukumar and Raghukumar 1991; Smith et al. 1998; Ravindran et al. 2001). Some are solely endolithic and can live in carbonate substrates without living coral and some appear to require living coral (Raghukumar and Raghukumar 1991; Ravindran et al. 2001). The interaction between these fungi and endolithic algae create the black band formations often seen in coral skeletons (Priess et al. 2000). SEM observations of coral skeletons indicate that fungal hyphae of these endoliths occasionally penetrate the coral skeleton and enter the coral tissue (Le Campion-Alsumard et al. 1995; Bentis et al. 2000). Corals respond or defend themselves by calcifying over the pore created by the hyphae, such that most of the observed hyphae are seen as nob-like protuberances and only a few have pores that indicate penetration into the coral tissue (Le Campion-Alsumard et al. 1995; Bentis et al. 2000). These observations indicate that corals and endolithic fungi live close together and may largely have a commensal, but it could range from a symbiotic to saprophytic relationship. The details of this relationship have not been examined fully and understood well, and remain an area of coral biology in need of further examination. Observations of necrosis associated with some coral and fungi (Raghukumar and Raghukumar 1991; Ravindran et al. 2001) indicate that a tenuous coexistence is at times replaced by a saprophytic relationship.

These studies indicate that the observed death may result from a local imbalance in the coral–fungi relationship that may simultaneously occur over wide areas due to similar oceanographic or physicochemical conditions as opposed to an infectious or invading organism that colonizes and kills the coral host. This reasoning would explain the uniform distribution in reefs, the rapid death, and simultaneous appearance and disappearance of this syndrome in disparate reefs. The final hypothesis may fit most of the observations, but many critical aspects of this relationship and the factors that influence the relationship need to be resolved before favoring any of the hypotheses. One way to test this hypothesis is to examine the coral skeletons for evidence of pores created by fungal hyphae. The frequency of these pores in the skeleton should increase when there is an imbalance between the coral and fungi and may be a direct measure of the health of corals. Given the close relationship between coral and fungi, it is likely that corals that are ill or dying are often consumed by fungi with penetration of hyphae into coral tissue occurring prior to actual cell death by other means. Consequently, it is likely to be complex and difficult to resolve the issue of whether coral death is caused by an imbalance in the coral-fungi relationship or if some external environmental factor created suboptimal conditions for corals, and this then allowed fungi to invade the ill coral. This dilemma is not unique to corals, but common to most studies on the influence of diseases on organisms.

7.3.3 Some Unresolved Questions and Future Research

The understanding of bleaching and diseases is still in its infancy in the region and nearly any investigation is likely to be a useful contributions to the region. A number of the areas for future investigation are suggested by the discussion above and I will briefly summarize some of these below. Bleaching has affected the reefs in this region at least four times since the first large-scale bleaching in 1987 and yet there are only a few persistent reef ecological monitoring programs or programs of study in coral bleaching and it consequences that are reporting their results in the scientific literature (although more in the gray literature, such as Linden et al. 2002). Clearly, this is still a priority as many of the important unresolved issues such as the resilience of reefs in the face of global climate change will need these moderate to long-term studies to answer these questions. The questions that need to be answered should focus on the acclimatization and adaptation issue of corals and reef organisms and the recovery rates and processes of the ecosystems. Further, how will the various forms of management interact with the disturbance and recovery processes and which management institutions are likely to maintain the resilience and diversity of these ecosystems? There is, therefore, a need to monitor reefs within an experimental management system such that the detriments and benefits of management interacting with reef and climate change become evident in the coming decades. Finally, there is a need to determine those species most vulnerable to this era of rapid climate change and develop management systems that will give these species special focus and management that will insure their persistence.

Disease research has yet to classify most of the diseases or to document their distributions, etiology and virulence. There is much to do in the face of our vast ignorance on this subject and as described above, the issues are complex and difficult to study. They will require a good deal of flexibility and opportunism as unusual sources of mortality often appear and disappear inexplicably and seldom lend themselves to standard field or experimental methods. Their appearance and cursory study often leave more questions unanswered than answered as the environment, cause and effect, and ecological relationships are intricately intertwined in ways that are poorly understood. To what extent are diseases an imbalance in the ecological relationships of microorganisms with their host and how will opportunism and ecological relationships change in the face of changing environmental conditions associated with climate change and increasing human influences? Are diseases truly novel or xenobiotic events that reef systems have not previously encountered, or are they simply a temporary imbalance or ecological advantage in a previously static relationship? What are the environmental conditions and changes that promote or prevent conditions for disease persistence and virulence and how will these change with changing climate patterns, land use and fishing effort? Investigations into all of these questions are likely to occupy coral reef scientists for some time and hopefully improve our understanding and ability to manage these complex ecosystems.

References

Anderson DM, Overpeck JT, Gupta AK (2002) Increase in the Asian Southwest Monsoon during the past four centuries. Science 297:596–599

Allison WR (1995) Changes in the Maldivian reef system. Coast Manage Trop Asia 4:6–8

Antonius A (1988) Distribution and dynamics of coral diseases in the eastern Red Sea. In: Choat JH, Barnes D, Borowizka MA et al. (eds) Proc 6th Int Coral Reef Symp, Townsville, Australia, 8–12 Aug 1988, pp 293–298

Aronson RB, Precht WF, Macintyre IG (1998) Extrinsic control of species replacement on a Holocene reef in Belize: the role of coral disease. Coral Reefs 17:223–230

Baker AC (2001) Reef corals bleach to survive change. Nature 411:765–766

Ben-Haim Y, Rosenberg E (2002) A novel *Vibrio* sp. pathogen of the coral *Pocillopora damicornis*. Mar Biol 141:47–55

Bentis CJ, Kaufman L, Golubic S (2000) Endolithic fungi in reef-building corals (Order: Scleractinia) are common, cosmopolitan, and potentially pathogenic. Biol Bull 198:254–260

Berkelmans R, Willis BL (1999) Seasonal and local spatial patterns in the upper thermal limits of corals on the inshore Central Great Barrier Reef. Coral Reefs 18:219–228

Brown BE (1997) Coral bleaching: causes and consequences. Coral Reefs 16:S129–S138

Buddemeier RW, Fautin DG (1993) Coral bleaching as an adaptive mechanism: a testable hypothesis. Bioscience 43:320–326

Charles CD, Hunter DE, Fairbanks RD (1997) Interaction between the ENSO and the Asian Monsoon in a coral record of tropical climate. Science 277:925–928

Cole J, Dunbar R, McClanahan T, Muthiga N (2000) Tropical Pacific forcing of decadal variability in SST in the western Indian Ocean. Science 287:617–619

Coles SL (1994) Extensive coral disease outbreak at Fahl Island, Gulf of Oman, Indian Ocean. Coral Reefs 3:242

Coles SL (1997) Reef corals occurring in a highly fluctuating temperature environment at Fahal Island, Gulf of Oman (Indian Ocean). Coral Reefs 16:269–272

Coles SL, Brown BE (2003) Coral bleaching – capacity for acclimatization and adaptation. Adv Mar Biol 46:183–223

Coles SL, Jokiel PL, Lewis CR (1976) Thermal tolerance in tropical versus subtropical reef corals. Pac Sci 30:159–166

D'Croz LD, Mate JL, Oke JE (2001) Responses to elevated sea water temperature and UV radiation in the coral *Porites lobata* from upwelling and non-upwelling environments on the Pacific Coast of Panama. Bull Mar Sci 69:203–214

Dunne RP, Brown BE (2001) The influence of solar radiation on bleaching of shallow water reef corals in the Andaman Sea, 1993–1998. Coral Reefs 20:201–210

Gates RD, Edmunds PJ (1999) The physiological mechanisms of acclimatization in tropical reef corals. Am Zool 39:30–43

Gleason DF, Wellington GM (1993) Ultraviolet radiation and coral bleaching. Nature 365:836–838

Glynn PW (1984) Widespread coral mortality and the 1982/83 El Nino warming event. Environ Cons 11:133–146

Glynn PW (1988) El Niño-Southern oscillation 1982–1983: Nearshore population, community, and ecosystem responses. Annu Rev Ecol Syst 19:309–345

Goreau T, McClanahan T, Hayes R, Strong A (2000) Conservation of coral reefs after the 1998 global bleaching event. Cons Biol 14:5–15

Goreau TJ, Hayes RL (1994) Coral bleaching and ocean "hot spot". Ambio 23:176–180

Hart AM, Klumpp DW, Guss GR (1996) Response of herbivorous fishes to crown-of-thorns starfish *Acanthaster planci* outbreaks. II. Density and biomass of selected species of herbivorous fish and fish-habitat correlations. Mar Ecol Prog Ser 132:21–30

Hoegh-Guldberg O, Salvat B (1995) Periodic mass-bleaching and elevated sea temperatures: bleaching of outer reef slope communities in Moorea, French Polysnesia. Mar Ecol Prog Ser 121:181–190

Houghton JT, Jenkins GJ, Ephraums JT (1990) Climate Change, The IPCC Scientific Assessment. Cambridge Univ Press, Cambridge, UK

Hughes TP (1994) Catastrophes, phase shifts, and large-scale degradation of a Caribbean coral reef. Science 265:1547–1551

Hughes TP, Tanner JE (2000) Recruitment failure, life histories, and long-term decline of Caribbean corals. Ecology 81:2250–2263

Huppert A, Stone L (1998) Chaos in the Pacific's coral bleaching cycle. Am Nat 152:447–459

Jokiel PL, Coles SL (1977) Effects of temperature on the mortality and growth of Hawaiian reef corals. Mar Biol 43:201–208

Kinzie RA (1999) Sex, symbiosis and coral communities. Amer Zool 39:80–91

Knowlton N, Lang JC, Keller BD (1988) Fates of staghorn coral fragments on hurricane-damaged reefs in Jamaica: The role of predators. In: Choat JH, Barnes D, Borowizka MA, et al. (eds) Proc 6th Int Coral Reef Symp, Townsville, Australia, 8–12 Aug 1988, pp 83–88

Korrubel JL (2000) Coral diseases in the Western Indian Ocean. In: McClanahan TR, Sheppard CS, Obura DO (eds) Coral reefs of the Indian Ocean: their ecology and conservation. Oxford Univ Press, New York, pp 279–283

Kuhnert H, Patzold J, Hatcher BG, Wyrwoll K-H, Eisenhauer A, Collins LB, Zhu ZR, Wefer G (1999) A 200-year coral stable oxygen isotope record from a high-latitude reef off Western Australia. Coral Reefs 18:1–12

Le Campion-Alsumard T, Golubic S, Priess K (1995) Fungi in corals: symbiosis or disease? Interaction between polyps and fungi causes pearl-like skeleton biomineralization. Mar Ecol Prog Ser 117:137–147

Lessios HA (1988) Mass mortality of *Diadema antillarum* in the Caribbean: what have we learned? Annu Rev Ecol Syst 19:371–393

Linden OL, Souter D, Wilhelmsson D, Obura D (2002) Coral reef degradation in the Indian Ocean: status report 2002. CORDIO, Kalmar, Sweden

Marcus J, Thorhaug A (1981) Pacific versus Atlantic responses of the subtropical hermatypic coral *Porites* spp. to temperature and salinity effects. Proc 4th Int Coral Reef Symp 2:15–20

McClanahan TR (2000) Bleaching damage and recovery potential of Maldivian coral reefs. Mar Poll Bull 40:587–597

McClanahan TR (2004) The relationship between bleaching and mortality of common corals. Mar Biol 144(6)

McClanahan TR, Maina J (2003) Response of coral assemblages to the interaction between natural temperature variation and rare warm-water events. Ecosystems 6:551–563

McClanahan TR, Mutere JC (1994) Coral and sea urchin assemblage structure and interrelationships in Kenyan reef lagoons. Hydrobiologia 286:109–124

McClanahan TR, Muthiga NA (1998) An ecological shift in a remote coral atoll of Belize over 25 years. Environ Cons 25:122–130

McClanahan TR, Sheppard CRC, Obura D (eds) (2000) Coral reefs of the Indian ocean: their ecology and conservation. Oxford University Press, New York, 525 pp

McClanahan TR, Muthiga NA, Mangi S (2001) Coral and algal response to the 1998 coral bleaching and mortality: interaction with reef management and herbivores on Kenyan reefs. Coral Reefs 19:380–391

McClanahan TR, Uku JN, Machano H (2002a) Effect of macroalgal reduction on coral-reef fish in the Watamu Marine National park, Kenya. Mar Freshwater Res 53:223–231

McClanahan TR, Maina J, Pet-Soede L (2002b) Effects of the 1998 coral mortality event on Kenyan coral reefs and fisheries. Ambio 31:543–550

McClanahan TR, McLaughlin SM, Davey JE, Wilson WH, Peters EC, Price KL, Maina J (2004) Observations of a new source of coral mortality along the Kenyan coast. (in press)

McPhaden MJ (1999) Genesis and evolution of the 1997–1998 El Nino. Science 283:950–954

Nakamura T, van Woesik R (2001) Differential survival of corals during the 1998 bleaching event is partially explained by water-flow rates and passive diffusion. Mar Ecol Prog Ser 212:301–304

Parker D, Folland CK, Jackson M (1995) Marine surface temperature: observed variations and data requirements. Climate Change 31:5559–5600

Pillay RM, Terashima H, Kawasaki H (2002) The extent and intensity of the 1998 mass bleaching event on the reefs of Mauritius, Indian Ocean. Galaxea 4:43–52

Priess K, Le Campion-Alsumard T, Golubic S, Gadel F, Thomassin BA (2000) Fungi in corals: black bands and density-banding of *Porites lutea* and P. *lobata* skeleton. Mar Biol 136:19–27

Raghukumar C, Raghukumar S (1991) Fungal invasion of massive corals. Mar Ecol 12:251–260

Ravindran J, Raghukumar C, Raghukumar S (2001) Fungi in *Porites lutea*: associations with healthy and diseased corals. Dis Aq Org 47:219–228

Reaka-Kudla ML, Feingold JS, Glynn PW (1996) Experimental studies of rapid bioerosion of coral reefs in the Galapagos Islands. Coral Reefs 15:101–109

Rosen R (1971) The distribution of reef coral genera in the Indian Ocean. Symp Zool Soc Lond 28:263–299

Rowan R, Knowlton N, Baker A, Jara J (1997) Landscape ecology of algal symbionts creates variation in episodes of coral bleaching. Nature 388:265–269

Saji NH, Goswami BN, Vinayachandran PN, Yamagata T (1999) A dipole mode in the tropical Indian Ocean. Nature 401:360–363

Sano M (2000) Stability of reef fish assemblages: responses to coral recovery after catastrophic predation by *Acanthaster planci*. Mar Ecol Prog Ser 198:121–130

Sheppard CRC (2000a) The Chagos Archipelago. In: McClanahan TR, Sheppard CRC, Obura DO (eds) Coral reefs of the Indian ocean: their ecology and conservation. Oxford Univ Press, New York, pp 445–470

Sheppard CRC (2000b) Coral reefs of the western Indian Ocean: an overview. In: McClanahan TR, Sheppard CRC, Obura DO (eds) Coral reefs of the Indian ocean: their ecology and conservation. Oxford Univ Press, New York, pp 3–38

Smith GW, Harvell CD, Kim K (1998) Response of sea fans to infection with Aspergillus sp. (Fungi). Rev Biol Trop 46 [Suppl 5]:205–208

Tanner JE (1995) Competition between scleractinian corals and macroalgae: an experimental investigation of coral growth, survival and reproduction. J Exp Mar Biol Ecol 190:151–168

Ware JR, Fautin DG, Buddemeier RW (1995) Modeling the adaptive bleaching hypothesis: insights into patterns of coral bleaching. Ecol Mod 84:199–214

Wilkinson C, Linden O, Cesar H, Hodgson G, Rubens J, Strong AE (1999) Ecological and socioeconomic impacts of 1998 coral mortality in the Indian Ocean: an ENSO impact and a warning of future change? Ambio 28:188–199

Williams ID, Polunin NVC (2001) Large-scale associations between macroalgal cover and grazer biomass on mid-depth reefs in the Caribbean. Coral Reefs 19:358–366

8 Symbiont Diversity on Coral Reefs and Its Relationship to Bleaching Resistance and Resilience

ANDREW C. BAKER

8.1 Introduction

8.1.1 Overview

Mass coral reef bleaching and mortality as a result of prolonged seawater warming following the 1997–1998 El Niño-Southern Oscillation forced a change in conservation priorities in assessing threats to the health of coral reefs worldwide. By some estimates, approximately one sixth of the world's coral reefs was destroyed over a single 9-month period during the 1997–1998 bleaching event (Wilkinson 2000). Most of this destruction occurred in the Indian Ocean, where prolonged elevations of sea surface temperature were maintained by prevailing currents that pooled warm water in the western Indo-Pacific. In most cases, coral reef destruction equated to a dramatic reduction in live coral cover on these reefs (e.g., McClanahan 2000; Loch et al. 2002), but it is noteworthy that even the most severely affected reefs maintained significant pockets of live coral scattered throughout their original distributions. Moreover, many coral reef ecosystems that suffered extensive bleaching (e.g., parts of the Caribbean and Great Barrier Reef) did not experience significant eventual mortality (Wilkinson 2002). Consequently, coral reef recovery has in many places been more rapid than initially expected, particularly in the western Pacific. Although 1997–1998 clearly represents an annus horribilis for many coral reefs worldwide, the destruction it witnessed may not be as irreversible or as cumulative as originally thought. How resistant and/or resilient were reef corals (and coral reefs) to this event? How might resistance and resilience change over time in response to rising temperatures and recurrent bleaching episodes? To what extent can we expect the destruction and recovery patterns of 1997–1998 to be common features of reef bleaching and mortality in the years to come?

This chapter presents an overview of some characteristics of reef corals as *symbioses* that might contribute to our understanding of current and future patterns of bleaching resistance and resilience. Because the algal symbionts of reef corals (and many other marine invertebrates and protists) are so extraordinarily diverse, and because the coral–algal symbiosis is a relatively flexible association, the possibility exists that the detrimental effects of rising sea surface temperatures may be mitigated to some extent by: (1) the persistence and spread of reef coral hosts that contain bleaching-resistant symbionts; and/or (2) the possibility that corals may switch or shuffle their symbionts in response to rising temperatures and/or recurrent bleaching. Investigation of these questions has only become possible within the last decade, as molecular DNA techniques to identify, classify and document the distribution of different types of algal symbiont have been developed. This chapter reviews the findings of these studies, presents some perspectives on how they contribute to our understanding of how reef corals (and the ecosystems they build) will respond to continued climate change, and identifies some research priorities that will test these perspectives and help resolve some current controversy over the role of symbiont diversity in understanding the past, present and future of coral reefs.

8.2 Diversity of *Symbiodinium*

8.2.1 Historical Perspective on Diversity

In his 1883 treatise *On the morphological and physiological importance of chlorophylls in animals*, Karl Brandt suggested that the mysterious "animal chlorophylls" found in radiolarians were in fact mutualistic algae living in association with their protist hosts (Brandt 1883). By correctly identifying the role of these "yellow cells", he initiated a line of scientific inquiry that would ultimately lead to the recognition of these algae to be vastly more important than anyone had previously suspected. Until then, generally considered parasitic, these algal mutualists were destined to become the key to understanding the existence and potential fate of the most biodiverse marine ecosystems on the planet: coral reefs.

Brandt coined the genus *Zooxanthella* (Gr. *zoo* animal + Gr. *xanth* yellow + Gr. *ella* diminutive = "yellow animal cells") to refer to these algae (Brandt 1883). However, the phylogenetic diversity of these algae soon became apparent (reviewed in Taylor 1973, 1984; Lee et al. 1980) and the new genus was quickly ignored, although the term "zooxanthellae" persisted in the vernacular to the present day[1]. Some of Brandt's "yellow cells" were recognized as dinoflagellates as early as 1884 (Klebs 1884), but members of this group were not successfully cultured and described until the 1950s, eventually resulting in a new genus

[1] The term "zooxanthellae" properly applies to a paraphyletic group consisting of all golden-colored endosymbiotic microalgae, but is now commonly (but incorrectly) used to refer exclusively to symbiotic dinoflagellates (and in particular the most common marine algal symbionts of all, those in the genus *Symbiodinium*). Because this term lacks useful taxonomic meaning, its use should generally be discouraged (Droop 1963; Taylor 1973; Blank and Trench 1986; Trench 1992).

Symbiodinium (Gr. *symbion* living together + Gr. *dinos* whirling) and a new species *microadriaticum* to describe symbionts isolated from the upside-down jellyfish *Cassiopeia xamachana* (Freudenthal 1962). The apparent uniformity of the symbionts subsequently isolated from a variety of hosts provoked Taylor (1973) to remark that "our knowledge of these organisms is sufficient to enable us to classify them with a reasonable degree of certainty". Despite cautions to the contrary (e.g., McLaughlin and Zahl 1966), this view laid the foundations for a general assumption that all symbiotic dinoflagellates belonged to a single pandemic species *Symbiodinium microadriaticum* (e.g., Taylor 1974).

8.2.2
Current Understanding of Diversity

We now know that the genus *Symbiodinium* is exceptionally diverse. Research undertaken over the last three decades demonstrates that different symbionts show considerable behavioral, morphological, biochemical, physiological and genetic variation (see Trench 1992 for review). Despite this, there has been remarkably little direct investigation of how symbiont diversity affects the biology of their hosts. One explanation for this is that, until recently, symbiont distributions have been assumed to be highly host-specific, with individual host species exclusively containing only one algal type. Such an assumption allowed host species (e.g., scleractinian corals) to be considered the fundamental units of biodiversity, and ignored how symbiont diversity within these hosts might shape the phenotypic characteristics of the resultant symbiosis (Buddemeier and Fautin 1993; Rowan and Knowlton 1995; Rowan et al. 1997; see also reviews by Rowan 1998; Baker 2003; Knowlton and Rohwer 2003).

The genus *Symbiodinium* is currently recognized from molecular evidence as consisting of at least seven principal clades (*A–G*, see Fig. 8.1), all of which contain multiple "taxa" (Pawlowski et al. 2001; Pochon et al. 2001; Baker 2003; LaJeunesse et al. 2003). It is currently difficult to draw species boundaries in *Symbiodinium*, not only because these organisms have not yet been observed to undergo sexual reproduction, but also because the unusual genetic organization of these organisms may invalidate reliable estimates of species diversity from sequence divergence (Rizzo 1987; Santos and Coffroth 2003). Nevertheless, it is clear that the genus is exceptionally diverse and contains numerous members that are undoubtedly different from one another in ways that are meaningful for our understanding of the symbioses they form with invertebrate and protist hosts (Rowan 1998). It is also evident that most diversity in *Symbiodinium* has yet to be discovered: the rate of discovery of novel molecular types continues to increase since the initial documentation of seven distinct types in three (or perhaps more) clades only 12 years ago (Rowan and Powers 1991a, b, 1992; see Fig. 8.1). A more complete review of *Symbiodinium* diversity, systematics and nomenclature is presented in Baker (2003).

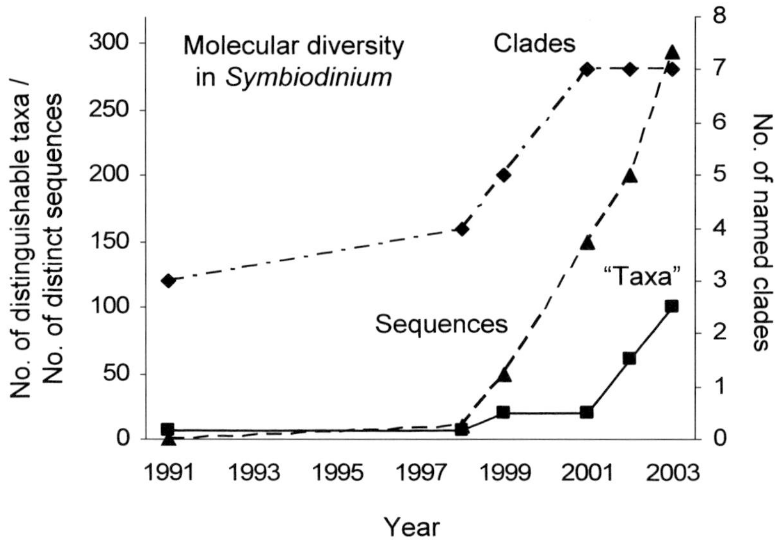

Fig. 8.1. The expanding appreciation of molecular diversity in *Symbiodinium* from 1991 to 2003. Shown are estimates of the number of named clades, the number of rapidly distinguishable "taxa", and the number of distinct large subunit rDNA sequences reported (included as a proxy measure of fine-scale diversity). Although criteria for defining different "clades", "taxa" or "sequences" are only arbitrary at this point in time, these data nevertheless indicate that additional molecular diversity in *Symbiodinium* remains to be discovered

8.3 Distribution of *Symbiodinium*

8.3.1 Host-Systematic Distribution of *Symbiodinium*

Symbioses involving *Symbiodinium* show clear patterns of host specificity (sensu Dubos and Kessler 1963; see also Schoenberg and Trench 1980) while simultaneously exhibiting a range of flexibility within these confines (see Baker 2003 for review). Some host species are highly specific, appearing to exclusively contain only one symbiont taxon (strict specificity); others are less specific, containing multiple symbiont types that are nevertheless a nonrandom subset of the full range of possibilities. Symbionts also exhibit a range of specificities, although to a generally lesser extent than their hosts (LaJeunesse 2002; LaJeunesse et al. 2003). Some appear strictly specific, for example, the described species *Symbiodinium microadriaticum* (a rare type of clade *A*) which is apparently exclusive to the jellyfish *Cassiopeia xamachana* (LaJeunesse 2002); others show cosmopolitan distributions, for example, a common type of *Symbiodinium* in clade *C* which is found in numerous species of scleractinians, jellyfish, gorgonians, zoanthids, anemones and corallimorphs (LaJeunesse

2002), and a common *B*-type which is found in most Caribbean gorgonian corals (Santos et al. 2003b). In general, reef ecosystems appear to be dominated by a few types of very common *Symbiodinium*, with numerous rarer types showing restricted ecological distributions and/or strict host specificity (Baker 1999; LaJeunesse 2002). These rare types may prove to be disproportionately important in understanding the resilience of coral reefs to future environmental change and recurrent bleaching events (Baker 2003).

8.3.2 Symbiont Diversity Within Single Species of Host

Recent research (within the last decade or so) indicates that single species of coral host can contain multiple symbiont types (Rowan and Powers 1991b; Rowan and Knowlton 1995; Baker and Rowan 1997; Baker et al. 1997; Rowan et al. 1997; Baker 1999, 2001; LaJeunesse 2002; LaJeunesse et al. 2003). This phenomenon is not confined to the scleractinian corals, but has now been documented in other cnidarians, including anemones (Santos et al. 2003b), gorgonians (Coffroth et al. 2001; Santos et al. 2003a; Santos et al. 2004) and hydrocorals (Baker 1999; LaJeunesse 2002). It has also been documented in unrelated (non-Cnidarian) host taxa that contain *Symbiodinium*, such as foraminiferans (Pawlowski et al. 2001; Pochon et al. 2001) and giant clams (Rowan 1998). Symbiont diversity involving *Symbiodinium* and other algae (both dinoflagellate and non-dinoflagellate) has also been documented: the foraminiferan *Globigerinoides ruber* contains both *Symbiodinium* (Lee et al. 1995) and *Amphidinium* (Taylor 1974), and the anemone *Anthopleura xanthogrammica* contains both *Symbiodinium* (LaJeunesse and Trench 2000) and green algae (Muscatine 1971).

In the scleractinian corals, many cases of intraspecific symbiont diversity also show clear patterns of photic zonation (both between individuals over a depth gradient and within individuals over a colony "landscape"; Rowan and Knowlton 1995; Baker et al. 1997; Rowan et al. 1997; Baker 1999, 2001; but see Billinghurst et al. 1997; Diekmann et al. 2002). These distributions suggest that symbionts vary in their photic optima, resulting in the partitioning of irradiance habitats between different symbionts. Experimental manipulations of these distributions support this interpretation (Rowan et al. 1997; Baker 2001). These studies also show that reef corals can host dynamic multi-species communities of symbionts, and that (in at least some cases) the colony represents a landscape over which these symbionts compete for space (Rowan et al. 1997). However, it is not currently known to what degree other environmental gradients (e.g., temperature) also affect symbiont distribution (but see LaJeunesse and Trench 2000; Rodriguez-Lanetty et al. 2001; Baker 2003; Baker, submitted).

The fact that closely related symbionts belonging to the same clade can be found in different hosts and environments indicates that physiological variability within-clades may be comparable to that found among them (Savage et al. 2002a; Douglas 2003). This precludes any robust generalizations about the physiological properties of particular clades, particularly in the highly diverse clades *A*, *B* and *C*. For this reason, physiological studies of symbionts where "clade" has been the highest taxonomic criterion are likely to fail in their attempts to explore the underlying basis for the ecological distribution of symbionts (e.g., Savage et al. 2002b). Finer molecular resolution will be required to understand these distributions.

8.3.3 Recombinant Symbioses and the Return of the Hopeful Monster

Why is intraspecific symbiont diversity (albeit to different degrees and involving variously diverse symbiont taxa) such a persistent feature of these invertebrate and protist hosts? The ability of different symbionts to exhibit photic zonation suggests that symbiont diversity within single host species increases the physiological range of the resultant symbiosis and the number of ecological niches they can inhabit (Buddemeier and Fautin 1993; Rowan and Knowlton 1995; Rowan et al. 1997). Symbioses that are flexible over ecological (as opposed to evolutionary) time scales may, therefore, provide reef corals (and other hosts) with the ability to quickly adapt to environmental change by "recombination" between different partners. This ability differs from Darwinian adaptation because it does not involve natural selection of each partner in isolation, nor any necessary change in genotype frequencies within their populations (see Chap. 24, Buddemeier et al., this Vol.).

Many invertebrate hosts, when sampled across their biogeographic, ecological and ontogenetic range, contain highly diverse symbionts that are often present only at low abundance ("cryptic" symbionts of LaJeunesse 2002; "moonlighting" symbionts of Baker 2003). This suggests the range of possible symbiotic associations is in fact very much larger than that commonly documented in ecological surveys, and suggests the "combinatorial" diversity characterizing these symbioses may be much greater than the sum of the number of host and symbiont taxa (Rowan and Knowlton 1995; Knowlton and Rohwer 2003).These otherwise rare and/or unusual symbionts may play an important role in understanding the long-term resilience of coral reefs to environmental change because their differential success and rapid proliferation in host tissues during times of dramatic change can result in the rapid appearance of a novel phenotype that is better suited to the new conditions (Buddemeier and Fautin 1993). Therefore, recombination functionally resembles the rapid macroevolution of "hopeful monsters" (Goldschmidt 1940).

8.4 Resistance and Resilience to Coral Bleaching

How might symbiont identity, diversity and/or flexibility affect the response of reef corals (and other hosts) to bleaching? By extension, what can this tell us about how coral reefs (as communities of diverse *Symbiodinium*-containing invertebrates and protists) might vary in their bleaching response? Two characteristics of bleaching response are distinguished for the purposes of this discussion: (1) bleaching resistance (defined here as a measure of how readily corals (or reefs) lose their symbiotic algae in response to environmental stress); and (2) bleaching resilience (defined here as a measure of how quickly bleached corals, or reefs, recover from bleaching). Resilience, as discussed here, includes recovery of individual colonies and of entire reefs (cf., West 2001).

8.4.1 Bleaching Resistance

If different symbionts vary in their physiological optima in hospite, as evidenced by their different ecological (and host-systematic) distributions, it follows that symbionts might also be expected to vary in their susceptibility to bleaching caused by environmental extremes. In fact, evidence does exist that the distribution of symbionts within scleractinian coral hosts can explain the variability characteristic of coral bleaching: during a high temperature bleaching event in the Caribbean in 1995, enigmatic patterns of bleaching in two closely related species of *Montastraea* containing three types of *Symbiodinium* in different clades were explained by the differential loss of *Symbiodinium* in clade *C* from the brightest end of its irradiance distribution (Rowan et al. 1997; see Fig. 8.2a, b). Similarly, during the 1997–1998 El Niño event, colonies of *Pocillopora* in the Panamanian Pacific that contained members of *Symbiodinium C* were susceptible to bleaching, while conspecific colonies (or parts of colonies) containing *Symbiodinium* in clade *D* were unusually resistant to bleaching (Glynn et al. 2001; see Fig. 8.2c–e). In the far eastern Pacific *Symbiodinium D* is found principally in the genus *Pocillopora*, perhaps explaining the ecological dominance of this coral genus in this region.

In both of these cases, the apparent susceptibility of *Symbiodinium* in clade *C* suggests this clade may contain some of the most thermally sensitive symbiont taxa found in scleractinian corals. Consequently, the dominance of clade *C* in the tropical Indo-Pacific (especially prior to the 1997–1998 El Niño) may be of particular significance in understanding future episodes of bleaching following recurrent episodes of high temperature stress (see below). However, *Symbiodinium* clade *C* is very diverse and likely to exhibit significant physiological variation that might include some taxa that are bleaching resistant: for example, an unusual type of *Symbiodinium C* in Indo-Pacific *Montipora digitata* and *Porites* spp. might explain their bleaching resistance (LaJeunesse et al. 2003).

Symbiont diversity has also been suggested as a proxy measure for explaining bleaching response when the specific identities and/or responses of different symbiont taxa are unknown. This approach has been applied to explaining differences in the patterns of bleaching incidence and variability between the Caribbean and far eastern Pacific (Baker and Rowan 1997), as well as the widespread bleaching of shallow *Acropora* (containing only one symbiont taxon) on the Great Barrier Reef during the 1997–1998 El Niño event (LaJeunesse et al.

Fig. 8.2A–E. Variability in coral bleaching explained by preferential loss of bleaching-susceptible *Symbiodinium* genotypes in clade *C*. Shown are Caribbean *Montastraea* (**A**, **B**) with normally pigmented tissue containing *Symbiodinium* in clades *A* and *B* (Rowan et al. 1997) and eastern Pacific *Pocillopora* (**C**, **D**, **E**) with normally pigmented tissue containing *Symbiodinium* in clade *D* (Glynn et al. 2001, images included from Bulletin of Marine Science with permission). See text for details

2003). Differential loss of symbiont taxa during bleaching has also been supported from studies of symbiont diversity: for example, fewer symbiont genotypes were found in Japanese coral species that were severely affected by El Niño bleaching in 1998 compared to species that were not (Loh et al. 2000).

In general, these findings suggest that the initial incidence and severity of bleaching may depend on the specific identities of symbionts present in affected hosts. They also demonstrate how symbiont distributions might be applied to interpreting regional patterns of bleaching involving multiple species.

Therefore, symbiont *identity* and *diversity* should be explicitly incorporated into models of bleaching resistance (and its inverse, susceptibility). This is equally true for species (or colonies) as it is for reefs (or regions), since the latter consist of assemblages of the former.

8.4.2 Bleaching Resilience

Coral reefs can recover from bleaching through a number of different mechanisms whose relative importance may vary significantly between reefs, regions and events. Because any given recovery event usually involves multiple mechanisms, the recovery histories of coral reefs usually represent a patchwork of different responses. If bleaching is severe and results in significant mortality, recovery occurs at the "reef level" – through asexual propagation and growth of surviving colonies and/or by recruitment and growth of sexually produced larvae ("ecological resilience"). If not all bleached colonies die, then reefs can also recover at the "colony level" – by symbiont proliferation within bleached tissue and re-growth of dead or necrotic tissue ("physiological resilience"). Both processes contribute to understanding bleaching resilience on coral reefs.

Recovery at the reef level has the potential to bring about significant change in symbiont community structure because surviving coral colonies are more likely to contain symbionts that are resistant to bleaching (see Sect. 8.4.1, above). Consequently, the persistence and spread of hosts containing these symbionts will change the bleaching resistance ("threshold") of that reef for future events. Some evidence suggests this process may have occurred during the 1982–1983 El Niño, which caused severe bleaching and mass mortality of reef corals in the far eastern Pacific, leading to high abundance of bleaching-resistant *Symbiodinium* in clade *D* on these reefs in the 1990s. As a result, bleaching and mortality in 1997–1998 was far less than in 1982–1983, despite the fact that temperature anomalies were comparable for the two events (Baker 2002; Baker and Glynn, unpubl. data). In general, recovery from local sources that have experienced recent bleaching might be expected to favor bleaching-resistant symbioses, while recovery from remote sources (usually by sexual recruitment) will depend on the bleaching history of the source reefs, but in any case will tend to introduce novel (and possibly bleaching-susceptible) combinations.

How reefs recover at the colony level will also have implications for their future bleaching resistance and/or resilience by virtue of the types of symbionts involved (Knowlton 2001). Colony recovery potentially depends on two (not mutually independent) processes: (1) *endogenous* recovery by internal proliferation of algal symbionts remaining in bleached tissue; and (2) *exogenous* recovery by acquisition of symbionts from external sources (surrounding water, benthic sediments, food intake, predation and/or contact transmission from other symbiotic organisms) followed by internal proliferation. In both cases, the source populations of symbionts may be genetically diverse, leading to a change in symbiont community structure following bleaching (symbiont "shuffling" and "switching" of Baker 2003). Because bleaching may facilitate symbiont community change, and because different symbiont communities may vary in their environmental optima, it has been suggested that bleaching may in fact represent an adaptive mechanism (Buddemeier and Fautin 1993) that ultimately helps them respond to environmental change. This process is independent of whether symbionts recover from endogenous or exogenous sources.

The "adaptive bleaching hypothesis" (ABH) that explores this process (Buddemeier and Fautin 1993; Ware et al. 1996) has generated controversy because the detrimental effects of bleaching are well documented and there appears to be abundant prima facie evidence that mass bleaching can lead to mass mortality of reef corals and other hosts. The use of the term "adaptive" has also resulted in some confusion; although it was correctly used in its Darwinian sense by Buddemeier and Fautin (1993) to refer to bleaching as a trait that has "adaptive value" (i.e., a trait upon which natural selection could act), it has been used in the vernacular to refer to a process that is more correctly described by the term "acclimatization". Accurately expressed, the ABH states that bleaching is adaptive because it helps corals acclimatize to new environments.

Because the ABH does not rely on the de novo acquisition of exogenous symbionts while in the bleached state (see Chap. 24, Buddemeier et al., this Vol.), the presence of significant reservoirs of symbiont diversity within corals can have important repercussions for their bleaching resilience. The ABH is important because it outlines a process by which these "minor" symbionts, usually found at low abundance within the host colony, can become disproportionately important during bleaching, leading to significant shifts in symbiont community structure (see Baker 2001 for an example). For this reason, the infrequent infection of some coral hosts by novel symbionts (only occasionally picked up by molecular surveys) may be critical in understanding which corals survive bleaching events, and why (when the symbionts are particularly unusual) the analogy of a "hopeful monster" is truly apt. Consequently, symbiont diversity and flexibility may indeed have adaptive value for corals, and bleaching may promote the change from one dominant population to another.

In summary, symbiont diversity will increase ecological resilience (at the level of the reef) because it introduces critical variation in the incidence and/or severity of bleaching. However, the ecological resilience of a reef is also dependent on the physiological resilience of its constituent colonies, which in turn depends on symbiont flexibility, or how often novel combinations appear in the roster of available partnerships. Symbiont "switching" (Baker 2003) will depend on how frequently individual symbioses acquire "new" (exogenous) populations of symbionts from external pools, but this need not occur during a bleaching event per se (in fact, it may be unreasonable to expect significant acquisition at a time of dramatic expulsion). A more likely hypothesis is that bleaching, by removing competing populations of symbionts, facilitates the internal proliferation of pre-existing novel types that would otherwise be unlikely to become dominant (see Chap. 24, Buddemeier et al., this Vol.).

8.4.3
Environmental History, Time Delays and Naïveté

Reef coral colonies represent shifting communities of different symbiont taxa (Rowan et al. 1997). These diverse symbionts can sometimes co-occur at similar abundance within the colony (e.g., in Caribbean *Montastraea*), but probably occur more often as dominant populations of one taxon, with other types occurring at low abundance (LaJeunesse 2002). Because symbiont change (in the absence of bleaching) occurs relatively slowly in these communities (the "competitive home advantage" of Baker 2001), significant time lags may hinder our ability to directly relate symbiont distributions to current environmental conditions. Accordingly, we should not expect perturbed communities to revert quickly to resemble their original communities before disturbance, and we should not expect to find similar symbiont communities in reefs with different bleaching histories.

One prediction of the processes outlined here is that reefs that have recently recovered from bleaching should be more resistant to subsequent bleaching (as a result of change in their symbiont communities) than naïve reefs that have not recently experienced bleaching. Data exist that support this idea: reefs in the tropical far eastern Pacific experienced severe coral bleaching during the 1982–1983 El Niño (suffering eventual mortality of 50–99%; Glynn 1984, 1988), but escaped significant bleaching or mortality during the 1997–1998 El Niño (despite the fact that temperature anomalies were at least as high in 1997–1998 as they were in 1982–1983; Enfield 2001; Podesta and Glynn 2001; Baker 2002). In contrast, reefs in the tropical Indo-west Pacific did not experience significant bleaching in 1982–1983, but were devastated by the 1997–1998 event. Symbiont distributions in both areas support the interpretation that an abundance of usually rare members of *Symbiodinium* clade *D* in eastern Pacific corals (as a result of recovery following previous El Niño bleaching) protected these corals from subsequent bleaching in 1997–1998 (Glynn et al. 2001; Baker and Glynn, unpubl. data).

8.4.4
Regional Differences in Bleaching Response

To what extent can these themes help us understand differences in the bleaching resistance and/or resilience of different coral reef regions? Until more detailed information is available on precise distributions and susceptibilities of different symbiont taxa, we are probably restricted for the time being to using symbiont diversity as a proxy for exact community composition. An illustration of how these data might be used to model the frequency and severity of past bleaching events is shown in Fig. 8.3. Low symbiont diversity in the tropical Indo-west Pacific correlates with infrequent, but severe episodes of bleaching, while relatively high symbiont diversity in the Caribbean corresponds to more frequent, but less severe bleaching (Baker and Rowan 1997). We urgently

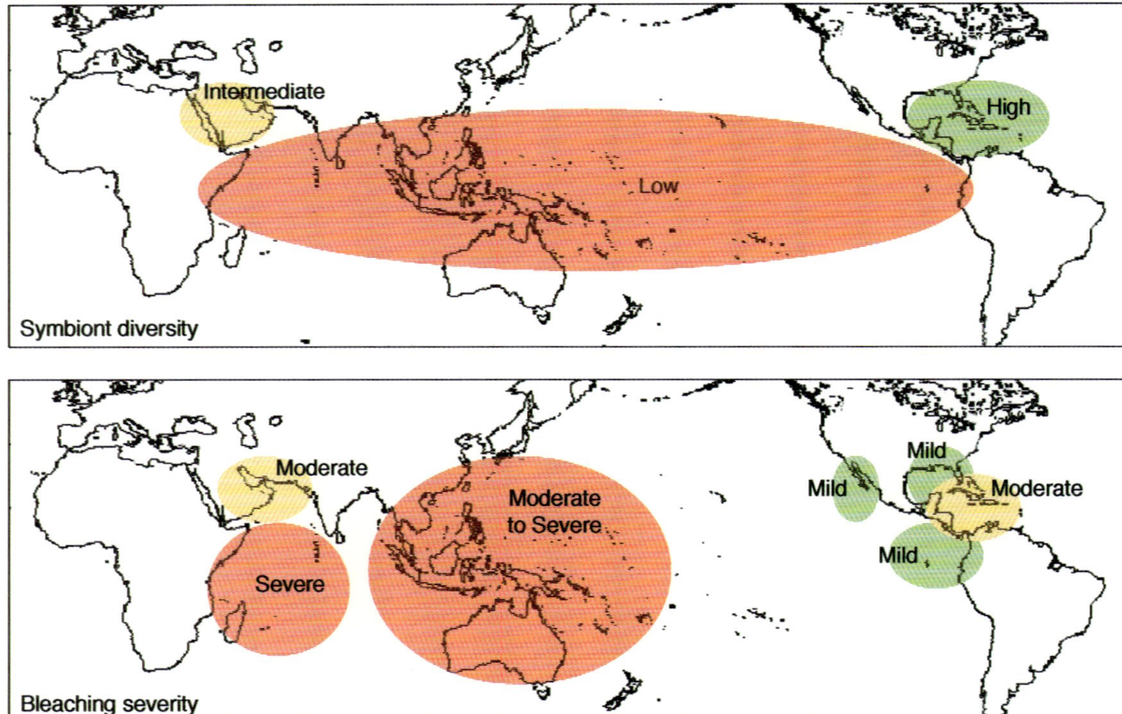

Fig. 8.3. Global patterns of distribution of *Symbiodinium* in scleractinian reef corals (*upper panel*) and observed severity of bleaching following the 1997–1998 El Niño event (*lower panel*). *Symbiodinium* data adapted from Baker 2003, bleaching data from volunteer surveys coordinated by Reef Check (www.reefcheck.org). Low symbiont diversity predicts infrequent episodes of severe bleaching (*red*), intermediate diversity predicts bleaching of moderate frequency and severity (*yellow*) and high symbiont diversity predicts frequent episodes of mild bleaching (*green*). Observable color correlations exist. A notable exception is the tropical Pacific coast of the Americas, which has low symbiont diversity (*red*), but only mild bleaching (*green*). However, these reefs are not naïve to severe bleaching and mass mortality (see text)

need to begin global surveys and monitoring of *Symbiodinium* diversity in reef corals and other hosts to understand how shifting mosaics of hosts and symbionts can be used to better understand the continued effects of climate change, global warming and coral bleaching.

8.4.5 Symbiont Diversity and Bacterial Bleaching

Recent research demonstrating the role of bacterial pathogens as the causative agents of coral bleaching (Kushmaro et al. 1996; Rosenberg and Ben-Haim 2002) does not exclude the possibility that algal symbionts are targets for infection, or that these symbionts determine sensitivity of the coral to the pathogen. Algal symbionts play a key role in the adhesion of bacterial pathogens to the host coral (Banin et al. 2001) and these pathogens may in fact be specific to particular symbiont taxa, making the discussion here regarding symbionts that are bleaching "resistant" or "susceptible" equally valid for bacterial bleaching as for conventional bleaching due to environmental stress. Of course, the possibility remains that both hypotheses are true (for different hosts, regions and/or events), in which case symbiont diversity may be an important theme in unifying the discoveries of both fields.

8.5 Understanding the Past History and Future Fate of Coral Reefs

To understand how symbiont diversity might contribute to our understanding of how reef corals respond to environmental change we need to distinguish: (1) responses due to natural selection ("adaptation") – represented by higher mortality of hosts containing bleaching-susceptible symbionts (Rowan et al. 1997; Glynn et al. 2001); and (2) responses due to symbiosis recombination ("acclimatization") – represented by recovery of bleached colonies with different symbionts ("adaptive" bleaching; Buddemeier and Fautin 1993; Ware et al. 1996; Kinzie et al. 2001; Baker 2001). How important will each of these mechanisms be in understanding future responses of reef corals (and coral reefs) to environmental change?

The answer probably depends on the scale at which the question is being framed: reefs versus colonies, mortality versus extinction, long-term versus short-term. If our perspective concerns the single colony over the short term, then we should probably not expect symbiosis recombination to play a significant role, and the fate of the colony is probably more tied to the specific identity of the symbionts it contains (which determines whether or not the colony bleaches). If, on the other hand, our perspective is at the level of the reef over the long-term, then symbiosis recombination following bleaching may play a significant role, because some fraction of the hosts will always contain unusual types that may be favored following bleaching. This increase in the recovery potential of bleached reef communities means that reefs might continue to experience extensive bleaching and mortality, but may recover fairly rapidly because they do not suffer complete loss of coral cover. Over time, unusual symbionts that promoted survival of bleached colonies may become replaced by the original symbiont communities if environmental conditions no longer favor the new combination, but this process may be much slower than symbiont change promoted by bleaching (see Sect. 8.4.3, above).

The symbiotic characteristics of reef corals discussed here may help resolve the apparent paradox of corals reefs as environmentally fragile, yet geologically persistent ecosystems (Smith and Buddemeier 1992; Buddemeier et al. 1997; Buddemeier and Smith 1999). Of course, symbiont diversity is not the only consideration likely to determine resistance and/or resilience of corals to bleaching (see West and Salm 2003, for a review of other considerations). The purpose of this chapter is to outline a conceptual basis for exploring symbiont diversity in this context, rather than provide a review of details, many of which are still unknown. Symbiont diversity may also be important in fully understanding other phenomena that may accompany environmental change, for example, "strategy shifts" (Done 1999) and accompanying changes in host species composition (Knowlton 2001) that may characterize reefs of the future (e.g., Loch et al. 2002).

In general, we can conclude that models of bleaching-related mortality that use climate model predictions to predict widespread demise of coral reefs by 2020 or 2030 (e.g., Hoegh-Guldberg 1999) are probably over-simplistic. Bleaching thresholds are unlikely to remain constant over time, and the shifting composition of hosts that accompanies mortality events, combined with associated time delays, make predictions of bleaching incidence, severity and recovery problematic without a detailed understanding of how these factors influence the symbiotic composition of affected reefs. As is becoming increasingly clear, the presence (or absence) of particular stress-resistant algal symbionts can structure the surviving community of reef-building corals following a bleaching event and determine successional processes and/or phase shifts at an ecosystem level (Done 1999). Regardless of how diversity in *Symbiodinium* influences the future fate of coral reefs, in an era of climate change and global warming, the continued success of these ecosystems is dependent on the stable association of these symbionts with the reef-building organisms which depend upon them. Karl Brandt would no doubt be pleased to discover how important his "animal chlorophylls" have become.

8.6 Research Priorities

1. Field surveys: there is an urgent need to document symbiont diversity not only from well-studied hosts, habitats or regions, but also from unusual or rare host taxa, latitudinal and environmental extremes, and early ontogenetic stages in the life cycle of various hosts. This will this provide a more complete picture of the full range of extant symbioses, as well as help

in the understanding of the role of symbionts found in extremely low abundance in some hosts.

2. Field monitoring: to better understand the time scales over which symbiont community change might occur, and relate these changes to environmental parameters, we need to begin large-scale monitoring of symbiont populations worldwide, especially in areas that are expected to experience severe bleaching in the near future.
3. Integrative activities: where possible, the results of these surveys should be incorporated into the design of coral reef Marine Protected Areas (MPAs) where resistance or resilience to bleaching is a consideration, as well as into geographically explicit models of climate change and seawater warming. Can we use emergent patterns to identify *Reefs of Hope* that might help us establish conservation priority areas?
4. Mechanisms of symbiont acquisition: further research needs to be undertaken on the diversity and distribution of free-living *Symbiodinium* and the ability and/or frequency with which adult hosts (particularly reef corals) can acquire these symbionts from free-living pools.
5. Functional diversity of *Symbiodinium*: the physiological basis for variation among different symbiont taxa in their bleaching resistance needs to be properly synthesized with our emerging understanding of symbiont distributions and bleaching resistance in the field.

Acknowledgments. I would like to thank E. Rosenberg and Y. Loya for providing the opportunity to discuss these ideas at the meeting in Eilat in April 2003, and other workshop participants, particularly R.W. Buddemeier, for useful discussion there. This paper was written with funding from NSF (OCE 0099301), NOAA (NURC 2002–15B) and the Wildlife Conservation Society (Tiffany and Co. Foundation, McBean Family Foundation, Peretti Foundation, Gimbel Foundation and Mr. H. Railton, contributors).

References

Baker AC (1999) The symbiosis ecology of reef-building corals. PhD Thesis, University of Miami, 120 pp

Baker AC (2001) Reef corals bleach to survive change. Nature 411:765–766

Baker AC (2002) Is bleaching really adaptive? Reply to Hoegh-Guldberg et al. Nature 415:602

Baker AC (2003) Flexibility and specificity in coral–algal symbiosis: diversity, ecology and biogeography of *Symbiodinium*. Annu Rev Ecol Syst 34:661–689

Baker AC, Rowan R (1997) Diversity of symbiotic dinoflagellates (zooxanthellae) in scleractinian corals of the Caribbean and eastern Pacific. Proceedings of the 8th International Coral Reef Symposium, Panama, 2, pp 1301–1305

Baker AC, Rowan R, Knowlton N (1997) Symbiosis ecology of two Caribbean acroporid corals. Proceedings of the 8th International Coral Reef Symposium, Panama, 2, pp 1295–1300

Banin E, Israely T, Fine M, Loya Y, Rosenberg E (2001) Role of endosymbiotic zooxanthellae and coral mucus in the adhesion of the coral bleaching pathogen *Vibrio shiloi* to its host. FEMS Microbiol Lett 199:33–37

Billinghurst Z, Douglas AE, Trapido-Rosenthal H (1997) On the genetic diversity of the symbiosis between the coral *Montastraea cavernosa* and zooxanthellae in Bermuda. Proceedings of the 8th International Coral Reef Symposium, Panama, 2, pp 1291–1294

Blank RJ, Trench RK (1986) Nomenclature of endosymbiotic dinoflagellates. Taxon 35:286–294
Brandt K (1883) Über die morphologische und physiologische Bedeutung des Chlorophylls bei Tieren. Mitt Zool Stat Neapol 4:191
Buddemeier RW, Fautin DG (1993) Coral bleaching as an adaptive mechanism – a testable hypothesis. BioScience 43:320–326
Buddemeier RW, Smith SV (1999) Coral adaptation and acclimatization: a most ingenious paradox. Am Zool 39:1–9
Buddemeier RW, Fautin DG, Ware JR (1997) Acclimation, adaptation and algal symbiosis in reef-building scleractinian corals. Proceeding of the 6th International Conference on Coelenterate Biology, pp 71–76
Coffroth MA, Santos SR, Goulet TL (2001) Early ontogenetic expression of specificity in a cnidarian-algal symbiosis. Mar Ecol Prog Ser 222:85–96
Diekmann OE, Bak RPM, Tonk L, Stam WT, Olsen JL (2002) No habitat correlation of zooxanthellae in the coral genus *Madracis* on a Curaçao reef. Mar Ecol Prog Ser 227:221–232
Done TJ (1999) Coral community adaptability to environmental change at the scales of regions, reefs and reef zones. Am Zool 39:66–79
Douglas AE (2003) Coral bleaching – how and why? Mar Pollut Bull 46:385–392
Droop MR (1963) Algae and invertebrates in symbiosis. Proceedings of the 13th Symposium of the Society for General Microbiology, Cambridge, pp 171–99
Dubos R, Kessler A (1963) Integrative and disintegrative factors in symbiotic associations. Symp Soc Gen Microbiol 13:1–11
Enfield DB (2001) Evolution and historical perspective of the 1997–1998 El Nino-Southern oscillation event. Bull Mar Sci 68:1–19
Freudenthal HD (1962) *Symbiodinium* gen. nov. and *Symbiodinium microadriaticum* sp. nov., a zooxanthella: Taxonomy, life cycle, and morphology. J Protozool 9:45–52
Glynn PW (1984) Widespread coral mortality and the 1982–1983 El Nino warming event. Environ Conserv 11:133–46
Glynn PW (1988) El Niño-Southern Oscillation 1982–1983 – nearshore population, community, and ecosystem responses. Annu Rev Ecol Syst 19:309–345
Glynn PW, Maté JL, Baker AC, Calderón MO (2001) Coral bleaching and mortality in Panamá and Ecuador during the 1997–1998 El Niño-Southern Oscillation event: spatial/temporal patterns and comparisons with the 1982–1983 event. Bull Mar Sci 69:79–109
Goldschmidt RB (1940) The material basis of evolution. Yale Univ Press, New Haven, CT
Hoegh-Guldberg O (1999) Climate change, coral bleaching and the future of the world's coral reefs. Mar Freshwater Res 50:839–866
Kinzie RA, Takayama M, Santos SR, Coffroth MA (2001) The adaptive bleaching hypothesis: Experimental tests of critical assumptions. Biol Bull 200:51–58
Klebs G (1884) Ein kleiner Beitrag zur Kenntnis der Peridineen. Bot Z 46–47:10
Knowlton N (2001) The future of coral reefs. Proc Natl Acad Sci USA 98:5419–5425
Knowlton N, Rohwer F (2003) Multispecies microbial mutualisms on coral reefs: the host as a habitat. Am Nat 162:51–62
Kushmaro A, Loya Y, Fine M, Rosenberg E (1996) Bacterial infection and coral bleaching. Nature 380:396
LaJeunesse TC (2002) Diversity and community structure of symbiotic dinoflagellates from Caribbean coral reefs. Mar Biol 141:387–400
LaJeunesse TC, Trench RK (2000) Biogeography of two species of *Symbiodinium* (Freudenthal) inhabiting the intertidal sea anemone *Anthopleura elegantissima* (Brandt). Biol Bull 199:126–134
LaJeunesse TC, Loh WKW, van Woesik R, Hoegh-Guldberg O, Schmidt GW, Fitt WK (2003) Low symbiont diversity in southern Great Barrier Reef corals relative to those of the Caribbean. Limnol Oceanogr 2046–2054
Lee JJ, McEnery ME, Lee MJ, Reidy JJ, Garrison JR, Rottger R (1980) Algal symbionts in larger Foraminifera. In: Schwemmler W, Schenk HEA (eds) Endocytobiology, vol I. De Gruyter, Berlin, pp 113–124

Lee JJ, Wray CG, Lawrence C (1995) Could foraminiferal zooxanthellae be derived from environmental pools contributed to by different coelenterate hosts? Acta Protozool 34:75–85
Loch K, Loch W, Schuhmacher H, See WR (2002) Coral recruitment and regeneration on a Maldivian reef 21 months after the coral bleaching event of 1998. Mar Ecol 23:219–236
Loh W, Sakai K, Hoegh-Guldberg O (2000) Coral zooanthellae diversity in bleached reefs. 9th Int Coral Reef Symp Abstr 33
McClanahan TR (2000) Bleaching damage and recovery potential of Maldivian coral reefs. Mar Pollut Bull 40:587–597
McLaughlin JJA, Zahl PA (1966) Endozoic algae. In: Henry SM (ed) Symbiosis. Academic Press, New York, pp 257–297
Muscatine L (1971) Experiments on green algae coexistent with zooxanthellae in sea anemones. Pac Sci 25:13–21
Pawlowski J, Holzmann M, Fahrni JF, Pochon X, Lee JJ (2001) Molecular identification of algal endosymbionts in large miliolid foraminifera: 2. Dinoflagellates. J Eukaryot Microbiol 48:368–373
Pochon X, Pawlowski J, Zaninetti L, Rowan R (2001) High genetic diversity and relative specificity among *Symbiodinium*-like endosymbiotic dinoflagellates in soritid foraminiferans. Mar Biol 139:1069–1078
Podesta GP, Glynn PW (2001) The 1997–1998 El Nino event in Panama and Galapagos: an update of thermal stress indices relative to coral bleaching. Bull Mar Sci 69:43–59
Rizzo PJ (1987) Biochemistry of the dinoflagellate nucleus. In: Taylor FJR (ed) The biology of dinoflagellates, vol 21. Blackwell, Oxford, pp 143–173
Rodriguez-Lanetty M, Loh W, Carter D, Hoegh-Guldberg O (2001) Latitudinal variability in symbiont specificity within the widespread scleractinian coral *Plesiastrea versipora*. Mar Biol 138:1175–1181
Rosenberg E, Ben-Haim Y (2002) Microbial diseases of corals and global warming. Environ Microbiol 4:318–326
Rowan R (1998) Diversity and ecology of zooxanthellae on coral reefs. J Phycol 34:407–417
Rowan R, Knowlton N (1995) Intraspecific diversity and ecological zonation in coral–algal symbiosis. Proc Natl Acad Sci USA 92:2850–2853
Rowan R, Powers DA (1991a) A molecular genetic classification of zooxanthellae and the evolution of animal–algal symbiosis. Science 251:1348–1351
Rowan R, Powers DA (1991b) Molecular genetic identification of symbiotic dinoflagellates (zooxanthellae). Mar Ecol Prog Ser 71:65–73
Rowan R, Powers DA (1992) Ribosomal RNA sequences and the diversity of symbiotic dinoflagellates (zooxanthellae). Proc Natl Acad Sci USA 89:3639–3643
Rowan R, Knowlton N, Baker AC, Jara J (1997) Landscape ecology of algal symbiont communities explains variation in episodes of coral bleaching. Nature 388:265–269
Santos SR, Coffroth MA (2003) Molecular genetic evidence that dinoflagellates belonging to the genus *Symbiodinium* Freudenthal are haploid. Biol Bull 204:10–20
Santos SR, Gutiérrez-Rodríguez C, Lasker HR, Coffroth MA (2003a) *Symbiodinium* sp. associations in the gorgonian *Pseudopterogorgia elisabethae* in the Bahamas: high levels of genetic variability and population structure in symbiotic dinoflagellates. Mar Biol 143:111–120
Santos SR, Gutiérrez-Rodriguez C, Coffroth MA (2003b) Phylogenetic identification of symbiotic dinoflagellates via length heteroplasmy in domain V of chloroplast large subunit (cp23S)-ribosomal DNA sequences. Mar Biotechnol 5:130–140
Santos SR, Shearer TL, Hannes AR, Coffroth MA (2004) Fine-scale diversity and specificity in the most prevalent lineage of symbiotic dinoflagellates (*Symbiodinium,* Dinophyceae) of the Caribbean, Molecular Ecology 13:459–469.
Savage AM, Goodson MS, Visram S, Trapido-Rosenthal H, Wiedenmann J, Douglas AE (2002a) Molecular diversity of symbiotic algae at the latitudinal margins of their distribution: dinoflagellates of the genus *Symbiodinium* in corals and sea anemones. Mar Ecol Prog Ser 244:17–26

Savage AM, Trapido-Rosenthal H, Douglas AE (2002b) On the functional significance of molecular variation in *Symbiodinium,* the symbiotic algae of Cnidaria: photosynthetic response to irradiance. Mar Ecol Prog Ser 244:27–37

Schoenberg DA, Trench RK (1980) Genetic variation in *Symbiodinium* (=*Gymnodinium microadriaticum* Freudenthal, and specificity in its symbiosis with marine invertebrates. III. Specificity and infectivity of *Symbiodinium microadriaticum.* Proc R Soc Lond Ser B Biol Sci 207:445–460

Smith SV, Buddemeier RW (1992) Global change and coral reef ecosystems. Annu Rev Ecol Syst 23:89–118

Taylor DL (1973) The cellular interactions of algal–invertebrate symbiosis. Adv Mar Biol 11:1–56

Taylor DL (1974) Symbiotic marine algae: taxonomy and biological fitness. In: Vernberg CBW (ed) Symbiosis and the sea. Univ South Carolina Press, Columbia,pp 245–262

Taylor DL (1984) Autotrophic eukaryotic marine symbionts. In: Pirson A, Zimmerman MH, Linskens HF, Heslop-Harrison J (eds) Encyclopedia of plant physiology, vol 17. Springer, Berlin Heidelberg New York, pp 75–90

Trench RK (1992) Microalgal-invertebrate symbiosis, current trends. Encyclopedia of microbiology, vol 3. Academic Press, London, pp 129–142

Ware JR, Fautin DG, Buddemeier RW (1996) Patterns of coral bleaching: Modeling the adaptive bleaching hypothesis. Ecol Modelling 84:199–214

West JM (2001) Environmental determinants of resistance to coral bleaching: implications for management of Marine Protected Areas. Proceedings of the Coral Bleaching and Marine Protected Areas. Proceedings of the Workshop on Mitigating coral bleaching impact through MPA design, Bishop Museum, Honolulu, Hawaii, 102, pp 40–52

West JM, Salm RV (2003) Resistance and resilience to coral bleaching: implications for coral reef conservation and management. Conserv Biol 17:956–967

Wilkinson CR (2002) Status of coral reefs of the world: 2000. Australian Institute of Marine Science, Cape Ferguson, Queensland, Australia, 363 pp

Wilkinson CR (2002) Status of coral reefs of the world: 2002. Australian Institute of Marine Science, Cape Ferguson, Queensland, Australia, 378 pp

9 Stress Effects on Metabolism and Photosynthesis of Hermatypic Corals

Noga Stambler and Zvy Dubinsky

9.1 Introduction

Coral reef ecosystems are one of the largest remaining reservoirs of biodiversity and among the most diverse in the world. Nevertheless, over the last few decades they have been undergoing a rapid decline resulting from exposure to increasing major stress agents (e.g., marine pollution; Dubinsky and Stambler 1996). One of the main initial visible expressions of such stress is the bleaching phenomenon that can lead to the death of entire reefs (see Chaps. 23 and 26, this Vol.).

The metabolism of coral colonies reflects the energy budget of the association, or the balance of the combined respiration of both the zooxanthellae and the coral animal, against photosynthesis by the zooxanthellae. This budget is supplemented to different species-specific degrees by predation on zooplankton. Even within a single species, the autotrophy/heterotrophy ratio strongly depends on ambient light.

While photosynthesis is the biological conversion of light energy to potential chemical energy stored as organic carbon compounds, respiration is the use of this chemical energy to drive all ongoing life processes in the coral colony. Although both photosynthesis and respiration rates are affected by stress, respiration is usually far more sensitive, and as such, a good indicator of stress.

Wilkinson and Buddemeier (1994) state: "Coral reefs are being seriously and increasingly stressed by exploitation and anthropogenic environmental changes, such as sedimentation, nutrient loading and pollution, physical destruction, and over-fishing."

While any particular stress can affect a variant of different biological processes, the present chapter focuses on their effect on the two main components of coral metabolism: photosynthesis and respiration. Furthermore, we confine this discussion to zooxanthellae hermatypic corals sensu (Schumacher and Zibrowius 1985). We will focus on the effect of the following different stressors on the respiration rates of the coral and the zooxanthellae, as well as the photosynthesis of the zooxanthellae light extremes: high light intensity, UV radiation, high and low temperatures, pollution (eutrophication, herbicides, metal), pathogens, and sedimentation.

9.2 Methods Used for Measuring Stress on Coral Photosynthesis and Respiration

The most frequently used methods for quantifying photosynthetic rates of corals are based either on laboratory or in situ measurements of ^{14}C incorporation or O_2 evolution, and more recently, also variable fluorescence. Photosynthesis, as measured by an oxygen electrode, is net photosynthesis (Pn), which is the balance of gross photosynthesis (Pg) and respiration (R; Falkowski et al. 1990). Laboratory studies were mainly done on freshly isolated zooxanthellae (FIZ) and colonies brought into the laboratory, while in situ measurements were done on intact colonies.

In situ measurements of respiration and photosynthesis, based on O_2 evolution, have been done by submersible respirometry (McCloskey et al. 1985; Dubinsky et al. 1990). These measurements require that small coral colonies or branches be placed in a submersible respirometer chamber for several hours or diel cycles. From these, photosynthesis versus energy curves can be generated, and all their parameters, such as α, the initial slope of the curve, P_{max}, the light-saturated rate of photosynthesis, I_k, the light of incipient saturation, and R, the dark respiration, are determined.

To avoid destructive sampling, in situ moored or SCUBA-based Fast Repetition Rate (FRR) fluorometers, and underwater Pulse Amplitude Modulated (PAM) fluorometers, (Diving-PAM) were developed and have been used in situ to evaluate changes in photosynthetic parameters. The advantage of these instruments is their being totally nonintrusive, since the organisms examined remain attached to their natural growth site (Beer and Ilan 1999; Gorbunov et al. 2000, 2001). Their inherent disadvantage is that fluorescence is a proxy, not a direct measure of photosynthetic rate, which is also affected by processes such as non-photochemical quenching (NPQ). Therefore, the interpretation of all fluorescence-based data is relatively complex, and occasionally ambiguous.

Fluorometers determine the variable of chlorophyll fluorescence, which is a relative measure of the rate at which photosystems II (PS II) can use light to process electrons flowing from the water-splitting reactions of photosynthesis. It is a sensible indicator of the redox state of PSII. In photosynthesis, the energy of absorbed light may be used for photochemical conversion, dissipated to heat, or reemitted as fluorescence. The quantum efficiencies of the three pathways are strongly controlled by ambient irradiance. The maximum quantum yield of photochemistry in PSII, its potential, is determined in a dark-adapted state as the ratio $Fv/Fm=(F_m-F_0)/Fm$, where F_0, F_m are the minimum and maximum yields of chlorophyll fluorescence, respectively, measured after a dark period (relative units). F_0 is measured in the dark by a weak probe flash, whereas F_m is measured by the same probe flash, but following a saturating "pump" flash (in the PAM) or a series of cumulative "flashlets" (in the FRRF). From these, Fv, the variable fluorescent is determined, $Fv=(F_m-F_0)$.

As ambient irradiance increases, the quantum yield of ongoing photochemistry in PSII decreases because of a gradual increase in thermal energy dissipation, and can be assessed from $\Delta F'/Fm'$ (Gorbunov et al. 2000, 2001), where F_m' and F_0' are determined under ambient light, rather than after dark relaxation.

FRR and PAM fluorometers have been used for estimating effective quantum yields of photosystem II in situ in various photosymbiont (zooxanthellae and zoocyanellae)-containing marine invertebrates (among others, Beer and Ilan 1999; Brown et al. 1999; Hoegh-Guldberg and Jones 1999; Gorbunov et al. 2000, 2001).

The parameters measured by the fluorometers are derived from the yields and kinetics of chlorophyll fluorescence. These include σ_{PSII}, the functional absorption cross section of PSII, ϕ, the quantum yield of photochemistry in PSII, τ, the rate of electron transport through PSII (ETR), energy transfer between photosynthetic units, coefficient of photochemical and non-photochemical quenching, and parameters of the photosynthesis versus irradiance curve.

Neither respiration nor net photosynthesis can be measured by fluorescence-based methods since they detect only PS II function. This remains the greatest disadvantage of these methods. Fluorometers are becoming increasingly popular since they are now available commercially and are increasingly user-friendly, in spite of the difficulties in interpretation of their results.

An additional methodology that has been successfully applied to metabolic studies of corals is that of microsensors (Kuhl et al. 1995; de Beer et al. 2000). These allow a uniquely detailed spatial resolution on the submillimeter scale, compared to the 1–2 cm^2 of the fluorometers and the entire colonies used in respirometry.

9.3 Photoacclimation

Corals are exposed to different light regimes, depending, besides depth, on their location on the reef and proximity to shade-casting colonies. Photoacclimation to decreasing light results in a marked change in the concentrations of the photosynthetic pigments of the endosymbiotic algae.

Light-harvesting pigments such as chlorophylls *a* and *c*, and peridinin concentrations per zooxanthella, increase with decreasing irradiance, whereas photoprotective pigments like β-carotene show an opposite trend (review by Barnes and Chalker 1990; Falkowski et al. 1990). These changes are associated with a general increase in the size, rather than the number of photosynthetic units (PSUs) within the zooxanthellae. Thus, P700 content per cell is constant while the molar ratio of chlorophyll-*a*/P700 increases with decreasing irradiance (Falkowski and Dubinsky 1981).

These photoacclimative changes in the photosynthetic apparatus of zooxanthellae within coral tissues produce profound changes in the shapes of P–I curves for corals growing under different light regimes. These changes include:

(1) the initial slopes of P–I curves (α) increase with decreasing intensity of growth irradiance; (2) I_k, which is the irradiance at initial slope of the curve that intersects the horizontal asymptote, increases with increasing growth irradiance; (3) the compensation intensity irradiance (Ic), which is the irradiance at which the rates of gross photosynthesis and respiration are exactly equal, decreases with decreasing growth irradiance (see Barnes and Chalker 1990 and Falkowski et al. 1990); (4) The maximum gross rate of photosynthesis when expressed in terms of chlorophyll *a* content, the assimilation number, decreases with decreasing growth irradiance (Wethey and Porter 1976; Zvalinskii et al. 1980; Falkowski and Dubinsky 1981). All of these photosynthesis parameters, as well as the areal density of the zooxanthellae themselves, may change under stress conditions.

Dark-adapted F_v/F_m showed clear diurnal changes, decreasing to a minimum at solar noon, and increasing in the afternoon. However, its value also drifted downwards at night or in prolonged darkness, and increased rapidly during the early morning twilight (Jones and Hoegh-Guldberg 2001; Lesser and Gorbunov 2001).

F_v/F_m also increased when the oxygen concentration surrounding the corals increased. These changes probably reflect state transitions associated with PQ pool reduction via chlororespiration and night-time tissue hypoxia that occurs in corals. During full sunlight, F_v/F_m of shade-acclimated (shade) colonies of *Stylophora pistillata* was lower than in light-acclimated (sun) colonies due to lowering of F_m and F_o, suggesting photoprotection and no lasting damage. Decrease in F_v/F_m is caused by a decrease in F_m and an increase in F_o in shade colonies, suggesting photoinactivation and long-term cumulative photoinhibition. Shade colonies rapidly lost their symbiotic algae (bleached) during exposure to full sunlight (Jones and Hoegh-Guldberg 2001).

9.4 Photoinhibition

Increasing light intensity will lead to an increased photosynthetic rate up to the point, I_k, at which the relationship between photosynthesis and light saturates. At relatively high light intensities, increasing light levels lead to the over-reduction of the light reactions and the production of potentially harmful products, such as various free radicals. Oxygen-free radicals, if not detoxified by several enzyme systems found in zooxanthellae (Lesser 1996), will lead to cellular damage and failure of the dark reactions of photosynthesis. Jones et al. (1998) and Hoegh-Guldberg and Jones (1999) concluded that coral bleaching is due to an increase in the susceptibility of zooxanthellae to photoinhibition. Hoegh-Guldberg and Jones (1999) and Ralph et al. (2002) showed that photoinhibition in corals results from damage to photosystem II of the zooxanthellae.

Photoinhibition may be separated into (1) dynamic photoinhibition, a reversible photoprotective process which diverts excess excitation energy away from the photosystem II (PS II) reaction center, probably a daily phenomenon taking place at noontime under high light intensity; and (2) chronic photoinhibition causing irreversible photodamage to PSII, so that de novo synthesis of the 34 kDa primary reaction center proteins is required before normal photochemistry can resume.

At depths of approximately 1 m, unexpectedly low rates of photosynthesis in corals are sometimes observed (Barnes and Chalker 1990). This phenomenon may be the result of photochemical damage caused by high light intensity, in the PAR region (photosynthetic available radiation, 400–700 nm), and ultraviolet radiation (UV, 290–400 nm), which has been demonstrated in corals transplanted from depths of 10–2 m (see Barnes and Chalker 1990). Solar UV inhibits coral growth in very shallow water and enhances bleaching (Jokiel and York 1982; Lesser 1996, 1997), whereas PAR assumes an important, but secondary role to that of UVR (Hoegh-Guldberg 1999).

Photosynthesis leads to elevated pO_2 within coral tissues (Shashar et al. 1993). Absorption of excitation energy in the presence of oxygen leads to the production of reactive oxygen species, such as singlet oxygen ($^1O_2$), superoxide radical (O_2^-), and hydrogen peroxide (H_2O_2). The effects of oxygen radicals on zooxanthellae probably include damage to photosynthesis II at the D1 protein (see Lesser 1997).

UV radiation reduces the carboxylase activity of Rubisco and decreases photosynthetic oxygen evolution. UV may be blocked by several cellular components such as mycosporine-like amino acids, MAAs, which act as sunscreens (see Shibata 1969; Shick et al. 1996). The concentration of these compounds which absorb solar ultraviolet light in the UV-B region (286–340 nm) increases with increasing solar-UV irradiance (Shibata 1969; Dunlap et al. 1986; Lesser 2000).

In addition, a range of antioxidant enzymes, superoxide dismutase (SOD) and catalase (CAT), present both in coral host and the algae, as well as ascorbate peroxidase (ASPX) in zooxanthellae, protect the coral from oxidative stress. These enzymes function together to inactivate O_2^- and H_2O_2 and by preventing the formation of HO· (Lesser et al. 1990; Lesser 1997; Levy et al. 2003).

The pocilloporine pigments and GFP (green fluorescent protein) that fluoresce under ultraviolet and blue light, may also function as photoprotectors, shielding the protein from UV damage and excessive light intensity (see Salih et al. 1998, 2000; Dove et al. 2001). It has been suggested that besides their photoprotective function, under low light conditions pocilloporine pigments may support photosynthesis by converting UV radiation into PAR (Schlichter et al. 1985). Mazel et al. (2003) found no correlation between GFP concentration and depth; moreover, these authors have shown that GFP absorption, emission and reflection have negligible impact on the solar radiation reaching the zooxanthellae. Based on that, Mazel et al. (2003) concluded that GFPs play no role in coral photosynthesis by either addition or removal of photons. It was suggested by Mazel et al. (2003) that GFPs are involved in the nonenzymatic scavenging of reactive oxygen species (ROS).

Freshly isolated zooxanthellae (FIZ) always show photoinhibition (see Muller-Parker 1984; Stambler 1992). However, from O_2-based measurements, this phenomenon can rarely be observed in entire coral colonies or whole reefs (Barnes and Chalker 1990). Absence of photoinhibition in the intact colonies or reef systems is due to the following: (1) only the zooxanthellae on the external branch tips or surface of massive colonies are ever exposed to the measured, incident irradiance and likely to suffer photoinhibition. A variable majority of symbionts resides among branches, deep within calices, or within reef crevices, where they still find themselves within the linear, light-limited domain of photosynthesis. Since the various respirometers integrate photosynthetic rates of whole colonies or reefs, any fractional photoinhibition will be masked; (2) photosynthetically derived free radicals may move to the animal tissue and do not damage the algae (Brown et al. 2002; Levy et al. 2003). Naturally, due to their small footprint, fluorometers and microelectrodes can easily detect photoinhibition by focusing on exposed colony areas.

Deep-water coral communities and species with deeper seated zooxanthellae are more resistant to bleaching caused by photoinhibition. Hence, the "deep" tissues of *Porites* spp. are less susceptible to bleaching than the "shallow" tissue configuration of *Acropora* spp. or *Pocillopora* spp. (Hoegh-Guldberg and Salvat 1995). Tissue retraction may be an important mechanism that some species use to reduce damage during solar bleaching (Brown et al. 1994, 2002).

Concentrations of yellow xanthophylls such as dinoxanthin and diadinoxanthin, and β-carotene are often, but not always, high in corals growing in shallow water (Titlyanov et al. 1980). The xanthophyll cycle is an efficient plant mechanism whereby excess energy may be dissipated harmlessly, whereas β-carotene may perform the photoprotective role of sacrificial carotenoids and thus reduce or prevent the advent of photoinhibition. According to their function, photosynthetic carotenoids may be divided into light-harvesting ones, and photoprotective ones. In zooxanthellae, peridinin transfers excitation energy to chlorophyll while β-carotene prevents some of the energy it absorbs from reaching, and potentially harming, reaction centers. The alternative energy-dissipating pathway known as the xanthophyll cycle is activated under high light. Since it does not produce any photochemistry products, it is termed non-photochemical quenching (NPQ). In this process, excitation energy is dissipated as heat (see Falkowski and Raven 1997). An active xanthophyll cycle in corals was described by Ambarsari et al. (1997), Brown et al. (1999), and Levy et al. (2003), with diurnal cycling of xanthophylls being closely correlated with photochemical efficiency (F_v/F_m) and quantum yield of photosynthesis of Photosystem II (PS II), ($\Delta F/Fm$). The photochemical efficiency (F_v/F_m) and quantum yield ($\Delta F/F_m$) of symbiotic dinoflagellates within shallow-water corals from Thailand, declined after dawn to a minimum between midday and early afternoon, recovering to former dawn levels by early evening. The xanthophylls diadinoxanthin (Dn) and diatoxanthin (Dt), and their inter-conversion, also revealed a strong diurnal pattern as well as inverse correlation between the xanthophylls ratio Dt/(Dn+Dt) and F_v/F_m and $\Delta F/F_m$.

During high light, diadinoxanthin is converted to diatoxanthin, with the resultant dissipation of excess light energy, while at low light levels diatoxanthin is converted back to diadinoxanthin. Therefore, these pigments fulfill a photoprotective function (Brown et al. 1999).

Brown et al. (2002) blocked the xanthophyll cycle by the inhibitor dithiothreitol (DTT). As a result, corals in full sunlight showed a significant increase in oxidative damage and cnidarian antioxidant enzyme concentrations, compared with untreated controls. Interestingly, antioxidant enzyme concentrations did not increase within the symbiotic algae. In addition, the expected reduction in steady state chlorophyll fluorescence (F_t) and maximum fluorescence (F_m') in response to increased irradiance, was initially suppressed in the DTT-treated corals, but after 6 h, both parameters recovered to levels similar to controls with a functional xanthophyll cycle. Extreme host retraction was observed in corals treated with DDT in full sunlight compared with untreated corals in full sunlight and DTT-treated corals in dim irradiance. These results suggest that the coral host is capable of enhancing behavioral (tissue retraction) and biochemical (antioxidant enzymes) defenses to protect the symbiotic algae under high natural irradiance when the xanthophyll cycle is absent. While these defenses could not prevent oxidative damage, nonetheless algal numbers and algal chlorophyll levels were not affected (Brown et al. 2002).

Several additional photoprotective mechanisms employed by corals include some or all of the following:

1. Modification of colony morphology, resulting in reducing photon fluxes under high light by profuse branching or columnar architecture being replaced by planar fanning and flattening under dim light.
2. Behavioral responses such as expansion and contraction (Salih et al. 2000; Brown et al. 2002).
3. Chemical adjustments of concentration of xanthophylls (Brown et al. 1999), fluorescent pigments (Hoegh-Guldberg and Jones 1999; Salih et al. 2000) and mycosporine-like amino acids which act as sunscreens (see Shick et al. 1996).
4. Activation and increase in levels of antioxidant enzymes present in both coral host and algae that also protect the coral from oxidative stress (Lesser et al. 1990; Lesser 1997; Levy et al. 2003).

High light conditions alone may lead to chronic damage to the symbiotic dinoflagellates and their eventual removal from reef-building corals in the course of bleaching events (Jones and Hoegh-Guldberg 2001).

The photosynthetic capacity of PS II, or the ratio of variable fluorescence to maximal fluorescence F_v/F_m of shallow-water reef-building corals, decreases in correlation to the daily course of irradiance (Brown et al. 1999; Gorbunov et al. 2000). This diel decrease in photosynthetic capacity has been attributed to several biochemical pathways, including dynamic photoinhibition of photosynthesis via the interconversion of xanthophyll pigments in the antenna complexes of the symbionts, as well as downregulation by damaged reaction centers of PS II (i.e., chronic photoinhibition; Brown et al. 1999; Gorbunov et al. 2001).

9.5 Depth Effects on Metabolism and Photosynthesis

Maximum photosynthesis per chlorophyll decreases with depth, along with ambient irradiance, but increases slightly with depth when expressed per biomass such as protein content (Barnes and Chalker 1990), or zooxanthellae cell number. The rate of colony dark respiration (R) decreases as well with decreasing growth irradiance and depth (see Barnes and Chalker 1990; Falkowski et al. 1990).

P_{max} per protein increased with increasing depth from 1 to 80 m for *Acropora* spp. from the Great Barrier Reef corals, while at the same depths, I_k was lowered. Respiration per protein decreased. Diel rates of gross photosynthesis remained relatively constant from 1–35 m for *Acropora* spp. Diel P/R ratios were maximal at 30 m and declined rapidly below 40 m. The diel compensation depth was 60 m (Chalker and Dunlap 1983).

F_v/F_m of corals increases with depth, reflecting increased PS II efficiency related to photoacclimation at depth as well as reduced PS II efficiency in shallow corals due to photoinhibition (Lesser and Gorbunov 2001). Warner et al. (2003) found that dinoflagellates in corals from 14 m and 3–4 m had higher F_v/F_m values than symbionts in corals at a shallower site (1–2 m). Photoinhibition occurred only in very shallow water (1–2 m).

Seasonal alterations in protein and pigment content influence the patterns of dynamic photoinhibition originating from the light-harvesting antennae. Photoinhibition shows seasonal patterns (see Warner et al. 2003), such as fluctuations in the quantum yield of charge separation (Fv/Fm) of PS II that were found for *Montastraea faveolata* (during 5 years) and *Montastraea franksi* (over 2 years) in the Bahamas, at all depths. The highest photosynthetic capacity was consistently recorded between mid-winter and early spring, and the lowest photosynthetic capacity occurred in mid- to late summer. Corals residing in shallow depths of 1–2 m showed the greatest seasonal fluctuations in F_v/F_m, whereas deeper corals (3–4 and 14 m depths) had consistently higher and less variable values of F_v/F_m. The seasonality of photosynthetic capacity showed a strong correlation with the patterns of water temperature and light. Such seasonal shifts in photosynthetic capacity are most likely due to several biochemical processes in the algae that lead to changes in both photoprotection and photodamage (Warner et al. 2003).

The increased winter photosynthetic efficiency may be controlled by reduced light levels that lead to seasonal photoacclimation at all depths, whereas the summertime decline in PS II yield is most likely the result of simultaneous photodamage and photoprotective processes. High temperatures can lead to decreases in photosynthesis/respiration ratios and destabilization of the reaction centers of PS II (Coles and Jokiel 1977; Fitt and Warner 1995; Warner et al. 1999, 2003). Seasonality can be especially important in corals residing in shallow habitats where xanthophyll cycling contributes to photoprotection (Brown et al. 1999; Warner et al. 1999). Gorbunov et al. (2001) suggested that a proportion of PS II reaction centers in symbiotic dinoflagellates are in a state of chronic photoinhibition, thus contributing to lower photosynthetic capacity.

9.6 Temperature Stress

As the so-called dark reactions of photosynthetic carbon assimilation are enzyme controlled, they are temperature-dependent processes (Falkowski and Raven 1997). The reaction's temperature dependence can be described by the Arrhenius equation and Q_{10} (the change of the rate constants for a 10 °C interval). The temperature effect is not linear, temperatures above 30–35 °C generally leads to decreases in activity (Falkowski and Raven 1997).

High temperature causes an increase in respiration up to a maximum (Coles and Jokiel 1977), while very high temperatures may cause a reduction in its rate (Jokiel and Coles 1990). Prolonged elevated temperature (days) reduced both the respiration rate of *Montastrea annularis* and to a lesser degree, its photosynthesis rate (Porter et al. 1989).

Photosynthetic activity of heat-stressed corals was drastically reduced (Coles and Jokiel 1977; Hoegh-Guldberg and Smith 1989). This reduction was partly due to the reduced population density of zooxanthellae in the heat-stressed coral (Coles and Jokiel 1977), and mainly due to the decrease of the photosynthetic rate of the zooxanthellae (Hoegh-Guldberg and Smith 1989; Iglesias-Prieto et al. 1992; Fitt and Warner 1995; Warner et al. 1996; Iglesias-Prieto 1997). Ralph et al. (2001) found that expelled zooxanthellae were largely unaffected in their photosynthesis and could be heated to 37 °C before showing temperature-induced photosynthetic impairment. Their results suggest that the early events that trigger temperature-induced expulsion of zooxanthellae involve a dysfunction in the interaction of the zooxanthellae and the coral host tissue, and not a dysfunction in the zooxanthellae per se (Ralph et al. 2001). It should be noticed that the loss of zooxanthellae occurred in response to heat stress without a decrease in algal chlorophyll concentration (Hoegh-Guldberg and Smith 1989; Fitt and Warner 1995)

Inhibition of photosynthetic oxygen evolution and a loss of variable fluorescence have been reported in cultured zooxanthellae exposed to temperatures of 34–36 °C (Iglesias-Prieto et al. 1992), and in zooxanthellae within Caribbean corals exposed to temperatures of 32 and 34 °C (Fitt and Warner 1995; Warner et al. 1996). These studies demonstrated a decrease in the efficiency of PS II when corals and their zooxanthellae were exposed to heat and led to the suggestion that the primary effect of temperature was to cause a malfunction of the light reactions of photosynthesis. Photoinhibition occurs at high temperatures as a result of the reduction in photosynthetic electron transport (Osmond 1981). Such reduction, as additional light is absorbed, will lead to production of oxygen radicals that target the photosynthetic unit. Iglesias-Prieto (1995) and Warner et al. (1996) proposed that the initial site of action of elevated water temperature in precipitating bleaching events affects the oxygen evolving complex and the reaction center of PSII, possibly at the Dl protein site (Warner et al. 1996). Iglesias-Prieto (1997) suggested that in *Symbioudinum kawagutii*, heat causes damage in both the acceptor and donor sides of PS II, resulting in the degradation of the Qb binding D1 protein. The recovery is a slow process which requires de novo synthesis of the protein. However, reduction of the PS II efficiency may protect the PSU units and may not indicate a degradation of D1.

Fitt and Warner (1995) and Warner et al. (1996) reported that corals with zooxanthellae which were able to develop strong non-photochemical quenching of chlorophyll fluorescence were more temperature-tolerant. Non-photochemical quenching occurs when the rate of light-driven electron transport exceeds the rate of ADP/Pi recycling by the dark reactions of photosynthesis. It is considered a photoprotective mechanism which prevents the over-reduction of the photosynthetic electron transport chain by dissipating excess absorbed light energy in the PSII antenna system as heat (see Jones et al. 1998).

Stylophora pistillata responded to exposure to heat stress at temperatures of 33 and 34 °C for 4 h by (1) development of strong non-photochemical quenching (qN) of the chlorophyll fluorescence signal (measured by PAM); (2) decreases in optimal quantum yield (F_v/F_m) of photosystem II (PSII); and (3) decreases in photosynthetic oxygen evolution (Jones et al. 1998). Jones et al. (1998) suggested that the first site of damage in corals from GBR due to thermal stress in zooxanthellae, were the dark reactions of photosynthesis and not the light reactions as previously thought. Jones et al. (1998) also noted that light amplified the extent of damage caused by thermal stress, almost perfectly replicating the many field reports of corals bleaching on their upper, most-sunlit surfaces. Quantum yield decreased to a greater extent on the illuminated surfaces of coral branches than on lower (shaded) surfaces, and also when high irradiance intensities were combined with elevated temperature (Jones et al. 1998). This decreased was interpreted as the initiation of photoprotective dissipation of excess absorbed energy as heat (qN) and O_2-dependent electron flow through the Mehler-Ascorbate-Peroxidase (MAP) cycle following the point at which the rate of light-driven electron transport exceeds the capacity of the Calvin cycle (Jones et al. 1998). The MAP cycle (Mehler 1951) and photorespiration can reduce photodamage to the photosynthetic apparatus (in the MAP cycle, oxygen and H_2O_2 act as electron acceptors for PSI (Asada and Takahashi 1987; Asada 1996, 1999). Thereby, the MAP cycle serves two functions: first, it consumes excess electrons, preventing an accumulation of light-generated reductant, and secondly, it establishes a pH gradient which induces an increase of energy dissipation into heat, as reflected by non-photochemical fluorescence quenching (Schreiber and Neubauer 1990; Jones et al. 1998).

Jones et al. (1998) observed the collapse of the non-photochemical quenching (qN) in heat-stressed zooxanthellae during quenching analysis under nitrogen (in the absence of oxygen). Such dysfunction of the qN further aggravates stress effect and implies that in heat-stressed coral, non-photochemical quenching partially results from O_2-dependent electron flow. O_2-dependent electron flow in the MAP cycle is mainly responsible for the energization of the trans-thylakoid membrane, when the rate of light-driven electron transport exceeds the rate of enzymatic reactions of the Calvin cycle (Schreiber and Neubauer 1990). Removal of molecular oxygen is likely to suppress respiration in both host and endosymbiont cells. This could lead to a drop in ATP/ADP and an increase in NAD(P)H/NAD(P), both of which could have an effect on photosynthetic electron transport. The drop in ATP/ADP may be related to the observed decrease in energy-dependent quenching, and the increase in NAD(P)H/NAD(P) could explain part of the apparent lack of electron acceptors after oxygen removal (Jones et al. 1998).

Jones et al. (1998) proposed that during heat stress in zooxanthellae of *Stylophora pistillata*, electron flow to NADP reductase is blocked. This causes (1) increased electron flow to the MAP cycle, thereby maintaining some electron flow; and (2) a rate of charge separation in PSII that exceeds, under high light, the capacity of electron flow through the MAP cycle. The resulting increase in ΔpH across the thylakoid membrane initiates photoprotective dissipation of excess absorbed light energy as heat, which is reflected by energy-dependent, non-photochemical quenching. Jones et al. (1998) also suggested that the maintenance of a sustained non-photochemical quenching of the fluorescence signal is supported by O_2-dependent electron flow in the MAP cycle. There is, only a limited capacity for photoprotection, beyond which the electron transport chain becomes over-reduced, leading to significant damage to PSII. Based on that, Jones et al. concluded that under temperature stress, damage to PSII is considered a secondary effect, which occurs after a reduction of the dark electron flow and the advent of redox imbalance. If PSII was the initial site of action during temperature stress, then it would seem unlikely that reduced rates of electron flow could support the ΔpH and, hence, the high levels of non-photochemical quenching observed in their study. The observed decrease in net oxygen evolution was accompanied by stimulated qN, without decrease in qP. This interpretation strongly depends on a type of electron flow such as MAP cycle which does not lead to net O_2 evolution, but which still involves the existence of functional PSII (Jones et al. 1998).

In the model for coral bleaching suggested by Jones et al. (1998), temperature stress begins with an impairment of the dark reaction of the zooxanthellae (sink limitation), ultimately resulting in photodamage to PSII or to the carbon concentrating mechanism. Potential causes of this are the effect of temperature on Calvin cycle enzymes of zooxanthellae and the unusual status of the zooxanthellar Rubisco. *Symbiodinium* Rubisco is completely different from other eukaryotic (form I) Rubiscos, since it is a form II enzyme (Rowan et al. 1996). Based on their model, Jones et al. (1998) concluded that (1) damage to

PSII in heat-stressed zooxanthellae is a secondary effect that follows damage to assimilatory electron flow, possibly to the first steps of the Calvin cycle; and (2) light plays a key secondary role in the initiation of the bleaching phenomena.

The xanthophyll cycle photoprotective mechanism described above is relevant not only in corals subjected to high irradiance, but also to those exposed to elevated sea temperatures, which increases susceptibility to the damaging effects of irradiance and which may ultimately result in loss of algae and/or their pigments (Hoegh-Guldberg 1999). It has been claimed that those corals more capable of dissipating excess excitation energy through non-photochemical quenching are less prone to temperature-induced bleaching (Warner et al. 1996).

Cold temperature is yet another well-known phenomenon that may cause release of zooxanthellae from sea anemones, and increased respiration rates of the algae (Steen and Muscatine 1987). Cold temperature stress had a negative impact on *Montipora digitata* colonies indicated by decreased photosynthetic efficiency (F_V/F_m), loss of symbiotic dinoflagellates and changes in photosynthetic pigment concentrations (Saxby et al. 2003). *Montipora digitata* in higher light regimes were more susceptible to cold temperature stress than low light ones. Moderate cold stress resulted in photoacclimatory responses, but severe cold stress resulted in photodamage, bleaching and increased mortality. Responses to cold temperature stress of *Montipora digitata* appeared similar to those observed in corals exposed to above-normal temperatures, suggesting a similar mechanism (Saxby et al. 2003). Decreased temperatures increase photoinhibition in corals, probably due to a reduction in the rate at which the quenching of PSII develops (Saxby et al. 2003), similar to the observation in corals exposed to elevated temperature as discussed above (Jones et al. 1998).

Corals become sensitive to photoinhibition with both decreasing and increasing temperatures, probably due to damage of the Calvin-Benson cycle (Jones et al. 1998; Hoegh-Guldberg 1999; Saxby et al. 2003). As enzyme rate activity is temperature-dependent, the rate of action of enzymes catalyzing the Calvin-Benson cycle will decrease with temperature. This reduction in photosynthetic electron transport combined with continued high absorption of light energy leads to a reduced capacity to process excitations and, hence, over-reduction of the light reactions. This, in turn, leads to damage or inactivation of photosynthetic components (including PSII) as a result of the production of toxic oxygen species (Lesser 1996).

The resulting cellular damage to the symbiotic dinoflagellates caused by oxidative stress results in an energetic cost to the coral host, either in terms of decreased translocation of photosynthate or exposure to highly reactive oxygen radicals (Lesser and Shick 1989). As a result, the dinoflagellates are expelled from the host tissues, probably as a protective mechanism against further oxidative stress (Lesser 1997; Saxby et al. 2003).

Light, which is essential for the high productivity of coral reefs under normal conditions, becomes a liability under conditions of higher or lower than normal temperatures. Although temperature has to be higher than normal for a mass bleaching event to occur, light will cause damage to the photosystems even at normal intensities, when water temperature is elevated above a critical maximum. Most coral bleaching starts on the upper, more-sunlit surfaces of corals and the extent of damage is light dependent (Salih et al. 1998, 2000).

9.7 Anthropogenic Effects

9.7.1 Eutrophication

Additional supplies of nutrients either as inorganic compounds, such as ammonium and phosphate, or via zooplankton consumption by the host animal, led to an increase in zooxanthellae growth rates, hence an increase in their density (see Dubinsky et al. 1990; Dubinsky and Stambler 1996; Stambler 1998). The photosynthetic rates of the algae were significantly reduced due to carbon limitation in the super-dense, multilayered algal population (Dubinsky et al. 1990). Light limitation as a result of the very high density, multilayered arrangement of cells, can also contribute to the lower P_{max}. Alternatively, photoinhibition of the zooxanthellae in the upper layer may occur as the colony becomes darker and absorbs much more light (Dubinsky et al. 1990).

9.7.2 Herbicide

The antifouling s-triazine herbicide Irgarol 1051 (2-methylthio-4-tertbutyl-amino-6-cyclopropylamino-s-triazine), was detected in marinas, harbors and coastal waters of the Florida Keys, Bermuda and St. Croix, with concentrations ranging between 3 and 294 ng/l. Isolated zooxanthellae from the coral *Madracis mirabilis* showed no incorporation of $H^{14}CO_3^-$ from the seawater medium after 4–8 h exposure to Irgarol 1051 concentrations as low as 63 ng/l. Reduction in net photosynthesis of intact corals was found at concentrations of 100 ng/l with little or no photosynthesis at concentrations exceeding 1000 ng/l after 2–8 h exposure, at all irradiances. These data suggest that Irgarol 1051 is a potent inhibitor of coral photosynthesis at environmentally relevant concentrations (Owen et al. 2002). Diuron (DCMU, dichlorophenyl dimethyl urea) another antifouling agent, often used and known as an inhibitor of PSII, inhibits photosynthetic electron transport. Based on another study, Owen et al. (2002) suggested that chronic exposure will cause photosynthesis inhibition even at much lower concentrations (250 ng/l).

9.7.3 Cyanide

Cyanide is used in some regions to capture live fish in coral reefs. *Stylophora pistillata* and *Acropora aspera* responded to exposure to sodium cyanide by decreasing variable fluorescence (F_V) indicating a reduction in photochemical capacity (Jones et al. 1999). NaCN, under a saturating light intensity of 250 μmol quanta m^{-2} s^{-1}, also caused a long-term decrease in variable fluorescence of zooxanthellae from the hard coral *Plesiastrea versipora*. The effect of cyanide on dark-adapted F_V/F_m was light-dependent. These corals lost their zooxanthellae, and became bleached.

High levels of non-photochemical quenching [q(N)] develop upon exposure of corals to cyanide. This result is consistent with the known property of cyanide as an inhibitor of the dark reactions of the Calvin cycle, specifically of Rubisco. (Jones and Hoegh-Guldberg 1999; Jones et al. 1999). Thus, cyanide exposure results in a series of responses similar to those seen during temperature-related bleaching and appears to conform to the same model (Jones and Hoegh-Guldberg 1999).

9.7.4 Copper

Porites cylindrica was exposed to copper (11 μg/l Cu) and to an increase of 4 °C above ambient. Both heat and the combination of heat and copper, significantly reduced the production rate and respiration. Copper alone had no effect on production, unless the colonies were pre-exposed to increased temperature. Even though no synergistic effects were observed, it showed that copper will act as a stressor only when combined with another stressor (Nystrom et al. 2001).

9.8 Starvation

Colonies of *Stylophora pistillata* that were fed and subjected to 90% PAR showed lower frequencies of zooxanthellae degradation, higher photosynthesis and respiration rates and higher chlorophyll concentrations than corals at the same light regime under starvation. Starvation also interfered with acclimation to low light (20 and 3% PAR), which was observed only in fed corals (Titlyanov et al. 2001).

Based on oxygen isotopic composition, Reynaud et al. (2002) found that net photosynthetic rate was higher in fed corals, but that respiration was not affected by feeding. The observed lack of change in respiration rate resulted probably from the technique and the delay in the measurement, which took place 1 day after feeding (Reynaud et al. 2002). Food capture and metabolism are expected to increase the respiration rate, as was found for the coral *Astrangia danae* (Jacques and Pilson 1980).

9.9 Sedimentation Effects

Sedimentation leads to an increase of respiration (Edmunds and Davies 1989), due to the energy expenditure needed for sediment removal and increased mucus production.

Maximal quantum yield in *Montipora peltiformis* declined linearly in relation to both the amount of sediment deposited per unit surface area and the duration of exposure. F_v/F_m decreased from 0.67–0.71 in control colonies to levels below 0.1 in most *Montipora peltiformis* colonies after 36 h coverage with sediment. Zooxanthellae density and chlorophyll concentration also decreased in response to sedimentation stress. The colonies recovered only from short-term exposure (less than 36 h) or low-level sedimentation (Philipp and Fabricius 2003).

9.10 Pathogens

Vibro shiloi secretes extracellular compounds that inhibit photosynthesis and lyse zooxanthellae isolated from coral, and lead to colony bleaching. The material responsible for inhibition of photosynthesis was heat-stable and was produced only when the bacteria were grown at elevated seawater temperatures (Ben-Haim et al. 1999). It has been identified as a peptide, referred to as toxin P, which inhibits photosynthesis of coral symbiotic algae (zooxanthellae). Toxin P was produced during the stationary phase when the bacterium was grown on peptone or Casamino Acids media at 29 °C, and was inhibited by glycerol. Toxin P was purified to homogeneity, yielding the following 12-residue peptide: PYPVYAPPPVVP (molecular weight, 1295.54). The inhibition was proportional to the toxin concentration, which bound avidly to zooxanthellae, and required the addition of NH_4Cl. In the presence of 12.5 mM NH_4Cl, pure natural or synthetic toxin P (10 mM) caused a 64% decrease in the photosynthetic quantum yield of zooxanthellae within 5 min.

When zooxanthellae were incubated in the presence of NH_4Cl and toxin P, there was a rapid decrease in the pH (pH 7.8 to 7.2) of the bulk liquid, suggesting that the toxin P facilitates transport of NH_3 into the cell, destroying the trans-membranal pH gradient, thereby blocking photosynthesis (Banin et al. 2001).

9.11 Zooxanthellae Genetic Variation

Different types of zooxanthellae often show strong zonal patterns, shallow – high light – or deep – low light – algae within given coral hosts, corresponding to light intensity (Baker 2001). In *Montastrea* spp., light-adapted Clade (Phylotype) A zooxanthellae are better able to resist thermal stress than shade-adapted Clade C genotypes (Rowan et al. 1997).

Fresh isolates of zooxanthellae (FIZ) from nine species of Bermuda corals exhibited significant variation in photosynthesis versus irradiance (P vs. I) parameters, with no phylotype connection. *Symbiodinium* of phylotype B isolated from *Montastraea franksi* from 4 to 16 m were acclimated to high light conditions. The photoacclimatory response was slight. It included an increase in the dark respiration rate by algae from all collection depths, increased light use efficiency and decreased saturating irradiance of algae from 4 m, and increased maximal photosynthetic rate per unit chlorophyll, but not per cell, for algae from 8 and 16 m (Savage et al. 2002). Savage et al. (2002) concluded that generalizations about differences between the photosynthetic traits between *Symbiodinium* phylotypes are not valid, and that the wide depth distribution of *Montastrea franksi* cannot be attributed to either depth-dependent association with multiple *Symbiodinium* phylotypes with different photosynthetic properties, or strong photoacclimatory capabilities of its complement of *Symbiodinium*. They suggested that *Symbiodinium* phylotype is not generally an important determinant of the abundance and distribution of symbioses on coral reefs. Different types of *Symbiodinium* may only partly explain the interspecific differences in seasonal variation of photosynthetic capacity.

Patterns associated with bleaching will be complicated by genotype, acclimatory state, and environment interactions. (see Chap. 8, this Vol)

9.12 Coral Bleaching

Bleaching events have been primarily associated with warming of waters surrounding reefs by ~2 °C for a period exceeding a few weeks (see detailed discussion in Chap. 23 by Jokiel, and Chap. 26 by Hoegh-Guldberg, this Vol.). High irradiance, UV, and any other stressors such as pollution and osmotic shock, have been reported to precipitate bleaching, either alone or synergistically, with elevated temperature.

Bleaching is the widespread loss of photosynthetic symbionts from zooxanthellate corals and other cnidarians, which, unless reverted, leads to the death of entire reefs (see below). Even though there have been reports of several different cellular mechanisms of zooxanthellae expulsion (see Gates et al. 1992), it has not been established which of these dominates in bleaching episodes. An additional, possibly related, phenomenon is that of "paling", where loss of pigmentation does not involve symbiont expulsion.

Bleaching of corals is a generalized response to stress; elevated seawater temperatures increase the susceptibility of the zooxanthellae to high irradiance (Fitt and Warner 1995; Warner et al. 1996; Brown 1997b; Iglesias-Prieto 1997; Coles and Brown 2003). Rowan et al. (1997) suggested that shallow populations of *Montastrea* spp. had more light-tolerant zooxanthellae (found in the tips), which were more resistant to thermal stress than shade-adapted genotypes living in less exposed places within the same colonies (Rowan et al. 1997). Hence, bleaching on different sunlight-exposed surfaces of corals was proposed to be due to the presence of a light/temperature-sensitive clade of zooxanthellae (Rowan et al. 1997). Differential clades may explain bleaching patterns in some corals, but not why mono-cladal corals like *Stylophora pistillata* bleach on the upper surfaces first (Hoegh-Guldberg and Smith 1989).

Current thinking associates the onset of bleaching and paling to photoinhibition, precipitated by high irradiance and/or UV, with any stress agent acting as sensitizer. Bleaching results partly from destabilization of the symbiotic relationship between the host and the zooxanthellae. The photosynthetic capacity of the zooxanthellae is reduced and their energy contribution to the animal is minimum (Iglesias-Prieto 1997).

Bleached corals during a natural bleaching event showed a significant decline in PS II function (Warner et al. 1999). This decline in F_v/F_m correlated with the loss of a primary PS II reaction-center protein, D1, thereby indicating that dinoflagellates in naturally bleached corals suffer from chronic photoinhibition.

Coral and zooxanthellae species that are better able to photoacclimatize may be better able to resist bleaching stress. Differences in the ability to resist bleaching stress will be related to the ability to produce and regulate accessory pigments such as the xanthophylls (Brown 1997b; Brown et al. 1999; Hoegh-Guldberg and Jones 1999). Any stress that blocks the dark reactions before the light reactions of photosynthesis will result in similar bleaching phenomena.

Photoprotective mechanisms are likely to play an important role in the way that corals and their zooxanthellae may be able to limit the effect of bleaching stresses arising from a combination of increased temperature and irradiance in the field (Brown 1997a, b).

9.13 Unresolved Questions and Future Research

1. What is the role of stress in activating "dormant" pathogens present in coral mucus or in the reef environment, and how do these affect photosynthetic performance of the zooxanthellae?
2. What are the cellular mechanisms involved in translating the impairment of photosynthetic capacity of the zooxanthellae into subsequent bleaching?
3. What is the chain of events leading from stress to loss of zooxanthellae photosynthesis?
4. In what ways do different stress agents sensitize zooxanthellae to thermal damage?
5. What are the genes involved in differential susceptibility of zooxanthellae clades to stress-induced damage to photosynthesis?

References

Ambarsari I, Brown BE, Barlow RG, Britton G, Cummings D (1997) Fluctuations in algal chlorophylls and carotenoid pigments during solar bleaching in the coral *Goniastrea aspera* at Phuket, Thailand. Mar Ecol Prog Ser 159:303–307

Asada K (1996) Radical production and scavenging in the chloroplasts. In: Baker NR (ed) Photosynthesis and the environments. Kluwer, Dordrecht, pp 123–150

Asada K (1999) The water-water cycle in chloroplasts: scavenging of active oxygen and dissipation of excess photons. Annu Rev Plant Phys 50:601–639

Asada K, Takahashi M (1987) Production and scavenging of active oxygen in photosynthesis. In: Kyle DJ, Osmond CB, Amtzen CJ (eds) Photoinhibition. Elsevier, Amsterdam, pp 227–287

Baker AC (2001) Reef corals bleach to survive change. Nature 411:765–766

Banin E, Khare SK, Naider F, Rosenberg E (2001) Proline-rich peptide from the coral pathogen *Vibrio shiloi* that inhibits photosynthesis of zooxanthellae. Appl Environ Microb 67:1536–1541

Barnes D, Chalker BE (1990) Calcification and photosynthesis in reef building corals and algae. In: Dubinsky Z (ed) Coral reefs. Ecosystems of the world, vol 25. Elsevier, Amsterdam, pp 109–131

Beer S, Ilan M (1999) A new technique for non-intrusive in situ measurements of symbiotic photosynthesis. Coral Reefs 18:74

Behrenfeld MJ, Prasil O, Kolber ZS, Babin M, Falkowski PG (1998) Compensatory changes in photosystem II electron turnover rates protect photosynthesis from photoinhibition. Photosynth Res 58:259–268

Ben-Haim Y, Banim E, Kushmaro A, Loya Y, Rosenberg E (1999) Inhibition of photosynthesis and bleaching of zooxanthellae by the coral pathogen *Vibrio shiloi*. Environ Microbiol 1:223–229

Brown BE (1997a) Adaptations of reef corals to physical environmental stress. Adv Mar Biol 31:221–299

Brown BE (1997b) Coral bleaching: causes and consequences. Coral Reefs 16:129–138

Brown BE, Dunne RP, Scoffin TP, Le Tissier MDA (1994) Solar damage in intertidal corals. Mar Ecol Prog Ser 105:219–230

Brown BE, Ambarsari I, Warner ME, Fitt WK, Dunne RP, Gibb SW, Cummings DG (1999) Diurnal changes in photochemical efficiency and xanthophyll concentrations in shallow water reef corals: evidence for photoinhibition and photoprotection. Coral Reefs 18:99–105

Brown BE, Downs CA, Dunne RP, Gibb SW (2002) Preliminary evidence for tissue retraction as a factor in photoprotection of corals incapable of xanthophyll cycling. J Exp Mar Biol Ecol 277:129–144

Chalker BE, Dunlap WC (1983) Primary production and photoadaptation by corals on the Great Barrier Reef. In: Baker JT, Carter RM, Sammarco PW, Stark KP (eds) Proceedings of the inaugural Great Barrier Reef Conference, 28 Aug–2 Sept 1983. James Cook University, Townsville, Queensland, Australia, pp 293–298

Coles SL, Jokiel PL (1977) Effects of temperature on photosynthesis and respiration in hermatypic corals. Mar Biol 43:209–216

Coles SL, Brown BE (2003) Coral bleaching – capacity for acclimatization and adaptation. Adv Mar Biol 46:186–223

De Beer D, Kuhl M, Stambler N, Vaki L (2000) A microsensor study of light enhanced Ca^{2+} uptake and photosynthesis in the reef-building hermatypic coral *Favia* sp. Mar Ecol Prog Ser 194:75–85

Dove SG, Hoegh-Guldberg O, Ranganathan S (2001) Major colour patterns of reefbuilding corals are due to a family of GFP-like proteins. Coral Reefs 19:197–204

Dubinsky Z, Stambler N (1996) Marine pollution and coral reefs. Global Change Biol 2:511–526

Dubinsky Z, Stambler N, Benzion M, Mccloskey LR, Muscatine L, Falkowski PG (1990) The effect of external nutrient resources on the optical properties and photosynthetic efficiency of *Stylophora pistillata*. Proc R Soc Lond Ser B Biol Sci 239:231–246

Dunlap WC, Chalker BE, Oliver JK (1986) Bathymetric adaptation of reef building corals at Davies Reef, Great Barrier Reef, Australia. III. UV-absorbing compounds. J Exp Mar Biol Ecol 104:239–248

Edmunds PJ, Davies PS (1989) An energy budget for *Porites porites* (Scleractinia) growing in a stressed environment. Coral Reefs 8:37–43

Falkowski PG, Dubinsky Z (1981) Light-shade adaptation of *Stylophora pistillata*, a hermatypic coral from the Gulf of Eilat. Nature 289:172–174

Falkowski PG, Raven JA (1997) Aquatic photosynthesis. Blackwell Science, Boston, 375 pp

Falkowski PG, Jokiel PL, Kinzie RAI (1990) Irradiance and corals. In: Dubinsky Z (ed) Coral reefs. Ecosystems of the world, vol 25. Elsevier, Amsterdam, pp 89–107

Fitt WK, Warner ME (1995) Bleaching patterns of four species of Caribbean reef corals. Biol Bull 189:298–307

Gates RD, Baghdasarian G, Muscatine L (1992) Temperature stress causes host cell detachment in symbiotic cnidarians: implications for coral bleaching. Biol Bull 182:324–332

Gorbunov MY, Falkowski PG, Kolber ZS (2000) Measurement of photosynthetic parameters in benthic organisms in situ causing a SCUBA-based fast repetition rate fluorometer. Limnol Oceanogr 45:242–245

Gorbunov MY, Kolber ZS, Lesser MP, Falkowski PG (2001) Photosynthesis and photoprotection in symbiotic corals. Limnol Oceanogr 46:75–85

Hoegh-Guldberg O (1999) Climate change, coral bleaching and the future of the world's coral reefs. Mar Freshwater Res 50:839–866

Hoegh-Guldberg O, Smith GJ (1989) The effect of sudden changes in temperature, light and salinity on the population density and export of zooxanthellae from the reef corals *Stylophora pistillata* Esper and *Seriatopora hystrix* Dana. J Exp Mar Biol Ecol 129:279–303

Hoegh-Guldberg O, Salvat B (1995) Periodic mass-bleaching and elevated sea temperatures – bleaching of outer reef slope communities in Moorea, French Polynesia. Mar Ecol Prog Ser 121:181–190

Hoegh-Guldberg O, Jones R (1999) Photoinhibition and photoprotection in symbiotic dinoflagellates from reef-building corals. Mar Ecol Prog Ser 183:73–86

Iglesias-Prieto R (1995) The effects of elevated temperature on the photosynthetic responses of symbiotic dinoflagellates. In: Mathis P (ed) Photosynthesis from light to biosphere, vol 4. Kluwer, Dortrecht, pp 793–796

Iglesias-Prieto R (1997) Temperature-dependent inactivation of photosystem II in symbiotic dinoflagellates. Proc 8th Int Coral Reef Symp (Panama) 2:1313–1318

Iglesias-Prieto R, Matta JL, Robins WA, Trench RK (1992) Photosynthetic response to elevated temperature in the symbiotic dinoflagellate *Symbiodinium microadriaticum* in culture. Proc Natl Acad Sci USA 89:10302–10305

Jacques TG, Pilson MEQ (1980) Experimental ecology of the temperate scleractinian coral *Astrangia danae.* I. Partition of respiration, photosynthesis and calcification between host and symbionts. Mar Biol 60:167–178

Jokiel PL, York RH Jr (1982) Solar ultraviolet photobiology of the reef coral *Pocillopora damicornis* and symbiotic zooxanthellae. Bull Mar Sci 32:301–315

Jokiel PL, Coles SL (1990) Response of Hawaiian and other Indo-Pacific reef corals to elevated temperature. Coral Reefs 8:155–162

Jones RJ, Hoegh-Guldberg O (1999) Effects of cyanide on coral photosynthesis: implications for identifying the cause of coral bleaching and for assessing the environmental effects of cyanide fishing. Mar Ecol Prog Ser 177:83–91

Jones RJ, Hoegh-Guldberg O (2001) Diurnal changes in the photochemical efficiency of the symbiotic dinoflagellates (Dinophyceae) of corals: photoprotection, photoinactivation and the relationship to coral bleaching. Plant Cell Environ 24:89–99

Jones RJ, Hoegh-Guldberg O, Larkum AWD, Schreiber U (1998) Temperature-induced bleaching of corals begins with impairment of the CO_2 fixation mechanism in zooxanthellae. Plant Cell Environ 21:1219–1230

Jones RJ, Kildea T, Hoegh-Guldberg O (1999) PAM chlorophyll fluorometry: a new in situ technique for stress assessment in scleractinian corals, used to examine the effects of cyanide from cyanide fishing. Mar Pollut Bull 38:864–874

Kuhl M, Cohen Y, Dalsgaard T, Barker, Jorgensen B, Revsbech NP (1995) Microenvironment and photosynthesis of zooxanthellae in scleractinian corals studies with microsensors for O_2, pH and light. Mar Ecol Prog Ser 117:159–172

Lesser MP (1996) Exposure of symbiotic dinoflagellates to elevated temperatures and ultraviolet radiation causes oxidative stress and inhibits photosynthesis in symbiotic dinoflagellates. Limnol Oceanogr 41:271–283

Lesser MP (1997) Oxidative stress causes coral bleaching during exposure to elevated temperatures. Coral Reefs 16:187–192

Lesser MP (2000) Depth-dependent photoacclimatization to solar ultraviolet radiation in the Caribbean coral *Montastraea faveolata.* Mar Ecol Prog Ser 192:137–151

Lesser MP, Shick JM (1989) Photoadaptation and defenses against oxygen toxicity in zooxanthellae from natural populations of symbiotic cnidarians. J Exp Mar Biol Ecol 134:129–141

Lesser MP, Gorbunov MY (2001) Diurnal and bathymetric changes in chlorophyll fluorescence yields of reef corals measured in situ with a fast repetition rate fluorometer. Mar Ecol Prog Ser 212:69–77

Lesser MP, Stochaj WR, Tapley DW, Shick JM (1990) Bleaching in coral reef anthozoans: effects of irradiance, ultraviolet radiation, and temperature on the activities of protective enzymes against active oxygen. Coral Reefs 8:255–232

Levy O, Achituv Y, Yacobi Y, Dubinsky Z, Stambler N (2003) Diurnal changes in metabolism, protective enzymes and pigments in three Red Sea hermatypic corals (in press)

Mazel CH, Lesser MP, Gorbunov MY, Barry TM, Farrell JH, Wyman KD, Falkowski PG (2003) Green-fluorescent proteins in Caribbean corals. Limnol Oceanogr 48:402–411

McCloskey LA, Aamodt LD, Hazelton WD (1985) A computer-controlled respirometer for monitoring production and respiration of symbiotic organisms in situ. Proceedings of the 5th international coral reef congress, Tahiti 6:137–142

Mehler AH (1951) Studies on the reaction of illuminated chloroplasts. I. Mechanisms of the reduction of oxygen and other Hill reagents. Arch Biochem Biophys 33:65–67

Muller-Parker G (1984) Photosynthesis-irradiance responses and photosynthetic periodicity in the sea anemone *Aiptasia pulchella* and its zooxanthellae. Mar Biol 82:225–232

Nystrom M, Nordemar I, Tedengren M (2001) Simultaneous and sequential stress from increased temperature and copper on the metabolism of the hermatypic coral *Porites cylindrical.* Mar Biol 138:1225–1231

Osmond CB (1981) Photo-respiration and photoinhibition, some implications for the energetics of photosynthesis. Biochim Biophys Acta 639:77–98

Owen R, Knap A, Toaspern M, Carbery K (2002) Inhibition of coral photosynthesis by the antifouling herbicide Irgarol 1051. Mar Pollut Bull 44:623–632

Philipp R, Fabricius K (2003) Photophysiological stress in scleractinian corals in response to short-term sedimentation. J Exp Mar Biol Ecol 287:57–78

Porter JW, Fitt WK, Spero HJ, Rogers CS, White MW (1989) Bleaching in reef corals: physiological and stable isotopic responses. Proc Natl Acad Sci USA 86:9342–9346

Ralph PJ, Gademann R, Larkum AWD (2001) Zooxanthellae expelled from bleached corals at 33 °C are photosynthetically competent. Mar Ecol Prog Ser 220:163–168

Ralph PJ, Gademann R, Larkum AWD, Kuhl M (2002) Spatial heterogeneity in active chlorophyll fluorescence and PSII activity of coral tissues. Mar Biol 141:639–646

Reynaud S, Ferrier-Pages C, Sambrotto R, Juillet-Leclerc A, Jaubert J, Gattuso J-P (2002) Effect of feeding on the carbon and oxygen isotopic composition in the tissues and skeleton of the zooxanthellate coral *Stylophora pistillata.* Mar Ecol Prog Ser 238:81–89

Rowan R, Whitney SM, Fowler A, Yellowlees D (1996) Rubisco in marine symbiotic dinoflagellates: form II enzymes in eukaryotic oxygenic phototrophs encoded by a nuclear multigene family. Plant Cell 8:539–553

Rowan R, Knowlton N, Baker A, Jara J (1997) Landscape ecology of algal symbionts creates variation in episodes of coral bleaching. Nature 388:265–269

Salih A, Hoegh-Guldberg O, Cox G (1998) Photoprotection of symbiotic dinoflagellates by fluorescent pigments in reef corals. In: Greenwood JP, Hall NJ (eds) ACRS, Aus Coral Reef Soc Proc 75th Annu Conf 1:217–230

Salih A, Larkum A, Cox G, Kuhl M, Hoegh-Guldberg O (2000) Fluorescent pigments in corals are photoprotective. Nature 408:850–853

Savage A, Trapido-Rosenthal MH, Douglas AE (2002) On the functional significance of molecular variation in *Symbiodinium*, the symbiotic algae of cnidaria: photosynthetic response to irradiance. Mar Ecol Prog Ser 244:27–37

Saxby T, Dennison WC, Hoegh-Guldberg O (2003) Photosynthetic responses of the coral *Montipora digitata* to cold temperature stress. Mar Ecol Prog Ser 248:85–97

Schlichter D, Weber W, Fricke HW (1985) A chromatophore system in the hermatypic, deep-water coral *Leptoseris fragilis* (Anthozoa: Hexacorallia). Mar Biol 89:143–147

Schreiber U, Neubauer C (1990) O_2-dependent electron flow, membrane energization and the mechanism of non-photochemical quenching of chlorophyll fluorescence. Photosynthesis Res 25:279–293

Schumacher H, Zibrowius H (1985) What is hermatypic? A redefinition of ecological groups in corals and other organisms. Coral Reefs 4:1–9

Shashar N, Cohen Y, Loya Y (1993) Extreme diel fluctuations of oxygen in diffusive boundary layers surrounding stony corals. Biol Bull 185:455–461

Shibata K (1969) Pigments and a UV-absorbing substance in corals and a blue-green alga living in the Great Barrier Reef. Plant Cell Physiol 10:325–335

Shick JM, Lesser MP, Jokiel PL (1996) Effects of ultraviolet radiation on corals and other organisms. Glob Change Biol 2:527–545

Stambler N (1992) Harvesting and utilization of light by hermatypic corals, 221 pp (Hebrew with English abstract). 1992 Thesis, Bar-Ilan Univ, Ramat-Gan, Israel

Stambler N (1998) Effects of light intensity and ammonium enrichment on the hermatypic coral *Stylophora pistillata* and its zooxanthellae. Symbiosis 24:127–145

Steen RG, Muscatine L (1987) Low temperature evokes rapid exocytosis of symbiotic algae by a sea anemone. Biol Bull 172:246–263

Titlyanov EA, Shaposhnikova MG, Zvalinskii VI (1980) Photosynthesis and adaptation of corals to irradiance. 1. Contents and native state of photosynthetic pigments in symbiotic microalgae. Photosynthesis 74:413–421

Titlyanov EA, Titlyanov TV, Yamazato K, van Woesik R (2001) Photo-acclimation of the hermatypic coral *Stylophora pistillata* while subjected to either starvation or food provisioning. J Exp Mar Biol Ecol 257:163–181

Warner ME, Fitt WK, Schmidt GW (1996) The effects of elevated temperature on the photosynthetic efficiency of zooxanthellae *in hospite* from four different species of reef coral: a novel approach. Plant Cell Environ 19:291–299

Warner ME, Fitt WK, Schmidt GW (1999) Damage to photosystem II in symbiotic dinoflagellates: a determinant of coral bleaching. Proc Natl Acad Sci USA 96:8007–8012

Warner ME, Chilcoat GC, McFarland FK, Fitt WK (2002) Seasonal fluctuations in the photosynthetic capacity of photosystem II in symbiotic dinoflagellates in the Caribbean reef-building coral *Montastraea*. Mar Biol 141:31–38

Wethey DS, Porter IW (1976) Sun and shade differences in productivity of reef corals. Nature 262:281–282

Wilkinson CR, Buddemeier RW (1994) Report of the UNEP-IOC-ASPEI-IUCN global task team on the implications of climate change on coral reefs: implications for people and reefs. IUCN, Gland, 124 pp

Zvalinskii VI, Leletkin VA, Titlyanov EA, Shaposhnikova MG (1980) Photosynthesis and adaptation of corals to irradiance. 2. Oxygen exchange. Photosynthetica 14:422–430

10 What Can Regeneration Processes Tell Us About Coral Disease?

ESTI KRAMARSKY-WINTER

10.1 Introduction

Life history processes of scleractinian corals play a crucial role in the development and maintenance of coral reefs. Factors such as recruitment, survival, and longevity of the corals determine the reef characteristics. These factors depend on environmental parameters as well as on physiological intrinsic traits of the primary reef builders, the scleractinian corals. Despite intensive studies, we still lack a good understanding of the mechanisms of stress response and recovery of corals at the organism and community levels (Buddemeier and Smith 1999). Buddemeier and Smith (1999) proposed that "in any environment there exists a "hyperspace" where interacting variables create a multidimensional space of optimal or acceptable conditions in terms of organism response". As long as all or most of the variables are within the central core of this hyperspace, the organisms may display resilience. An organism or community may remain viable or seem to thrive even in response to environmental change that reduces its volume of optimal hyperspace (Buddemeier and Smith 1999). Further change, however slight, may then push them over the threshold causing a disintegration of adaptive responses. If we perceive a coral as a unit where resources are limited and partitioned to a multitude of life functions (see Fig 10.1), we can envision the mechanism in which the "hyperspace" may act upon it. Corals that suffer injury to tissue, suffer damage to reproduction, growth and the ability to withstand stress, and by definition exist in a shrunken hyperspace. These corals will also ultimately be more susceptible to disease.

In nature, corals suffer physical damage from a multitude of sources, including storms (Scoffin 1993), sedimentation (Loya 1976a, b; Gilmour 2002), emersion at low tide (Loya 1976a), anthropogenically induced breakage (Zakai et al. 2000), competitive interactions, predation, and disease (Bak and Criens 1981; Cameron et al. 1991; Richardson 1998). The ability of these corals to repair damage lies in their capacity for regeneration. Regeneration is defined as the capacity of a fully developed organism to replace lost parts by growth or remodeling of somatic tissue, and dictates the organism's ability to repair tissue (Goss 1992). Although this process is guided by a genetic blueprint (Alvarado 2000), the capacity for repair is determined by a multitude of responses to environmental parameters. This chapter will review some of the environmental, physiological, and cellular parameters that enable corals to regenerate lost tis-

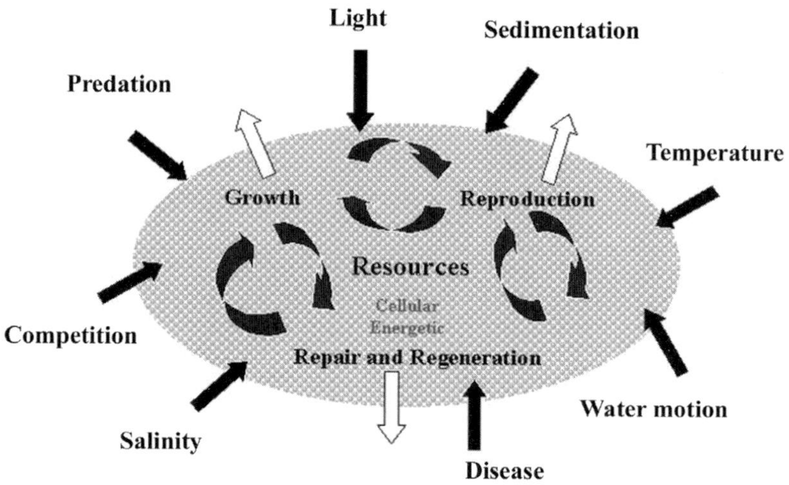

Fig. 10.1. Environmental effect on coral resource pool and reserves and links between growth, regeneration and reproduction. *Black arrows* indicate the environmental effects on the hyperspace of the coral and the *white arrows* indicate the resources a coral has to enlarge this hyperspace

sues and attempt to extrapolate these processes to the corals' ability to ward off pathogens and recuperate from disease.

10.2 Environmental Conditions Affecting Regeneration

Different coral growth forms and/or species are known to be better adapted to different environmental conditions. Environments with differing sedimentation loads, turbidity, and wave action are characterized by different species (Loya 1976a, b; Rosenfeld et al. 1999; Gilmour 2002). Physical disturbances can cause damage to coral tissue and or skeleton. For example, exposure to high sedimentation is often cited as an important cause of coral tissue damage due to abrasion, smothering by blocking oxygen-carrying water, bleaching due to reduced light, energy drain due to self-cleaning effort, or tissue necrosis due to microbial action (Loya 1976b; Hodgson 1990; Meesters et al. 1992; Rosenfeld et al. 1999; Gilmour 2002). Thus, coral survival under these conditions depends on the ability to remove it via costly energy processes such as ciliary action, tissue swelling, and the production of copious amounts of mucus. The duration of these reactions depends on the extent of the sedimentation event. High sedimentation may cause the corals to become more susceptible to damage, on the one hand, and when damaged may make them less capable of regeneration, on the other (Meesters et al. 1992; Gilmour 2002). When comparing regeneration rates of artificially induced lesions in *Acropora palmata* from sites with different sedimentation loads, Meesters et al. (1992) found that regeneration was slower in colonies from the high sediment site than in those from the low sediment site. It is possible, however, that physical abrasion by sediment alone may not have been the only parameter involved in repair. The microbial community, and particularly pathogens found in the suspended sediment, may have had additional detrimental effects on the corals' ability for repair. This is supported by a study by Hodgson (1990) showing that tetracycline-treated seawater reduced the rate of tissue necrosis and colony mortality during sedimentation experiments.

It is likely, however, that the rates of regeneration also depend on other innate processes that differ between species (Bak and Steward-Van Es 1980; Hall 1997, 2001). Meesters et al. (1992) found that when comparing repair rates in different species under the same sediment load, the branching coral *Acropora palmata* was slower to repair than the more massive corals such as *Porites asteroides* or *Siderastrea sidera*. Meesters et al. (1992) concluded that this was due to poorer or more inefficient sediment rejection in the former. However, it is also possible that the mechanisms of response to injury differ with species and may be tied to the growth strategies of the coral, i.e., depend on the type of budding. Coral colonies that grow through intratentacular budding will inevitably regenerate differently from those undergoing intertentacular budding, and solitary corals will differ in that all the tissue reserves are found in the single polyp (Meesters et al. 1992; Ruesnik 1997; Kramarsky-Winter and Loya 2000).

Seasonal factors, such as changes in light intensity, temperature, food availability and water quality that are known to affect normal growth in corals, also affect regeneration (Lester and Bak 1985; Guzman and Cortes 1989; Meesters et al. 1994; Lough and Barnes 1997; Kramarsky-Winter and Loya 2000). The multivariate changes in environmental conditions have made it difficult to isolate each effect on coral regeneration. Very few controlled experiments have been carried out to test how isolated parameters may affect regeneration. For example, the effect of irradiance on calcification has been well studied (Barnes and Chalker 1990), though virtually nothing is known of its effect on tissue repair. Hidaka et al. (1981) showed that following removal of the oral portion of a polyp, light affects the appearance and position of newly regenerating polyp buds in *Galaxea fascicularis*. Meesters et al. (1997) showed that water depth had an important effect on regeneration as well. Our preliminary experiments using the solitary free-living coral *Fungia granulosa* as a model showed that light enhanced tissue regeneration. Following breakage of six polyps into two halves, and maintenance of one half of each in the light and the other half in the dark showed that the fragment maintained in the dark for over 1 month only partially regenerated their tissues, while the fragment maintained in the light underwent complete regeneration (Fig. 10.2).

In order to ascertain the effect of water temperature on tissue regeneration in this coral, a set of controlled laboratory experiments using *F. granulosa*, which had been previously acclimated to two temperature regimes (Kramarsky-Winter and Loya 2000), was compared with complement experiments in situ. Thirty percent of the corals' oral surface tissues was removed using an air pick and the total number of corals that underwent regeneration was recorded.

Fig. 10.2. Regeneration in complementary fragments of the same *Fungia granulosa* individual maintained in the laboratory. Fragments were maintained for a period of 3 weeks either under a normal light regime or in the dark. **A** Fragment maintained in the light. Note tissue covering most of the broken edge. **B** Fragment maintained in the dark. Only mouth tissues are repaired

These experiments demonstrated that overall, corals maintained at warmer temperatures undergo repair faster and more completely than those kept at lower temperatures (Kramarsky-Winter and Loya 2000; Fig. 10.3). Interestingly, this effect was particularly marked during the fall (nongametogenic season) and almost negligible during spring when the corals undergo gametogenesis (Fig. 10.3). The discrepancy in the effect of temperature on regeneration rates between the seasons may be explained by an additional parameter (reproductive state) on tissue repair (Kramarsky-Winter and Loya 2000). These experiments may be regarded as evidence that environmental parameters such as water temperature and physiological thresholds act synergistically or antagonistically in affecting regeneration abilities.

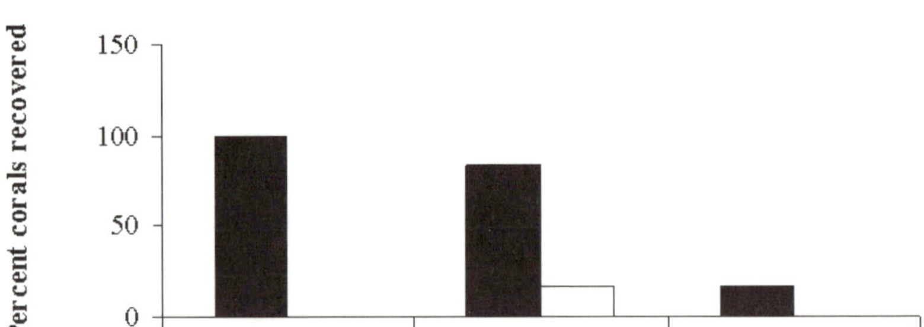

	Field	25°C	21°C
■ Fall	100	83.3	16.7
□ Spring	0	16.7	0

Fig. 10.3. Percent *Fungia granulosa*, maintained at two temperature regimes, that underwent full recovery following removal of 30% of the oral surface tissues. Parallel experiments were carried out in the field and in the laboratory during fall and spring months with new sets of corals each time. n=10 for each treatment in the field and n=6 for each treatment in the laboratory

10.3 Physiological Parameters

The ability to undergo repair of tissue and skeleton is dependent on a number of physiological parameters. Energy reserves provided by zooxanthellae, ingestion of dissolved organic matter (DOM), bacteriovory and predation will ultimately also have effects on the rate and ability to regenerate (Rinkevich 1997). Assuming a finite energy reserve, the amount of energy available for repair will inevitably also depend on trade-offs between other life functions such as reproduction, defense against disease, competition, etc. For example, the extent of damage dictates the amount of resources the coral will invest in these processes. The amount of tissue lost, or size of the wound and the amount of tissue available for repair (i.e., coral size) will dictate the coral's ability to recuperate. A number of studies showed that the ability to undergo regeneration depends largely on the extent of damage incurred (Bak and Steward-Van Es 1980; Oren et al. 1997a). In general, the smaller the lesion, the faster the coral undergoes repair (Bak and Steward-Van Es 1980). This is true for colonial species as well as solitary ones (Bak and Steward-Van Es 1980; Meesters et al. 1994; Oren et al. 1997a; van Woesik 1998; Kramarsky-Winter and Loya 2000). Despite this, even large lesions might be repaired quickly if there is a high perimeter to surface area ratio of the lesion (Oren et al. 1997a; van Woesik 1998). Oren et al. (1997a), for example, reported that, in the colonial coral *Favia favus*, "large" elongate lesions might recover as quickly as round lesions with a small surface area and faster than in round lesions where the perimeter to surface ratio is smaller. Using ^{14}C Oren et al. (1997b, 2001) showed that, in this coral, lesions with large perimeters caused large portions of the colony to become involved in translocation of photosynthates toward the injury, indicating involvement of larger portions of the colony.

Coral size also plays an important role in the ability of a coral to recuperate from damage. In general, larger corals undergo repair faster and more completely than small corals (Loya 1976c; Bak and Steward-Van Es 1980; Kramarsky-Winter and Loya 2000; Oren et al. 2001; Gilmour 2002). Loya (1976c) found that regeneration of tissue and skeleton in the colonial branching coral *Stylophora pistillata* was faster in large colonies than in small ones. Oren et al. (2001) showed that small faviid colonies underwent repair more slowly than larger colonies. Moreover, they showed that in *Favia favus*, small lesion repair was unaffected by colony size, while the repair of larger lesions was. The extent and rate of tissue repair in the solitary coral *Fungia granulosa* also depended on coral size; small immature corals (<5 cm diameter) (Kramarsky-Winter and Loya 2000) did not undergo tissue regeneration as well as larger polyps. An explanation for the discrepancy in repair between large and small corals, may be that the actual amount of tissue removed from smaller and, consequently, thinner corals was relatively greater (in terms of total tissue volume available for repair) in small corals than in large specimens, indicating that total percent tissue removed may be a better index than lesion size when comparing corals of different sizes. Another not mutually exclusive explanation may be that in this coral there may be a difference in energy allocation during two phases of its life history. Individuals of *F. granulosa* smaller than 5 cm diameter are sexually immature (Kramarsky-Winter and Loya 1998). Small immature corals were previously found to allocate energetic and cellular investment to growth while large mature ones also invest in reproduction (Loya 1976b). Thus, small colonies or individuals may not have the reserves necessary for regeneration. The hypothesis is further supported by the fact that following breakage in *F granulosa*, once tissue coverage was complete, ensuing skeletal accretion in smaller individuals was faster than in large ones (Chadwick and Loya 1990). Thus, overall growth in terms of skeletal accretion and tissue investment following damage may be such as to quickly return the coral to a threshold size where it may then be able to transfer its energy toward reproduction. This leads to the suggestion that a size threshold of repair may be indicative of the availability of cellular and energetic reserves for tissue maintenance and growth. These studies resulted in the presentation of a colony activation model that asserts that the capacity for repair depends on coral size as well as the amount of tissue directly involved in the repair process, i.e., the number of polyps or amount of tissue (see Oren et al. 2001).

A number of coral species have revealed energy trade-offs occurring between regeneration and reproduction (Kojis and Quinn 1981, 1985; Rinkevich and Loya 1989; van Veghel and Bak 1994; Rinkevich 1997; Kramarsky-Winter and Loya 2000; Lirman 2000; Zakai et al. 2000; Oren et al. 2001). Generally speaking, in many corals, reproduction begins once they pass a certain size threshold (Harrison and Wallace 1990; Kramarsky-Winter and Loya 1998). Presumably, this is due to the fact that larger corals have greater energy and tissue reserves. Since physical damage to coral colony integrity also affects its growth rate, it is assumed that damage to reproductively active corals will result in less reserve for repair and growth. Thus, damaging corals at such a critical stage may affect not only their reproductive effort for that season, but their ability to undergo repair and to grow (Loya 1976c; Rinkevich and Loya 1989; Meesters et al. 1994; Kramarsky-Winter and Loya 2000).

Damage also affects coral growth. Comparison of the *Stylophora*'s growth capacity between physically damaged colonies and intact controls showed that damaged corals grew at a faster rate than nondamaged controls (Rinkevich and Loya 1987, 1989). This indicated that damage may have actually stimulated repair and growth in this coral. On the other hand, the lower growth rate in damaged corals during the winter months (reproductive months) may have been a reflection of the corals' reproductive state. Indeed, Rinkevich and Loya (1986) found that growth was suppressed in mature, healthy *Stylophora pistillata* colonies during the reproductive months when compared with growth in these corals in nonreproductive months. Thus, the damaged colonies may have been allocating resources toward repair rather than the energetically costly development of oocytes, resulting in faster growth in these corals, but at the cost of reproduction. Since physical damage to coral colony integrity also affects its growth rate, damage to reproductively active corals that have less reserves for repair and growth causes harm at a critical stage, thus affecting not only their ability to undergo repair and to grow, but also their reproductive effort for that season (Loya 1976c; Rinkevich and Loya 1989; Meesters et al. 1994; Kramarsky-Winter and Loya 2000).

Whale (1983) examined the effect of reproductive state on tissue repair rate in the gorgonian *Plexeura homomalla*, and showed that in this octocoral the regeneration rate of small lesions was independent of reproductive phase. He reasoned that this coral possesses a permanent and independent energetic reserve for repair (Whale 1983). On the other hand, Kramarsky-Winter and Loya (2000) showed that tissue regeneration was severely reduced in reproducing *F. granulosa*, indicating that in this scleractinian coral there is a strong relationship between the reserves for repair and reproduction. Rinkevich and Loya (1989) showed that fecundity and even sexuality were reduced in physically damaged colonies of *S. pistillata*. They found that reproductive effort in regenerating colonies was an order of magnitude lower than in undamaged controls even 19 months following pruning of branch tips. Damage caused a loss of synchrony in reproductive effort between branches of the same colony. Colonies that were mature hermaphrodites prior to breakage reverted to maleness or became sterile following damage, with some of the colonies remaining affected even 2 years after damage (Rinkevich and Loya 1989). Similarly, Lirman (2000) showed that severe fragmentation in *Acropora palmata* resulted in loss of reproductive capacity 4 years following damage and Kojis and Quinn (1981, 1985) showed that damaged *Goniastrea* colonies had an overall lower fecundity than nondamaged controls.

The extent of the change in fecundity was examined at the polyp level and as a function of distance from the damage in a variety of corals. Van Veghel and Bak (1994) found that in damaged *Montastrea annularis*, newly regenerated polyps did not contain gonads. They further noted that both the polyps surrounding the lesion and those further from it had greatly reduced fecundities when compared to polyps in control corals (van Veghel and Bak 1994). Oren et al. (2001) showed that in regenerating *Favia favus* colonies, the reproductive effort of polyps following damage to a colony depended on the polyp's distance as well as on the size of the damage inflicted. These results suggested that there is a hierarchy in resource allocation; polyps closest to the damage investing more than those distant from the lesion. In addition, they showed that there was an overall colonial priority of energy allocation toward repair over reproduction (Oren et al. 2001).

In the solitary *F. granulosa*, all repair and reproductive reserves are found in the single polyp and prioritizing must take place at the polyp or organism level. Tissue loss (30% of surface tissue) resulted in a closing off of damaged tissue with no ensuing tissue growth and a reduction of fecundity by half. Interestingly, no significant difference (*t*-test, $P>0.05$) was found between the oocyte sizes in regenerating and control 1 month following damage (Kramarsky-Winter and Loya 2000). One explanation for this may be that stem cells already committed to gametogenesis continue to differentiate, while those cells which have not entered the pathway of differentiation are diverted toward repair (Kramarsky-Winter and Loya 2000). Thus, in contrast to some of the colonial corals, in the solitary *Fungia*, if there are not enough energetic and cellular reserves necessary for both repair and sexual reproduction to take place, reproduction seems to take priority and germ cells that are already committed will continue developing. It is possible, however, that a greater loss of tissue would have resulted in a cessation of reproduction as well, illustrating the importance of ascertaining what levels of damage a coral may be able to withstand before long-lasting or irreversible effects take place. Further studies are necessary to reveal at what stage this is reversible and at what stage commitment is final.

In a previous study, we showed that in fungiids that had extensive damage to tissues, deep tissue remnants formed new buds that eventually developed into new fungiids (Kramarsky-Winter and Loya 1996). Thus, it is likely that in these corals there is a critical mass of tissue necessary for regeneration to occur and when regeneration is not possible, the clone may survive by budding new polyps from tissue remnants. Understanding the parameters affecting the regeneration size threshold is particularly important when studying coral tissue degenerative diseases as they will certainly affect the ability of young coral communities to survive following disease flare-up.

10.4 Bleaching and Regeneration

A number of studies have evaluated the ability of corals to heal damage to tissue and/or skeleton following bleaching events (Meesters and Bak 1993; Mascarelli and Williams 1999; Fine et al. 2002). These studies showed that bleaching has a marked effect on the ability of tissues to undergo regeneration. Szmant and Gassman (1990) reported a decrease in tissue biomass associated with bleaching in the coral *Montastrea annularis*. Fine et al. (2002) reported that a hierarchy in regeneration capacities ranges from high capacity for non-bleached corals, to intermediate capacity in partly bleached colonies to no repair in completely bleached ones. Similarly, Mascarelli and Bunkley-Williams (1999) showed that unbleached fragments from a single coral colony underwent tissue repair faster than experimentally bleached ramets. In all cases, the ramets recovered pigmentation, but most (8/12) were not able to rid themselves of sediment or colonized algae, suggesting that reduction in zooxanthellae may decrease the resources that enable the corals to repel colonizers. Importantly, it should be noted that the amount of time in which the coral is stressed, or in this case bleached, would likely affect its ability to recover injury. A recent study by Edmunds et al. (2003) has shown that the depression in zooxanthellae number might be long-standing before any effects to tissue composition become obvious. A profound and measurable effect of tissue composition may be considered a delayed response of bleaching. Thus, it is likely that very short-lived bleaching may have little effect on the corals' ability to recuperate since they may have enough energy reserves to overcome damage. Corals that undergo either repeated bleaching (e.g., *Oculina*, see Fine et al. 2002; Fine and Loya, Chap. 6, this Vol.) or long-term bleaching might not have enough reserves to be able to regenerate. Indeed, Fine et al. (2002) recently showed that partially bleached colonies (less than 40% bleaching) were still able to regenerate tissue and translocate photosynthates to regenerating tissue, while those with a stronger degree of bleaching could not (discussed by Fine and Loya, Chap. 6, this Vol.).

10.5 Cellular Aspect of Regeneration

In general, very little is known about cellular aspects of regeneration in the Anthozoa. What is known is primarily from studies on the process of small wound healing in sea anemones (Young 1974; Patterson and Landolt 1979) and in gorgonians (Whale 1983; Meszaros and Bigger 1999; Olano and Bigger 2000). These studies showed that specific amoeboid cells termed wound cells or mesogleal cells were the first cell types to show movement toward the injury site. In *F. granulosa* there is a clear rise in amoebocytic cells adjacent to the wound site (Fig 10.4). This is particularly evident in regenerating mouth tissue (Fig. 10.4C). In the few published studies on cellular processes involved in these stages of recuperation, one of the first steps of small wound healing seems to be the involvement of "amoebocytes" that act in cleaning up cellular debris (Young 1974; Patterson and Landolt 1979; Olano and Bigger 2000). Moreover, additional studies showed that cells not normally phagocytic might be activated to become so, thus acting as clean-up cells as well (Hutton and Smith 1996; Olano and Bigger 2000). In corals undergoing pathogen-induced

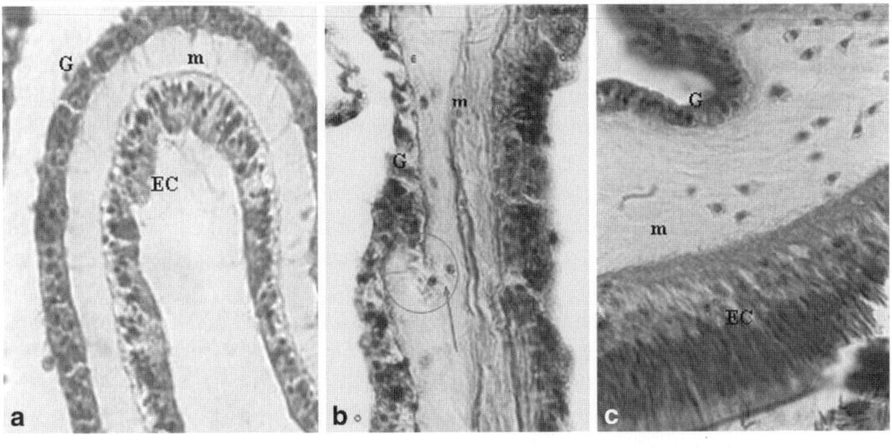

Fig. 10.4. The rise in amoebocytes occurring during tissue regeneration in the solitary coral *Fungia granulosa* (A), distal tissues (B), regenerating tissues (C). *Arrows* in regenerating mouth tissues indicate amoebocytes found in the mouth mesoglea. *EC* Ectoderm, *G* gastroderm

disease, it is possible that there is a reduction of amoebocytes that may be engaged in tissue repair.

One of the first outcomes to damage is a cessation of calcification at the site of repair. Corals stained with alizarine red S 1 day following breakage showed no staining at the site of breakage (Meesters et al. 1994; Kramarsky-Winter, unpubl. results). This depression is short-lived and indicates that there may be a depression in translocation occurring immediately following damage ("shock response"). In *F. granulosa* during this period, necrotic tissues and cellular debris were sloughed off. Following this step, the tissues surrounding the wound stretched to cover as much of the wound as possible. This occurs in fragments of *F. granulosa* maintained in the light as well in those maintained in the dark, indicating the use of available localized reserves. The cells in these tissues changed shape, resulting in a thinning of these tissues and, consequently, in the coverage of skeletal elements. Torn epithelial tissues sealed themselves off and formed lip-like edges. All these processes may be defined as local processes and when the damage is slight they may be adequate enough to quickly overcome it. Similarly, Meesters et al. (1992) showed that during early regeneration of 1-cm-diameter lesions in *Acropora palmata*, the bare skeleton is quickly covered by a transparent, undifferentiated tissue veneer. If damage is extensive, following the period of stretching, processes of cell proliferation, movement and differentiation occur. In *Acropora palmata*, polyp formation is only visible 8 days after damage (Meesters et al. 1992). This is also the period of renewed calcification. The length of these processes may then affect the ability of the coral to regrow tissue at a rate fast enough to compete with unwanted settlers.

This physical stretching of tissue to cover the wound may then explain the quick repair of lesions with large perimeter/surface area. If the lesion is too large to be covered by stretching tissues, much slower processes such as tissue growth, i.e., cell proliferation, and translocation must take place. Studies using both ^{3}H thymidine and BrdU (a thymidine analog) showed no rise in cell proliferation during the first phase of regeneration in *Favia favus* and *F. granulosa* (Oren et al. 2001; Kramarsky-Winter, unpubl. results). A rise in proliferation was evident in gastrodermal tissues 6 days following damage in *F. granulosa* (Kramarsky-Winter, unpubl. results). Thus, in *Fungia*, following the stretching and sealing stages of recovery (occurring during the first week of recovery), the tissues grow through cell proliferation and translocation. This process may also entail transdifferentiation (Olano and Bigger 2000) of existing cells and/or of stem cells though, unfortunately, virtually nothing is understood of this process.

At the cellular level, mechanisms of repair also require thresholds for different stages to be activated. Results presented by Oren et al. (1997b) showed that 3 weeks following damage there was a significant rise in translocation of ^{14}C only in regenerating large lesions. Fine et al. (2002; Fine and Loya, Chap. 6, this Vol.) report oriented translocation of ^{14}C toward damaged tissues in healthy and partly bleached colonies. Moreover, they found that corals with over 40% bleaching translocation toward damage stopped indicating a physiological threshold according to which the coral regulates its resource allocation (Fine et al. 2002). In addition to translocation of photosynthates, it is likely that there is also a translocation of cells toward the lesion. Gateño et al. (1998) showed that there is significant translocation of ^{14}C into the octocoral *Parerythropodium fulvum fulvum* from labeled grafts to intact colonies. They showed that coral gastrodermal cells with or without healthy zooxanthellae are transported in two interconnected canals of the gastrovascular system. In addition, a study by Meszaros and Bigger (1999) showed that during regeneration in the octocoral *Plexaurella fusifera* there is an overall increase in zooxanthella concentrations at the site of injury. This indicated that following early repair processes there is a translocation of photosynthates and cells to the damaged area.

10.6 What Regeneration Mechanisms May Tell Us About Disease and Recovery?

To date, no study has been carried out to try and determine if and to what extent corals that lose tissue as a result of pathogenic infection recuperate and regenerate lost tissue. Understanding what parameters affect regeneration following tissue lesion may aid in the elucidation of how diseased corals may or may not recuperate once the pathogen has been removed or as a result of acquired immunity. What can be inferred from some of the studies presented above, is that small corals are likely to be affected more by disease and less likely to recover than larger ones.

The prevalence of coral disease in the summer reproductive month has far-reaching consequences. Damage to tissue, whether due to the disease process or other parameters, may affect the coral over a long period of time. This results in corals affected by disease being less likely to grow and reproduce. The depression in reproductive effort is likely to extend into ensuing seasons even if the disease process is stopped. Moreover, even in those corals where the disease process is arrested, or causes only minimal tissue damage, depletion of cellular and energy resources may be such as to cause irreversible damage causing corals affected by repeated bouts of disease to be less likely to recover.

10.7 Unresolved Questions and Future Research

Despite intensive research, the process of regeneration remains poorly understood. One of the major questions to be resolved is what is the threshold beyond which corals cannot regenerate? This threshold may vary in different species and following different types of damage, and will depend on how limited the coral's resources are. These questions have only begun to be addressed (see Fine et al. 2002). Much more work needs to be done to ascertain the physiological and environmental thresholds that make up each coral species "hyperspace" as well as the balances between these parameters.

Much of our comprehension of these thresholds will be clarified once we understand the cellular mechanisms driving the regeneration process. Unfortunately, we still do not know what role the different cell types may play in the regeneration process. Although undoubtedly "amoebocytes" have a role in tissue repair, virtually nothing is known of what mechanisms are involved. Do these cells only act in cleaning up the tissues and phagocytizing debris, or are there subsets of these cells that act as reserve cells? Is there a reservoir of these cells that becomes activated upon injury or is this a manifestation of a cellular pluripotency? Do these cells really have a role as "immunocytes" acting against invading pathogens? If so, what is the mechanism of action? What is the role of zooxanthellae in the repair process? Is the translocation of photosynthates evident through ^{14}C carried out via dissolved materials or in actual circulating algal cells or in both? Since nonzooxanthellate corals also undergo regeneration, is it possible that their role is secondary and corals will be able to regenerate if provided with enough nutrients (e.g., prey, DOM)? Other major questions include what are the molecular mechanisms responsible for regeneration processes and how may they be influenced by environmental change?

References

Alvarado AS (2000) Regeneration in the metazoans: why does it happen? BioEssays 22:578–590

Bak RPM, Criens SR (1981) Survival after fragmentation of colonies of *Madracis mirabilis*, *Acropora palmata* and *A. cervicornis* (Scleractinia) and the subsequent impact of coral disease. Proc 4th Int Coral Reef Symp 2:221–227

Bak RPM, Steward-Van Es Y (1980) Regeneration of superficial damage in the scleractinian corals *Agaricia agaricites f. pupurea* and *Porites asteroides*. Bull Mar Sci 30:883–887

Barnes DJ, Chalker BE (1990) Calcification and photosynthesis in reef building corals and algae. In: Dubinsky Z (ed) Ecosystems of the world, vol 25. Coral reefs. Elsevier, Oxford, pp 109–131

Buddemeier RW, Smith SV (1999) Coral adaptation and acclimatization: a most ingenious paradox. Am Zool 39:1–9

Cameron AM Endean R, deVantieer LM (1991) The effect of *Acanthaster plancii* predation on populations of two species of massive corals. Hydrobiologia 216/217:257–262

Chadwick NE, Loya Y (1990) Regeneration after experimental breakage in the solitary reef coral *Fungia granulosa* Klutzinger 1879. J Exp Mar Biol Ecol 142:221–234

Fine M, Oren U, Loya Y (2002) Bleaching effect on regeneration and resource translocation in the coral *Oculina patagonica*. MEPS 234:119–125

Edmunds PJ, Gates RD, Gleason DF (2003) The tissue composition of *Montastrea franksi* during a natural bleaching event in the Florida Keys. Coral Reefs 22(1):54–62

Gateño D, Israel A, Barki Y, Rinkevich B (1998) Gastrovascular circulation in an Octocoral: evidence of significant transport of a coral and symbiotic cells. Biol Bull 194:178–186

Gilmour JP (2002) Acute sediment deposition causes size specific mortality and asexual budding in the mushroom coral *Fungia fungites*. Mar Freshwater Res 53(4):805–812

Goss RJ (1992) The evolution of regeneration: adaptive or inherent? J Theor Biol 159:241–260

Guzman HM, Cortes J (1989) Growth rates of eight species of scleractinian corals in the eastern Pacific (Costa Rica). Bull Mar Sci 44:1186–1194

Hall V (1997) Interspecific differences in the regeneration of artificial injuries on scleractinian corals. J Exp Mar Biol Ecol 212:9–23

Hall V (2001) The response of *Acropora hyacynthus* and *Montipora tuberculosa* to three different types of colony damage: scraping injury, tissue mortality and breakage. J Exp Mar Biol Ecol 264:209–223

Harrison PL, Wallace CC (1990) Coral reproduction. In: Dubinsky Z (ed) Ecosystems of the world: Coral reefs. Elsevier, Amsterdam, pp 133–208

Hidaka M, Uechi A, Yamazato K (1981) Effects of certain factors on budding of isolated polyps of a scleractinian coral *Galaxea fascicularis*. Proc 4th Int Coral Reef Symp 2:229–231

Hodgson G (1990) Tetracycline reduces sedimentation damage to corals. Mar Biol 104:493–496

Hutton DMC, Smith VJ (1996) Antibacterial properties of isolated amoebocytes from the sea anemone *Actinia equina*. Biol Bull 191:441–451

Kojis BL, Quinn NJ (1981) Aspects of sexual reproduction and larval development in the shallow water hermatypic coral, *Goniastrea australensis*. Bull Mar Sci 31:558–573

Kojis BL, Quinn NJ (1985) Puberty in *Goniastrea favulus* age or size limited? Proc 5th Int Coral Reef Congress, Tahiti, 4, pp 289–293

Kramarsky-Winter E, Loya Y (1996) Regeneration versus budding in fungiid corals: a trade-off. Mar Ecol Prog Ser 134:179–185

Kramarsky-Winter E, Loya Y (1998) Reproductive strategies of two fungiid species *Fungia granulosa* and *Fungia scutaria* from the Gulf of Eilat. Mar Ecol Prog Ser 174:175–182

Kramarsy-Winter E, Loya Y (2000) Tissue regeneration in the coral *Fungia granulosa*: the effect of extrinsic and intrinsic factors. Mar Biol 137:867–873

Krupp DA, Jokiel PL, Chartrand TS (1993) Asexual reproduction by the solitary scleractinian coral *Fungia scutaria* on dead parent coral in Kanehoe Bay, Oahu, Hawaiian Islands. Coral Reefs 12:527–534

Lester RT, Bak RPM (1985) Effect of environment on regeneration of tissue lesions in the reef coral *Montastrea annularis* (Scleractinia). Mar Ecol Prog Ser 24:183–185

Lirman D (2000) Fragmentation in the branching coral *Acropora palmata* (lamarck): growth, survivorship, and reproduction of colonies and fragments. J Exp Mar Biol Ecol 251:41–57

Lough JM, Barnes DJ (1997) Several centuries of variation in skeletal extension, density and calcification in massive *Porites* colonies from the Great Barrier Reef: a proxy for seawater temperature and a background of variability against which to identify unnatural change. J Exp Mar Biol Ecol 211:29–67

Loya Y (1976a) Recolonization of the Red Sea corals affected by natural catastrophes and man-made perturbations. Ecology 57:278–289

Loya Y (1976b) Effect of water turbidity on the growth and sedimentation on the community structure of Puerto Rican reefs. Bull Mar Sci 26:450–466

Loya Y (1976c) Skeletal regeneration in a Red Sea scleractinian coral population. Nature 261:490–491

Mascarelli PE, Bunkley-Williams L (1999) An experimental field evaluation in damaged, unbleached and artificially leached star coral *Montastrea annularis*. Bull Mar Sci 2:577–586

Meesters EH, Bak RP (1993) Effects of coral bleaching on tissue regeneration potential and colony survival. Mar Ecol Prog Ser 96:189–198

Meesters EH, Bos A, Gast CG (1992) Effect of sedimentation and lesion position on coral tissue regeneration. Proc 7th Int Coral Reef Symp 2:671–678

Meesters EH, Noordeloos M, Bak RPM (1994) Damage and regeneration: links to growth in the reef-building coral *Montastrea annularis*. Mar Ecol Prog Ser 112:119–128

Meesters EH, Wesseling I, Bak RPM (1997) Coral colony tissue damage in six species of reef-building corals: partial mortality in relation with depth and surface area. J Sea Res 37:131–134

Meszaros A Bigger CH (1999) Qualitative and quantitative study of wound healing processes in the coelenterate, *Plexaurella fusifera*: spatial, temporal and environmental (light attenuation) influences. J Invert Pathol 73:321–331

Nugues MM (2002) Impact of a coral disease outbreak on coral communities in St Lucia: what and how much has been lost? Mar Ecol Prog Ser 229:61–71

Olano CT, Bigger CH (2000) Phagocytic activities of the Gorgonian coral *Swifta exserta*. J Invert Pathol 76:176–184

Oren U, Benayahu Y, Loya Y (1997a) Effect of lesion size and shape on regeneration of the Red Sea coral *Favia favus*. Mar Ecol Prog Ser 146(1):101–107

Oren U, Rinkevich B, Loya Y (1997b) Oriented intra-colonial transport of ^{14}C labeled material during coral regeneration. MEPS 161:117–122

Oren U, Benayahu Y, Lubinevsky H, Loya Y (2001) Colony integration during regeneration in the stony coral *Favia favus*. Ecology 82(3):802–813

Patterson MJ, Landolt ML (1979) Cellular reaction to injury in the anthozoa *Anthopleura elegantissima*. J Invert Pathol 33(2):189–196

Richardson LL (1998) Coral disease: what is really known? Tree 13:438–443

Rinkevich B (1997) Do reproduction and regeneration in damaged corals compete for energy allocation? Mar Ecol Prog Ser 143:297–302

Rinkevich B, Loya Y (1986) Senescence and dying signals in a reef building coral Experientia 42:320–322

Rinkevich B, Loya Y (1987) Variability in the pattern of sexual reproduction in the coral *Stylophora pistillata*. at Eilat the Red Sea: a long term study. Biol Bull 173:474–488

Rinkevich B, Loya Y (1989) Reproduction in regenerating colonies of the coral *Stylophora pistillata*. In: Spanier E, Steinberger Y, Luria M (eds) Environmental quality and ecosystem stability. Hebrew University, Jerusalem, pp 257–265

Rosenfeld M, Bresler V, Abelson A (1999) Sediment as a possible source of food for coral. Ecol Lett 2:345–348

Ruesnik JL (1997) Coral injury and recovery: matrix model link process to pattern. J Exp Mar Biol Ecol 210:187–208

Scoffin TP (1993) The geological effects of hurricanes on coral reefs and interpretation of storm deposits. Coral Reefs 12:203–221

Szmant A, Gassman NJ (1990) The effect of prolonged "bleaching" on the tissue biomass and reproduction of the reef coral *Montastrea annularis*. Coral Reefs 8:217–224

Van Veghel MLJ, Bak RPM (1994) Reproductive characteristics of the polymorphic Caribbean reef building coral *Montastrea annularis*. III. Reproduction in damaged and regenerating colonies. Mar Ecol Prog Ser 109:229–233

Van Woesik R (1998) Lesion healing on massive *Porites* spp. corals. Mar Ecol Prog Ser 164:213–220

Whale CM (1983) Regeneration of injuries among Jamaican Gorgonians: the role of colony physiology and environment. Biol Bull 165:778–790

Young JC (1974) The nature of tissue regeneration after wounding in the sea anemone Calliactis parasitica (Couch). J Mar Biol Assoc UK 54:599–617

Zakai D, Levy O Chadwick-Furman NE (2000) Experimental fragmentation reduces reproductive output by the reef building coral *Pocillopora damicornis*. Coral Reefs 19:185–188

11 Bacteria as a Source of Coral Nutrition

ARIEL KUSHMARO and ESTI KRAMARSKY-WINTER

11.1 Introduction

Coral reefs are diverse and important communities in tropical and subtropical marine environments. Hermatypic corals play a key role in forming the structure of coral reefs and in providing substrata and shelter for a wide variety of organisms. The symbiotic association between corals and their photosynthetic algae (zooxanthellae) is one of the primary keys to coral success in this ecosystem (Fallowski et al. 1984; Sebens 1994). Hermatypic corals attain a large part of their nutrition from photosynthetic products of the zooxanthellae, which are located in their gastrodermal cell layer. However, corals must utilize other particulate matter to gain enough nitrogen and phosphorus to grow (Sorokin 1990; Sebens 1994; Schlichter and Brendelberger 1998; Rosenfeld et al. 1999; Anthony and Fabricius 2000).

11.2 Bacterial Suspension Feeding

Hermatypic corals gain the majority of their carbon requirement through their symbiotic association with zooxanthellae (Fallowski et al. 1984). They are also passive suspension feeders with the potential for utilizing a wide range of food sources (Sebens 1997; Anthony 1999, 2000). Previous studies demonstrated the ability of these corals to capture and ingest various sources of particulate food, including zooplankton and bacteria (Sorokin 1973a, b; Sebens 1997, 1994; Bak et al. 1998). Replenishment of nitrogen, phosphorus, and other nutrients that cannot be supplied by symbiotic algae must come from suspension feeding (Sebens et al. 1998). Coral morphology can affect their ability to capture plankton and other particulate materials (Schlichter and Brendelberger 1998). Abelson et al. (1993) demonstrated that branching forms are more efficient in suspension feeding than the massive growth forms. On the other hand, corals with flattened morphology such as *Mycedium elephantotus* or *Mycetophyllia reesi* may act as traps that capture particulate matter on their surface via their mucus layer (Schlichter and Brendelberger 1998; Goldberg 2002). The massive coral *Mycetophyllia reesi* was reported to capture particulate food by mucus entanglement which was then collected by the mesenterial filaments (Goldberg 2002).

Heterotrophic bacteria have been recognized as important sources of carbon and nitrogen cycling in coral reefs (Sorokin 1978). Because of their small size, free-living bacteria are thought to escape capture by most filter feeders in the reef. However, suspended bacteria attached to detritus particles or to each other as aggregates are effectively consumed (Ducklow 1990). Sorokin (1978) showed that 9–23% of the bacterial activity in Majuro Atoll reef was in a form of aggregates. Experiments using ^{14}C-labeled bacteria as food source demonstrated that filter feeders including corals can feed and digest bacterioplankton, and that the consumption is facilitated by the presence of 20–30% of the total bacterioplankton in aggregates larger than 5 μm. In these experiments, massive and solitary corals, *Porites compressa* and *Fungia scutaria*, were less efficient in bacterial feeding than the branching corals *Pocillopora damicornis* and *Montipora verrucosa* (Sorokin 1973a).

Bacteria and detritus found attached to particulate matter in reef sediment are also believed to play a role as a source of food (Hansen et al. 1992; Rosenfeld et al. 1999). By using fluorescently labeled organic material in the sediment, Rosenfeld et al. (1999) demonstrated the transfer of organic matter from the sediment into the cells of the solitary coral *Fungia horrida*. They also suggested that this phenomenon may be valid for other corals with common morphological features such as low slenderness ratio and large polyp size.

11.3 Endolithic Community and Nitrogen Fixation in the Coral Skeleton

Scleractinian coral skeletons are porous structures inhabited by a variety of organisms, mainly microorganisms (e.g., bacteria, fungi and algae). Endolithic microorganisms within the coral skeleton are a potential source of nutrients to the overlying coral tissue (Le Campion-Alsumard et al. 1995; Fine and Loya 2002). This endolithic community may satisfy 55–65% of the nitrogen required by the coral (Ferrer and Szmant 1998) Recent studies using ^{14}C-carbon dioxide have shown that endolithic algae transfer photoassimillates to the coral tissue, and that this is particularly significant during bleaching (Schlichter et al. 1995; Fine and Loya 2002).

It is well known that nitrogen fixation is an important feature of the marine food web and is associated with bacterial activity (Capone 2001). Williams et al. (1987) detected nitrogen fixation occurring in a living coral. Shashar et al. (1994) measured nitrogen fixation by acetylene reduction, in coral skeletons, and in bacteria isolated from the skeleton of the massive coral *Favia favus*. Indeed, these bacteria were found to contain a *nifH* gene known for its role in nitrogen fixation (Shashar et al. 1994). Thus, some of the bacterial populations associated with corals may play an additional nutritive role as a means of nitrogen fixation in a nitrogen-limited environment (Shashar et al. 1994). It is possible that in the coral–algal system the symbiotic zooxanthellae provide the high amount of carbon substrate supplying ATP and reducing power required for nitrogen fixation by the associated bacteria. It is still unclear if and how the fixed nitrogen is transferred from the bacteria to the coral or algae.

11.4 Coral Surface and the Mucus Microlayer

The coral surface microlayer is highly productive, mucus-rich and extends a few millimeters above the surface tissue of the coral (Paul et al. 1986). In this layer, physical and chemical changes are rapid and dynamic. For example, the mucus layer is supersaturated with oxygen during the day when zooxanthellae carry out photosynthesis. In the dark, coral respiration depletes the oxygen levels to near anaerobic conditions (Kuhl et al. 1995). These changes must affect the diversity of microorganisms in this layer. Indeed, we have observed that most of the bacteria isolated from coral mucus are oxidase-positive. Specialized mucus cells present in the polyp epidermis secrete mucus. Intact mucus contains polymers that form a highly hydrated viscoelastic polymeric gel, consisting of fucose, arabinose, mannose, galactose and small amounts of glucose residues (Meikle et al. 1988). Since arabinose is not a common constituent of animal cells, its presence in coral mucus samples suggests the involvement of zooxanthellae in contributing material to the mucus. The differences in the ratio of protein to carbohydrate, glucose content, and amino acid composition of mucus from different corals demonstrate that there is no universal chemical composition to coral mucus (Meikle et al. 1988). Lubbock (1979) suggested that the differences in mucus composition may be species-specific and may be one of the factors used by sea anemones and corals in allorecognition.

One of the functions of coral mucus is to clean their surface from sediment (Krupp 1985). Some corals also use mucus as a mechanism for capturing prey and for feeding (Ducklow and Mitchell 1979b; Schlichter and Brendelberger 1998). Lewis and Price (1975) found that most of the coral species they examined used mucus to entrap food. In addition, coral mucus is believed to constitute a major nutrient source in the reef (Ducklow 1990). It has been reported by Ducklow and Mitchell (1979a) that coral mucus sustains high bacterial growth that presumably results from the degradation of mucus constituents. They hypothesized that mucus degradation products or the bacteria themselves can be used as nutrients by other organisms.

Previous research showed that bacterial lawns cover the tissue surface of corals (Ducklow and Mitchell 1979a). Sorokin (1978) indicated that bacterial growth on coral mucus enhanced their aggregation. Palincsar et al. (1989) reported the presence of bacterial aggregates 10–100 μm in diameter on the coral surface and in the epidermis of the sea anemone *Aiptasia pallida*. These aggregates were often observed proximately to the oral disc and tentacles. Following treatment with antibiotics, the aggregate population decreased (Palincsar et al. 1989). Unfortunately, the role of these aggregates was not discerned in that study and no further work was published.

11.5 Interaction Between Corals and Bacteria

During the last two decades, most of the research in coral microbiology has been restricted to studying bacteria that are associated with coral pathogenesis (e.g., Mitchell and Chet 1975; Edmundes 1991; Ritchie et al. 1994; Ritchie and Smith 1995; Kushmaro et al. 1996, 1997, 2001; Santavy and Peters 1997; Harvell et al. 1999; Patterson et al. 2002; Frias-Lopez et al. 2002; Rosenberg and Ben-Haim 2002). In addition, the general contribution of bacteria to the biomass and energy budget of the coral reef has been the subject of a number of studies (Sorokin 1973a, b; Schiller and Herndl 1989). These studies have shown that there is a dynamic microbial biota living on the surface and in the tissue of many coral species (Ritchie and Smith 1997; Kushmaro et al. 1999; Rohwer et al. 2001), though their role in the reef ecosystem remains, for the large part, a mystery.

Studies on the ecological role of coral mucus and skeleton bacterial composition have been limited by the fact that these microorganisms were poorly defined (Ducklow and Mitchell 1979a; Rublee et al. 1980; Segel and Duklow 1982; Wahbeh and Mahasneh 1988). In general, the viable counts, determined by colony-forming units (CFU), indicate that mucus from a variety of hermatypic corals contain large numbers of bacteria. The viable counts of bacteria from coral mucus from a variety of corals (Table 11.1) showed a large increase in number compared to the surrounding seawater viable counts. This represents only a very small portion of the bacteria found in the marine environment since only approximately 1% of the total bacteria form colonies on standard agar medium. Recent research using molecular tools, e.g., PCR and denaturing gradient gel electrophoresis (DGGE), has begun to explore the uncultivable bacteria (Rohwer et al. 2001, 2002), though the specific functions and associations of these bacteria are still poorly understood.

Many lines of evidence show that less than 1% of the microorganisms in a particular niche are readily cultured using current techniques (Amann et al. 1995). Molecular microbial ecology allowed the development of tools to address a central dogma of microbial ecology: an inability to cultivate more than

Table 11.1. Viable counts of bacteria in the coral mucus microlayer

Eilat coral	Mucus (cfu/ml)
Fungia granulosa	3.6×10^7
Platygyra sp.	3.0×10^5
Favia favus	1.3×10^7
Stylophora pistillata	1.4×10^7
Acropora sp.	7.0×10^6
Seawater	2×10^3

Viable counts were performed by diluting mucus samples, plating them on 50% marine agar (Difco) plates. Colony number was determined after 4 days following incubation at 25 °C.

a small proportion of the bacteria that can be visualized by direct count procedures. The culture-independent approach based on direct recovery of bacterial 16S rDNA indicated the predominance of many different uncultured species. It is feasible that the yet uncultured types of bacteria might be grown under laboratory conditions if just the right nutrients are applied.

A new method for isolating uncultivable microorganisms using a diffusion chamber has been recently developed (Kaeberlein et al. 2002). In this method, microorganisms were separated from sediment particles, serially diluted, mixed with warm agar made with seawater, and placed in the diffusion chamber. Two membranes allow the exchange of chemicals between the chamber and the environment, but restrict movement of cells. After the first membrane was affixed to the base of the chamber, the agar with microorganisms was poured in, and the top was sealed with another membrane (Kaeberlein et al. 2002). Recently, we have developed a novel encapsulation method for isolating and culturing "uncultivable" microorganisms in agar spheres of varied sizes. In this method, an environmental sample (e.g., from coral mucus or seawater) is collected and the bacterial number is estimated by DAPI staining. Dilution of the sample is carried out in order to entrap approximately one bacterium in one agar sphere. The agar sphere is coated with a polymeric membrane that allows exchange of molecules between the sphere and the environment, but restricts movement of cells. After incubating the coated spheres on the coral surface, the spheres are cut and placed between a glass slide and cover slip, and the entire volume of agar scanned for microbial colonies at magnifications of 400× and 1000×. This method allows the isolation of the microcolonies into pure culture by successive re-inoculation of individual colonies into new spheres followed by incubation on the coral. Spheres with pure culture clones are then further analyzed using molecular biology, biochemical tests and subculturing.

11.6 Aggregates as a Possible Nutrition Source

Recently, we have observed rounded bacterial-like aggregates in the mucus and on the surface of a variety of scleractinian corals (Fig. 11.1). Preliminary in situ

Fig. 11.1A–B. Two examples of corals covered by the aggregates. A *Fungia granulosa* and B *Favia favus*; the white area covering the coral surface is mucus packed with bacterial-like aggregates

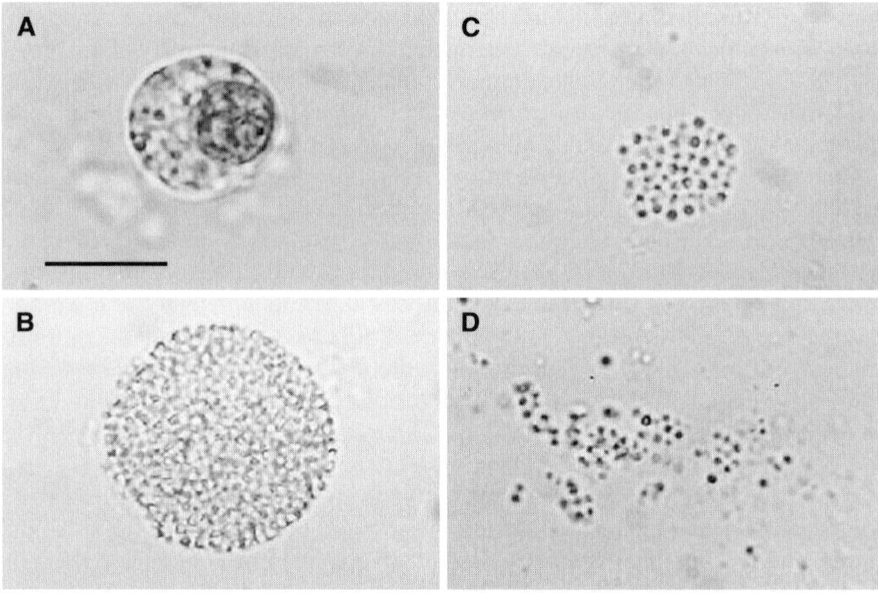

Fig. 11.2. Photomicrographs showing bacterial-like aggregate (**A**), zooxanthella (**B**) bacterial-like aggregate following mechanical dissociation (**C, D**). Bacterial-like aggregate, 20 µm in diameter; zooxanthella, 10 µm in diameter. Single coccoid-like bodies approximately 1 µm

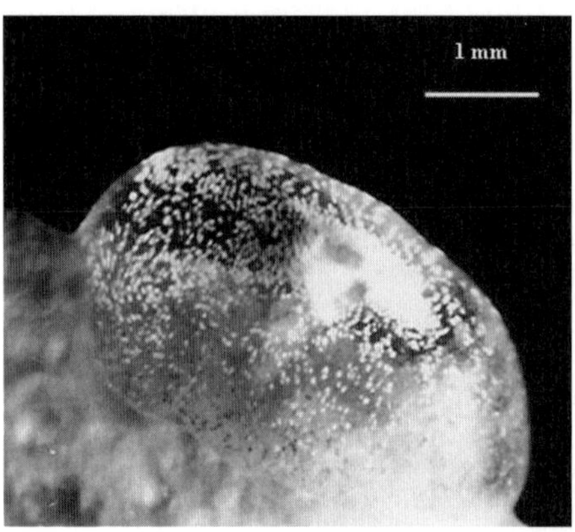

Fig. 11.3. Primary polyp developing from tissue fragment of *Fungia granulosa* (enlargement 40×). Aggregates are dispersed on the body surface as fluorescing bodies. Note the high density surrounding the developing mouth

belt transects (Loya 1978) in the Gulf of Eilat showed that approximately 32% of shallow water corals are infected with dense populations of these aggregates. Coral families with a high incidence of aggregate coverage include fungiids and faviids. The aggregates are round bodies of approximately 20 µm in diameter (Fig. 11.2) and are composed of unique coccoid bacterial-like bodies of approximately 1 µm (Fig. 11.2.). They are dispersed on the coral surface in a patchy distribution, with the highest density in the area of the polyp mouth (Fig. 11.3), suggesting that they may be transported toward it. These preliminary observations suggest the possibility that the corals may utilize these aggregates as food. This is supported by observations of Sorokin (1973a, 1978) in which, (1) bacterial growth on coral mucus enhanced their aggregation and (2) 20–30% of bacterioplankton suspension feeding was in the form of bacterial aggregates.

11.7 Conclusion

Microorganisms may contribute to the total coral energy budget through a number of pathways summarized in Fig. 11.4 (adapted from Rohwer et al. 2002). Until recently, the symbiotic associations between the coral and their zooxanthellae constituted the major source of carbon to corals in oligotrophic environments (Fallowski et al. 1984; Sebens 1994). Recently, it has been shown that additional carbon sources and nutrients (e.g., nitrogen and phosphorus) may also be translocated in forms of photoassimilates and fixed nitrogen by the coral's endolithic community (Shashar et al. 1994; Fine and Loya 2002). Suspension feeding and detritus are additional sources of nutrition to the corals (Sorokin 1973a, b; Rosenfeld et al. 1999). Studies have shown that the mucous provided by the coral acts as an enriched surface for microbial growth. This microbial community may also be consumed by the coral. These communities may be specific for certain corals and environmental conditions. It is also possible that these communities provide the coral with protection from pathogens by interspecific competition and secretion of antibiotic substances (Rohwer et al. 2002).

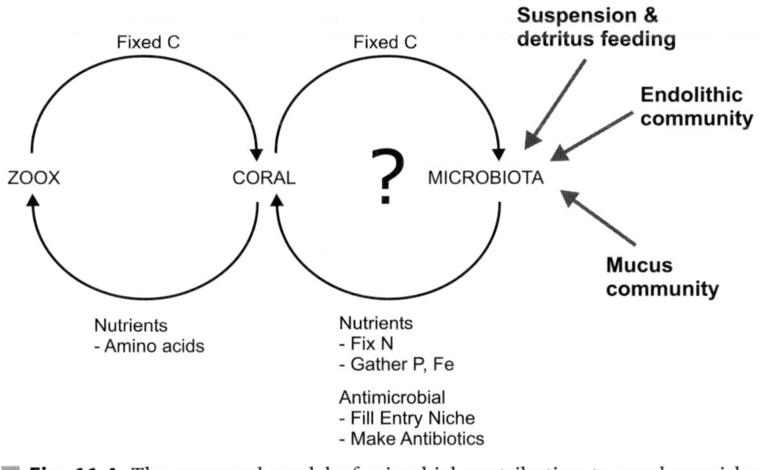

Fig. 11.4. The proposed model of microbial contribution to coral nourishment (adapted from Rowher et al. 2002). This includes contribution by symbiotic algae, endolithic community, suspension and detritus feeding, and mucus microbial community

11.8 Unanswered Questions and Further Research Directions

The limited number of studies related to coral–bacteria associations and their contribution to coral nourishment undermines our ability to fully understand coral microbial ecology. Additional research efforts must be directed to furthering our knowledge and understanding of the coral–bacteria association and their contribution to coral nutrition. More specifically, what is the contribution of bacterial consumption to the energy budget of corals? What are the mechanisms of consumption of bacteria from the different sources (i.e., sediment, suspension and mucus)? How much of a coral's energy budget depends on bacterivory? One of the questions that arises is, do the corals utilize bacteria singly or in aggregates? Understanding the mechanism of bacterial aggregation and ascertaining if the coral has a role in this process will be a step toward understanding the relationship between the coral and the aggregates. Another important question is, do the bacteria have a role in nitrogen fixation and if so, what are the mechanisms of nitrogen transport in the system?

One of the major issues to be considered is, do corals use alternate energy sources when they undergo bleaching? Is it possible that recovery of some of the corals following bleaching events may be due to bacterivory? Thus, is it possible that bacteria play a role as alternative energy sources during stress conditions such as bleaching events? It is important to understand how the coral–mucus environment influences the bacterial community and how in return this bacterial community influences the mucus layer composition. Do ingredients in the coral mucus layer promote growth of beneficial bacteria to the coral? To understand this it is of utmost importance to examine the composition of the mucus secreted by corals. It is likely that the mucus layer is a result of a combination of compounds secreted by the coral together with compounds secreted by or transformed by the bacterial community. If so, are there differences in composition between species, at different locations and seasons?

Acknowledgments. This work was supported by ISF grant no. 511/02-1.

References

Abelson A, Miloh T, Loya Y (1993) Flow patterns induced by substrata and body morphologies of benthic organisms, and their roles in determining availability of food particles. Limnol Oceanogr 38:1116–1124

Amann RI, Ludwig W, Schleifer K-H (1995) Phylogenetic identification and in situ detection of individual microbial cells without cultivation. Microbiol Rev 59:143–169

Anthony KRN (1999) Coral suspension feeding on fine particulate matter. J Exp Mar Biol Ecol 232:85–106

Anthony KRN (2000) Enhanced particle-feeding capacity of corals on turbid reefs (Great Barrier Reef, Australia). Coral Reefs 19:59–67

Anthony KRN, Fabricius EF (2000) Shifting roles of heterotrophy and autotrophy in coral energetics under varying turbidity. J Exp Mar Biol Ecol 252:221–253

Bak RPM, Joenje M, de Jong I, Lambrechts DYM, Nieuwland G (1998) Bacterial suspension feeding by coral reef benthic organisms. Mar Ecol Prog Ser 175:285–288

Ben-Haim Y, Rosenberg E (2002) A novel *Vibrio* sp. pathogen of the coral *Pocillopora damicornis.* Mar Biol 141:47–55

Capone DG (2001) Marine nitrogen fixation: what's the fuss? Curr Opin Microbiol 4(3):341–348

Ducklow WH (1990) The biomass, production and fate of bacteria in coral reefs. In: Dubinsky Z (ed) Coral reefs. Elsevier, Amsterdam, pp 265–289

Ducklow WH, Mitchell R (1979a) Bacterial populations and adaptations in the mucus layers on living corals. Limnol Oceanogr 24(4):715–725

Ducklow WH, Mitchell R (1979b) Composition of mucus released by coral reef coelenterates. Limnol Oceanogr 24:706–714

Edmundes JP (1991) Extent and effect of black band disease. Coral Reefs 10:161–165

Fallowski PG, Dubinsky Z, Muscatine L, Porter JW (1984) Light and the bioenergetics of symbiotic coral. Bioscience 34(11):705–709

Ferrer LM, Szmant AM (1988) Nutrient regeneration by the endolithic community in coral skeletons. Proceeding of the 6th International Coral Reef Symposium, Australia, 3, pp 1–4

Fine M, Loya Y (2002) Endolithic algae – an alternative source of energy during coral bleaching. Proc R Soc Lond B 269:1205–1210

Frias-Lopez J, Zerkle LA, Bonheyo T, Fouke WB (2002) Partitioning of bacterial communities between seawater and healthy, black-band diseased, and dead coral surfaces. Appl Environ Microbiol 68:2214–2228

Goldberg MW (2002) Feeding behavior, epidermal structure and mucus cytochemistry of the scleractinian *Mycetophyllia reesi,* a coral without tentacles. Tissue Cell 34(4):232–245

Hansen JA, Klumpp DW, Alongi DM, Dayton PK, Riddle MJ (1992) Detrital pathways in a coral reef lagoon II. Detritus deposition, benthic microbial biomass and production. Mar Biol 113:363–372

Harvell CD, Kim K, Burkholder JM, Colwell RR, Epstein PR, Grimes DJ, Hofmann EE, Lipp EK, Ostrerhaus A, Overstreet RM, Porter JW, Smith GW, Vasta GR (1999) Review: marine ecology – emerging marine diseases – climate links and anthropogenic factors. Science 285(5433): 1505–1510

Kaeberlein T, Lewis K, Epstein SS (2002) Isolating "uncultivable" microorganisms in pure culture in a simulated natural environment. Science 296:1127–1129

Krupp AD (1985) An immunochemical study of the mucus from the solitary coral *Fongia scutaria.* Bull Mar Sci 36(1):163–176

Kuhl M, Cohen Y, Dalsgaard T, Jorgensen BB, Revsbech PN (1995) Microenvironment and photosynthesis of zooxanthellae in scleractinian corals studied with microsensors for O_2, pH and light. Mar Ecol Prog Ser 117:159–172

Kushmaro A, Loya Y, Fine M, Rosenberg E (1996) Bacterial infection and coral bleaching. Nature 380:396

Kushmaro A, Rosenberg E, Fine M, Loya Y (1997) Bleaching of the coral *Oculina patagonica* by *Vibrio* AK-1. Mar Ecol Prog Ser 147:159–165

Kushmaro A, Rosenberg E, Loya Y (1999) Species-specific properties of the coral mucus microbial communities. 7th International Conference of the Israel Society for Ecology And Environmental Quality Sciences on Environmental Challenges for the Next Millennium

Kushmaro A, Banin A, Loya Y, Stackebrandt E, Rosenberg E (2001) *Vibrio shiloi* sp. nov, the causative agent of bleaching of the coral *Oculina patagonica.* Int J Syst Evol Microbiol 51:1383–1388

Le Campion-Alsumard T, Golubic S, Hutchings P (1995) Microbial endoliths in skeletons of live and dead corals: *Porites lobata* (Moorea, French Polynesia). Mar Ecol Prog Ser 117:149–157

Lewis JB, Price WS (1975) Feeding mechanisms and feeding strategies of Atlantic reef corals. J Zool Lond 176:527–544

Loya Y (1978) Plotless and transect methods. In: Stoddart DR, Johannes RE (eds) Monographs on oceanic methodology. Coral reefs: research methods, vol 5. UNESCO Press, Paris, pp 197–218
Lubbock R (1979) Mucus antigenicity in sea anemones and corals. Hydrobiologia 66:3–6
Meikle P, Richards NG, Yellowlees D (1988) Structural investigations on the mucus from six species of coral. Mar Biol 99:187–193
Mitchell R, Chet I (1975) Bacterial attack of corals in polluted seawater. Microb Ecol 2:227–233
Palincsar EE, Warren RJ, Palincsar SJ, Glogowski AM, Mastro LJ (1989) Bacterial aggregates within the epidermis of the sea anemone *Aiptasia pallida*. Biol Bull 177(30):140
Patterson KL, Porter JW, Ritchie KE et al. (2002) The etiology of white pox, a lethal disease of the Caribbean elkhorn coral, *Acropora palmata*. Proc Natl Acad Sci USA 99(13):8725–8730
Paul JH, DeFlaun MF, Jeffrey WH (1986) Elevated levels of microbial activity in the coral surface microlayer. Mar Ecol Prog Ser 33:29-40
Ritchie KB, Smith WG (1995) Preferential carbon utilization by surface bacterial communities from water mass, normal, and white-band diseased *Acropora cervicornis*. Mol Mar Biol Biotechnol 4(4):345–352
Ritchie KB, Smith WG (1997) Physiological comparison of bacterial communities from various species of scleractinian corals. Proc 8th Int Coral Reef Symp 1:521–526
Ritchie KB, Dennis JH, McGrath T, Smith GW (1994) Bacteria associated with bleached and nonbleached areas of *Montastrea annularis*. In: Kass L (ed) Bahamian field station. Proc 5th Symp Nat Hist Bahamas, San Salvador, Bahamas, pp 75–79
Rohwer F, Breitbart M, Jara J, Azam A, Knowlton N (2001) Diversity of bacteria associated with the Caribbean coral *Montastraea franksi*. Coral Reefs 20:85–91
Rohwer F, Seguritan V, Azam A, Knowlton N (2002) Diversity and distribution of coral-associated bacteria. Mar Ecol Prog Ser 243:1–10
Rosenfeld M, Bresler V, Abelson A (1999) Sediment as a possible source of food for corals. Ecol Lett 2(6):345–348
Rublee AP, Lasker RH, Gottfried M, Roman RM (1980) Production and bacterial colonization of mucus from the soft coral *Briarium asbestinum*. Bull Mar Sci 30(4):888–893
Santavy DL, Peters EC (1997) Microbial pests: coral disease in the western Atlantic. Proc 8th Int Coral Reef Symp 1:607–612
Schiller C, Herndl JG (1989) Evidence of enhanced microbial activity in the interstitial space of branched corals: possible implication for coral metabolism. Coral Reefs 7:179–184
Schlichter D, Brendelberger H (1998) Plasticity of the scleractinian body plan: functional morphology and trophic specialization of *Mycedium elephantotus* (Pallas, 1766). FACIES 39:227–242
Schlichter D, Zscharnack B, Krisch H (1995) Transfer of photoassimilates from endolithic algae to coral tissue. Naturwissenschaften 82(12):561–564
Sebens PK (1994) Biodiversity of coral reefs: what are we losing and why? Am Zool 34:115–133
Sebens PK (1997) Adaptive responses to water flow: morphology, energetics, and distribution of reef corals. Proc 8th Int Coral Reef Symp 2:1053–1058
Sebens PK, Grace PS, Helmuth B, Maney JE Jr, Miles SJ (1998) Water flow and prey capture by three scleractinian corals, *Madracis mirabilis*, *Montastrea cavernosa* and *Poritas porites*, in a field enclosure. Mar Biol 131:347–360
Segel AL, Ducklow WH (1982) A theoretical investigation into the influence of sublethal stresses on coral-bacterial ecosystem dynamics. Bull Mar Sci 32(4):919–935
Shashar N, Cohen Y, Loya Y, Sar N (1994) Nitrogen fixation (acetylene reduction) in stony corals: evidence for coral bacteria interactions. Mar Ecol Prog Ser 11(3):259–264
Sorokin YI (1973a) Trophical role of bacteria in the ecosystem of coral reef. Nature 242:415–417
Sorokin YI (1973b) On the feeding of some scleractinian corals with bacteria and dissolved organic matter. Limnol Oceanogr 18:380–385
Sorokin YI (1978) Microbial production in the coral-reef community. Arch Hydrobiol 83:281–323
Sorokin YI (1990) Aspects of trophic relation, productivity and energy balance in coral-reef ecosystem. In: Dubinsky Z (ed) Coral reefs. Elsevier, Amsterdam, pp 401–410

Wahbeh IM, Mahasneh MA (1988) Composition and bacterial utilization of mucus of corals from Aqaba (Red Sea), Jordan. Proceeding of the 6th International Coral Reef Symposium, Australia, 3, pp 53–57

Williams WM, Viner AB, Broughton WL (1987) Nitrogen fixation (acetylene reduction) associated with the living coral *Acropora variabilis*. Mar Biol 94:531–535

12 Antimicrobial Activity of Sponges and Corals

Dovi Kelman

12.1 Introduction

Bacteria and other microorganisms are ubiquitous in the marine environment. They are taxonomically diverse, biologically active and colonize all marine habitats, from the deep oceans to the shallowest estuaries (Austin 1988; Rheinheimer 1992), as well as coral reefs (Ducklow 1990). Living benthic marine organisms such as sponges and corals are frequently colonized by bacteria. Sponges are known to harbor a diverse range of microorganisms (Bergquist 1978; Wilkinson 1978). In some cases, up to 50% of the sponge weight was attributed to their symbiotic bacteria (Wilkinson 1978). The surface of living corals is covered by mucus (Ducklow and Mitchell 1979a). This mucus layer is colonized by bacteria, allowing the establishment of a bacterial community that can be characteristic to a particular coral species (Mitchell and Chet 1975; Ducklow and Mitchell 1979b; Rublee et al. 1980; Segal and Ducklow 1982; Ritchie et al. 1994; Rohwer et al. 2002). Some of these bacteria can be pathogenic to sponges and corals, and may initiate disease, such as tissue necrosis in sponges (Webster et al. 2002), and several diseases in corals, such as black band disease (Antonius 1985; Carlton and Richardson 1995), white plague type II (Smith et al. 1996; Richardson et al. 1998), tissue necrosis (Hodgson 1990; Ben-Haim and Rosenberg 2002), and even bleaching of the Mediterranean scleractinian coral *Oculina patagonica* (Kushmaro et al. 1996, 1997). On the other hand, bacteria could serve as beneficial symbionts or as benign associates. For example, Gil-Turnes et al. (1989) showed that bacteria on the surface of externally held eggs of the shrimp *Palaeman macrodactylus* produce a metabolite that inhibits fungal infections that are lethal to the eggs. Therefore, sponges and corals need the ability to regulate the bacteria they encounter and to resist microbial colonization and the invasion of potential pathogens in order to prevent possible detrimental effects. One method of combating microbial attack is by chemical defense.

Sponges and corals offer a rich source of unique and diverse secondary metabolites (for a recent review, see Faulkner 2002). Many of these compounds have potent pharmacological activities, including anti-tumor, anti-fungal, anti-viral, and anti-bacterial properties that have generated great pharmaceutical interest (Wallace 1997). However, natural products were also found to play important biological and ecological roles to the producing organisms, such as defense against predators, competition for space, prevention of fouling, roles in reproduction and antimicrobial activity (see reviews by Sammarco and Coll 1988, 1992; Coll 1992; Paul 1992; Pawlik 1993; Hay 1996). Antimicrobial activity has been extensively reported for extracts of various groups of marine organisms, such as sponges (Burkholder and Ruetzler 1969; Bergquist and Bedford 1978; Amade et al. 1982, 1987; McCaffrey and Endean 1985; Becerro et al. 1994; Newbold et al. 1999; Kelman et al. 2001), bryozoans (Walls et al. 1993), ascidians (Wahl et al. 1994), scleractinian corals (Koh 1997), gorgonian octocorals (Burkholder and Burkholder 1958; Kim 1994; Jensen et al. 1996), and alcyonacean soft corals (Slattery et al. 1995; Kelman et al. 1998). Most of the above-mentioned studies examined the activity of secondary metabolites against biomedically relevant microbes, such as human pathogens or marine bacteria that have no ecological relevance to the studied organism. A few of the reports have looked at the activity of secondary metabolites against ecologically relevant bacteria in order to elucidate their function in the mediation of marine sessile invertebrate interactions with bacteria (Becerro et al. 1994; Wahl et al. 1994; Slattery et al. 1995; Jensen et al. 1996; Kelman et al. 1998, 2001; Newbold et al. 1999). In these studies, the antimicrobial activity assays were conducted on bacteria isolated from the natural environment of the organism being studied.

This chapter will review several examples of antimicrobial activity of sponges and corals as well as antibiotics that were subsequently isolated from these organisms. This review will also include our recent research on antimicrobial activity and antibiotics of Red Sea sponges and corals. This will be followed by discussions on some unresolved questions and future research in the field.

12.2 Antimicrobial Activity of Sponges

The first discovery that some sponges possess antimicrobial activity was made by Nigrelli et al. (1959). They observed the antibiotic effects of some subtropical sponges by placing fresh sponge fragments on Petri dishes seeded with a test bacterium. Later, they found that extracts of freshly collected sponges exhibited a relatively high degree of antimicrobial activity (Nigrelli et al. 1959; Jakowska and Nigrelli 1960). Burkholder and Ruetzler (1969) tested the antimicrobial activity of several sponges from different geographical areas, and found that similar levels of activity were present in extracts from sponges sampled from the Caribbean and the Great Barrier Reef (Australia). They also found that in both oceans, the sponge extracts inhibited Gram-positive bacteria more than Gram-negative bacteria or yeast. At that time, it was assumed that antimicrobial activity was significantly higher in the tropics than in temperate regions. Bergquist and Bedford (1978) contradicted this assumption when they showed a higher percentage of activity in temperate sponges from New Zealand than reported for tropical sponges. They also found that sponge extracts inhibited the growth of marine bacteria more than non marine bacteria and that Gram-negative bacteria were more susceptible to inhibition by sponge extracts than were Gram-positives. Subsequently, many research groups reported the occurrence of antimicrobial activity of extracts from marine sponges from various geographical areas. Amade et al. (1982) found antimicrobial activity in several marine sponges from French Polynesia and the Mediterranean coast of Brittany. McCaffrey and Endean (1985) showed that extracts of 19 out of 24 species of Australian tropical and subtropical sponges possess potent antimicrobial activities, especially against Gram-positive bacteria. Amade et al. (1987) tested the antimicrobial activity of 28 species of marine sponges from the French Mediterranean coast and found that antibacterial activity against Gram-positive bacteria is more frequent than against Gram-negative bacteria. Green et al. (1990) found antimicrobial activity in marine sponges from the Gulf of Mexico, and McClintock and Gauthier (1992) discovered antimicrobial activity in Antarctic marine sponges. Antimicrobial activity was also found to be widespread in extracts of marine sponges from the coast of Brazil (Muricy et al. 1993; Monks et al. 2002).

Several different antibiotics were isolated from marine sponges. For example, the lipid-soluble antibiotic, plakortin (Fig. 12.1a) was isolated from the Carib-

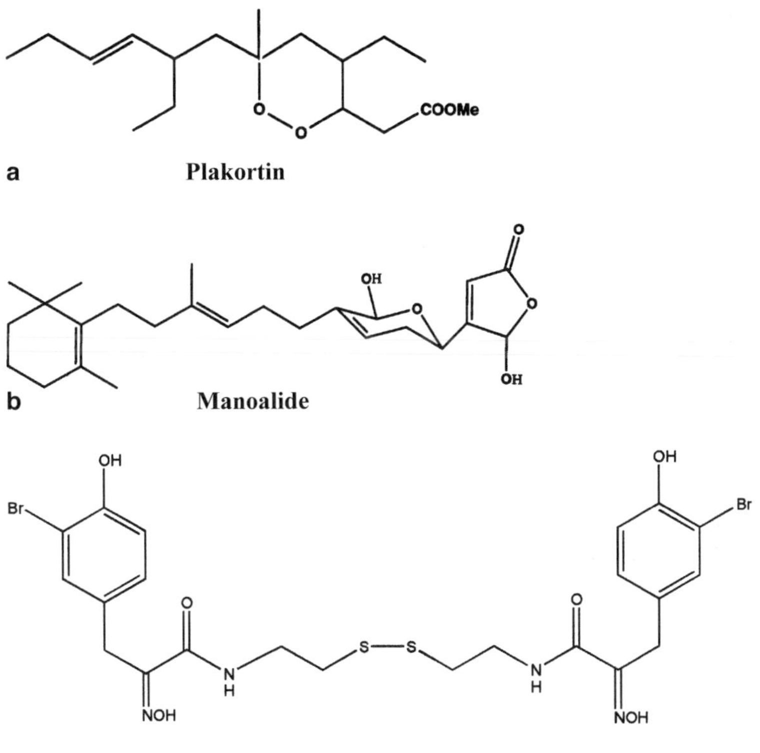

Fig. 12.1a–c. Antibiotics isolated from marine sponges. **a** Plakortin, **b** manoalide, and **c** psammaplin A

bean sponge *Plakortis halichondriodes* (Higgs and Faulkner 1978). The sesterterpenoid antibiotic manoalide (Fig. 12.1b) was isolated from the marine sponge *Luffariella variabilis* from Palau (de Silva and Scheuer 1980). Kim et al. (1999) isolated the bromotyrosine derivative psammaplin A (Fig. 12.1c) from an association of two Korean sponges, *Poecillastra* sp. and *Jaspis* sp. They found that this compound possesses significant antimicrobial activity on Gram-positive bacteria, especially against the methicillin-resistant *Staphylococcus aureus*.

These antibiotics and others isolated from marine sponges, as well as the above-mentioned studies on antimicrobial activity of marine sponges, were tested against human pathogens or against marine bacteria that have no obvious ecological relevance to the producing organism. Several recent reports on antimicrobial activity of marine sponges have tested the extracts and the subsequent purified antibiotics against ecologically relevant bacteria isolated from the natural environment of the sponge being studied. This approach of using ecologically relevant bacteria gives a better understanding of the potential chemical mediation of interactions between marine sponges and bacteria. Becerro et al. (1994) isolated several bacterial strains from the vicinity of the sponges they investigated. They found that the Mediterranean encrusting sponge, *Crambe crambe*, featured the strongest antimicrobial activity against these bacteria, while the other two sponges they investigated from the same location, *Ircinia fasciculata* and *Spongia officinalis*, exhibited lower antimicrobial activity. They also showed that Gram-positive and Gram-negative bacteria were equally affected by the sponge extracts. Newbold et al. (1999) performed a comprehensive investigation of the antimicrobial activity of extracts from 33 species of Caribbean sponges against a panel of bacterial strains that included isolates from the seawater near the sponges, as well as from the tissues of healthy and necrotic sponges. They also tested the activity of the extracts at concentrations that were volumetrically equivalent to those found in the sponge tissues (whole-tissue concentrations). Their results revealed that 48% (16 species) of the sponges investigated exhibited antimicrobial activity. Among them, the sponges *Amphimedon compressa*, *A. erina*, *Aplysina lacunosa*, and *Ptilocaulis spiculifera* exhibited the strongest antimicrobial activity.

In our investigation of the antimicrobial activity of Red Sea sponges (Kelman et al. 2001), similar approaches were used in testing the sponge extracts and subsequent antibiotics against ecologically relevant bacteria and with concentrations volumetrically equivalent to those found in the sponge tissues. Antimicrobial assays were performed with organic extracts of 11 dominant Red Sea reef sponges against a panel of Red Sea marine bacteria isolated from the natural environment of the sponges investigated. The results revealed considerable variability in antimicrobial activity (Table 12.1). Eight out of 11 (73%) sponge species inhibited at least one bacterial isolate. Among them, *Amphimedon viridis* exhibited the highest antimicrobial activity. Observation of surfaces of *A. viridis* by scanning electron microscopy (SEM) failed to reveal any bacteria (Fig. 12.2), suggesting that sponges possessing antimicrobial activity

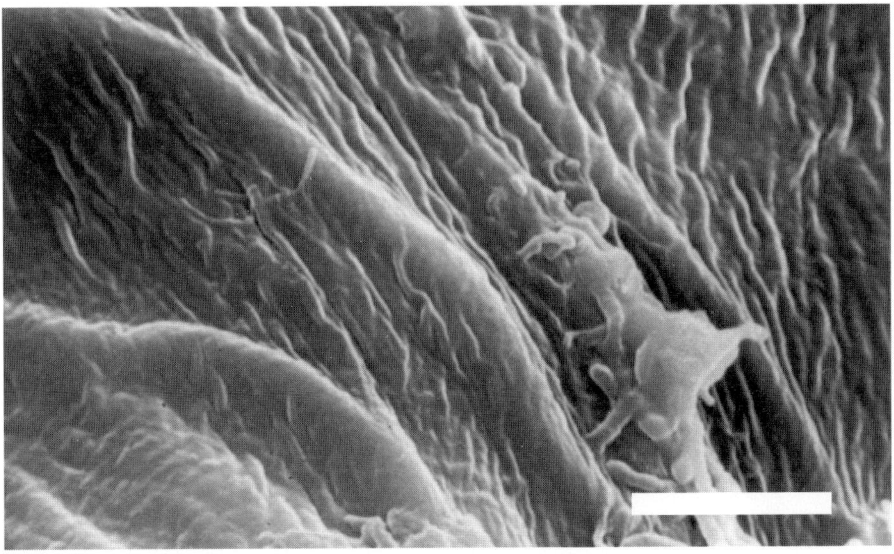

Fig. 12.2. A SEM micrograph of a typical surface of the Red Sea sponge *Amphimedon viridis* demonstrating a surface virtually free of epibiotic bacteria. *Scale bar* 5 µm

Table 12.1. Antimicrobial activity of extracts of 11 dominant Red Sea reef sponges, applied at natural concentrations, against various Red Sea marine bacteria (for more details on these strains see Kelman et al. 2001)

Sponge	Test microorganism							
	RSW-2	RSW-3[a]	RSW-14	ST-1	P-1	D-1	S-15	S-13
Amphimedon viridis	+++	++++	+	++++	+++	+++	–	+
Callyspongia siphonella	+	–	–	+	–	–	–	–
Callyspongia sp.	+	nd	–	–	–	–	–	–
Petrosia sp.	–	nd	–	–	–	–	–	–
Epipolasis sp.	++	++	–	+	–	–	–	–
Suberites clavatus	+	–	–	++	–	–	++	–
Theonella swinhoei	+	nd	–	–	–	–	–	–
Grayella cyatophora	–	–	–	–	–	–	–	–
Biemna erhenbergi	+	–	–	–	–	–	–	–
Diacarnus erythraenus	–	++	–	+++	–	–	–	–
Negombata magnifica	–	–	–	–	–	–	–	–

Size of inhibition zones is expressed as: + 0–2 mm, ++ 2–4 mm, +++ 4–6 mm, ++++ 6–8 mm, – no inhibition, nd not determined.
[a]Gram-positive, all the remaining strains are Gram-negative.

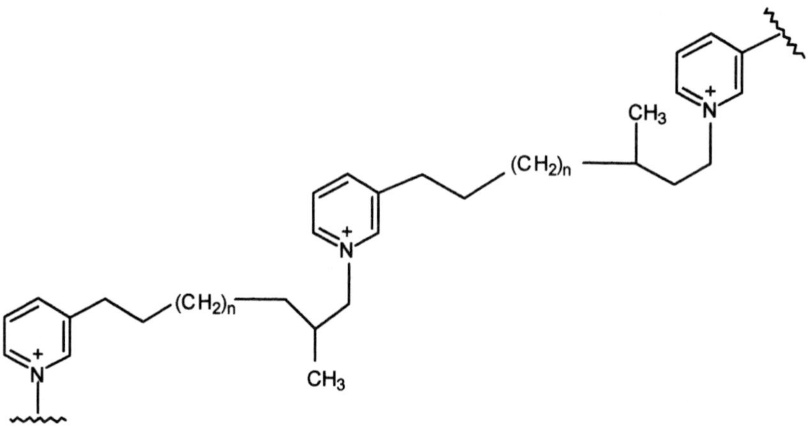

(1) n = 2,3,4,5

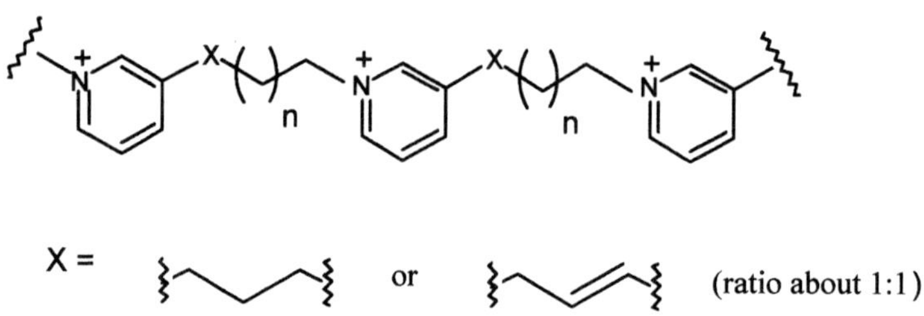

(2) n =5

Fig. 12.3. The chemical structure of halitoxin (1) and amphitoxin (2), the antimicrobial compounds present in the tissues of *Amphimedon viridis*

maintain a surface relatively clean of fouling microorganisms. A similar observation was made by Becerro et al. (1994) with Mediterranean sponges.

Bioassay-directed fractionation of the crude extract of *Amphimedon viridis* resulted in the isolation of an active fraction containing an inseparable mixture of two closely related toxin homologues. These metabolites were identified as a mixture of halitoxin and amphitoxin (Fig. 12.3), which are complex pyridinium alkaloids, by NMR (nuclear magnetic resonance) spectroscopy and comparison of the spectral data reported in the literature (Schmitz et al. 1978; Albrizio et al. 1995; Berlinck et al. 1996). Amphitoxin differs from halitoxin in having an additional carbon–carbon double bond in the alkyl chain linking the pyridinium rings. The estimated volumetric concentration of the amphitoxin/halitoxin mixture in tissues of *A. viridis* was ca. 6.2 mg/ml (assuming 100% recovery).

In previous studies, it was shown that halitoxin exhibits general cytotoxicity to various cell lines, and was toxic to mice and sea urchin eggs, as well as possessing hemolytic and neurotoxic activity (Berlinck et al. 1996). Amphitoxin deterred the feeding of a predatory fish in laboratory experiments (Albrizio et al. 1995). In the present study, however, these highly bioactive pyridinium alkaloids showed selective activity to specific bacteria rather than being of a broad spectrum. The amphitoxin/halitoxin mixture isolated from *A. viridis* was highly active against eight strains of bacteria isolated from the seawater surrounding these sponges, whereas six strains associated with the sponge were resistant to these compounds. This pattern of selectivity can be seen in Table 12.2, where the minimum inhibitory concentrations (MICs) are presented for the purified mixture of halitoxin and amphitoxin, crude extracts of *A. viridis*, as well as selected commercial antibiotics against a number of seawater bacteria and sponge-associated bacteria isolated from *A. viridis*. It can be seen that five of the six *A. viridis*-associated bacteria were not inhibited by 250 μg/ml of purified amphitoxin/halitoxin mixture and crude extracts of *A. viridis*. Strain AV-2 was inhibited at an MIC of 125 μg/ml of the purified amphitoxin/halitoxin mixture and 63 μg/ml of crude extracts of *A. viridis*. In contrast, the purified amphitoxin/halitoxin mixture and crude extracts of *A. viridis* inhibited the growth of eight out of eight seawater bacteria in MICs ranging from 4 to 250 μg/ml. The eight seawater bacteria consisted of equal numbers of Gram-negative and Gram-positive strains. This se-

Table 12.2. Minimum inhibitory concentrations (μg/ml) of amphitoxin/halitoxin mixture, crude extract of *Amphimedon viridis* and selected commercial antibiotics against growth of bacterial isolates from *A. viridis* and surrounding seawater (for more details on these strains see Kelman et al. 2001)

Source	Isolate	Crude *A. viridis*	Amphitoxin/ halitoxin	Poly-B	Pen-G	Chl	Nal	Strep	Tet
A. viridis	AV-1	>250	>250	>250	1	2	16	31	2
	AV-2	63	125	>250	63	2	16	31	8
	AV-3	>250	>250	>250	1	2	16	63	2
	AV-4	>250	>250	>250	1	1	31	31	2
	AV-5	>250	>250	>250	>250	2	16	31	8
	AV-6	>250	>250	>250	1	2	16	31	4
Seawater	RSW-2	31	31	1	1	4	>125	8	2
	RSW-3	31	31	250	1	8	>125	16	8
	RSW-16	31	31	1	1	31	63	16	8
	RSW-13	31	31	8	1	4	>125	16	63
	RSW-17	125	250	2	1	4	1	16	8
	RSW-18	250	250	4	63	4	16	16	8
	RSW-14	125	8	4	4	8	nd	63	nd
	RSW-1	31	4	125	8	31	nd	63	nd

Poly-B Polymixin B; *Pen-G* penicillin-G; *Chl* chloramphenicol; *Nal* nalidixic acid; *Strep* streptomycin; *Tet* tetracycline; *nd* not determined.

lective toxicity may be important in enabling certain bacteria to live in close association with their sponge host while it maintains a chemical defense against microbial pathogenesis.

The selected commercial antibiotics exhibited variable MICs against *Amphimedon viridis*-associated and seawater bacteria (Table 12.2). It is interesting to note that, as a group, the six strains of *A. viridis*-associated bacteria were resistant to the amphitoxin/halitoxin mixture, as well as to polymixin B. The latter is a bactericidal cyclic peptide antibiotic, which acts by binding to the bacterial cytoplasmic membrane. The positively charged peptide ring is thought to bind electrostatically with the anionic phosphate head groups of the membrane, thus affecting the normal organization of the membrane and altering its permeability characteristics (Franklin and Snow 1981). The fact that amphitoxin/halitoxin are also positively charged and all isolated *A. viridis*-associated bacteria were resistant to them as well as to polymixin B, led us to suggest that amphitoxin/halitoxin mimics the mode of action of this class of antibiotics. Furthermore, it also led us to hypothesize that the resistant bacteria probably have special membrane properties that interfere with the antibiotic action. Further investigation is still required in order to determine the mode of action of amphitoxin/halitoxin and the mode of resistance of *A. viridis*-associated bacteria to these highly potent antibiotics.

12.3 Antimicrobial Activity of Corals

The first study on antimicrobial activity of corals was made by Burkholder and Burkholder (1958) on gorgonian octocorals (horny corals) from the Caribbean. They tested small pieces of corals, or various extracts from different species of corals, that were placed on agar plates, which had been seeded with a test bacterium. They tested the activity against various marine bacteria that were isolated from the same region where the corals were collected, as well as a panel of human pathogenic microorganisms. They found that several gorgonian species caused a zone of inhibition around the coral fragments and around paper discs containing coral extracts. The corals that were found to possess antibacterial activity were *Antillogorgia turgida*, *A. americana*, *Rhipidogorgia flabellum*, *Briareum asbestinum*, *Plexaura homomalla*, *Plexaurella dichotoma*, and *Plexauropsis crassa*. In contrast to the results with gorgonian corals, little or no antimicrobial activity could be detected in the various species of stony corals that were tested. Since this report, a number of screening programs for bioactivity in various marine invertebrate taxa demonstrated antimicrobial activity in gorgonian corals (Burkholder 1973; Rinehart et al. 1981; Ciereszko and Guillard 1989). Kim (1994) tested the antimicrobial activity of eight gorgonian octocorals from San Blas Islands, Panama, against a panel of five bacteria from marine and nonmarine sources. Extracts from all eight coral species demonstrated moderate levels of antimicrobial activity. Nonpolar fractions were found to be much more active than polar fractions. A more comprehensive study, testing the antimicrobial activity of gorgonians as a taxonomic group, was performed by Jensen et al. (1996) for Caribbean gorgonian corals. The antimicrobial activity of extracts of 39 species of gorgonians against 15 strains of marine bacteria was tested. The test bacteria consisted of ecologically relevant bacteria that included marine opportunistic pathogens and strains isolated from healthy and decayed gorgonians. Their data indicated that organic extracts of most Caribbean gorgonians do not possess potent, broad-spectrum antibacterial activity against co-occurring and potentially pathogenic marine bacteria. They concluded that the inhibition of bacterial growth is not the primary ecological function of gorgonian natural products. However, several species of gorgonian corals did possess the ability to inhibit the growth of bacteria they encounter.

As opposed to the wealth of evidence of antimicrobial activity in gorgonian corals, a limited number of studies have shown antimicrobial activity in other groups of corals. Slattery et al. (1995) found antimicrobial activity in two species of Antarctic soft corals against marine bacteria isolated from the same habitat. Koh (1997) performed a comprehensive antimicrobial screening of extracts from 100 Australian scleractinian corals (representing 44 genera and 13 families) against a panel of 7 microbes. These microbes consisted of six marine bacteria isolated from Australian waters as well as one human pathogen. All of the coral species inhibited the growth of the marine cyanobacterium *Synechococcus* sp., while only eight of these coral species inhibited the growth of marine heterotrophic bacteria.

Antimicrobial compounds have also been isolated from corals, although to a lesser extent than marine sponges. One such example is the diterpenes, sinulariolide and flexibilide (Fig. 12.4), isolated from the Australian soft coral *Sinularia flexibilis* (Aceret et al. 1998).

The first example of antimicrobial activity in Red Sea corals was shown for the encrusting soft coral *Parerythropodium fulvum fulvum* (Kelman et al. 1998). This species reproduces by a unique mode of reproduction, which involves the surface brooding of embryos in a layer of mucus (Benayahu and Loya 1983). In

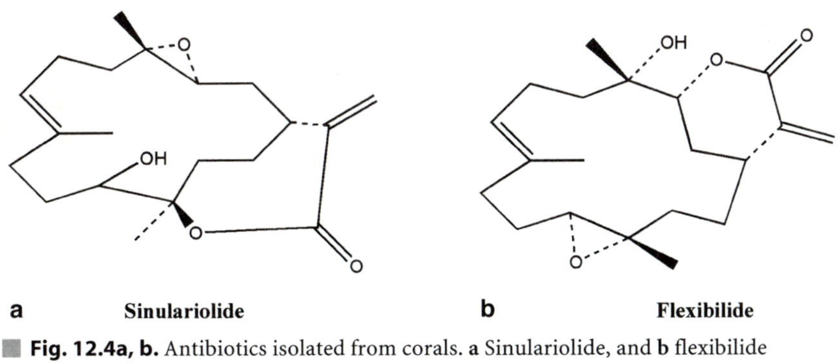

Fig. 12.4a, b. Antibiotics isolated from corals. **a** Sinulariolide, and **b** flexibilide

this study, it was interesting to investigate the antimicrobial activity of extracts from various reproductive and developmental stages of this coral. Organic extracts were prepared from colonies with no gonads, male and female colonies with gonads, 1–2 day-old embryos with mucus, 3–6 day old embryos with mucus, and 3–6 day old embryos without mucus. These extracts, which represent different reproductive and developmental stages of *Parerythropodium fulvum fulvum*, were tested for antimicrobial activity against a panel of 12 bacteria isolated from the environment of the coral. The results showed appreciable, but variable antimicrobial activity in all reproductive and developmental stages against several of the co-occurring and potentially pathogenic marine bacteria. High activity was found particularly against a *Vibrio* sp., isolated from necrotic coral tissue. However, no antimicrobial activity was observed against coral-associated bacteria isolated from healthy coral tissues and their mucoid surface. It was concluded that antimicrobial activity detected in the extracts of *P. f. fulvum* is specific rather than broad-spectrum. This specificity may be important in order to enable certain bacteria to live in close association with their coral host, while maintaining a chemical defense against microbial pathogenesis.

A more comprehensive study was performed to compare the antimicrobial activity of extracts of several of the most dominant stony (scleractinian) and soft (alcyonacean) coral species from the coral reef of Eilat (northern Red Sea) against bacteria isolated from the waters surrounding the corals. The results revealed considerable variability in antimicrobial activity. Extracts of five out of six (83%) of the Red Sea alcyonacean soft corals tested exhibited appreciable antimicrobial activity against the test bacteria (Table 12.3), while none of the extracts of the six stony corals tested inhibited the test bacteria (Table 12.4). The results suggested that soft and hard corals have developed different means to combat potential microbial infections. Alcyonacean soft corals use chemical defense through the production of antibiotic compounds to combat microbial attack, whereas stony corals seem to rely on other means.

Among the active soft coral species examined, *Xenia macrospiculata* exhibited the highest and most potent antimicrobial activity. Bioassay-directed fractionation indicated that the antimicrobial activity was due to the presence of a range of compounds of different polarities. One of these antibiotic compounds was isolated and identified as desoxyhavannahine (Fig. 12.5), by NMR spectroscopy and comparison with the spectral data reported in the literature (Almourabit et al. 1989; König et al. 1989). The estimated volumetric concentration of desoxyhavannahine in tissues of *Xenia macrospiculata* was ca. 590 μg/ml (assuming 100% recovery). The MIC of purified desoxyhavannahine was 48 μg/ml against a marine bacterium, which is about tenfold lower than its estimated natural concentration, while the MIC of the crude extract of *X. macrospiculata* was 25 μg/ml. This may suggest that the extract of this coral contains additional antimicrobial compounds. Further work is still required in order to determine the nature of these compounds, and to show whether these metabolites act in an additive or a synergistic fashion towards potentially harmful bacteria.

Table 12.3 Antimicrobial activities of extracts of Red Sea soft corals against various Red Sea marine bacteria

Coral species	Test microorganisms										
	Seawater strains			Coral isolates							
	1	2	3	1	2	3	4	5	6	7	8
Litophyton arboreum	+	++	++	++	+++	+++	+	++++	++++	+++	+
Parerythropodium f. fulvum	++	+++	++	+++	+++	++	++	+++	+++++	+++	++
Heteroxenia fuscesence	+	+	–	++	++	+	–	+++	++++	++	+
Sarcophyton glaucum	–	++	+	+	+	+	+	++	+++	+	+
Dendronephtia hemprichi	–	++	–	+	–	–	–	+	++	–	–
Xenia macrospiculata	+++	++++	++	+++++	+++++	+++++	++	+++++	+++++	+++++	+++++

Size of inhibition zones is expressed as: + 0–2 mm, ++ 2–4 mm, +++ 4–6 mm, ++++ 6–8 mm, +++++ 8–12 mm, – no inhibition.

Table 12.4 Antimicrobial activities of extracts of Red Sea stony corals against various Red Sea marine bacteria

Coral species	Test microorganisms										
	Seawater strains			Coral isolates							
	1	2	3	1	2	3	4	5	6	7	8
Stylophora pistillata	–	+	–	–	–	–	–	–	–	–	–
Favia favus	–	+	–	–	+	–	–	–	–	++	–
Pocillopora damicornis	–	–	–	–	–	–	–	–	–	–	–
Fungia fungites	–	–	+	–	–	–	–	–	–	–	–
Fungia scutaria	–	+	–	–	–	–	–	–	–	–	–
Acropora variabillis	–	+	+	–	–	–	–	–	–	–	–

Size of inhibition zones is expressed as: + 0–2 mm, ++ 2–4 mm, +++ 4–6 mm, ++++ 6–8 mm, +++++ 8–12 mm, – no inhibition.

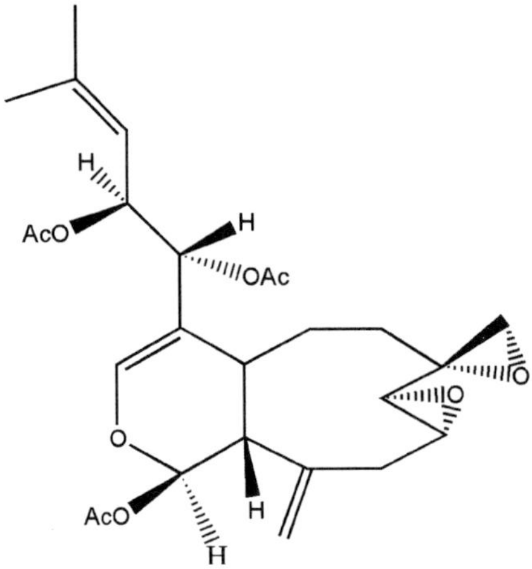

■ **Fig. 12.5.** Chemical structure of desoxyhavannahine, the antimicrobial compound present in the tissues of *Xenia macrospiculata*

12.4 Some Unresolved Questions and Future Research

12.4.1 Nontoxic Antimicrobial Activity

The use of the antibiotic disc susceptibility test can rapidly identify active metabolites and, therefore, is especially useful in the initial screening for antimicrobial activity and as a means to follow activity during chemical purification (Jenkins et al. 1998). However, since this assay measures toxicity (cell death or inhibition of cell growth), the absence of antimicrobial activity in laboratory assays does not necessarily indicate a lack of antimicrobial chemical defense. Chemicals produced by higher organisms against co-occurring microorganisms may not simply kill or inhibit the growth of the target microorganism, but can act selectively against particular phenotypes or characteristics that are expressed by the bacteria. Maximilien et al. (1998) showed the effects of halogenated furanones from the Australian red alga *Delisea pulchra* on the colonization phenotypes of co-occurring marine bacteria. These specific phenotypes represent different stages of the bacterial fouling process that includes swimming, chemotaxis, attachment, growth, swarming, or even interference with bacterial signaling systems such as acylated homoserine lactone (AHL) regulatory system as was shown by Givskov et al. (1996) for the macroalga *D. pulchra*. Therefore, investigations to determine if sponges or corals can defend themselves against microbes by targeting functions essential for pathogenesis, but not survival in culture (i.e., nontoxic) will greatly extend our knowledge of the extent by which these organisms are chemically defended against microorganisms.

12.4.2 Localization of Antimicrobials

The simulation of natural concentrations of extracts on a volumetric basis for the testing of antimicrobials assumes that the extracts are equally distributed throughout the volume of the organism being tested. In corals, algae and in some sponges it is reasonable to assume that higher concentrations of bioactive compounds accumulate on their surfaces, thus providing greater defense against bacterial colonization. However, since sponges pass large volumes of water through their tissues (Bergquist 1978), which contain potentially harmful bacteria, as well as encountering them on their external surfaces, it may be advantageous for sponges to distribute their antimicrobials throughout their tissues. Several studies provided evidence for the localization of natural products within the tissues of sponges (e.g., Bewley et al. 1996; Uriz et al. 1996). In the marine alga, *Delisea pulchra*, de Nys et al. (1998) determined the naturally occurring surface concentrations of the antimicrobial furanone. Further investigations on the localization of antibiotics in tissues of marine sponges and corals are therefore warranted.

12.4.3 Role of Associated Bacteria in the Production of Antimicrobials

Certain symbiotic marine bacteria were shown to be responsible for the production of secondary metabolites that were previously thought to be derived from their host (e. g., Elyakov et al. 1991; Mikki et al. 1996). Studies that will assign the production of antimicrobials in corals or sponges to associated bacteria will be highly interesting and will provide us with a unique perspective on coral/sponge-microbe interactions.

References

Aceret TL, Coll JC, Ychio Y, Sammarco PW (1998) Antimicrobial activity of the diterpenes flexibilide and sinulariolide derived from *Sinularia flexibilis* Quoy and Gaimard 1833 (Coelenterata: Alcyonacea, Octocorallia). Comp Biochem Physiol C 120:121–126

Albrizio S, Ciminiello P, Fattorusso E, Magno S, Pawlik JR (1995) Amphitoxin, a new high molecular weight antifeedant pyridinium salt from the Caribbean sponge *Amphimedon compressa.* J Nat Prod 58:647–652

Almourabit A, Gillet B, Ahond A, Beloeil JC, Poupat C, Potier P (1989) Invertébrés marins du lagon Néo-Calédonien, XI. Les desoxyhavannahines, nouveaux métabolites de *Xenia membranacea.* J Nat Prod 52:1080–1087

Amade P, Pesando D, Chevolot L (1982) Antimicrobial activities of marine sponges from French Polynesia and Brittany. Mar Biol 70:223–228

Amade P, Charroin C, Baby C, Vacelet J (1987) Antimicrobial activities of marine sponges from the Mediterranean Sea. Mar Biol 94:271–275

Antonius A (1985) Black band disease infection experiments on hexacorals and octocorals. Proc 5th Int Coral Reef Symp, Tahiti 6:155–160

Austin B (1988) Marine microbiology. Cambridge Univ Press, New York

Becerro MA, Lopez NI, Turon X, Uriz MJ (1994) Antimicrobial activity and surface bacterial film in marine sponges. J Exp Mar Biol Ecol 179:195–205
Benayahu Y, Loya Y (1983) Surface brooding in the Red Sea soft coral *Parerythropodium fulvum fulvum* (Forskål, 1775). Biol Bull 165:353–369
Ben-Haim Y, Rosenberg E (2002) A novel *Vibrio* sp. pathogen of the coral *Pocillopora damicornis*. Mar Biol 141:47–55
Bergquist PR (1978) Sponges. Univ California Press, Berkeley
Bergquist PR, Bedford JJ (1978) The incidence of antibacterial activity in marine Demospongiae; systematic and geographic considerations. Mar Biol 46:216–221
Berlinck RGS, Ogawa CA, Almeida AMP, Sanchez MAA, Malpezzi ELA, Costa LV, Hajdu E, de Freitas JC (1996) Chemical and pharmacological characterization of halitoxin from *Amphimedon viridis* (Porifera) from the southeastern Brazilian coast. Comp Biochem Physiol C 115: 155–163
Bewley CA, Holland ND, Faulkner DJ (1996) Two classes of metabolites from *Theonella swinhoei* are localized in distinct populations of bacterial symbionts. Experientia 52:716–722
Burkholder PR (1973) The ecology of marine antibiotics and coral reefs. In: Jones OA, Endean R (eds) Biology and geology of coral reefs, vol II. Biology. Academic Press, New York, pp 117–176
Burkholder PR, Burkholder LM (1958) Antimicrobial activity of horny corals. Science 127: 1174–1175
Burkholder PR, Ruetzler K (1969) Antimicrobial activity of some marine sponges. Nature 222: 983–984
Carlton RG, Richardson LL (1995) Oxygen and sulfide dynamics in a horizontally migrating cyanobacterial mat: black band disease of corals. FEMS Microbiol Ecol 18:155–162
Ciereszko LS, Guillard RRL (1989) The incidence of some cembranolides from gorgonian corals on motility of marine flagellates. J Exp Mar Biol Ecol 127:205–210
Coll JC (1992) The chemistry and chemical ecology of octocorals (Coelenterata, Anthozoa, Octocorallia). Chem Rev 92(4):613–631
De Nys R, Dworjanyn SA, Steinberg PD (1998) A new method for determining surface concentrations of marine natural products on seaweeds. Mar Ecol Prog Ser 162:79–87
De Silva ED, Scheuer PJ (1980) Manoalide, an antibiotic sesterterpenoid from the marine sponge *Luffariella variavilis* (Polejaeff). Tetrahedron Lett 21:1611–1614
Ducklow HW (1990) The biomass, production and fate of bacteria in coral reefs. In: Dubinsky Z (ed) Ecosystems of the world: coral reefs. Elsevier, New York, pp 265–289
Ducklow HW, Mitchell R (1979a) Composition of mucus released by coral reef coelenterates. Limnol Oceanogr 24:706–714
Ducklow HW, Mitchell R (1979b) Bacterial populations and adaptations in the mucus layers on living corals. Limnol Oceanogr 24:715–725
Elyakov GB, Kuznetsova T, Mikhailov VV, Maltsev II, Voinov VG, Fedoreyev SA (1991) Brominated diphenyl ethers from a marine bacterium associated with the sponge *Dysidea* sp. Experientia (Basel) 47:632–633
Faulkner DJ (2002) Marine natural products. Nat Prod Rep 19:1–48
Franklin TJ, Snow GA (1981) Biochemistry of antimicrobial action. Chapman and Hall, London
Gil-Turnes MS, Hay ME, Fenical W (1989) Symbiotic marine bacteria chemically defend crustacean embryos from a pathogenic fungus. Science 246:116–118
Givskov M, de Nys R, Manefield M, Gram L, Maximilien R, Eberl L, Molin S, Steinberg PD, Kjelleberg S (1996) Eukaryotic interference with homoserine lactone-mediated prokaryotic signalling. J Bacteriol 178:6618–6622
Green G, Gomez P, Bakus GJ (1990) Antimicrobial and ichthyotoxic properties of marine sponges from Mexican Waters. In: Rutzler K (ed) New perspectives in sponge biology. Smithsonian Institution Press, Washington, DC, pp 109–114
Hay ME (1996) Marine chemical ecology: what's known and what's next? J Exp Mar Biol Ecol 200:103–134
Higgs MD, Faulkner DJ (1978) Plakortin, an antibiotic from *Plakortis halichondroides*. J Org Chem 43:3454–3457

Hodgson G (1990) Tetracycline reduces sedimentation damage to corals. Mar Biol 104:493–496
Jakowska S, Nigrelli RF (1960) Antimicrobial substances from sponges. Ann NY Acad Sci 90:913–916
Jenkins KM, Jensen PR, Fenical W (1998) Bioassays with marine microorganisms. In: Haynes KF, Millar JG (eds) Methods in chemical ecology, vol 2. Bioassay methods. Chapman and Hall, New York, pp 1–38
Jensen PR, Harvell CD, Wirtz K, Fenical W (1996) Antimicrobial activity of extracts of Caribbean gorgonian corals. Mar Biol 125:411–419
Kelman D, Kushmaro A, Loya Y, Kashman Y, Benayahu Y (1998) Antimicrobial activity of a Red Sea soft coral, *Parerythropodium fulvum fulvum*: reproductive and developmental considerations. Mar Ecol Prog Ser 169:87–95
Kelman D, Kashman Y, Rosenberg E, Ilan M, Ifrach I, Loya Y (2001) Antimicrobial activity of the reef sponge *Amphimedon viridis* from the Red Sea: evidence for selective toxicity. Aquat Microb Ecol 24:9–16
Kim D, Lee IS, Jung JH, Yang SI (1999) Psammaplin A, a natural bromotyrosine derivative from a sponge, possesses the antibacterial activity against methicillin-resistant *Staphylococcus aureus* and the DNA gyrase-inhibitory activity. Arch Pharm Res 22(1):25–29
Kim K (1994) Antimicrobial activity in gorgonian corals (Coelenterata, Octocorallia). Coral Reefs 13:75–80
Koh EGL (1997) Do scleractinian corals engage in chemical warfare against microbes? J Chem Ecol 23(2):379–398
König GM, Coll JC, Bowden BF, Gulbis JM, MacKay MF, La Barre SC, Laurent D (1989) The structure determination of xenicane diterpene from *Xenia garciae*. J Nat Prod 52:294–299
Kushmaro A, Loya Y, Fine M, Rosenberg E (1996) Bacterial infection and coral bleaching. Nature 380:396
Kushmaro A, Rosenberg E, Fine M, Loya Y (1997) Bleaching of the coral *Oculina patagonica* by *Vibrio* AK-1. Mar Ecol Prog Ser 147:159–165
Maximilien R, de Nys R, Holmström C, Gram L, Givskov M, Crass K, Kjelleberg S, Steinberg PD (1998) Chemical mediation of bacterial surface colonisation by secondary metabolites from the red alga *Delisea pulchra*. Aquat Microb Ecol 15:233–246
McCaffrey EJ, Endean R (1985) Antimicrobial activity of tropical and subtropical sponges. Mar Biol 89:1–8
McClintock JB, Gauthier JJ (1992) Antimicrobial activities of Antarctic sponges. Antarct Sci 4(2):179–183
Mikki W, Otaki N, Yokoyama A, Kusumi T (1996) Possible origin of zeaxanthin in the marine sponge, *Reniera japonica*. Experientia (Basel) 52:93–96
Mitchell R, Chet I (1975) Bacterial attack of corals in polluted seawater. Micro Ecol 2:227–233
Monks NR, Lerner C, Henriques AT, Farias FM, Schapoval EES, Suyenaga ES, da Rocha AB, Schwartsmann G, Mothes B (2002) Anticancer, antichemotactic and antimicrobial activities of marine sponges collected off the coast of Santa Catarina, southern Brazil. J Exp Mar Biol Ecol 281:1–12
Muricy G, Hajdu E, Araujo FV, Hagler AN (1993) Antimicrobial activity of southwestern Atlantic shallow-water marine sponges (Porifera). Sci Mar 57(4):427–432
Newbold RW, Jensen PR, Fenical W, Pawlik JR (1999) Antimicrobial activity of Caribbean sponge extracts. Aquat Microb Ecol 19:279–284
Nigrelli RF, Jakowska S, Calventi I (1959) Ectyonin, an antimicrobial agent from the sponge *Microciona prolifera* Verril. Zoologica (NY) 44:173–175
Paul VJ (1992) Ecological roles of marine natural products. Cornell Univ Press, New York
Pawlik JR (1993) Marine invertebrate chemical defenses. Chem Rev 93:1911–1922
Rheinheimer G (1992) Aquatic microbiology, 4th edn. Wiley, New York
Richardson LL, Goldberg WM, Kuta KG (1998) Florida's mystery coral-killer identified. Nature 392:557–558
Rinehart KL et al. (25 co-authors) (1981) Marine natural products as sources of antiviral, antimicrobial, and antineoplastic agents. Pure Appl Chem 53:795–817

Ritchie KB, Smith GW, Gerace DT (1994) Grouping of bacterial heterotrophs from scleractinian corals using metabolic potentials. Proc 26th Meet Assoc Mar Lab Caribbean. San Salvador, Bahamas, Bahamian field stations, pp 224–236

Rohwer F, Seguritan V, Azam F, Knowlton N (2002) Diversity and distribution of coral-associated bacteria. Mar Ecol Prog Ser 243:1–10

Rublee AP, Lasker RH, Gottfriend M, Roman RM (1980) Production and bacterial colonization of mucus from the soft coral *Briarium asbestinum*. Bull Mar Sci 30(4):888–893

Sammarco PW, Coll JC (1988) The chemical ecology of alcyonarian corals (Coelenterata: Octocorallia). In: Scheuer PJ (ed) Bioorganic marine chemistry. vol 2. Springer, Berlin Heidelberg New York, pp 87–116

Sammarco PW, Coll JC (1992) Chemical adaptations in the Octocorallia: evolutionary considerations. Mar Ecol Prog Ser 88:93–104

Schmitz FJ, Hollenbeak KH, Campbell DC (1978) Marine natural products: Halitoxin, toxic complex of several marine sponges of the genus Haliclona. J Org Chem 43:3916–3922

Segel AL, Ducklow WH (1982) A theoretical investigation into the influence of sublethal stresses on coral-bacterial ecosystem dynamics. Bull Mar Sci 32(4):919–935

Slattery M, McClintock JB, Heine JN (1995) Chemical defenses in Antarctic soft corals: evidence for antifouling compounds. J Exp Mar Biol Ecol 190:61–77

Smith GW, Ives LD, Nagelkerken IA, Ritchie KB (1996) Caribbean sea-fan mortalities. Nature 383:487

Uriz MJ, Turon X, Galera J, Tur JM (1996) New light on the cell location of avarol within the sponge *Dysidea avara* (Dendroceratida). Cell Tissue Res 285:519–527

Wahl M, Jensen PR, Fenical W (1994) Chemical control of bacterial epibiosis on ascidians. Mar Ecol Prog Ser 110:45–57

Wallace RW (1997) Drugs from the sea: harvesting the results of aeons of chemical evolution. Mol Med Today 3:291–295

Walls JT, Ritz DA, Blackman AJ (1993) Fouling, surface bacteria and antibacterial agents of four bryozoan species found in Tasmania, Australia. J Exp Mar Biol Ecol 169:1–13

Webster NS, Negri AP, Webb RI, Hill RT (2002) A spongin-boring alpha-proteobacterium is the etiological agent of disease in the Great Barrier Reef sponge *Rhopaloeides odorabile*. Mar Ecol Prog Ser 232:305–309

Wilkinson CR (1978) Microbial associations in sponges. II. Numerical analysis of sponge and water bacterial populations. Mar Biol 49:169–176

13 Microbial Communities of Coral Surface Mucopolysaccharide Layers

Kim B. Ritchie and Garriet W. Smith

13.1 Introduction

Microbes associated with corals form communities that differ significantly from water mass communities. Physiological and metabolic activities occurring within these close, mutualistic relationships are very different from the surrounding environment and tend to be very specific. Coral-associated microbial communities are likely to resemble microbial mats in structure. Vertical stratification of populations, from proximal toward distal regions (with respect to the surface of the coral), results in fluxes of organic and inorganic nutrients. The presence of an actively metabolizing microbiota has significant influences on overall population structure. Theoretical and observed consequences of environmental changes lead to the disruption of the 'normal' communities associated with healthy corals. Because it is likely that the normal microbiota protects the coral animal from invading microbes, changes in the community structure may result in the development of disease.

13.2 Structure of the Coral Surface Mucopolysaccharide Layer

All corals secrete mucus. Much of the fixed carbon that makes up the surface mucopolysaccharide layer (SML) originates from the zooxanthellae (Patton et al. 1977), but is secreted from epidermal mucus cells. The basic structure of coral mucus is secreted as an insoluble, hydrated, glycoprotein, which forms a viscoelastic gel (Ducklow and Mitchell 1979; Meikle et al. 1988). The chemical composition of the SML from different coral species varies qualitatively and quantitatively (Meikle et al. 1988). In addition, the chemical nature and quantity of mucus can change when corals are stressed (Ritchie and Smith 1995a), thus changing the SML environment. It is not known if the normal differences in the chemical composition of the SML are due to variations in the compliment of zooxanthellae (Rowan et al. 1997) or variations in the metabolism of the coral host. Figure 13.1 illustrates that bacterial communities associated with corals are largely coral species-specific. In addition, Fig. 13.1 shows that these bacterial communities appear to reflect the phylogenetic grouping of the corals at the genus level, where bacterial communities are more similar within genera (data for Fig. 13.1 compiled from Ritchie et al. 1994, Ritchie and Smith 1995a, b and Ritchie and Smith, unpubl. data). Although microbial community

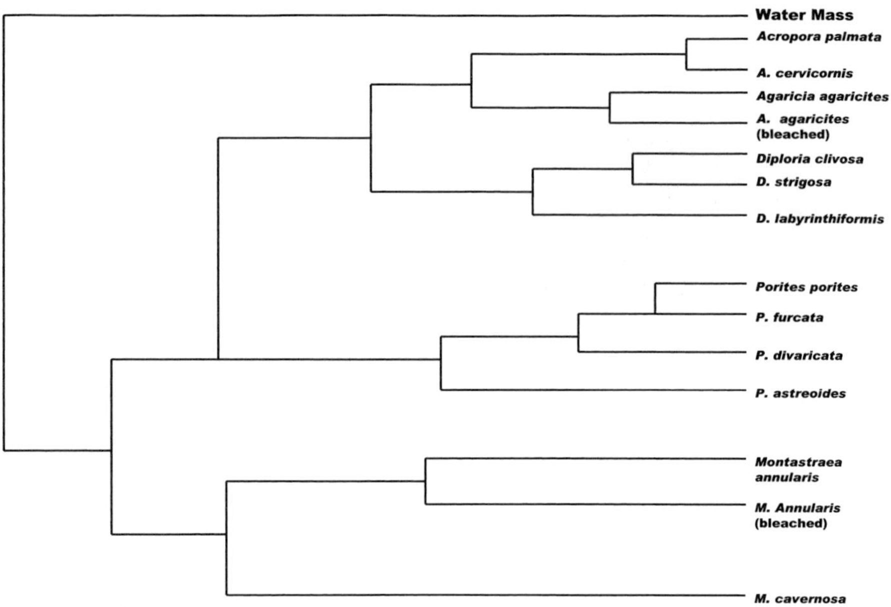

Fig. 13.1. Dendrogram based on metabolic characteristics of microbial communities from a number of different coral species. On average, 230 distinct metabolic groups were isolated from each coral species. When these communities are compared using cluster analysis of carbon source utilization patterns of each bacterium within the community, we see that the bacterial communities are different in different corals. In addition, these communities appear to reflect the phylogenetic grouping of the corals at the genus level (data compiled from Ritchie et al. 1994, Richie and Smith 1995a, b and Ritchie and Smith, unpubl. data)

profiles appear to reflect phylogenetic relationships among coral species, these community profiles are known to change when corals bleach (Ritchie et al. 1994, 1995, 1996a, b; Ritchie and Smith 1995b; Jindal et al. 1996), or become diseased (Ritchie and Smith 1995a).

The most obvious feature of the coral SML is that it is a medium high in organic compounds. This is in contrast to the pelagic bacterioplankton, which exist in a low organic (oligotrophic) environment. Most bacterioplankton are very small (even for bacteria), exist in a state of starvation (which may account for the inability to culture many of them), and are basically dependent on the products of photoautotrophy for metabolism (Morita 1997; Giovannoni and Rappe 2000). Comparatively, culturable SML bacterial populations are 100 times higher than those of the surrounding water mass and are many orders of magnitude more metabolically active (Ritchie et al. 1996). Although both communities consist of primarily organoheterotrophs, the importance of photoautotrophy in the water mass community is much higher than in SML communities. This is true even though cyanobacteria can be detected in the SML using PCR (Rohwer et al. 2001).

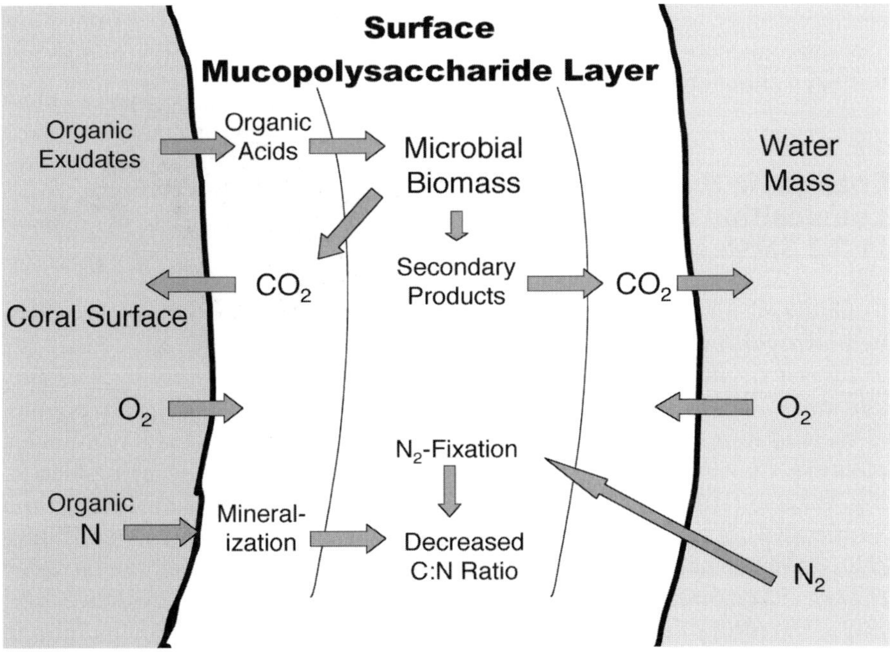

Fig. 13.2. Theoretical stratification of the coral surface mucopolysaccharide layer. *Top* to *bottom* Flux of carbon, oxygen and nitrogen

The physical structure of SML microbial communities has not been studied in detail, but it is likely to be as complex as the seagrass rhizosphere, which is stratified with respect to nitrogen-fixing bacteria from aerobic (root surface) to anaerobic (sediment) environments (Smith and Hayasaka 1982a, b; Smith et al. 1982). The width of the SML can vary from less than 1 mm in some scleractinians, to a few centimeters in some gorgonians. Easily metabolized coral exudates may first be oxidized either completely to CO_2, or incompletely to organic acids (Fig. 13.2). These, in turn, may be broken down toward the center of the SML, which is likely to be microaerophilic or even anaerobic due to increased activity at the surface of the coral coupled with retarded O_2 diffusion rates through the mucus and microbial utilization of O_2 diffusing in from the water mass.

Figure 13.2 indicates probable fluxes through the coral SML. Organic carbon, in the form of structural glycoproteins and carbohydrates, is secreted into the SML from the host coral. More labile compounds, like sugars and amino acids, could be exuded and/or produced in situ as more complex polymers are hydrolyzed by microbial exoenzymes. Organic acids could then be used as electron donors or substrates for fermenting processes, depending on the O_2 concentration in the middle of the SML. Although these conditions have not been shown in the coral SML, they are well known in microbial mats (Fenchel et al. 2000). Nitrogen relationships are equally unknown in the coral SML, but if the matrix becomes microaerophilic to anaerobic, nitrogen fixation becomes a real possibility. Regardless, the microbial population is known to have a low C:N ratio, and the use of the SML bacterial community as a coral food source has been shown (Sorokin 1973).

13.3 Changes in the Coral Surface Mucopolysaccharide Layer Composition Lead to Changes in the Bacterial Population Structure

Ritchie and Smith (1995a, b) have shown that shifts in the SML organo-heterotrophic microbial populations occur when corals are stressed, either due to disease or during bleaching. Often corals recover from bleaching and specific microbial populations also recover. Figure 13.3 shows that *Vibrio* populations tend to increase during bleaching, but return to previous levels during recovery. Conversely, populations of pseudomonads decrease during bleaching, but also return to previous levels during recovery (Fig. 13.3; McGrath and Smith 1999). It is likely that since qualitative changes (and often quantitative changes) have been reported in coral mucus during bleaching, the bacterial change is a response to available carbon sources. Selective carbon source utilization by culturable SML communities was shown to differ among 11 different species of scleractinian corals (Ritchie and Smith 1997).

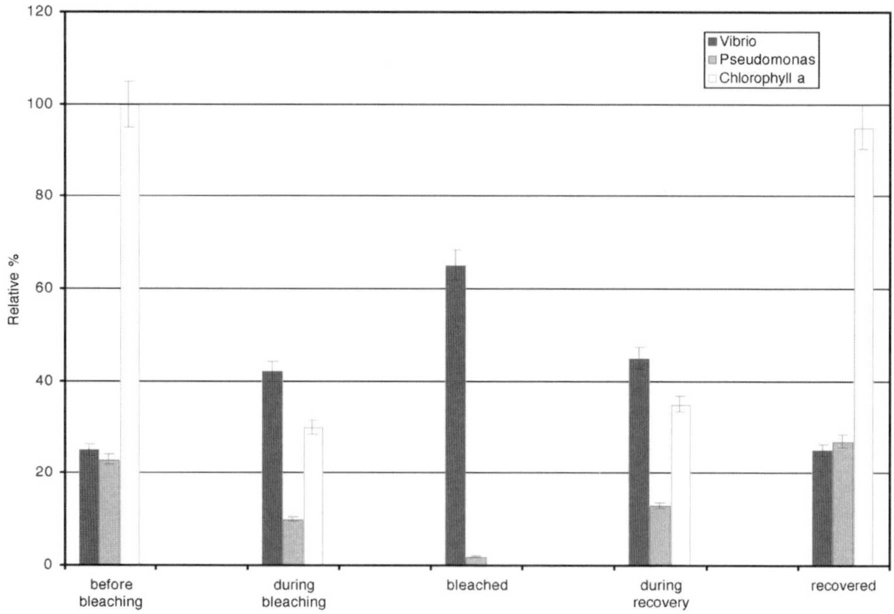

Fig. 13.3. Microbial changes during bleaching and recovery. The percentage of *Vibrio* metabolic groups, *Pseudomonas* and chlorophyll

Although the example in Fig. 13.3 demonstrates population changes in only two metabolic groups, overall microbial community changes would be expected. Such physically close microbial populations typically show a high degree of syntropy or co-metabolism. The organic by-products of one metabolic group are often the organic substrates of another. This is particularly the case in complex communities where biochemical activity can lead to changes in the surroundings (i.e., pH, O_2 tensions).

13.4 Unresolved Questions and Future Research

The establishment of specific microbial communities on larger organisms can be shown to be adaptive mutualism. In nature, selective pressures tend to favor mutually beneficial relationships. This is undoubtedly true for microbial communities in association with corals. These selective pressures led to the type of specific metabolic interactions referred to in Fig. 13.2. To date, these relationships have been inferred based on the composition of culturable heterotrophs. Actual measurements of biogeochemical cycles within the mucus layers should reveal known and unique pathways that will likely be specific for various coral groups. There is a recent increase in information concerning total microbial communities associated with healthy (Rohwer et al. 2001, 2002; Frias-Lopez et al. 2002) and diseased corals (Frias-Lopez et al. 2002). Due to an inability to culture the vast majority of environmental microbes, this information will be invaluable to our understanding of the overall structure of coral microbial communities. However, the importance of culturable heterotrophic bacteria associated with the surface mucus of corals cannot be underestimated, as it is this subset of microbes that will ultimately reveal essential information concerning adaptive mutualistic and nutrient interrelationships.

References

Ducklow HW, Mitchell R (1979) Bacterial populations and adaptations in the mucus layers on living corals. Limnol Oceanogr 24:715–725

Fenchel T, King GM, Backburn TH (2000) Bacterial biogeochemistry: the ecophysiology of mineral cycling. Academic Press, San Diego

Frias-Lopez J, Zerkle AL, Bonheyo GT, Fouke BW (2002) Partitioning of bacterial communities between seawater and healthy, black-band diseases, and dead coral surfaces. Appl Environ Microbiol 68:2214–2228

Giovannoni S, Rappe M (2000) Evolution, diversity, and molecular ecology of marine prokaryotes. In: Kirchman DL (ed) Microbial ecology of the oceans. Wiley-Liss, New York, pp 47–80

Jindal A, Ritchie KB, Hayes RL, Goreau TJ, Smith GW (1996) Bacterial ecology of selected corals following the south central pacific bleaching event. Proc Assoc Mar Labs Caribb 27:26–32

McGrath TA, Smith GW (1999) Community shifts in the surface mucopolysaccharide layer microbiota of *Agaricia* sp. during the 1995/6 and 1998/9 bleaching events on patch reefs of San Salvador Island, Bahamas. 29th meeting of the Association of Marine Laboratories of the Caribbean. Cumana, Venezuela, July

Meikle P, Richards GN, Yellowlees D (1988) Structural investigations on the mucus from six species of coral. Mar Biol 99:187–193

Morita RY (1997) Bacteria in oligotrophic environments. Chapman and Hall, New York

Patton JS, Abraham S, Benson AA (1977) Lipogenesis in the intact coral *Pocillopora clapitata* and its isolated zooxanthellae: evidence for a light-driven carbon cycle between symbiont and host. Mar Biol 44:235–247

Ritchie KB, Smith GW (1995a) Carbon-source utilization of coral-associated marine heterotrophs. J Mar Biotechnol 3:107–109

Ritchie KB, Smith GW (1995b) Preferential carbon utilization by surface bacterial communities from water mass, normal and white-band diseased *Acropora cervicornis*. Mol Mar Biol Biotechnol 4:345–352

Ritchie KB, Smith GW (1997) Physiological comparisons of bacterial communities from various species of scleractinian corals. Proc 8th Int Coral Reef Symp 1:521–526

Ritchie KB, Dennis JH, McGrath T, Smith GW (1994) Bacteria associated with bleached and nonbleached areas of *Montastrea annularis*. Proc Symp Nat Hist Bahamas 5:75–80

Ritchie KB, Gerace DT, Smith GW (1995) Grouping of bacterial heterotrophs from scleractinian corals using metabolic potentials. Proc Assoc Mar Labs Caribb 26:224–236

Ritchie KB, Holley RA, McGrath T, Smith GW (1996) INT-linked dehydrogenase activity in the surface mucopolysaccharide layers of Bahamian scleractinian corals. Proc Symp Nat Hist Bahamas 6:18–27

Rohwer F, Breitbart M, Jara J, Azam F, Knowlton K (2001) Diversity of bacteria associated with the Caribbean coral *Montastraea franksi*. Coral Reefs 20:85–91

Rohwer F, Seguritan V, Azam F, Knowlton N (2002) Diversity and distribution of coral-associated bacteria. Mar Ecol Prog Ser 243:1–10

Rowan R, Knowlton K, Baker A, Jara J (1997) Landscape ecology of agal symbionts creates variation in episodes of coral bleaching. Nature 388:265–269

Smith GW, Hayasaka SS (1982a) Nitrogenase activity associated with *Zostera marina* from a North Carolina estuary. Can J Microbiol 28:448–451

Smith GW, Hayasaka SS (1982b) Nitrogenase activity of bacteria associated with *Halodule wrightii* roots. Appl Environ Microbiol 43:1244–1248

Smith GW, Kozuchi AM, Hayasaka SS (1982) Heavy metal sensitivity of the seagrass rhizosphere and sediment bacteria. Bot Mar 25:19–24

Sorokin YI (1973) On the feeding of some scleractinian corals with bacteria and dissolved organic matter. Limnol Oceanogr 18:380–385

14 Culture-Independent Analyses of Coral-Associated Microbes

FOREST ROHWER and SCOTT KELLEY

14.1 Introduction

The influence of microbes on coral reefs has been underappreciated, even though it is widely recognized that Bacteria, Archaea, and unicellular eukaryotes are vital components of all marine ecosystems. Several studies have applied modern microbial ecology methods to investigate the microbes living with reef organisms. These new avenues of research are changing our view of how coral reefs function. This chapter is intended to make these new findings accessible to those interested in interdisciplinary studies of coral and microbial ecology. A model of how microbes are structured on healthy corals is proposed, as is the hypothesis that disrupting this structure leads to coral disease and death.

14.2 Coral Reefs and the Marine Microbial Food Web

The marine microbial food web (MMFW) is the consortium of heterotrophic and autotrophic prokaryotes, as well as their predators that inhabit the earth's oceans and seas (Fig. 14.1; Azam 1998). The MMFW regulates the transfer of energy and nutrients to higher trophic levels and influences global carbon and nutrient cycles (Pomeroy 1974; Azam et al. 1983; Joint and Pomeroy 1983; Kirchman 1994). For example, ~80% of the primary production in the equatorial and subtropical North Pacific Ocean is carried out by autotrophic bacteria (Liu et al. 1997), and secondary production by MMFW heterotrophic prokaryotes accounts for ~50% of the carbon fixed by photosynthesis in the world's oceans (Azam 1998). Both auto- and heterotrophic microbes are controlled in a top-down fashion by two predator guilds: (1) protozoa and (2) viruses, also called as phages, that kill prokaryotes (Proctor and Fuhrman 1988; Fuhrman and Noble 1995). On average, there are 1 million prokaryotes and 10 million phages/ml of surface seawater (Fig. 14.2).

Microbes sequester essential nutrients within the MMFW (Azam 1998). By virtue of their high affinity transport systems and their large surface area to volume ratios, prokaryotes scavenge nutrients at much lower concentrations than eukaryotic cells (Geesey and Morita 1979; Nissen et al. 1984). In nutrient-poor waters, prokaryotes assimilate most of the limiting nutrients and restrict primary production [e.g., iron in the equatorial Pacific (Martin 1992) and phos-

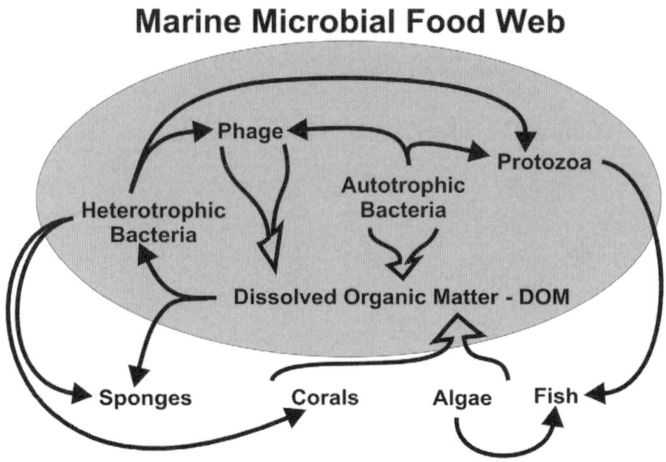

Fig. 14.1. Much of marine biogeochemistry in the oligotrophic waters surrounding the world's coral reefs is carried out by members of the marine microbial food web (MMFW). Dissolved organic matter (DOM), the food source for marine bacteria, is the largest biogenic sink of carbon in the ocean (Kennish 2001). The DOM pool is so large that heterotrophic bacterial populations are generally not energy- or-nutrient-limited. Instead, bacterial numbers are controlled by protozoa and phage predation. Each of these predator guilds accounts for 50% of the microbial mortality each day (Fuhrman and Noble 1995). In the open ocean, when a bacterium is ingested by a protozoa there is a possibility that the carbon can be transferred to the larger members of the marine food web. In contrast, when a bacterium is killed by a lytic phage, both the lysed host cell and the phage become part of the DOM pool (Middelboe et al. 1992, 1996) and the carbon never leaves the MMFW. On coral reefs, however, the "wall-of-mouths" may directly harvest bacteria, protozoa, phage, and DOM. This allows coral reefs to assimilate much of the incoming energy and nutrients from the surrounding water column

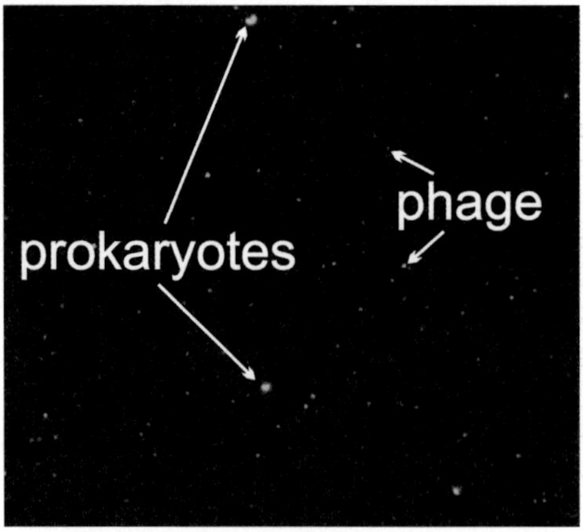

Fig. 14.2. Epifluorescent micrograph of seawater treated with SYBR-Gold (a dsDNA stain). On average, there are ~10^6 bacteria and ~10^7 viruses/ml of surface seawater. Most of the viruses are phage, which infect and kill bacteria. (Photo by Linda Wegley, San Diego State University)

phate in the Mediterranean (Thingstad et al. 1998)]. Nutrient concentrations on coral reefs are low (Odum and Odum 1955; Muscatine 1980; Rahav et al. 1989; Szmant et al. 1990; Gast et al. 1998; Gili and Coma 1998), and prokaryotes will be competing with metazoans to obtain these resources.

Corals and other sessile organisms may acquire necessary nutrients directly by harvesting microbes from the water column through mucus netting and indirectly via capture of protozoa that graze on bacteria (Sorokin 1973; Bak et al. 1998; Ferrier-Pages et al. 1998). Coral reefs derive a substantial amount of essential nutrients and energy by harvesting incoming phyto- and zooplankton from the oceanic waters that continuously flush onto reefs (Ayukai 1995). The reef community is so efficient at sweeping the incoming water of food that it is referred to as the "wall-of-mouths". Corals themselves gain up to 70% of their nitrogen budget from particulate sources, and bacteria from the water column account for 30–45% of the incorporated particulate matter (Bak et al. 1998). This process is one of the factors that permits coral reefs to support high biomass concentrations in oligotrophic waters. When considering how coral reefs function, it is essential to remember that they are surrounded by the MMFW. Corals live within and depend on this soup of microbes, some of which are potential pathogens.

14.3 Uncultured Microbes and the Three Domains of Life

Until recently, environmental microbiology has been limited by the ability to culture microorganisms in the laboratory (Pace et al. 1985). Direct microscopic counts have shown that only a small percentage of the microbes in an environment are readily cultured (Wagner et al. 1993; Amann et al. 1995). This observation has been made so consistently in so many different environments that it is known as the "Great Plate Anomaly". Currently, it is estimated that more than 99% of microorganisms have not been cultured (Ferguson et al. 1984; Staley and Konopka 1985; Wagner et al. 1993; Amann et al. 1995).

In 1977, Carl Woese and colleagues used ribosomal RNA (rRNA) sequences to show that all organisms belonged to three evolutionary domains of life: Bacteria, Eukarya, and Archaea (Balch et al. 1977; Woese and Fox 1977; Pace 1997). The Eukarya comprise all multi-cellular organisms, as well as many single-celled organisms such as protists. The Archaea are prokaryotes like Bacteria, but they are phylogenetically distinct and share core genetic machinery with the Eukarya (Harris et al. 2003; Pace 1997). Subsequently, the existence of these three domains has been reinforced by a great deal of phylogenetic and molecular research (Amann et al. 1995; Brown and Doolittle 1995; Pace 1997; Glansdorff 2000; Brown et al. 2001; Harris et al. 2003).

Pace et al. (1985) developed molecular methods, based on ribosomal DNA (rDNA), to directly search the environment for novel microorganisms without cultivation. Identification of microbial organisms using cloned rDNA sequences provided a powerful new means of discovering previously unknown microbial diversity, and a great number of molecular tools have been developed based on rDNA for studying microbes in the environment (Amann et al. 1995). Some of the most common techniques, and their application to coral-microbe communities, are discussed below.

Studies of uncultured microbial diversity require DNA that is suitable for analysis with the polymerase chain reaction (PCR) and/or cloning. The extraction technique also needs to include DNA from all of the microbes present in the original sample. Both Rohwer et al. (2001) and Frias-Lopez et al. (2002) have worked out protocols that appear to meet these conditions for corals. Once high quality DNA has been isolated from a coral, the 16S rDNA genes in the sample are amplified using universal PCR primers designed to hybridize to evolutionarily conserved regions of the gene (Amann et al. 1995). The resulting PCR products include a mixture of all the targets in the original sample and must be analyzed in such a way that the different 16S rDNAs, representing unique microbial species, are separated from each other. The most straightforward method is to clone the products and then sequence them. Once sequence data have been obtained from the microbial 16S rDNAs, they are compared to known sequences in databases such as GenBank to determine the phylogenetic affiliations of the sequences (Benson et al. 2000). The most commonly used software tools for these analyses are BLAST (Altschul et al. 1990, 1997) and global alignment algorithms like Clustal W (Thompson et al. 1994, 1997).

14.4 Uncultured Microbes Associated with Healthy Corals

The first culture-independent analyses of microbial 16S rDNA on corals showed that coral-associated bacteria are distinct from those in the water column and that culture-based studies underestimate both the types and diversity of microbes on corals (Rohwer et al. 2001). The majority of coral-associated microbes so far identified using 16S rDNA analyses are novel species (<93% identity at the 16S rDNA; Rohwer et al. 2001, 2002; Frias-Lopez et al. 2002). Furthermore, most of the coral-associated microbial 16S ribotypes have only been observed once and overall bacterial diversity is very high (Shannon Index = 4.95; Rohwer et al. 2002). In contrast, similar analyses on water column samples revealed that most of the clones are closely related to previously described 16S rDNA sequences in GenBank (Hagstrom et al. 2002; Rohwer et al. 2002). Together these results showed that corals have diverse microbiotas that are distinct from those in the water column.

Rohwer et al. (2002) used 16S rDNA cloning and sequencing to demonstrate that *Montastraea franksi*, *Porites astreoides*, *P. furcata*, and *Diploria strigosa* have bacterial communities that are distinct for each coral species. This study showed: (1) that the same coral species had similar microbial communities even when separated by hundreds of kilometers, and (2) the microbiotas associated with different coral species were distinct even when the colonies were located immediately adjacent to one another (e.g., ~10 cm). The types of coral-bacteria associations that were observed are summarized in Table 14.1. All *P. astreoides* colonies harbored PA1, a single γ-proteobacteria species. PA1 was never found on any other coral type. Similarly, all *P. furcata* colonies contained the bacteria PF1. Because PF1 and PA1 are closely related γ-proteobacteria, it is possible that all *Porites* species have an associated PA1-like bacterial symbiont. In the case of *M. franksi*, a group of closely related, but not identical, α-proteobacteria was found with all colonies studied. This finding confirmed earlier denaturing gradient gel electrophoresis (DGGE) results which showed that there was an α-proteobacteria ubiquitously associated with *M. franksi* in Panama (Rohwer et al. 2001). A broader type of association was found between the bacterial groups *Cytophaga–Flavobacter/Flexibacter–Bacterioides* (CFB) and *Bacillus–Clostridium* (BC) with *D. strigosa*. In this case, no specific CFB or BC was observed on more than one coral colony. There were, however, statistically more CFB and BC associated with *D. strigosa* than with any other coral species. Together these results showed that different coral species harbor distinct bacterial microbiotas.

Coral-associated bacteria can also be structured within colonies. For example, the white, fast growing tips of *Porites furcata* have a particular bacterial species that is not found in other parts of the same coral (Rohwer et al. 2002). While the significance of this finding is not known, it is reminiscent of zooxanthellae–coral interactions that have important ecological consequences (Rowan and Knowlton 1995; Rowan et al. 1997; Toller et al. 2001).

Kellogg, Rohwer, and Wegley (unpubl. data) have found that coral-associated Archaea are abundant and diverse. Direct counts with Archaea-specific peptide nucleic acid (PNA) probes have shown that coral contain ~10^8 Archaea/cm^2. Unlike the Bacteria, Archaea do not appear to be associated with specific corals. The same archaeal ribotypes were found on different coral species, as well as on corals from Florida, Bermuda, and Panama.

Table 14.1. Different types of coral–microbe relationships. CFB *Cytophaga-Flavobacter/Flexibacter-Bacterioides*, BC *Bacillus-Clostridium*

	One microbial species	Very closely related microbes	Microbial division
One coral species	PA1 and *P. astreoides*	*M. franksi* and at least three closely related α-proteobacteria	*D. strigosa* with CFB and BC
	PF1 and *P. furcata*		
One coral genera	No example	γ-Proteobacteria and *Porites* spp.	No example
Many coral genera	No example	*Rickettsia*-like bacteria and at least two Archaea with various Caribbean corals	No example

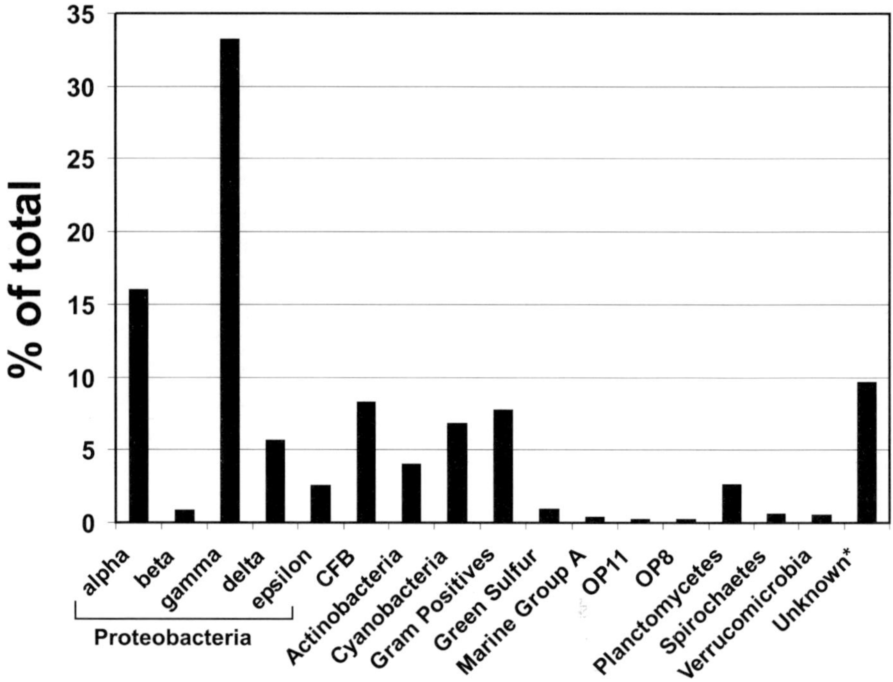

Fig. 14.3. Bacterial divisions of uncultured 16S rDNA found on healthy and diseased corals. All of the reported 16S rDNA in GenBank (April 2003) was classified into divisions using ARB. Most of the sequences come from Rohwer et al. 2001, 2002; Frias-Lopez et al. 2002. *Unknown** includes sequences that are probably derived from plastids

In summary, the association of microbes and corals is not random. Corals have microbiotas that are distinct from those on other coral species and from the surrounding water (Rohwer et al. 2001, 2002; Cooney et al. 2002; Frias-Lopez et al. 2002; Hagstrom et al. 2002). Coral-associated prokaryotes are extremely diverse, in terms of both species richness and abundance. Currently, uncultured analyses of coral-associated microbes have documented ~1000 distinct ribotypes. Most of the ribotypes come from bacteria that have never been observed before and include representatives from at least 16 bacterial divisions (Fig. 14.3; Rohwer et al. 2001, 2002; Cooney et al. 2002; Frias-Lopez et al. 2002; unpubl. data).

14.5 Culture-Independent Analyses of Coral Diseases

Cooney et al. (2002) and Frias-Lopez et al. (2002) have used culture-independent methods to investigate the microbes associated with black band diseased corals. Both groups found that *Phormidium corallyticum*, the presumed etiological agent of black band disease, was not present in the clone libraries or evident in DGGEs. Further studies have now shown that a variety of cyanobacteria are associated with black band disease (Frias-Lopez et al. 2003), but *P. corallyticum* is not one of them. In their study, Frias-Lopez et al. (2002) sampled the healthy tissue harvested from the same coral colony that was displaying symptoms of black band disease. They found that the bacterial diversity associated with this healthy coral tissue was lower than in the diseased tissue, dead coral surface, and water column. What could account for the differences between Frias-Lopez et al. (2002) and Rohwer et al. (2002), who observed high microbial diversity on healthy corals? Pantos et al. (2003) answered this question by showing that the microbial community associated with apparently healthy tissue taken from a diseased coral colony is much less diverse than that of a remote colony belonging to the same coral species. Therefore, the microbial community associated with a diseased coral is different even where symptoms of the disease are not apparent. This finding has important implications for how coral diseases develop and are detected. Pantos et al. (2003) also found that the microbial communities associated with both black band disease and white plague are very similar to each other. In particular, there is an α-proteobacteria associated with both coral diseases that is closely related to the causative agent of juvenile oyster disease (JOD). This JOD bacterium is a good candidate etiological agent of both black band disease and white plague.

14.6 Microbial Growth on Corals

Even though diverse microbial communities are present on corals, this does not necessarily mean that they are growing. Growth rates of uncultured bacteria can be determined by measuring incorporation of radiolabeled thymidine. This works because bacteria, but not eukaryotic cells, salvage exogenous thymidine at tracer concentrations. Samples are treated with radio-labeled thymidine, which is taken up by bacteria and incorporated into the newly synthesized DNA. The amount of incorporated radiolabel is proportional to the bacterial growth rates. Using this method, Paul et al. (1986) determined that coral-associated bacteria are growing approximately ten times faster than those in the overlying water column.

To determine whether certain microbial groups preferentially grow on corals, fragments of *Diploria strigosa* were incubated with tracer amounts of Br-dUTP, a thymidine analog (Rohwer, unpubl. data; based on method by Urbach et al. 1999). Actively growing bacteria incorporate the Br-dUTP into their genome. After the incubation, total DNA was harvested from the corals and one half was immunoprecipitated with anti-Br-dUTP antibodies. Clone libraries were made from both the immunoprecipitated DNA and the total DNA. The 16S rDNA in these libraries were sequenced and compared using the phylogenetic permutation tail probability tests (PTP) described by Martin (2002). The PTP test found little phylogenetic overlap between growers and non-growers (PTP test; $P<0.0001$). Almost all the actively growing bacterial sequences belonged to the Proteobacteria division. In contrast, 16S rDNA sequences found on the corals, but not in the actively growing fraction, came from many different bacterial divisions, including the CFB and BC groups that are normally numerically dominant on *D. strigosa*.

14.7 Conclusions

Though in its infancy, coral reef microbial ecology is showing that when viewing a coral reef, it is necessary to imagine the surfaces, crevices, and animals as landscapes of highly diverse, ecologically structured microbial communities. The associations of the coral animal, prokaryotes, zooxanthellae, fungi, and other undefined components help define the niche that a coral colony occupies on a reef. Disrupting the relationships between microbes and corals has the potential to collapse the whole community. Only by considering these invisible components will we understand how coral reefs live and die.

14.8 Major Unanswered Questions and Future Research

14.8.1 What Are Coral-Associated Prokaryotes Doing?

It is not known how coral–prokaryote associations affect the coral holobiont. By associating with specific prokaryotes, a coral colony may gain important metabolic functions such as photosynthesis or nitrogen fixation (Saffo 1992). Reef sponges, for example, can acquire fixed nitrogen via associated cyanobacteria (Wilkinson and Fay 1979). Corals also have associated nitrogen fixers (Shashar et al. 1994; Rohwer et al. 2001, 2002; Frias-Lopez et al. 2002) and some nitrogen fixation has been recorded on living corals (Williams et al. 1987; Shashar et al. 1994), though no one has shown that the corals assimilate the newly fixed nitrogen.

Corals may use microbes to "fish" for nutrients. In this scenario, corals would encourage certain microbes to grow by secreting fixed carbon in the form of mucus. The mucus-associated prokaryotes would then harvest limiting nutrients from the surrounding waters and the host coral could eat a proportion of the microbial community (see also Chap. 10, Kushmaro and Kramasky-Winter, this Vol.). In support of this model, corals are known to ingest microbes associated with their mucus (Sorokin 1973; Bak et al. 1998; Ferrier-Pages et al. 1998; Gast et al. 1998; Gili and Coma 1998), and there are large numbers of mucus-adapted microbes associated with corals (Ducklow and Mitchell 1979a; Herndl and Velimirov 1986; Paul et al. 1986).

A third possibility is that specialized microbiota may protect the coral animal from opportunistic pathogens by occupying entry niches and/or through the production of antibiotics. Figure 14.4 shows how coral-associated microbes may act as inhibitors of colonization by other organisms. Corals would encourage the

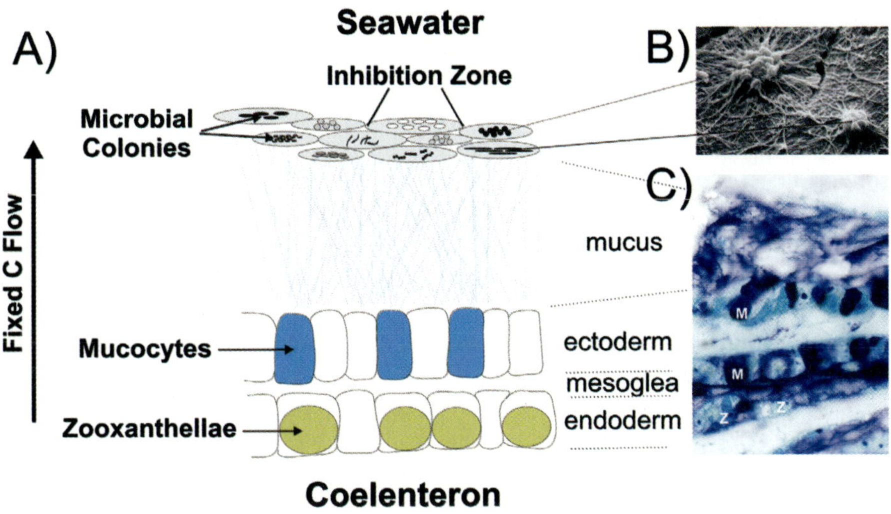

Fig. 14.4. A Coral-associated microbes as inhibitors of colonization by other organisms. In this model, corals encourage the growth of a specific microbiota by the secretion of mucus, which is ultimately derived from photosynthate produced by the zooxanthellae. The microbes form colonies on the mucus surface. In turn, these colonies produce inhibition zones that cannot be colonized by bacteria from the water column. Potential methods of inhibition include physical occupation of space, sequestering of nutrients, and production of antibiotics. **B** Scanning electron micrograph of microbial colonies on the surface of *Diploria strigosa*. Note that the colonies produce mucus networks. **C** Cross section of fixed and embedded coral stained with Trypan blue. Microbes are rarely observed below the mucus layer (Bythell et al. 2002). *M* Mucocytes, *Z* zooxanthellae

growth of specific microbes by the secretion of specialized mucus (Ducklow and Mitchell 1979a). In turn, the microbes form colonies on the mucus surface and create "inhibition zones" by physical occupation of space, sequestration of nutrients, and production of antibiotics (Koh 1997). These inhibition zones prevent colonization by outside microbes, which include opportunistic pathogens. In this scenario, the high microbial diversity observed on corals is a result of competition amongst these mucus-adapted microbes (Czaran et al. 2002) and by phage and protist predation (Thingstad and Lignell 1997). One consequence of this proposed model is that disrupting the mucus-associated microbial community will make the coral holobiont more susceptible to pathogens. For example, there may be stresses that cause "microbial bleaching" on corals, which leads to opportunistic infections and eventually coral death.

14.8.2 What Is the Mucus Made of?

Corals interact with the environment through their mucus layer. This layer forms a boundary through which nutrients and gases must diffuse. The mucus is a barrier against pathogens and should be considered the primary immune organ for corals. As proposed above, the mucus layer itself may be composed of different compounds designed to recruit specific microbiotas. The mucus is also an important source of particulate and dissolved organic matter on coral reefs (Benson and Muscatine 1974). Given that coral mucus is so important, it is surprising that very little is actually known about its composition or structure. Lipids, polysaccharides, and proteins are the major components of coral mucus (Benson and Muscatine 1974; Ducklow and Mitchell 1979b). More detailed analyses of the macromolecules have been performed (Krupp 1981; Meikle et al. 1988), but no general patterns are apparent. EM studies of the mucus layer (Fig. 14.4B) show that the microbes are probably contributing a significant amount of material to the mucus, or alternatively, the microbes are transforming the mucus with ectoenzymes or aggregation reactions (e.g., flocculation). Anecdotally, many cultured coral-associated bacteria are very "mucusy" (i.e., they form large mucus layers on agar plates). Given the high concentration of microbes found on corals, it is probably more appropriate to think of the mucus layer as a microbial biofilm rather than a pure product of the coral animal.

The most pressing questions for future studies are: does the mucus composition vary consistently among different coral species? How does the mucus change as a particular coral is moved between heterotrophic and autotrophic conditions? How much of the mucus layer is derived from the microbes versus the coral?

14.8.3
Culturing the Uncultured

Identifying important microbes, as described above, is just the beginning. Ultimately, we will have to culture coral animals, zooxanthellae, and microbes separately to determine the roles that each play in coral holobiont health and disease. Axenic coral cultures and coral tissue culture cell lines need to be developed. Tissue culture is essential for studies of coral viruses and the development of coral genetic systems.

Characterization of prokaryotes by 16S rDNA analyses narrows down which microbes are the most interesting. Once a particular bacterial species or group has been identified, culturing strategies targeting these microbes can be designed. This approach has been very successful with marine microbes (Rappe et al. 2002). Obvious candidates for culturing studies include PA1, the bacterium ubiquitously associated with *P. astreoides*, the JOD-like bacterium found with white plague and black band disease, and the α-proteobacteria group found on *M. franksi*. When the various components of the coral holobiont have been cultured, it will be possible to determine the roles each play in the greater whole holobiont by add-back experiments. Only then will we understand how corals have come to dominate tropical near-shore systems and how they will adapt to changing environmental conditions.

References

Altschul SF, Gish W, Miller W, Myers EW, Lipman DJ (1990) Basic local alignment search tool. J Mol Biol 215:403–410

Altschul SF, Madden TL, Schäffer AA, Zhang J, Zhang Z, Miller W, Lipman DJ (1997) Gapped BLAST and PSI-BLAST: a new generation of protein database search programs. Nucleic Acids Res 25:3389–3402

Amann RI, Ludwig W, Schleifer KH (1995) Phylogenetic identification and in situ detection of individual microbial cells without cultivation. Microbiol Rev 59:143–169

Ayukai T (1995) Retention of phytoplankton and planktonic microbes on coral reefs within the Great Barrier Reef, Australia. Coral Reefs 14:141–147

Azam F (1998) Microbial control of oceanic carbon flux: the plot thickens. Science 280:694–696

Azam F, Fenchel T, Field JG, Gray JS, Meyer-Reil LA, Thingstad F (1983) The ecological role of water-column microbes in the sea. Mar Ecol Prog Ser 10:257–263

Bak R, Joenje M, de Jong I, Lambrechts D, Nieuwland G (1998) Bacterial suspension feeding by coral reef benthic organisms. Mar Ecol Prog Ser 175:285–288

Balch WE, Magrum LJ, Fox GE, Wolfe RS, Woese CR (1977) An ancient divergence among the bacteria. J Mol Evol 9:305–311

Benson A, Muscatine L (1974) Wax in coral mucus: energy transfer from corals to reef fishes. Limnol Oceanogr 19:810–814

Benson DA, Karsch-Mizrachi I, Lipman DJ, Ostell J, Rapp BA, Wheeler DL (2000) GenBank. Nucleic Acids Res 28:15–18

Brown JR, Doolittle WF (1995) Root of the universal tree of life based on ancient aminoacyl-tRNA synthetase gene duplications. Proc Natl Acad Sci USA 92:2441–2445

Brown JR, Douady CJ, Italia MJ, Marshall WE, Stanhope MJ (2001) Universal trees based on large combined protein sequence data sets. Nat Genet 28:281–285

Bythell J, Barer M, Cooney R, Guest J, O'Donnell A, Pantos O, Le Tissier M (2002) Histopathological methods for the investigation of microbial communities associated with disease lesions in reef corals. Lett Appl Microbiol 34:359–364

Cooney R, Pantos O, Le Tissier M, Barer M, O'Donnell A, Bythell J (2002) Characterization of the bacterial consortium associated with black band disease in coral using molecular microbiological techniques. Environ Microbiol 4:401–413

Czaran T, Hoekstra R, Pagie L (2002) Chemical warfare between microbes promotes biodiversity. Proc Natl Acad Sci USA 99:786–790

Ducklow HW, Mitchell R (1979a) Bacterial populations and adaptations in the mucus layers on living corals. Limnol Oceanogr 24:715–725

Ducklow HW, Mitchell R (1979b) Composition of mucus released by coral reef coelenterates. Limnol Oceanogr 24:707–714

Ferguson RL, Buckley EN, Palumbo AV (1984) Response of marine bacterioplankton to differential filtration and confinement. Appl Environ Microbiol 47:49–55

Ferrier-Pages C, Allemand D, Gattuso J-P, Jaubert J, Rassoulzadegan F (1998) Microheterotrophy in the zooxanthellate coral *Stylophora pistillata*: effects of light and cilliate density. Limnol Oceanogr 43:1639–1648

Frias-Lopez J, Zerkle AL, Bonheyo GT, Fouke BW (2002) Partitioning of bacterial communities between seawater and healthy, black band diseased, and dead coral surfaces. Appl Environ Microbiol 68:2214–2228

Frias-Lopez J, Bonheyo G, Qusheng J, Fouke B (2003) Cyanobacteria associated with coral black band disease in Caribbean and Indo-Pacific reefs. Appl Environ Microbiol 69:2409–2413

Fuhrman JA, Noble RT (1995) Viruses and protists cause similar bacterial mortality in coastal seawater. Limnol Oceanogr 40:1236–1242

Gast GJ, Wiegman S, Wieringa E, Duyl FC, Bak RPM (1998) Bacteria in coral reef water types: removal of cells, stimulation of growth and mineralization. Mar Ecol 167:37–45

Geesey GG, Morita RY (1979) Capture of arginine at low concentrations by a marine psychrophilic bacterium. Appl Environ Microbiol 38:1092–1097
Gili J-M, Coma R (1998) Benthic suspension feeders: their paramount role in littoral marine food webs. TREE 13:316–321
Glansdorff N (2000) About the last common ancestor, the universal life-tree and lateral gene transfer: a reappraisal. Mol Microbiol 38:177–185
Hagstrom A, Pommier T, Rohwer F, Simu K, Svensson D, Zweifel UL (2002) Bio-informatics reveal surprisingly low species richness in marine bacterioplankton. Appl Environ Microbiol 67:3628–3633
Harris JK, Kelley ST, Spiegelman GB, Pace NR (2003) The genetic core of the universal ancestor. Genome Res 13:407–412
Herndl GJ, Velimirov B (1986) Microheterotrophic utilization of mucus released by the Mediterranean coral *Cladocora cespitosa*. Mar Biol 90:363–369
Joint IR, Pomeroy AJ (1983) Production of picoplankton and small nanoplankton in the Celtic Sea. Mar Biol 77:19–27
Kennish MJ (2001) Practical handbook of marine science. CRC Press, Boca Raton
Kirchman DL (1994) The uptake of inorganic nutrients by heterotrophic bacteria. Microbial Ecol 28:255–271
Koh EGL (1997) Do scleractinian corals engage in chemical warfare against microbes? J Chem Ecol 23:379–398
Krupp DA (1981) The composition of the mucus from the mushroom coral, *Fungia scutaria*. Proc 4th Int Coral Reef Symp 2:69–73
Liu H, Nolla H, Campbell L (1997) *Prochlorococcus* growth rate and contribution to primary production in the equatorial and subtropical North Pacific Ocean. Aquat Microb Ecol 12:39–47
Martin A (2002) Phylogenetic approaches for describing and comparing the diversity of microbial communities. Appl Environ Microbiol 68:3673–3682
Martin JH (1992) Iron as a limiting factor in oceanic productivity. In: Falkowski PG, Woodhead AD (eds) Primary productivity and biogeochemical cycles in the sea. Plenum Press, New York, pp 123–137
Meikle P, Richards GN, Yellowlees D (1988) Structural investigations on the mucus from six species of coral. Mar Biol 99:187–193
Middelboe M, Nielsen B, Søndergaard M (1992) Bacterial utilization of dissolved organic carbon (DOC) in coastal waters – determination of growth yield. Arch Hydrobiol Ergeb Limnol 37:51–61
Middelboe M, Jørgensen NOG, Kroer N (1996) Effects of viruses on nutrient turnover and growth efficiency of noninfected marine bacterioplankton. Appl Environ Microbiol 62:1991–1997
Muscatine L (1980) Uptake, retention, and release of dissolved inorganic nutrients by marine algae invertebrate associations. In: Cook CB, Pappas PW, Rudolph ED (eds) Cellular interactions in symbioses and parasitism. Ohio State Univ Press, Columbus, pp 229–244
Nissen H, Nissen P, Azam F (1984) Multiphasic uptake of D-glucose by an oligotrophic marine bacterium. Mar Ecol Prog Ser 16:155
Odum HT, Odum EP (1955) Trophic structure and productivity of a windward coral reef community on Eniwetok Atoll. Ecol Mono 25:291–320
Pace NR (1997) A molecular view of microbial diversity and the biosphere. Science 276:734–740
Pace NR, Stahl DA, Lane DJ, Olsen GJ (1985) Analyzing natural microbial populations by rRNA sequences. ASM News 51:4–12
Pantos O, Cooney R, Le Tissier M, Barer M, O'Donnell A, Bythell J (2003) The bacterial ecology of a plague-like disease affecting the Caribbean coral *Montastraea annularis*. Environ Microbiol 5:370–382
Paul J, DeFlaun M, Jeffery W (1986) Elevated levels of microbial activity in the coral surface monolayer. Mar Ecol Prog Ser 33:29–40
Pomeroy LR (1974) The ocean's food web, a changing paradigm. BioScience 24:499–504
Proctor LM, Fuhrman JA (1988) Marine bacteriophages and bacterial mortality. EOS 69:1111–1112

Rahav O, Dubinsky Z, Achituv Y, Falkowski PG (1989) Ammonium assimilation in the zooxanthelle coral, *Stylophroa pistillata*. Philos Trans R Soc Lond B 236:325–337
Rappe M, Connon S, Vergin K, Giovannoni S (2002) Cultivation of the ubiquitous SAR11 marine bacterioplankton clade. Nature 418:630–633
Rohwer F, Breitbart M, Jara J, Azam F, Knowlton N (2001) Diversity of bacteria associated with the Caribbean coral *Montastraea franksi*. Coral Reefs 20:85–95
Rohwer F, Serigutan V, Azam F, Knowlton N (2002) Diversity and distribution of coral-associated bacteria. Mar Ecol Prog Ser 243:1–10
Rowan R, Knowlton N (1995) The ecological significance of genetic diversity in corals and their algal symbionts. J Cell Biochem [Suppl] 19B:334
Rowan R, Knowlton N, Baker A, Jara J (1997) Landscape ecology of algal symbionts creates variation in episodes of coral bleaching. Nature 388:265–269
Saffo MB (1992) Invertebrates in endosymbiotic associations. Am Zool 32:557–565
Shashar N, Cohen Y, Loya Y, Sar N (1994) Nitrogen fixation (acetylene reduction) in stony corals: evidence for coral–bacteria interactions. Mar Ecol Prog Ser 111:259–264
Sorokin YI (1973) Trophical role of bacteria in the ecosystem of the coral reef. Nature 242:415–417
Staley JT, Konopka A (1985) Measurement of in situ activities of nonphotosynthetic microorganisms in aquatic and terrestrial habitats. Annu Rev Microbiol 39:321–346
Szmant AM, Ferrer LM, FitzGerald LM (1990) Nitrogen excretion and O:N ratios in reef corals: evidence for the conservation of nitrogen. Mar Biol 104:119–127
Thingstad TF, Lignell R (1997) Theoretical models for the control of bacterial growth rate, abundance, diversity and carbon demand. Aquat Microb Ecol 13:19–27
Thingstad FT, Zweifel UL, Rassoulzadegan F (1998) P limitation of heterotrophic bacteria and phytoplankton in the northwest Mediterranean. Limnol Oceanogr 43:88–94
Thompson JD, Higgins DG, Gibson TJ (1994) CLUSTAL W: improving the sensitivity of progressive multiple sequence alignment through sequence weighting, positions-specific gap penalties and weight matrix choice. Nucleic Acids Res 22:4673–4680
Thompson JD, Gibson TJ, Plewniak F, Jeanmougin F, Higgins DG (1997) The Clustal X windows interface: flexible strategies for multiple sequence alignment aided by quality analysis tools. Nucleic Acids Res 24:4876–4882
Toller WW, Rowan R, Knowlton N (2001) Repopulation of zooxanthellae in the Caribbean corals *Montastraea annularis* and *M. faveolata* following experimental and disease-associated bleaching. Biol Bull (Woods Hole) 201:360–373
Urbach E, Vergin K, Giovannoni S (1999) Immunochemical detection and isolation of DNA from metabolically active bacteria. Appl Environ Microbiol 65:1207–1213
Wagner M, Amann R, Lemmer H, Schleifer KH (1993) Probing activated sludge with oligonucleotides specific for proteobacteria: inadequacy of culture-dependent methods for describing microbial community structure. Appl Environ Microbiol 59:1520–1525
Wilkinson CR, Fay P (1979) Nitrogen fixation in coral reef sponges with symbiotic cyanobacteria. Nature 279:527–529
Williams WM, Viner AB, Broughton WJ (1987) Nitrogen fixation (acetylene reduction) associated with the living coral *Acropora variabilis*. Mar Biol 94:531–535
Woese CR, Fox GE (1977) Phylogenetic structure of the prokaryotic domain: the primary kingdoms. Proc Natl Acad Sci USA 74:5088–5090

15 Aspergillosis of Gorgonians

Garriet W. Smith and Ernesto Weil

15.1 Introduction and History

In the 1980s, certain areas in the Caribbean experienced mass mortalities of sea fans (*Gorgonia flabellum*). These areas included the coasts of Costa Rica (Guzman and Cortez 1984), Panama (Garzon-Ferreira and Zea 1992; Diaz et al. 1995), and Trinidad (Laydoo 1983). Laydoo (1983) suggested that the cause may be due to a species-specific pathogen, but microbiological studies were not performed. Although limited mortalities of gorgonian colonies were previously observed due to fish grazing (Kinzie 1973), *Cyphoma* sp. grazing (Harvell and Suchanek 1987), fouling and overgrowth (Wahle 1985) and tumors (Morse et al. 1977, 1981), the 1980's outbreak showed signs of an epizootic which had spread from the southwestern to the southeastern Caribbean.

In 1995, sea fans showing disease signs similar to those reported in the 1980s were observed near the island of Saba and at several other Caribbean locations (Nagelkerken et al. 1997a). Participating member institutions (24 locations throughout the Caribbean) of the Caribbean Coastal Marine Productivity Network (CARICOMP) made qualitative observations on sea fan tissue mortality, and collected samples of healthy and affected colonies for microbiological analysis in the same year. With the exception of Bermuda and Mexico, affected sea fans were found at all Caribbean locations. Later, diseased sea fans were also found in Bermuda and Mexico, extending the geographical range to include all of the Caribbean.

15.2 Infection Patterns

Observations made by members of the CARICOMP network revealed that the disease affected both *Gorgonia ventalina* and *G. flabellum* (Nagelkerken et al. 1997b). Smith et al. (1998) later showed that *G. ventalina* was somewhat more susceptible than *G. flabellum*, which, at least on one reef in the Bahamas, resulted in a change of dominance from *G. ventalina* to *G. flabellum* in a 3-year period.

The disease was found to increase in incidence (% of diseased sea fans) and virulence (% tissue loss per diseased colony) in the Bahamas from 1995 through 1998. Incidence, virulence and the number of lesions were positively correlated with water depth (Nagelkerken et al. 1997b).

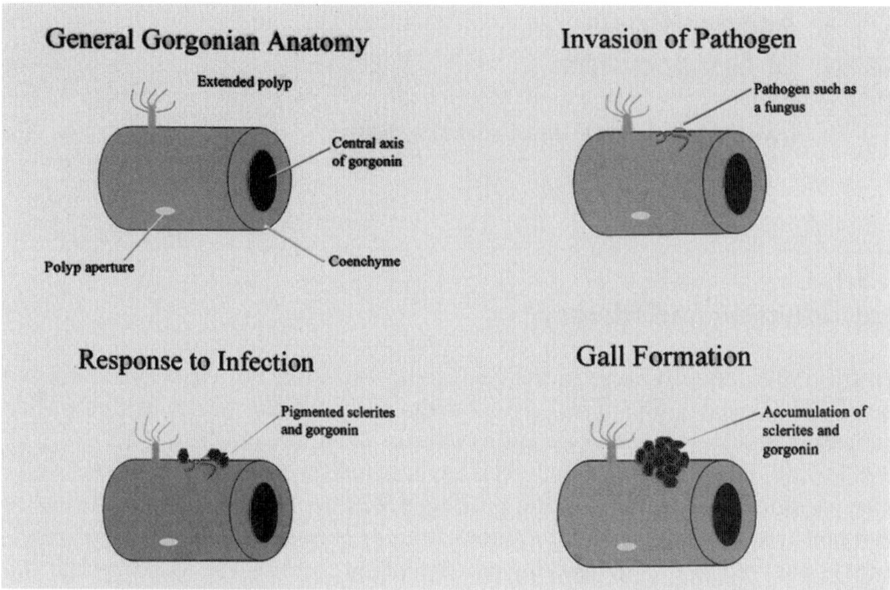

Fig. 15.1. *Top left* Generalized structure of a sea fan colony. *Top right* The initial invasion by a fungus. *Bottom left* The sclerite response to invasion. *Bottom right* The formation of a gall which can sequester the advance of the fungal pathogen

Infections are characterized by initial degradation and recession of the coenenchyme (the outer rind tissue containing the living polyps), and eventual exposure of the axial skeleton (central protein core). The exposed core often becomes colonized by a variety of cyanobacteria, algae and other organisms. Purpling of the tissue was common after the infection began and the eventual production of galls occurred with some, but not all affected sea fans (Fig. 15.1). Microscope comparisons of affected and unaffected tissue revealed the presence of fungal hyphae in all diseased samples.

15.3 Identity of the Pathogen

Evidence indicating that sea fan disease was actually an infectious disease came from a series of transfections (taking affected tissue from a diseased sea fan and attaching it to healthy sea fans) performed in the Bahamas (Fig. 15.2). In this study, 20 tissue squares (5×5 cm^2) were removed from healthy and affected *G. ventalina* colonies. Squares of each tissue type were grafted onto ten healthy *G. ventalina* colonies and ten healthy *G. flabellum* colonies. Some purpling of the tissue was observed on all of the *G. flabellum* grafts, but four developed disease lesions at the site of the disease grafts. Disease lesions were observed on six of the *G. ventalina* disease graft sites. This simple experiment indicated three things. Firstly, the disease was transmissible. Secondly, *G. ventalina* appeared

Fig. 15.2A–F. Transfections (grafting) of affected tissue onto healthy sea fans. **A–D** Affected *G. ventalina* transfected to healthy *G. ventalina*. Initial transfection (**A**), recession of tissue after 2 days (**B**), and 1 week (**C**). Close-up showing purpling response and secondary colonization (**D**). Infection of *G. mariae* by *A. sydowii* from *G. ventalina* (**E**), transfection of healthy *P. americana* from infected *P. americana* (**F**)

more susceptible than *G. flabellum*. Finally, not all exposures to the pathogen resulted in disease, indicating the presence of some resistance mechanisms.

Microscope and microbiological analysis of healthy and affected sea fan tissue throughout the Caribbean revealed that receding tissue always contained fungal hyphae (Fig. 15.3), while healthy tissue was free of hyphae (Smith et al. 1996). All affected tissue, when plated on fungal media, resulted in the growth of a specific

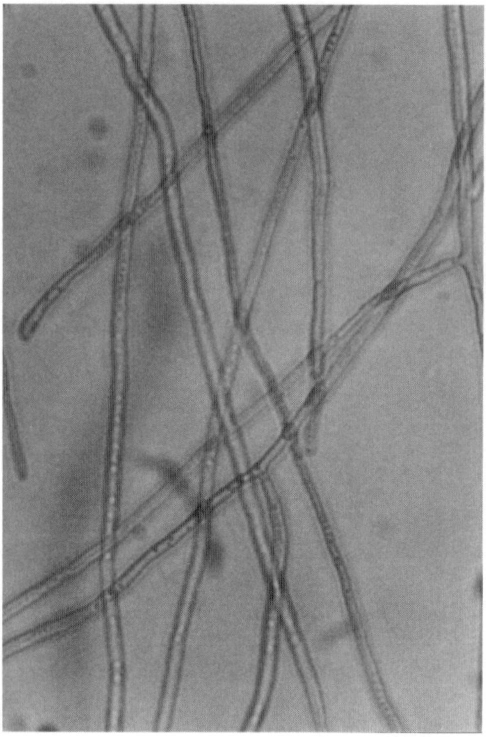

Fig. 15.3. Hyphae in affected *G. ventalina* tissue

bluish colony type. Microscope observations of the colonies showed the presence of echinulate asexual spores. Phenotypically, these isolates appeared to be in the *Aspergillus versicolor* group of fungi. Sequences of the 18S nuclear ribosomal small subunit genes from fungal isolates obtained from diseased sea fans in Trinidad, Saba, the Florida Keys and San Salvador, Bahamas were identical and a Blast search showed more identity to *Aspergillus fumigatus* (96%) than to other sequences in the GenBank database (Smith et al. 1996). Later, a 485-base-pair portion of the 5′ nontranslated region of the *trpC* genes from the same isolates were compared with those from four known *A. sydowii* strains, *A. versicolor*, and two *Emericella* species. These comparisons showed that the sea fan pathogens were indeed *A. sydowii* (Geiser et al. 1998).

Chemotaxonomy, based on the production of secondary metabolites (Malmstrom 2000), and carbon source utilization patterns (Alker et al. 2001), has also been used to determine species identity. Alker et al. (2001), using cluster analysis of the ability of strains to use 95 different sole carbon sources, showed that previously identified *A. sydowii* strains isolated from diseased sea fans were more similar to one another and reference strains than to other closely related species. Malmstrom (2000) showed that *A. sydowii* from diseased sea fans produced sydowinol, sydowinin A and B and sydowic acid. These are typical secondary metabolites produced by *A. sydowii*. Using a bioassay (Fig. 15.4), Malmstrom et

Fig. 15.4A–C. Bioassay for pathogenicity of *A. sydowii* strains. Different strains of *A. sydowii* are grown on solid media on which sterile strips of gauze are overlaid. The strips are then removed and inserted on healthy colonies (**A**). **B** The beginning of the purpling response. **C** The receding tissue (*lighter area*) around the inoculum

al. (2001) correlated pathogenicity with the production of these compounds. It is not yet known if these compounds play a roll in the actual pathogenic mechanism of the fungus.

15.4 Host Range

By 1999, other species of gorgonians were found to be degrading in the Bahamas, Puerto Rico and the Bahamas (see Chap. 2, Weil, this Vol.). The most obviously affected was *Pseudoterogorgia americana* (Smith et al., Weil et al., in prep). This species showed disease signs similar to the sea fans, including the recession of coenenchyme, purpling of adjacent tissue, production of galls and secondary colonization of exposed axial skeleton (Fig. 15.5). Transmissibility experiments (Fig. 15.4) showed that it was an infectious disease and that the pathogen resembled *A. sydowii* morphologically. Carbon source utilization patterns were consistent with *A. sydowii* and bioassays showed that the isolate caused lesions on *G. ventalina*. Furthermore, in 1999, a large stand of *P. acerosa* in St Croix USVI was found to be wasting away (Kaczmarsky et al., in prep.), and again, *A. sydowii* was isolated which also was positive in sea fan bioassays. Since this time, additional gorgonians have been observed showing signs of aspergillosis. Among them, *A. sydowii* was isolated from *Pseudoterogorgia americana, Gorgonia mariae, Plexaura flexuosa, Pseudoplexaura porosa and Plexaurella nutans* (Fig. 15.6; Weil et al., in prep.). It appears that *A. sydowii* is a general pathogen of Caribbean octocorals.

Fig. 15.5. *A. sydowii* infection of *Pseudoterogorgia americana*. Note the purpling response and galling previously described for *G. ventalina*

Fig. 15.6. Gorgonians from which *A. sydowii* has been isolated. Top right, *Gorgonia mariae*; top left, *Pseudoterogorgia americana*; center right, *Pseudoplexaura porosa*; center and bottom left, *Plexaura flexuosa*; bottom right, *Plexaurella nutans*

15.5 Future Work

Although there is little doubt that *A. sydowii* can infect living octocoral colonies, there is less known about the pathogenic mechanisms, the extent of susceptible host, the geographic distribution of the pathogen and the source/origin of the inoculum. Microscope observations give insight into the progression of the infection and secondary compounds (mycotoxins) have been identified from pathogenic strains. Assays using purified sydowinol, sydowinin A and B, and sydowic acid need to be evaluated. In addition, it is not known how the fungus can overcome natural defenses like the sclerite response (Smith et al. 1998) and naturally occurring anti-fungals (Kim et al. 2000). It is, however, likely that host defenses are weakened at the time of infection (Alker et al. 2001). So far, aspergillosis of gorgonians has only been reported from the Atlantic Ocean. Preliminary evidence exits (Garzon-Ferreira, pers. comm.) that the disease may also affect Pacific sea fans. The origin of the marine pathogenic strains of *A. sydowii* has been discussed and two theories have emerged. The first is that because *A. sydowii* is a ubiquitous soil saprophyte, it can be introduced into the marine environment through runoff. Large river systems could distribute the fungal spores throughout the Caribbean. Indeed, *A. sydowii* has been isolated from the benthic environments near the mouth of the Orinoco (Malmstrom 2000). On the other hand, pathogenic strains of *A. sydowii* have also been isolated during African dust events from the US Virgin Islands (Shinn et al. 2000; Garrison et al. 2003). These African dust strains produced mycotoxins and scored positive for pathogenicity in bioassays. Additional work is in progress to further characterize the pathogenic isolates genetically to help determine their origin.

Acknowledgments. This work was supported by NOAA CRES Grant 0648-0384, and the World Bank administered through GEF. We thank the Center for the Study of Emerging Diseases, Israel, for supporting the Coral Health and Disease Meeting. We are indebted to Eugene Rosenberg, Ruth Rossing and Andy Hooten for their support.

References

Alker AP, Smith GW, Kim K (2001) Characterization of *Aspergillus sydowii* (Thom et Church), a fungal pathogen of Caribbean sea fan corals. Hydrobiologia 460:105–111

Diaz JM, Garzon-Ferreira J, Zea S (1995) Los arrecifes coralinos de la isla de San Andres: estado actual y perspectivas para su conservacion. Academia Colombiana de Ciencias Exactas, Fisicas y Naturales, Coleccion Jorge Alvarez Lleras no 7, Santafe de Bogota, DC

Garrison VH, Shinn EA, Foreman WT, Griffin DW, Holmes CW, Kellogg CA, Majewski MS, Richardson LL, Ritchie KB, Smith GW (2003) African and Asian dust: from desert soils to coral reefs. BioScience 53:469–480

Garzon-Ferreira J, Zea S (1992) A mass mortality of *Gorgonia ventalina* (Cnidaria: Gorgoniidae) in the Santa Marta area, Caribbean coast of Colombia. Bull Mar Sci 50:522–526

Geiser DM, Taylor JW, Ritchie KB, Smith GW (1998) Cause of sea fan death in the West Indies. Nature 394:137–138

Guzman HM, Cortez J (1984) Mortand de *Gorgonia flabellum* Linnaeus (Octocorallia: Gorgoniidae) en la Costa Caribe de Costa Rica. Rev Biol Trop 32:305–308
Harvell CD, Suchanek TH (1987) Partial predation on tropical gorgonians by *Cyphoma gibbosum* (Gastropoda). Mar Ecol Prog Ser 38:37–44
Kim K, Harvell CD, Kim PD, Smith GW, Merkel SM (2000) Fungal disease resistance of Caribbean sea fan corals (*Gorgonia* spp.). Mar Biol 136:259–267
Kinzie RA (1973) The zonation of West Indian gorgonians. Bull Mar Sci 23:93–155
Laydoo R (1983) Recent mass mortality of gorgonians in Trinidad. Report Institute of Marine Affairs, Trinidad
Malmstrom J (2000) Marine fungal chemistry. PhD Diss, University of Copenhagen, Denmark
Malmstrom J, James SC, Polson SW, Smith GW, Frisvad JC (2001) Study of secondary metabolites associated with virulent and non-virulent strains of *Aspergillus sydowii*: sea fan pathogen. Proceedings of the 8th Symposium on the Natural history of the Bahamas, 8, pp 48–51
Morse DE, Morse A, Duncan H (1977) Algal tumors in the Caribbean sea fan, *Gorgonia ventalina*. Proc 3rd Int Coral Reef Symp 1:623–629
Morse DE, Morse A, Duncan H, Trench RK (1981) Algal tumors in the Caribbean octocorallian, *Gorgonia ventalina* II. Biochemical characterization of the algae, and first epidemiological observations. Bull Mar Sci 31:399–409
Nagelkerken I, Buchan K, Smith GW, Bonair K, Bush P, Garzon-Ferreira J, Botero L, Gayle P, Heberer C, Petrovic C, Pors L, Yoshioka P (1997a) Widespread disease in Caribbean sea fans I. Spreading and general characteristics. Proc 8th Int Coral Reef Symp 1:679–682
Nagelkerken I, Buchan K, Smith GW, Bonair K, Bush P, Garzon-Ferreira J, Botero L, Gayle P, Harvell CD, Heberer C, Kim K, Petrovic C, Pors L, Yoshioka P (1997b) Widespread disease in Caribbean sea fans: Patterns of infection and tissue loss. Mar Ecol Prog Ser 160:255–263
Shinn EA, Smith GW, Prospero JM, Betzer P, Hayes MA, Garrison V, Barber RT (2000) African dust and the demise of Caribbean coral reefs. Geophys Res Lett 27:3029–3032
Smith GW, Ives LD, Nagelkerken IA, Ritchie KB (1996) Caribbean sea fan mortalities. Nature 383:487
Smith GW, Harvell CD, Kim K (1998) Response of sea fans to infection with *Aspergillus* sp. (Fungi). Rev Biol Trop 46 [Suppl] 5:205–208
Wahle CM (1985) Habitat-related patterns of injury and mortality among Jamaican gorgonians. Bull Mar Sci 37:905–927

16 White Pox Disease of the Caribbean Elkhorn Coral, *Acropora palmata*

KATHRYN PATTERSON SUTHERLAND and KIM B. RITCHIE

16.1 Introduction

Populations of the most common Caribbean reef-building coral, *Acropora palmata,* are being decimated by white pox disease, with losses of living cover in the Florida Keys National Marine Sanctuary (FKNMS) averaging 88%. Elkhorn coral plays a significant role in the structural and functional integrity of Caribbean coral reef ecosystems. *A. palmata* is an important shallow water species, providing elevated rates of calcium carbonate deposition (Adey 1978) and a highly complex three-dimensional structure of the shallow water fore reef. This keystone species provides shelter and food for reef organisms and aids in the protection of coastal regions by serving as a buffer between land and sea. Severe population declines of *A. palmata* in the FKNMS and elsewhere in the Caribbean have led to the identification of this species as a candidate for inclusion on the Endangered Species List (Diaz-Soltero 1999).

White pox disease, also termed acroporid serratiosis (Patterson et al. 2002), and patchy necrosis (Bruckner and Bruckner 1997), exclusively affects *Acropora palmata* and was first documented in 1996 on reefs off Key West, Florida (Holden 1996). White pox has since been observed throughout the Caribbean (Porter et al. 2001; Rodríquez-Martínez et al. 2001; Santavy et al. 2001; Patterson et al. 2002). White pox disease is characterized by irregularly shaped, distinct white patches devoid of coral tissue (Fig. 16.1). Lesions can vary from a few to greater than 80 cm^2 and can develop simultaneously on all surfaces of the coral colony. Coral tissue loss occurs radially along the perimeter of the lesion at an average rate of 2.5 cm^2/day and is greatest during periods of seasonally elevated temperature. In heavily affected *A. palmata*, lesions can merge, resulting in tissue loss that spans the entire colony. White pox is highly contagious with nearest neighbors most susceptible to infection (Patterson et al. 2002). The disease spread rapidly within and between reefs in the FKNMS during the mid-1990s (Porter et al. 2001; Patterson et al. 2002).

The distinct white patches and the potential for tissue loss throughout the coral colony distinguishes white pox disease from white band disease, which also affects *Acropora palmata*, developing at the base of a coral branch and progressing upward toward the branch tip in a concentric ring. White band disease is characterized by a distinct white band of recently denuded coral skeleton (type I; Gladfelter 1982), or bleached tissue (type II; Ritchie and Smith 1998), followed by a line of necrosis. White pox disease signs also clearly differ

Fig. 16.1. A *Acropora palmata* colony affected with white pox disease. (Photograph by J.W. Porter). **B** White pox disease lesions on *A. palmata.* (Photograph by K.P. Sutherland)

from coral bleaching and predation scars produced by the corallivorous snail, *Coralliophila abbreviata* (Knowlton et al. 1990; Miller 2001).

16.2 The Coral Pathogen, *Serratia marcescens*

White pox disease is caused by *Serratia marcescens* (Patterson et al. 2002), a common Gram-negative bacterium classified as a coliform and a member of the Enterobacteriaceae family. *S. marcescens* is found in the intestines of humans, insects, and other animals, and in freshwater, soil, and plants (Grimont and Grimont 1994). *S. marcescens* is pathogenic to humans, cows, goats, chickens, fishes, insects, and plants (Baya et al. 1992; Grimont and Grimont 1994). *S. marcescens* is an opportunistic pathogen of humans associated with both waterborne infections in tropical waters (Hazen 1988) and hospital-acquired infections, including urinary tract infections, wound infections, pneumonia, and bacteremia (Grimont and Grimont 1994; Miranda et al. 1996; Shi et al. 1997).

The pathogenic mechanism of *Serratia marcescens* for elkhorn coral is currently unknown, thus our best available model for pathogenesis is derived from human clinical studies. *S. marcescens* is known to produce a number of virulence factors including serralysin (a peptidase), proteases, chitinases, extracellular lipases and nucleases (Hertle 2002). *S. marcescens* cell wall includes a lipopolysaccharide (LPS) that is responsible for endotoxin activity (Hejazi and Falkiner 1997). A hemophore secreted by *S. marcescens* was shown to be cytotoxic to cells other than red blood cells (Arnoux et al. 1999; Marty et al. 2002). Polson (2002) showed that the white pox isolate also produces lipases and nucleases, but their role in pathogenesis is presently unknown. There is an inverse relationship between pigment production (prodigiosin) and toxicity of *S. marcescens* strains against nematodes (Carbonelli et al. 2000). Likewise, human clinical isolates are rarely pigmented (Hejazi and Falkiner 1997). The white pox isolate is nonpigmented and, therefore, does not produce prodigiosin, but the role of the pigment in pathogenesis is also unclear. Hejazi and Falkiner (1997) described adherence and pathogenicity differences between *S. marcescens* strains exhibiting mannose-resistant and mannose-sensitive surface pili. Mannose-sensitive pili are important in bacterial colonization. In *Vibrio cholerae* mannose-sensitive hemagglutinin (MSHA) mediates the adherence to commensal zooplankton carapaces (Chiavelli et al. 2001).

Quorum sensing is a signaling process by which a number of marine (and other) bacteria turn on genes (including those for virulence factors) as a result of the production of the signal compound by other bacteria. Most known quorum sensing systems of Gram-negative bacteria involve the secretion of N-acyl-homoserine lactone or similar lactones and/or quinolones. Swarming and the production of protease (behaviors found in other quorum sensors) have been reported for *Serratia* species (Givskov et al. 1997; Lindum et al. 1998).

The prevalence of *Serratia marcescens* in the marine environment is unknown. However, this bacterium has been found in the marine environment in sewage-polluted estuaries. For example, *S. marcescens* has been linked to disease of white perch (*Morone americanus*) in the sewage-polluted Back River, Maryland (Baya et al. 1992). Other *Serratia* species are known to cause disease in marine fishes (Austin and Austin 1999) and to pose a serious threat as opportunistic pathogens of marine organisms (Ingles et al. 1993).

White pox is one of approximately 18 coral diseases documented worldwide (Sutherland et al. 2004), but *Serratia marcescens* is only the fifth pathogen to be confirmed as a coral disease agent through the fulfillment of Koch's postulates (Kushmaro et al. 1996; Smith et al. 1996; Richardson et al. 1998a, b; Ben-Haim and Rosenberg 2002; Patterson et al. 2002) and the first agent with a possible link to human sewage pollution (Patterson et al. 2002). Identification of *S. marcescens* as a coral pathogen marked the first time that a common member of the human gut microbiota was shown to be a marine invertebrate pathogen. While *S. marcescens* is ubiquitous, its noted association with human hosts prompts speculation that improperly treated sewage from the Florida Keys may be associated with white pox disease. Human sewage markers (e.g., human enteric bacteria and viruses) are prevalent on coral surfaces and in near-shore, offshore, and canal waters of the FKNMS (Paul et al. 1995a, b, 1997; Griffin et al. 1999; Lipp et al. 2002), suggesting land-based activities may affect reef coral health and coral reef survival.

16.3 Loss of *Acropora palmata* in the Florida Keys National Marine Sanctuary: 1994–2002

Annual monitoring of Eastern Dry Rocks Reef (24°27.715′N, 81°50.801′W) in the FKNMS began in 1994 (Patterson et al. 2002). Seven additional reef sites in the FKNMS have been monitored on an annual basis since 1996: Carysfort Reef (25°13.205′N; 80°12.628′W), Grecian Rocks Reef (25°06.450′N; 80°18.410′W), Molasses Reef (25°00.525′N; 80°22.589′W), Rock Key Reef (24°27.285′N; 81°51.589′W), Sand Key Reef (24°27.119′N; 81°52.650′W), Sombrero Reef (24°37.531′N; 81°06.624′W), and Western Sambo Reef (24°28.771′N; 81°42.970′W; Patterson et al. 2002; Porter et al. 2002).

Between 1994 and 2002, the percent cover of *Acropora palmata* at Eastern Dry Rocks declined by 97%, from 23.9% in 1994 to 0.82% in 2002 (Fig. 16.2). Between 1996 and 2001, the percent cover of *A. palmata* dramatically declined at each of the seven additional reef sites: Carysfort, 90%; Grecian Rocks, 89%; Molasses, 90%; Rock Key, 79%; Sand Key, 100%; Sombrero, 100%; Western Sambo, 90% (Fig. 16.3). The complete loss of *A. palmata* at Sombrero Reef and Sand Key Reef occurred by 1999 and 2001, respectively. Loss of *A. palmata* at all eight surveyed reefs in the FKNMS averaged 92% between 1996 and 2001 (Fig. 16.3). Between 2001 and 2002, the percent cover of *A. palmata* increased at Carysfort, Molasses, and Rock Key and continued to decline at Grecian Rocks and Western Sambo. The continued loss was especially pronounced at Grecian Rocks where an additional 53% of *A. palmata* cover was lost, with a total loss at this reef of 95% between 1996 and 2002 (Fig. 16.3).

Populations of *Acropora palmata* in the FKNMS sustained losses averaging 88% between 1996 and 2002. These losses are approximately double the loss caused by white plague type II disease on *Dichocoenia stokesii* colonies in the FKNMS (Richardson et al. 1998a, b) and are comparable to *A. palmata* losses at other locations throughout the Caribbean (Bythell and Sheppard 1993; Aronson and Precht 2001; Miller et al. 2002). While *A. palmata* population declines elsewhere in the Caribbean were attributed to hurricanes, bleaching, and white band disease (Gladfelter 1982; Porter and Meier 1992; Bythell and Sheppard 1993; Aronson and Precht 2001; Miller et al. 2002), losses in the

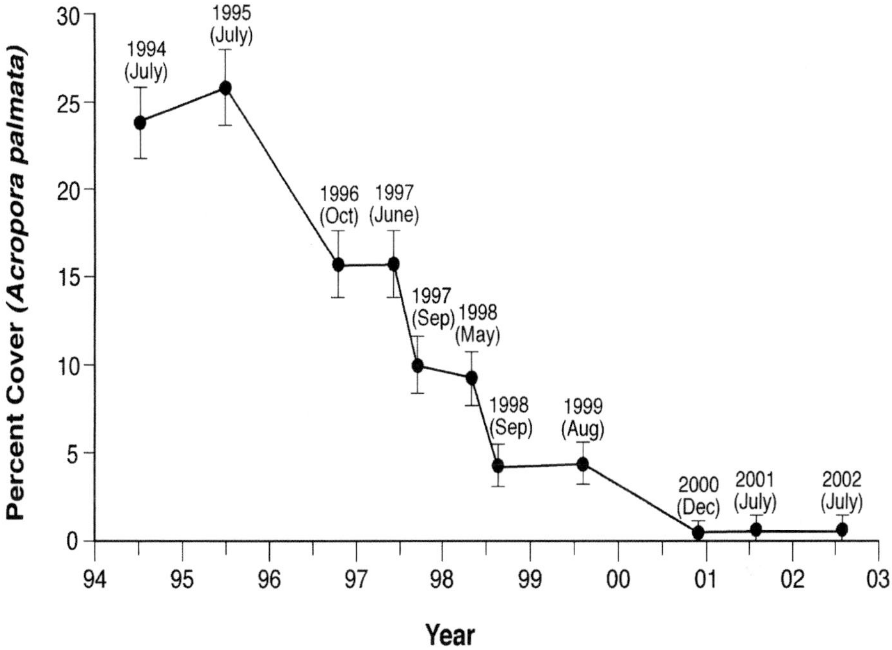

Fig. 16.2. Percent cover of *Acropora palmata* at Eastern Dry Rocks Reef, Key West, FL, 1994–2002. Data are presented as mean ± SD

FKNMS were primarily caused by white pox disease. By 1997, 1 year after the first documentation of the disease (Holden 1996), white pox was found at all surveyed reefs in the FKNMS (Patterson et al. 2002; Porter et al. 2002). Signs of active white pox disease were observed on *A. palmata* colonies at Eastern Dry Rocks every year between 1996 and 2000 (Patterson et al. 2002) and at each of the other reefs with living cover of *A. palmata* every year between 1997 and 2002. One exception was Carysfort Reef, where white pox disease signs were not observed at the time of the 2000 survey. Observations of white band disease, on the other hand, were rare at monitored reefs in the FKNMS between 1996 and 2002.

In late September 1998, reefs in the FKNMS were impacted by both Hurricane Georges and mass-bleaching (Wilkinson et al. 1999). However, the 82% loss of *Acropora palmata* recorded at Eastern Dry Rocks between 1994 and 1998 (Fig. 16.2) occurred prior to both the hurricane and the bleaching event. Hurricane or bleaching damage may have contributed to the further decline of *A. palmata* on this reef after 1998. However, it is important to note that the first post-hurricane/post-bleaching survey (August 1999) showed a 3% increase in percent live cover of *A. palmata* (Fig. 16.2).

Between 2000 and 2001, living cover of *Acropora palmata* at Eastern Dry Rocks increased 66.7% from 0.49 to 0.82% (Fig. 16.2). This increase in living cover was due to growth of remaining fragments of living tissue (Fig. 16.4A, B).

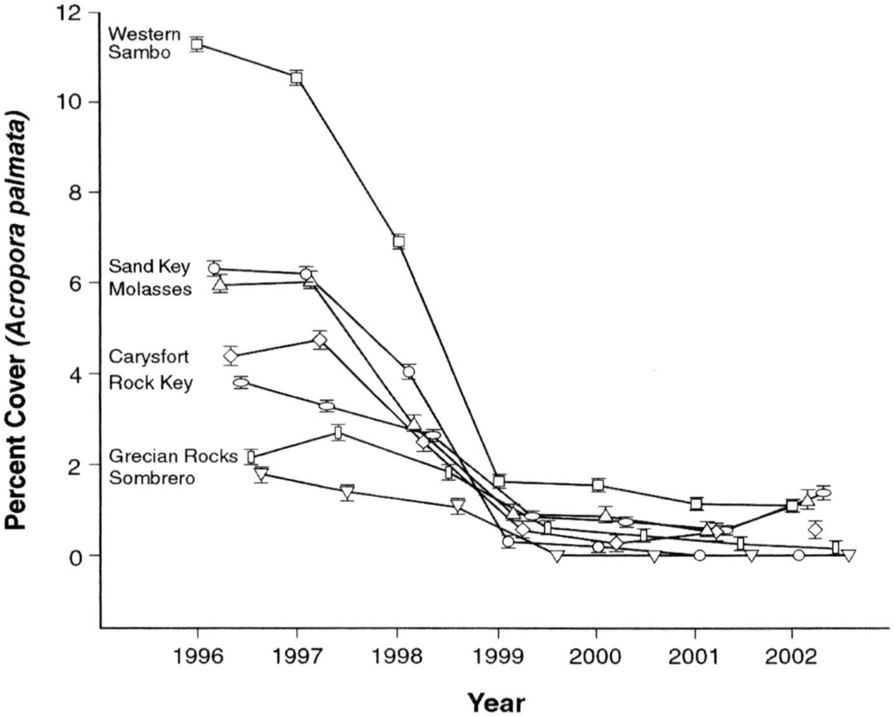

Fig. 16.3. Percent cover of *Acropora palmata* at seven reef sites in the Florida Keys National Marine Sanctuary, 1996–2002: Western Sambo Reef (*squares*), Sand Key Reef (*circles*), Molasses Reef (*triangles*), Carysfort Reef (*diamonds*), Rock Key Reef (*ovals*), Grecian Rocks Reef (*rectangles*), Sombrero Reef (*inverted triangles*). Data are presented as mean ± SD

Coral recruitment was not observed at Eastern Dry Rocks at any time during the 9-year survey. All seven elkhorn colonies that exhibited tissue gain in 2001 showed either complete or partial tissue loss in 2002 (Fig. 16.4C). This loss may be attributed to disease or predation. While signs of active white pox disease were not observed on this reef at the time of the 2001 and 2002 surveys, percent cover of *A. palmata* was only 0.82% during these two survey years, and therefore, minimal surface area of living tissue was available for the disease to affect. Declining population numbers may make *A. palmata* especially vulnerable to white pox disease and predation by *Coralliophila abbreviata*, which preferentially feeds on this coral species (Knowlton et al. 1990; Miller 2001). Decimation of *A. palmata* populations may also limit the reproductive capacity of this species which reproduces almost exclusively by fragmentation (Aronson and Precht 2001). While vegetative reproduction may be well adapted to recolonization following mechanical disturbances such as hurricanes, colony fragmentation is ineffectual following severe population declines due to disease, which frequently kills the entire coral colony.

Fig. 16.4. Photographic time series of gain and loss of percent cover *Acropora palmata* at Eastern Dry Rocks Reef, Key West, FL, 2000–2002. The *A. palmata* fragment was photographed on **A** 28 December 2000 and **B** 30 July 2001. Despite the growth observed between 2000 and 2001, the fragment had disappeared by **C** 29 July 2002. This same pattern of tissue gain between 2000 and 2001, followed by tissue loss between 2001 and 2002 was observed on seven *A. palmata* colonies at this reef. (Photographs by J.W. Porter and K.P. Sutherland)

Acropora palmata colonies at Eastern Dry Rocks exhibited reduced tissue loss in the winter followed by accelerated tissue loss in the summer. This seasonal trend in percent change of *A. palmata* is evidenced by the stair-step pattern of tissue loss (Fig. 16.2) and may be attributable to the effects of seasonal seawater temperatures and precipitation on white pox disease pathogenicity. Elevated seawater temperature is a stressor in corals, causing thermally induced breakdown in the coral–zooxanthellae host–symbiont relationship

(Brown 1997), promoting growth and virulence of pathogens (Kushmaro et al. 1996, 1998; Toren et al. 1998; Alker et al. 2001; Banin et al. 2001; Israely et al. 2001), and reducing immune response in host corals (Toren et al. 1998; Alker et al. 2001). Increased rainfall may increase seepage of sewage from septic tanks (Rose et al. 2001) and seed the marine environment with *Serratia marcescens* and other human fecal enteric bacteria and viruses.

The white pox disease epizootic has caused catastrophic losses of *Acropora palmata* in the FKNMS in just 9 years. These losses illustrate the impact that disease can have on coral communities. The susceptibility of *A. palmata* to disease and predation, combined with the reproductive strategies of this species, exacerbate the impact of declining populations on the stability of Caribbean coral reef ecosystems. The monumental losses of *A. palmata* in the FKNMS and elsewhere in the Caribbean signify an urgent need for the protection of this keystone species under the Endangered Species Act.

16.4 Potential Sources of the White Pox Pathogen

Potential sources of the white pox disease pathogen include wastewater influent, septic tank effluent, feces of reef fishes, canal water, reef water column, and white pox diseased and apparently healthy *Acropora palmata*. Accumulating evidence suggests that the health of reef organisms in the FKNMS and elsewhere in the Caribbean is affected by pollution of fecal origin. White pox disease is caused by a fecal enteric bacterium of possible human origin (Patterson et al. 2002) and human sewage markers are concentrated on coral surfaces in nearshore waters of the FKNMS (Lipp et al. 2002). Bacteria associated with human fecal contamination have been found within the microbial mat that causes black band disease of corals (Frias-Lopez et al. 2002). Since 1996, populations of the sewage consuming reef sponge *Cliona delitrix* have increased by a factor of 10 on reefs in the FKNMS (Ward-Paige 2003) while, concurrently, corals have declined by 37% (Porter et al. 2002).

It is tempting to speculate that poor waste disposal practices may be associated with the declining health of tropical marine invertebrates in the FKNMS. Full-scale sewage treatment plants service the communities of Key West, Key Colony Beach on Marathon, and Ocean Reef Resort on Key Largo. The remainder of the Florida Keys utilizes on-site waste disposal practices including septic systems, injection wells, and illegal cesspools. Keys-wide, there are at least 24,000 septic tanks, 600–700 injection wells, and 5000–10,000 cesspools (Lapointe et al. 1990; Shinn et al. 1994). On-site waste disposal contaminates the waters of the FKNMS with nutrients and microorganisms of human fecal origin (Lapointe et al. 1990; Paul et al. 1995a, b, 1997; Griffin et al. 1999). Fecal bacteria and viruses migrate quickly (0.57–140.9 mph) from the site of disposal through the porous limestone bedrock of the Florida Keys and finally to the marine environment (Paul et al. 2000). The general direction of contaminant flow from land is toward the reef tract (Paul et al. 1995a, 1997, 2000).

16.5
Unresolved Questions and Future Research

Although *Serratia marcescens* occurs in human sewage, the source of the *S. marcescens* strain that causes white pox disease is uncertain (Patterson et al. 2002). Given the magnitude of *Acropora palmata* loss in the FKNMS, research is needed to identify the source of this pathogen. A variety of potential sources are presently being screened for biotypes similar to the *Serratia* pathogen. These potential sources include wastewater influent, septic tank effluent, feces of reef fishes, seabird guano, canal water, reef water column, and white pox diseased and apparently healthy *A. palmata*. Coral reef managers and wastewater treatment engineers require certainty of the human origin of the white pox pathogen in order to recommend expensive sewage treatment upgrades in Florida and around the Caribbean.

Research is continuing to determine mechanisms of pathogenesis of *Serratia marcescens* against *Acropora palmata*. Future research will identify virulence genes in the coral pathogen and prevalence of these genes in environmental isolates of *S. marcescens*. In addition, future studies will reveal the host range of the *Serratia* pathogen. It is possible that this *Serratia* isolate does not exclusively affect *A. palmata*, but instead affects other coral species on which disease signs are manifested differently. Since the etiologies of the majority of the coral disease conditions described to date are unknown (Richardson 1998), the extent of the pathogenicity of the white pox pathogen warrants examination.

References

Adey WH (1978) Coral reef morphogenesis: a multidimensional model. Science 202:831–837

Alker AP, Smith GW, Kim K (2001) Characterization of *Aspergillus sydowii* (Thom et Church) a fungal pathogen of Caribbean sea fan corals. Hydrobiologia 460:105–111

Arnoux, P, Hasser R, Izadi N, Lecroisey A, Delepierre M, Wandersman C, Czjzek M (1999) The crystal structure of HasA, a hemophore secreted by *Serratia marcescens*. Nat Struct Biol 6:516–520

Aronson RB, Precht WF (2001) White-band diseases and the changing face of Caribbean coral reefs. Hydrobiologia 460:25–38

Austin B, Austin DA (1999) Bacterial fish pathogens: disease of farmed and wild fish. Praxis Publ, Chichester

Banin E, Khare SK, Naider F, Rosenberg E (2001) Proline-rich peptide from the coral pathogen *Vibrio shiloi* that inhibits photosynthesis of zooxanthellae. Appl Environ Microbiol 67:1536–1541

Baya AM, Toranzo AE, Lupiani B, Santos Y, Hetrick FM (1992) *Serratia marcescens:* a potential pathogen for fish. J Fish Dis 15:15–26

Ben-Haim Y, Rosenberg E (2002) A novel *Vibrio* sp. pathogen of the coral *Pocillopora damicornis*. Mar Biol 141:47–55

Brown BE (1997) Coral bleaching: causes and consequences. Coral Reefs 16:129–138

Bruckner AW, Bruckner RJ (1997) Outbreak of coral disease in Puerto Rico. Coral Reefs 16:260

Bythell J, Sheppard C (1993) Mass mortality of Caribbean shallow corals. Mar Pollut Bull 26:296–297

Carbonelli GV, Della Colleta HH, Yano T, Darini AL, Levy CE, Fonseca BA (2000) Clinical relevance and virulence factors of pigmented *Serratia marcescens*. FEMS Immunol Med Microbiol 28(2):143–149

Chiavelli DA, Marsh JW, Taylor RK (2001) The mannose-sensitive hemagglutinin of *Vibrio cholerae* promotes adherence to zooplankton. Appl Environ Microbiol 67(7):3220–3225

Diaz-Soltero H (1999) Endangered and threatened species: a revision of candidate species list under the Endangered Species Act. Fed Register 64:33466–33468

Frias-Lopez J, Zerkle AL, Bonheyo GT, Fouke BW (2002) Partitioning of bacterial communities between seawater and healthy black band diseased and dead coral surfaces. Appl Environ Microbiol 68:2214–2228

Givskov M, Eberl L, Molin S (1997) Control of exoenzyme production, motility and cell differentiation in *Serratia liquefaciens*. FEMS Microbiol Lett 148:115–122

Gladfelter WB (1982) White-band disease in *Acropora palmata*: implications for the structure and growth of shallow reefs. Bull Mar Sci 32:639–643

Griffin DW, Gibson CJ III, Lipp EK, Riley K, Paul JH III, Rose JB (1999) Detection of viral pathogens by reverse transcriptase PCR and of microbial indicators by standard methods in the canals of the Florida Keys. Appl Environ Microbiol 65:4118–4125

Grimont PA, Grimont F (1994) Genus VIII *Serratia* Bizio, 1823. In: Holt JG, Kreig NR, Sneath PHA, Staley JT, Williams ST (eds) Bergey's manual of determinative bacteriology, vol 4. Williams and Wilkins, Baltimore, pp 477–484

Hazen TC (1988) Fecal coliforms as indicators in tropical waters: a review. Tox Assess Int J 3:461–477

Hejazi A, Falkiner FR (1997) *Serratia marcescens*. J Med Microbiol 46:903–912

Hertle R (2002) *Serratia marcescens* hemolytic (ShlA) binds artificial membranes and forms pores in a receptor-independent manner. J Membr Biol 189:1–14

Holden C (1996) Coral disease hot spot in the Florida Keys. Science 274:2017

Ingles V, Roberts RJ, Bromage NR (eds) (1993) Bacterial diseases of fish. Halsted Press, New York

Israely T, Banin E, Rosenberg E (2001) Growth differentiation and death of *Vibrio shiloi* in coral tissue as a function of seawater temperature. Aquat Microb Ecol 24:1–8

Knowlton N, Lang JC, Keller BD (1990) Case study of natural population collapse: post-hurricane predation on Jamaican staghorn corals. Smithson Contrib Mar Sci 31:1–25

Kushmaro A, Loya Y, Fine M, Rosenberg E (1996) Bacterial infection and coral bleaching. Nature 380:396

Kushmaro A, Rosenberg E, Fine M, Ben-Haim Y, Loya Y (1998) Effect of temperature on bleaching of the coral *Oculina patagonica* by *Vibrio shiloi* AK-1. Mar Ecol Prog Se 171:131–137

Lapointe BE, O'Connell JD, Garrett GS (1990) Nutrient coupling between on-site sewage disposal systems groundwaters and nearshore surface waters of the Florida Keys Biogeochem 10:289–307

Lindum PW, Anthoni U, Chistopherson C, Eberl L, Molin S, Giskov M (1998) N-acyl-homoserine lactone autoinducers control production of an extracellular lipopeptide biosurfactant required for swarming motility of *Serratia liquefaciens* MG1. J Bacteriol 180:6384–6388

Lipp EK, Jarrell JL, Griffin DW, Lukasik J, Jacukiewicz J, Rose JB (2002) Preliminary evidence for human fecal contamination in corals of the Florida Keys, USA. Mar Pollut Bull 44:666–670

Marty KB, Williams CL, Guynn LJ, Benedik MJ, Blanke SR (2002) Characterization of a cytotoxic factor in culture filtrates of *Serratia marcescens*. Infect Immun 70:1121–1128

Miller MW (2001) Corallivorous snail removal: evaluation of impact on *Acropora palmata*. Coral Reefs 19:293–295

Miller MW, Bourgue AS, Bohnsack JA (2002) An analysis of the loss of acroporid corals at Looe Key, Florida USA: 1983–2000. Coral Reefs 21:179–182

Miranda G, Kelly C, Solorzano F, Leanos B, Coria R, Patterson JE (1996) Use of pulsed-field gel electrophoresis typing to study an outbreak of infection due to *Serratia marcescens* in a neonatal intensive care unit. J Clin Microbiol 34:3138–3141

Patterson KL, Porter JW, Ritchie KB, Polson SW, Mueller E, Peters EC, Santavy DL, Smith GW (2002) The etiology of white pox a lethal disease of the Caribbean elkhorn coral, *Acropora palmata*. Proc Natl Acad Sci USA 99:8725–8730

Paul JH, Rose JB, Brown J, Shinn EA, Miller S, Farrah SR (1995a) Viral tracer studies indicate contamination of marine waters by sewage disposal practices in Key Largo, Florida. Appl Environ Microbiol 61:2230–2234

Paul JH, Rose JB, Jiang S, Kellogg C, Shinn EA (1995b) Occurrence of fecal indicator bacteria in surface waters and the subsurface aquifer in Key Largo, Florida. Appl Environ Microbiol 61:2235–2241

Paul JH, Rose JB, Jiang SC, Zhou X, Cochran P, Kellogg C, Kang JB, Griffin D, Farrah S, Lukasik J (1997) Evidence for groundwater and surface marine water contamination by waste disposal wells in the Florida Keys. Water Res 31:1448–1454

Paul JH, McLaughlin MR, Griffin DW, Lipp EK, Stokes R, Rose JB (2000) Rapid movement of wastewater from on-site disposal systems into surface waters in the lower Florida Keys. Estuaries 23:662–668

Polson SW (2002) Isolation and characterization of bacteria associated with three coral diseases. Masters Thesis, Clemson University, Clemson, South Carolina

Porter JW, Meier OW (1992) Quantification of loss and change in Floridian reef coral populations. Am Zool 32:625–640

Porter JW, Dustan P, Jaap WC, Patterson KL, Kosmynin V, Meier OW, Patterson ME, Parsons M (2001) Patterns of spread of coral disease in the Florida Keys. Hydrobiologia 460:1–24

Porter JW, Kosmynin V, Patterson KL, Porter KG, Jaap WC, Wheaton JL, Hackett K, Lybolt M, Tsokos CP, Yanev G, Marcinek DM, Dotten J, Eaken D, Patterson M, Meier OW, Brill M, Dustan P (2002) Detection of coral reef change by the Florida Keys Coral Reef Monitoring Project. In: Porter JW, Porter KG (eds) The everglades Florida Bay and coral reefs of the Florida Keys. CRC Press, Boca Raton, pp 749–769

Richardson LL (1998) Coral diseases: what is really known? Trends Ecol Evol 13:438–443

Richardson LL, Goldberg WM, Carlton RG, Halas JC (1998a) Coral disease outbreak in the Florida Keys: plague type II. Rev Biol Trop 46:187–198

Richardson LL, Goldberg WM, Kuta KG, Aronson RB, Smith GW, Ritchie KB, Halas JC, Feingold JS, Miller SL (1998b) Florida's mystery coral killer identified. Nature 392:557–558

Ritchie KB, Smith GW (1998) Type II white-band disease. Rev Biol Trop 46:199–203

Rodríquez-Martínez RE, Banaszak AT, Jordán-Dahlgren E (2001) Necrotic patches affect *Acropora palmata* (Scleractinia: Acroporidae) in the Mexican Caribbean. Dis Aquat Org 47: 229–234

Rose JB, Epstein PR, Lipp EK, Sherman BH, Bernard SM, Patz JA (2001) Climate variability and change in the United States: potential impacts and foodborne diseases caused by microbiologic agents. Environ Health Perspect 109:211–221

Santavy DL, Mueller E, Peters EC, MacLaughlin L, Porter JW, Patterson KL, Campbell J (2001) Quantitative assessment of coral diseases in the Florida Keys: strategy and methodology. Hydrobiologia 460:39–52

Shi Z-Y, Liu PY-F, Lin Y-H, Hu B-S (1997) Use of pulsed-field gel electrophoresis to investigate an outbreak of *Serratia marcescens*. J Clin Microbiol 35:325–327

Shinn EA, Reese RS, Reich CD (1994) Fate and pathways of injection-well effluent in the Florida Keys. Department of the Interior/US Geological Survey, Washington, DC

Smith GW, Ives LD, Nagelkerken IA, Ritchie KB (1996) Caribbean sea-fan mortalities Nature 383:487

Sutherland KP, Porter JW, Torres T (2004) Disease and immunity in Caribbean and Indo-Pacific zooxanthellate corals. Mar Ecol Prog Ser 266:273–302

Toren A, Landau L, Kushmaro A, Loya Y, Rosenberg E (1998) Effect of temperature on adhesion of *Vibrio* strain AK-1 to *Oculina patagonica* and on coral bleaching. Appl Environ Microbiol 64:1379–1384

Ward-Paige CA (2003) Bioerosion surveys on the Florida reef tract suggest widespread land-based stress on reefs. MS thesis, McMaster University, Ontario

Wilkinson C, Lindén O, Cesar H, Hodgson G, Rubens J, Strong AE (1999) Ecological and socioeconomic impacts of 1998 coral mortality in the Indian Ocean: an ENSO impact and a warning of future change? Ambio 28:188–196

17 Temperature-Regulated Bleaching and Tissue Lysis of *Pocillopora damicornis* by the Novel Pathogen *Vibrio coralliilyticus*

Yael Ben-Haim Rozenblat and Eugene Rosenberg

17.1 Introduction

Coral reefs are the most diverse and dynamic tropical and subtropical ecosystems on earth, providing substrata and shelter for hundreds of thousands of species. They are renowned for their extraordinary natural beauty and high productivity. Coral reefs provide an essential source of food, tourist income and natural coastal protection from storm damage, erosion and flooding for many coastal communities, the majority of which are in developing countries. They also provide a potential source for the isolation of biologically active metabolites of potential medical value (Sammarco and Coll 1988, 1992; Coll 1992; Jensen et al. 1996; Koh 1997). Hermatypic corals play a key role in forming the structure of coral reefs. The symbiotic association between corals and their photosynthetic algae, the zooxanthellae, is one of the primary reasons for the success of corals in these ecosystems (Sebens 1994). This symbiosis allows the coral animal to exist and flourish in oligotrophic environments.

Coral bleaching and other diseases of corals have been increasing during the last few decades. This increase includes the appearance of new, previously undescribed diseases, the reemergence of more virulent forms of known diseases and an increase in the number of coral species affected (Brown 1997; Richardson 1998; Hoegh-Guldberg 1999; Rosenberg and Ben-Haim 2002). In many cases, these bleaching events resulted in significant losses of hard coral cover, and in some cases entire coral reef ecosystems were damaged (Glynn 1984; Peters 1984; Oliver 1985; Lang et al. 1992; Loya et al. 2001). Coral bleaching is the disruption of symbioses between the coral animals and their photosynthetic zooxanthellae (Hayes and Goreau 1998). The loss of the symbiotic algae and/or their photosynthetic pigments results in paling or bleaching of the coral due to the reduced concentration of algal pigments and the increased visibility of the white calcareous skeleton of the coral. Bleaching of corals results in malnutrition, since a major portion of a coral's nutrition comes from the photosynthetic products of the zooxanthellae (Glynn 1991), reduced growth rates, and suppression of sexual reproduction (Szmant and Gassman 1990; Fine et al. 2002b). Bleaching may also lead to impaired healing after mechanical damage, as well as increased susceptibility to secondary diseases (Harvell et al. 2001). If the bleaching process is not reversed within a few weeks or months, depending upon the specific coral species and conditions, bleaching would result in coral mortality (Szmant and Gassman 1990; Brown et al. 2002; Rosenberg and Ben-Haim 2002).

Considerable data have been accumulated suggesting that bleaching is triggered by environmental factors, which impose stress on the corals. The most frequently reported stress condition is increased sea temperature (Hoegh-Guldberg and Smith 1989; Glynn and D'Croz 1990; Jokiel and Coles 1977, 1990; Lesser et al. 1990; Iglesias-Prieto et al. 1992; Glynn 1993; Fitt and Warner 1995; Warner et al. 1996). In general, coral bleaching has coincided with the hottest period of the year (Glynn 1991; Kushmero et al. 1997), and is most severe at times of warmer-than-normal conditions (Hoegh-Guldberg 1999). In addition, a number of other environmental factors have been suggested to cause coral bleaching, e.g., decreased seawater temperature (Coles and Fadlallah 1991; Muscatine et al. 1991; Gates et al. 1992; Kobluk and Lysenko 1994), increased solar irradiation including ultraviolet radiation (Gleason and Wellington 1993; Brown et al. 1994; Fisk and Done 1985; Dunne and Brown 2001), decreased ultraviolet radiation (Lyons et al. 1998; Fine et al. 2002a), reduced salinity (Goreau 1964; Fang et al. 1995; van Woesik et al. 1995) and combinations of elevated temperature and irradiance (Harriott 1985; Brown and Suharsono 1990; Lesser et al. 1990; Williams and Bunkley-Williams 1990; Glynn et al. 1992).

In 1996, Kushmaro et al. reported for the first time that bleaching of *Oculina patagonica* corals in the Mediterranean Sea was the result of a bacterial infection. Bleaching of these corals has been recorded each summer when the seawater temperature reaches a maximum of 29–31 °C. The aetiological agent of bleaching *O. patagonica* is *Vibrio shiloi* (Kushmaro et al. 1996, 1997). In controlled aquaria experiments, it was shown that the infection and resulting bleaching disease was temperature-dependent, occurring only at summer temperatures (Kushmaro et al. 1998). The major effect of increased temperature is the induction of bacterial virulence factors, including an adhesin that binds to a β-galactose containing receptor on the coral surface (Toren et al. 1998), superoxide dismutase (Banin et al. 2003), and toxins that inhibit photosynthesis, bleach and lyse the zooxanthellae (Ben-Haim et al. 1999; Banin et al. 2001b). The virulence factors of *V. shiloi* and the effect of temperature are discussed by E. Rosenberg in Chapter 25. In principle, the correlation between increased seawater temperature and infectious disease could be the result of increased sensitivity of the host to the pathogen, increased virulence of the pathogen, higher frequency of transmission, or a combination of the three factors.

17.2 *Pocillopora damicornis*

Pocillopora damicornis is a major reef-building coral, widely distributed in coral reefs throughout the Pacific Ocean, Indian Ocean, Persian Gulf and the Red Sea. It is thought that the last genus to be extinct in the Caribbean Sea was *Pocillopora. P. damicornis* colonies appear in brown, pink or greenish morphs, while the growing tips are white. It occurs in all shallow water habitats, from exposed reef fronts to mangrove swamps and wharf piles. The colonies are compact and consist of thin irregular branching. Branches are fine and widely spaced in low current zones, such as sheltered reef flats or deep habitats, and become thick and compact under strong wave action, characteristic of the upper reef slopes. *P. damicornis* phylogenetic classification: *Family*: Pocilloporidae; *Order*: Scleractinia; *Subclass*: Zoantharia; *Class*: Anthozoa; *Phylum*: Cnidaria (Vernon 2000).

P. damicornis corals have been commonly affected by bleaching worldwide, and have often served as an experimental subject for studies on reef-coral physiology with relation to bleaching, e.g., the effect of increased temperature on photosynthesis and respiration, effect of temperature and UV radiation on bleaching and mortality, and coral reproduction (Coles and Jokiel 1977; Jokiel and Coles 1977; Jokiel and Guinther 1978; Jokiel and Morrissey 1986; Glynn and D'Croz 1990; Hoegh-Guldberg and Salvat 1995; Lesser and Lewis 1996; Stimson 1997).

17.3 How General Is Bacterial Bleaching of Corals?

The current consensus opinion amongst coral biologists is that bleaching is the result of environmental stress, primarily temperature and/or light, affecting directly the coral and resulting in the expulsion of the symbiotic algae (e.g., reviews by Brown 1997; Hoegh-Guldberg 1999). The discovery that bleaching of the coral *O. patagonica* in the Mediterranean Sea was the result of a bacterial infection by the pathogen *V. shiloi*, and that elevated temperature increased the virulence of the pathogen, raised an alternative hypothesis to explain coral bleaching, i.e., bacterial infection. One of the difficulties in extrapolating from bacterial bleaching of *O. patagonica* to other corals is that *O. patagonica* is a temperate stony coral, which is not found on warm-water coral reefs (Fine et al. 2001). Therefore, testing the generality of the bacterial hypothesis of coral bleaching has required an investigation of a more representative coral, *P. damicornis*, an abundant warm-water coral present in coral reefs, and which has been shown to being commonly affected by bleaching worldwide.

The data presented here demonstrate that *Vibrio coralliilyticus*, a newly isolated coral pathogen, is an etiological agent of bleaching and tissue lysis of *P. damicornis* on coral reefs in the Indian Ocean and the Red Sea, thus providing support for the bacterial hypothesis of coral bleaching. Seawater temperature is the critical environmental condition determining the outcome of the infection.

17.4 Isolation and Characterization of *Vibrio coralliilyticus*

In order to address the question regarding the generality of the bacterial hypothesis of coral bleaching, bacteria were isolated in March 1999 from bleached and healthy tissues of seven corals, from the reef off Mawi Island,

Zanzibar, in the Indian Ocean. Corals were collected from depths of 3–5 m, while seawater temperature was 29.5 °C. Streaking crushed tissue samples of the coral *P. damicornis* on *Vibrio*-selective thiosulfate-citrate-bile-sucrose (TCBS) agar yielded a relatively high concentration of *Vibrio*s compared to healthy corals. Comparison of colony morphologies obtained from the healthy and diseased samples led to isolation in pure culture of 15 *Vibrio* spp. isolates, which seemed to be present in bleached corals, but not in healthy ones. These isolates were subsequently tested for their pathogenic effect on *P. damicornis*. Preliminary infection experiments of healthy corals with the suspected strains led to one particular strain, referred to as YB1, originally isolated from bleached tissue of *P. damicornis*, which was highly pathogenic to these corals in controlled inoculation experiments. Strain YB1, which gave characteristic cream-colored and yellow colonies on Marine agar and *Vibrio*-selective TCBS agar media, respectively, and appeared dominant in diseased tissue, but absent from healthy tissue, was chosen for further study. Based on its 16S rDNA sequence, biochemical tests and carbon compound utilization studies, strain YB1 was found to be a novel species of the genus *Vibrio* that was named *Vibrio coralliilyticus* (Ben-Haim and Rosenberg 2002). *V. coralliilyticus* has been deposited in three international culture collections (American Type Culture Collection BAA-450^{T}, Ghent Bacterial Collection LMG 20984^{T}, and Pasteur Institute Collection CIP 107925T).

During the course of investigation, five additional strains of *V. coralliilyticus* were isolated and taxonomically classified as members of the same species as the type strain YB1 (Table 17.2). Three strains were isolated from different colonies of bleached and partially lysing *P. damicornis* corals from the Red Sea (Eilat) in August 2001. Two other strains were originally isolated from a diseased oyster in the Atlantic Ocean (Brazil) and from bivalve larvae in Europe (Kent region, UK), respectively. All five strains were phenotypically and genotypically classified together with the type species, and were shown to be *V. coralliilyticus* strains (Ben-Haim et al. 2003a).

V. coralliilyticus is a Gram-negative, rod-shaped and motile bacterium, which contains a single sheathed polar flagellum when grown on solid or liquid medium (Fig. 17.1a). Morphologically, the exponentially grown cells on Marine Broth liquid medium are 0.8×1.2–1.5 μm. Colonies are cream-colored on Marine Broth agar and yellow on TCBS agar after 48-h incubation at 25–30 °C, have smooth edges and do not luminescence (Fig. 17.1b). *V. coralliilyticus* conforms to the description of the family Vibrionaceae and of the genus *Vibrio* (Farmer and Hickman-Brenner 1992). Growth occurred over a wide range of salinity conditions (NaCl, 1–7%), but the bacterium failed to grow in media containing 0% NaCl or more than 8% salt. Table 17.1 summarizes the phenotypic characteristics of *V. coralliilyticus*. Biochemical tests and carbon utilization differences between the six strains of *V. coralliilyticus* are shown in Table 17.2. Comparisons revealed that of the 117 tests performed, the six strains gave the same qualitative results in 110 tests. One strain, isolated from a diseased *P. damicornis* coral in the Red Sea, gave identical results in all 117 tests compared to the type strain YB1. The other strains differed from *V. coralliilyticus* type strain in only 1–5 of the 117 tests (Ben Haim et al. 2003a).

A genetic analysis showed that the 16S rDNA sequence of *V. coralliilyticus* (NCBI database accession number of the type species: AJ440005) was allocated to the γ-proteobacteria. Figure 17.2 shows the phylogenetic tree of *V. coral-*

Table 17.1. Phenotypic characteristics of *Vibrio coralliilyticus*. (Adapted from Ben-Haim and Rosenberg 2002; Ben-Haim et al. 2003a)

Bacterial colony morphology	On Marine agar: cream-beige, smooth-edged, 3-mm diameter after 3 days	
	On TCBS agar: yellow, smooth-edged, 3-mm diameter after 3 days	
Cell morphology	Rod, 1.2–1.5×0.8 μm	
Gram stain	Negative	
Motility	Positive, polar sheathed flagellum	
Salinity tolerance (Marine Broth + NaCl)	Growth in 1–7% NaCl	
	No growth in 0 and 8% or higher	
Growth rate (doubling time in MBT medium)	20 °C – 140 min	
	25 °C – 36 min	
	30 °C – 25 min	
Biochemical tests		
Casein degradation, protease	+ (High activity)	
Gelatinase, protease	+	
Urease	+	
Oxidase	+	
Nitrate reduction	+	
Indole production	–	
Acidification of glucose medium	+	
Acidification of sucrose medium	+	
Arginine dehydrolase	–	
β-Galactosidase	+	
Utilization of carbon source	Positive	D-Mannose, D-galactose, D-fructose, D-mannitol, maltose, m-inositol, D-trehalose, D,L-serine, L-leucine, D-gluconic acid, propionic acid, D,L-lactic acid
	Negative	L-Arabinose, cellobiose, lactulose, α-D-lactose, gentiobiose, L-fucose, L-rhamnose, maltonic acid, urocanic acid, D-glucoronic acid, D-galacturonic acid
Antibiotic sensitivity	Sensitive to: erythromycin, tetracycline, chloramphenicol, gentamycin	
	Resistant to: kanamycin, ampicillin, penicillin	
Fatty acid composition profile	12:0, 2.2; 12:0 3-OH, 2.7; 13:0 iso, 2.6; 14:0, 6.5; 14:0 iso, 0.5; 15:0, 1.3; 15:0 iso, 2.8; 15:0 iso 3-OH, 1.7; 16:0, 14.1; 17:0, 2.0; 17:0 iso, 3.2; 17:1 ω8*c*, 1.3; 18:1 ω7*c*, 18.2	

Table 17.2. Strains of *V. coralliilyticus* – source of isolation and phenotypical differences. (Adapted from Ben-Haim et al. 2003a)

Vibrio coralliilyticus strains	Source and date of isolation
YB1^T (ATCC BAA-450^T, LMG 20984^T)	Diseased coral (*P. damicornis*), Zanzibar TZ, Indian Ocean, 1999
YB2 (LMG 21348)	Diseased coral (*P. damicornis*), Eilat Israel, Red Sea, 2001
YB3 (LMG 21349)	Diseased coral (*P. damicornis*), Eilat Israel, Red Sea, 2001
YB4 (LMG 21350)	Diseased coral (*P. damicornis*), Eilat Israel, Red Sea, 2001
LMG 10953 (NCIMB 2165)	Diseased oyster larvae (*C. gigas*), Kent UK, 1980s
LMG 20538 (INCO 83)	Bivalve larvae (*N. nodosus*), Brazil, Atlantic Ocean, 1998

Biochemical differences between *V. coralliilyticus* strains

Enzymatic activity or utilization	YB1^T	YB2	YB3	YB4	LMG 10953	LMG 20538
Arginine dehydrolase	+	–	–	+	–	–
Citrate	+	+	–	+	+	+
β-Methyl D-glucoside	–	+	+	–	–	–
Propionic acid	+	+	–	+	+	+
Methyl pyruvate	+	–	+	+	+	+
Glucose-1-phosphate	+	+	–	+	+	+
D-Mannitol	+	+	+	+	–	+

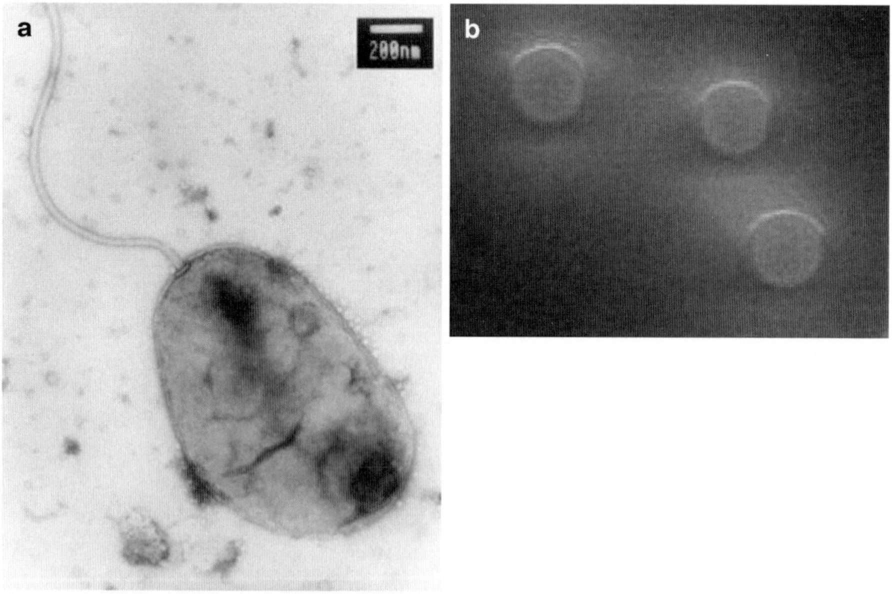

Fig. 17.1. Electron micrograph of negatively stained *V. coralliilyticus* (**a**). *V. coralliilyticus* colony morphology on TCBS agar (**b**). *Bar* 0.2 μm

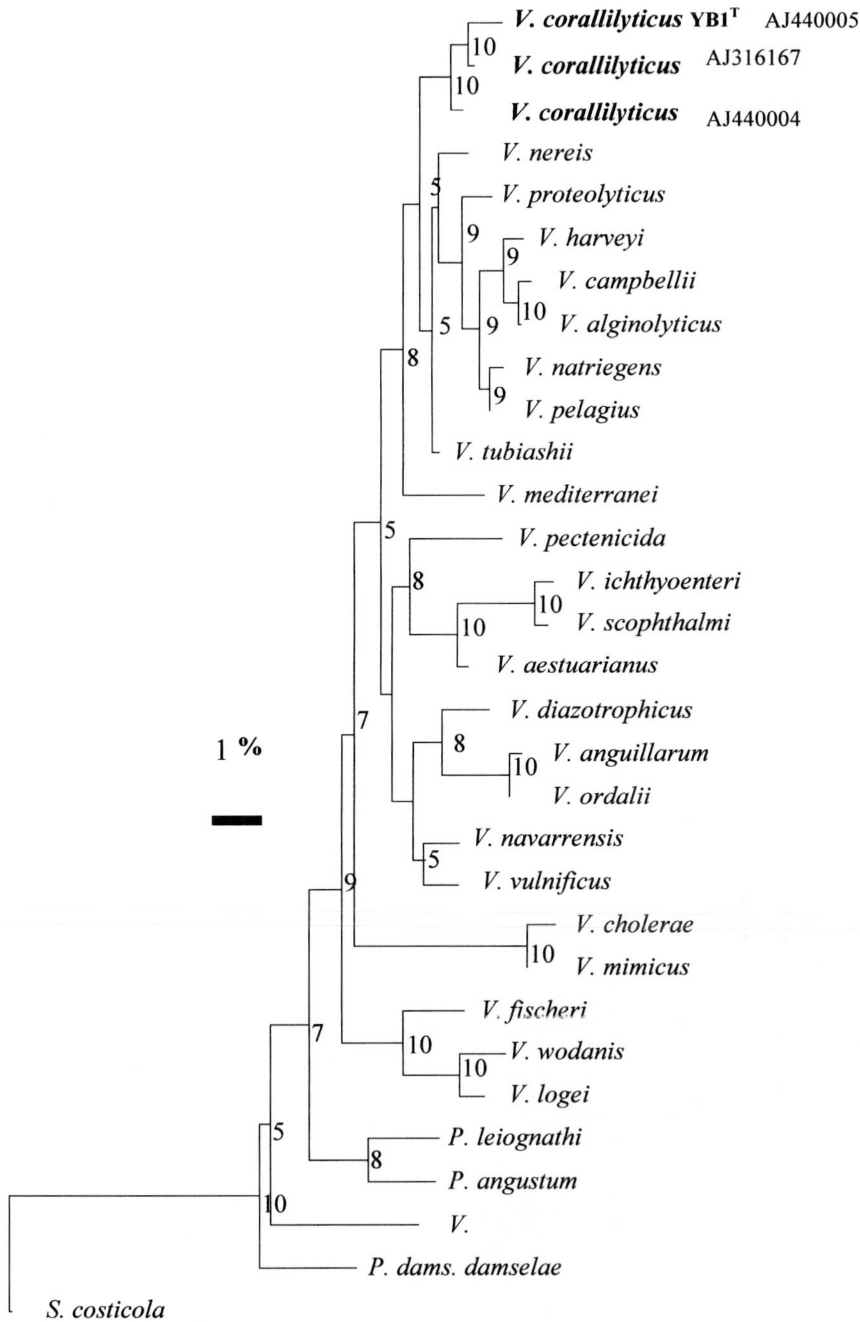

Fig. 17.2. Phylogenetic tree showing the relationships between *V. corallilyticus* and other representative type strains of the family Vibrionaceae, based on the neighbor joining method based on the almost complete 16S rDNA sequences. Bootstrap values after 500 simulations are shown. *Bar* 1% estimated sequence divergence

liilyticus and its relationship with other closely related *Vibrios*, based on 16S rDNA sequences. *V. coralliilyticus* strains form a tight cluster with more than 99% 16S rDNA identity. The closest phylogenetic neighbors of *V. coralliilyticus* are *V. tubiashii*, *V. nereis*, and *V. shiloi*, with 97.2, 96.8 and 96.6% identity to the 16S rDNA of *V. coralliilyticus*, respectively. DNA–DNA hybridization confirmed that *V. coralliilyticus* strains form a single genomic group with DNA-DNA similarities higher than 85% (Ben-Haim et al. 2003a, b).

17.5 Bleaching of *Pocillipora damicornis by Vibrio coralliilyticus*

Vibrio coralliilyticus is an etiological agent of bleaching of the coral *Pocillopora damicornis* (Ben-Haim et al. 2003b). At temperatures of 24.5 and 25 °C, inoculation of healthy *P. damicornis* corals with pure cultures of *V. coralliilyticus* in controlled aquaria experiments resulted in bleaching of the corals. *V. coralliilyticus*

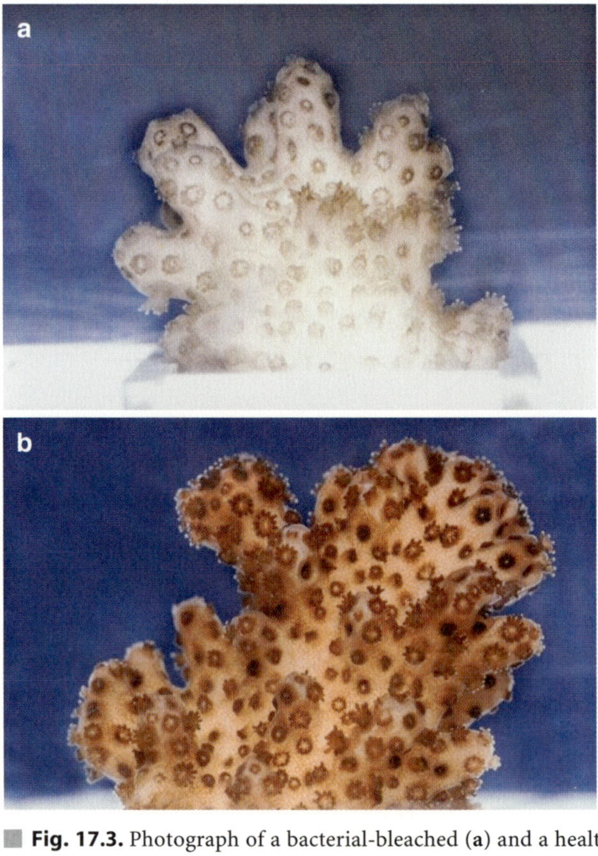

Fig. 17.3. Photograph of a bacterial-bleached (**a**) and a healthy control (**b**) of *P. damicornis* corals. Experimental temperature was 24.5 °C

was introduced to the healthy corals either by directly applying the cells to the coral surface, or by adding the bacteria to the water. All 16 inoculated corals appeared healthy and pigmented for 10 days after inoculation. Bleaching began at 12 days, and between 15 and 21 days, 15 out of the 16 infected corals bleached with a decrease of at least 50% of their pigmentation (compared to the control corals). By day 25, all the infected corals were bleached, with the coral tissue visually appearing totally transparent, but otherwise looking intact. The corals remained bleached, and showed no signs of tissue lysis for at least two additional weeks. The 14 control corals that were 'inoculated' with filtered seawater, and maintained at 24.5–25 °C under similar conditions, remained healthy and pigmented for at least 2 months, except for two controls that died after 3 weeks as a result of tissue lysis, but with no signs of bleaching. A photograph of *P. damicornis* bleached by *V. coralliilyticus* is presented in Fig. 17.3, alongside a healthy uninoculated control coral. At the end of the experiment, healthy and bacterial-bleached corals were crushed, diluted and plated on TCBS agar. Samples of bleached and control corals were also ultra-sectioned and examined by electron microscopy. After bleaching occurred, the pathogen was re-isolated from the coral tissues in high numbers. No *V. coralliilyticus* was detected in healthy corals, while the bleached corals showed bacteria inside the corals tissue (Fig. 17.4). No bacteria were seen in the tissue of the uninfected control corals (Ben-Haim et al. 2003b).

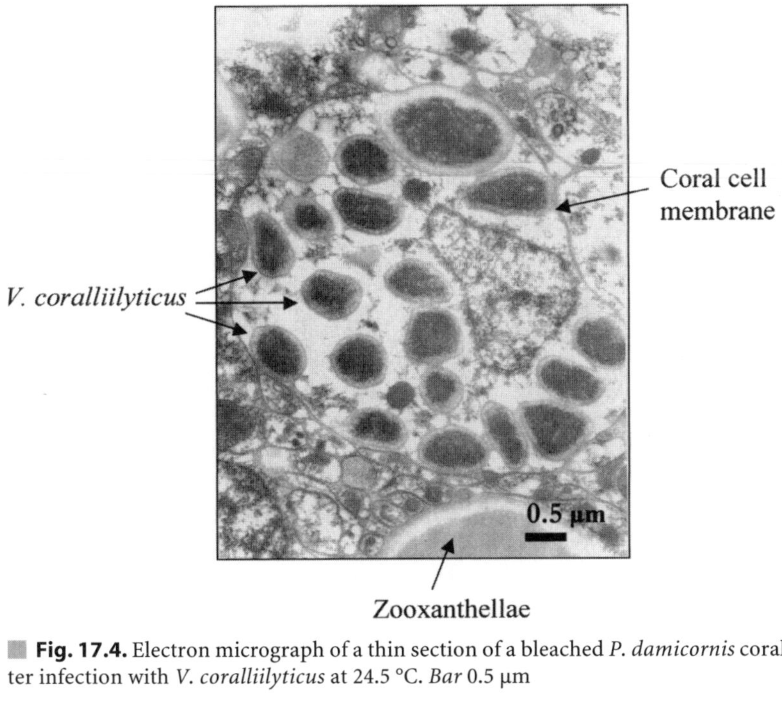

Fig. 17.4. Electron micrograph of a thin section of a bleached *P. damicornis* coral, 13 days after infection with *V. coralliilyticus* at 24.5 °C. *Bar* 0.5 µm

Table 17.3. Zooxanthellae, chlorophyll *a* content, and maximal fluorescence of healthy and bacterial-bleached coral tissue[a]

Experimental coral and condition		Zooxanthellae (cells/g×10^3)	Chlorophyll *a* (μg/g)	Maximal fluorescence (F_m)
Controls	Healthy 1	43.7	120	3540
	Healthy 2	45.5	90	3460
	Healthy 3	111.7	219	3630
	Healthy 4	62.6	183	3860
	Mean ± SEM	65.9±15.9	153±29	3630±90
Bacterial-infected	70% Bleached	7.2	55	1040
	70% Bleached	7.8	70	1430
	90% Bleached	5.8	24	820

[a]Corals were taken from the experiment described in Table 17.2 (performed at 24.5 °C). Pigmented zooxanthellae were enumerated microscopically after disruption of the coral tissue. Chlorophyll *a* was determined spectrophotometrically after extraction of the tissue with acetone. The number of algae and μg chlorophyll *a* are expressed per g wet weight tissue. Maximum fluorescence (F_m) was determined on the intact coral fragment with a mini-pulse-amplitude modulated fluorometer (adapted from Ben-Haim et al. 2003b).

In addition to visual determinations of bleaching, quantitative measurements of bacterial-induced bleaching were performed, and are summarized in Table 17.3. The four control corals in the experiment contained a mean of 66×10^3 zooxanthellae per gram wet weight coral tissue. Two of the infected corals that appeared visually as 70% bleached, contained 10.9 and 11.8% of the zooxanthellae concentration compared to the controls, whereas the infected coral which was 90% bleached, contained only 8.8% of the zooxanthellae concentration. Photosynthetic pigments were also reduced in the bleached corals compared to the controls, but to a lesser than algal concentration. For example, the 90% bleached coral contained 16 and 23% chlorophyll *a* and maximum fluorescence, respectively, compared to the unbleached controls.

17.6 The Pathogen Was Present in Bleached Corals in the Red Sea, Eilat

During the summer of 2001, the seawater surrounding the Eilat reef reached 26–27 °C, higher than the normal maximum temperature. During that time, bleaching and partial tissue lysis of *P. damicornis* were observed. In order to test whether these diseases were associated with the presence of *V. coralliilyticus*, coral fragments of healthy, bleached and partially lysed colonies were removed, rinsed, crushed and plated on TCBS agar (Table 17.4). *V. coralliilyticus* was present in all four bleached and five partially lysed corals, but was absent from the five healthy corals examined (limits of detection 100 CFU/cm^3 coral tissue). *V. coralliilyticus* was recognized by its characteristic colony morphol-

Table 17.4. Enumeration of *V. coralliilyticus* in healthy and diseased *P. damicornis* corals from Eilat, the Red Sea[a]

		V. coralliilyticus (CFU/cm^3)	
Coral condition	*n*	Mean ± SEM	Range
Healthy	5	$<10^2$	$<10^2$
Bleached	4	$1.8\pm1.2\times10^5$	$0.11–11\times10^5$
Partially lysed	5	$10.5\pm5.1\times10^5$	$1.1–24\times10^5$

[a]Coral fragments of ca. 2 cm^3 were sampled from 2–6 meters on the Eilat reef during the summer of 2001. Colony-forming units (CFU) were determined from the crushed fragments on TCBS agar (adapted from Ben-Haim et al. 2003b).

ogy. In addition, several colonies from both bleached and partially lysed corals were isolated and shown to be *V. coralliilyticus* by genotypic and phenotypic characterization, as discussed above (Ben-Haim et al. 2003a). It should be pointed out that the values in Table 17.4 are minimum numbers, because a significant number of *Vibrio* may be in the viable-but-not-culturable (VBNC) state, as has been shown for *V. shiloi* (Israely et al. 2001).

In order to examine the pathogenic effectiveness of each one of the additional *V. coralliilyticus* strains (three strains from diseased corals, and two strains from diseased oyster and bivalve larvae), controlled aquaria infection experiments were performed. All strains were pathogenic to *P. damicornis* corals, causing similar tissue damage, whereas control corals infected with *V. shiloi* under similar experimental conditions remained healthy (Ben-Haim et al. 2003a). Thus, virulence to *P. damicornis* appears to be a general property of the species *V. coralliilyticus.*

17.7 Lysis of Coral Tissue by *Vibrio coralliilyticus*

In comparison with the outcome of infection at the moderate temperatures of 24–25 °C, bleaching of the corals, and infection of *P. damicornis* with *V. coralliilyticus* at higher temperatures (26–29 °C) resulted in different symptoms. At 29 °C, 10 days after infection, most of the infected corals began to lyse, and by 15 days, all infected corals lysed and died (Ben-Haim and Rosenberg 2002). There was no sign of bleaching above 26 °C. Tissue lysis typically started in small spots, usually on the verrucae of the branch, which slowly united into white patches, progressing until the entire tissue was destroyed, leaving only the intact bare skeleton. A typical example showing macroscopically the progressive tissue lysis of infected corals is shown in Fig. 17.5. Inoculation of 13 different coral species from the Red Sea, using 10^8 *V. coralliilyticus* per coral at 29 °C, revealed that the infection is host-specific, i.e., *V. coralliilyticus* only infected *P. damicornis* and possibly a closely related species, *Stylophora pistillata,* which showed variable effects (Table 17.5).

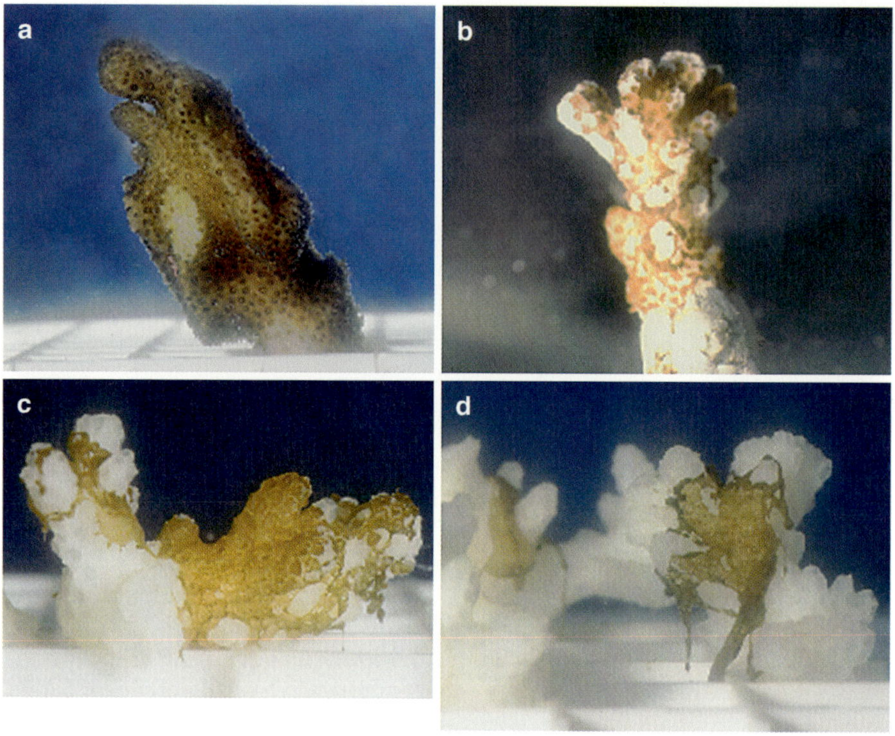

Fig. 17.5. Tissue lysis of *P. damicornis* by *V. coralliilyticus* at 26–29 °C. Series of photographs showing the progressive tissue lysis of infected corals at 29 °C. A healthy uninoculated coral, after 31 days (**a**), 15–20% lysis, 7 days after infection (**b**), 40–50% lysis after 10 days (**c**) and 90% lysis of the tissue 14 days after infection (**d**)

Additional infection experiments were conducted in order to examine the effect of inoculum size on the infectious process. Inoculation of the corals at 29 °C with 10^4–10^8 bacteria per coral (in 2.5 l aquaria) resulted in the same outcome of tissue degradation, except for a ca. 2-day delay in the disease progression at the lowest inoculum size. Thus, even with a relatively small inoculum size, 30 bacteria/ml, *V. coralliilyticus* was able to infect and lyse the corals at 29 °C. When healthy corals were placed in direct contact with infected corals undergoing tissue lysis, rapid (2–4 days) tissue degradation of the healthy coral was observed. The tissue of the newly infected coral lysed in a similar manner to the originally infected coral, and the pathogen was reisolated from these tissues. Furthermore, the corals infected by direct contact could also propagate the disease and the pathogen, further showing that the disease was contagious (Ben-Haim and Rosenberg 2002).

Table 17.5. Host-specificity of *V. coralliilyticus* to different Eilat corals[a]

Group	Coral inoculated	*n*		Virulence results
		Infected	Controls	
Branchy corals (Red Sea)	*Stylophora pistillata*	18	25	Variable
	Acropora sp. (1)	3	10	Resistant
	Acropora sp. (2)	3	7	Resistant
	Acropora sp. (3)	12	13	Resistant
	Acropora sp. (4)	3	3	Resistant
Massive corals (Red Sea)	*Platygyra lamellina*	3	9	Resistant
	Favia favus	3	7	Resistant
	Favites sp.	1	–	Resistant
	Goniastrea sp.	3	13	Resistant
	Acanthastrea sp.	1	2	Resistant
	Cyphastrea micropthalma	3	4	Resistant
	Cyphastrea chalcidicum	3	3	Resistant
Solitary corals (Red Sea)	*Fungia granulosa*	1	1	Resistant
	Fungia echinata	1	–	Resistant

[a]Corals were inoculated with 10^8 *V. coralllyticus* per coral at 29 °C in 3-l aquaria. Observations were conducted over 2 months.

17.8
The Effect of Temperature on the Infection

What is particularly important about this new coral pathogen is its temperature-regulated infective mode of action. In order to examine the effect of temperature on the outcome of the infection, a series of infection experiments at temperatures of 20 to 29 °C were performed. *V. coralliilyticus* pathogenesis is temperature-dependent, causing bleaching of the corals at moderate temperatures (24–25 °C), and coral tissue lysis at higher temperatures (26–29 °C). Figure 17.6 demonstrates the differences in progression of the infection at each of the temperatures examined. Corals infected and maintained at 29 °C showed the fastest rate of tissue lysis. At 27 °C the infection progression was a little slower, delayed by 1–2 days, but at 26 °C the infection process occurred significantly slower compared to 29 °C. Interestingly, a decrease of only 1 °C (26 to 25 °C) resulted in a major difference in the outcome of the infection. Inoculation of the coral at lower temperatures of 20–22.5 °C resulted in neither bleaching nor lysis of the corals. In control aquaria, where corals were maintained at each of the temperatures from 20 to 27 °C without inoculation, all the corals remained healthy. At 29 °C, the control corals showed no change for at least 3 weeks, except for 5 of the 31 corals that died after 3 weeks (the signs were different to the tissue lysis in the infected corals). Therefore, *V. coralliilticus* pathogenesis is regulated by seawater temperature.

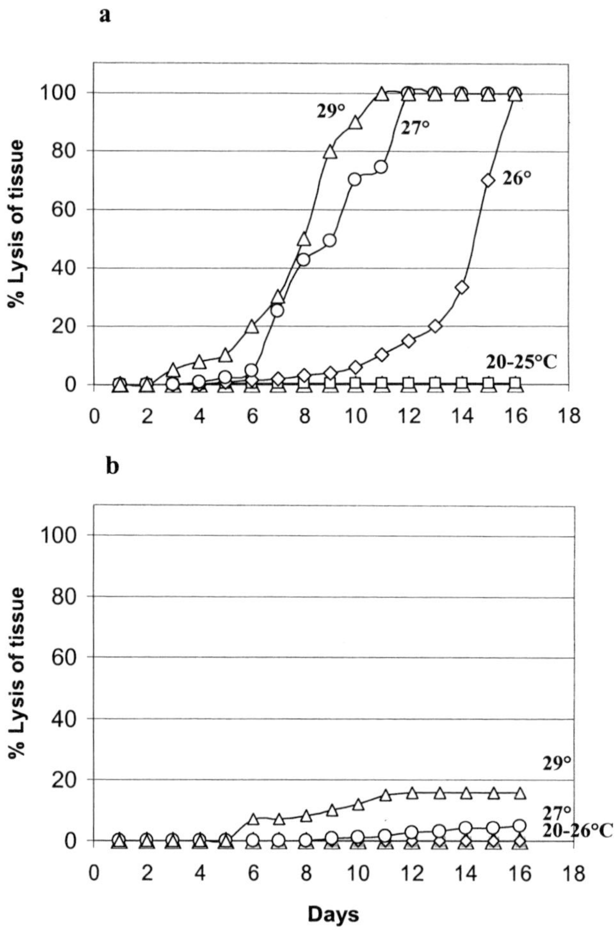

Fig. 17.6. Effect of temperature on the infection of *P. damicornis* by *V. coralliilyticus*. Average percentage of tissue lysis of *P. damicornis*, inoculated with *V. coralliilyticus* (**a**), compared to uninoculated control corals (**b**). The corals were inoculated with ca. 10^7 bacteria per coral, in 2–3 l aquaria, and maintained at 20, 22, 25, 26, 27 and 29 °C (*n*=2, 4, 13, 6, 9 and 5 inoculated corals, and *n*=2, 4, 16, 9, 10 and 31 uninoculated control corals, respectively)

17.9 Protease Activity of *Vibrio coralliilyticus*

During studies on the classification of *V. coralliilyticus*, it was observed that the pathogen produced a potent extracellular protease (Ben-Haim and Rosenberg 2002; Ben-Haim et al. 2003b). Since, in several cases, virulence of *Vibrio* pathogens is associated with protease activity (e.g., Hada et al. 1984; Wright and Morris 1991; Harrington 1996), the production and properties of the *V. coralliilyticus* protease was investigated. Extracellular protease activity was a

Table 17.6. Specific extracellular protease activity of *Vibrio coralliilyticus* as a function of growth temperature[a]

Temperature (°C)	Specific protease activity	
	Exponential phase (h)	Stationary phase (h)
16	0.1 (24)	0.5 (50)
20	0.1 (20)	0.5 (48)
22	0.2 (20)	0.7 (40)
24	0.3 (16)	0.8 (36)
26	1.0 (12)	4.7 (30)
28	1.1 (12)	5.9 (26)
30	1.2 (10)	6.1 (24)

[a] *V. coralliilyticus* was grown with aeration in MBT medium at the indicated temperature. Values for culture turbidity (A_{600}) and extracellular protease activity (units) were determined during the exponential phase (10–24 h depending on growth temperature) and stationary phase (24–50 h), as described in Ben-Haim et al. (2003b).

function of growth temperature and phase of growth (Table 17.6). Using a 0.5% inoculum, maximum culture turbidity was achieved within 24 h over the entire temperature range studied (16–30 °C) and maximum turbidity values differed by only 20% from 20–30 °C. However, the production of the extracellular protease was greatly influenced by the growth temperature with the largest change in specific activity occurring between 24 °C (0.8 U/A_{600}) and 28 °C (5.9 U/A_{600}). At all temperatures, the specific activity was three- to five-fold higher during stationary phase than exponential phase.

The *V. coralliilyticus* protease was purified to homogeneity by column chromatography (Ben Haim et al. 2003b). The molecular mass, estimated by SDS-PAGE was 36 kDa. The first 17 amino acids from the N-terminus of the protease were sequenced, and comparative sequence analyses showed high homology to a range of proteases within the family Vibrionaceae, including the well-studied hemagglutinin/protease (HA/P) of *Vibrio cholerae* (Finkelstein and Hanne 1982). The *V. coralliilyticus* protease showed maximum activity at pH 8–11 and 30–40 °C. The enzyme was inactivated by 1 mM EDTA and the apoenzyme reactivated by addition of 10 mM Ca^{2+} or Zi^{2+}, indicating it was a metallo-proteinase. Preliminary experiments indicated that addition of the purified protease to coral fragments caused tissue damage within 18 h at 27 °C, as measured by release of protein and pigments (A_{max}=320 nm) from the coral. The effect was proportional to enzyme concentration: in the presence of 2, 4.5 and 9 U/ml of the purified enzyme at 28 °C, the A_{320} in the aquaria water containing the corals were 0.5, 2.1 and 4.8, respectively after 18 h. EDTA inhibited the release of pigment (Ben-Haim et al. 2003b).

17.10 Effect of Higher Than Normal Temperatures on Different Eilat Corals

To obtain more general information on the effect of increased temperature on Eilat corals, 12 different corals were collected during the late summer of 2001 from the Eilat coral reef. Seawater temperature at the time of collection was 24 °C, the average maximal sea temperature being ca. 26 °C. Corals of different groups: (1) branchy corals, e.g., three different *Acropora* species, *Stylophora pistillata*, and *P. damicornis*, (2) massive corals, e.g., *Favia favus*, *Platygyra lamellina* and *Goniastrea* sp., and (3) solitary corals, e.g., two different *Fungia* species were slowly adjusted to 30 and 31 °C, in two temperature-controlled

Table 17.7. Effect of higher-than-normal seawater temperatures on different Eilat corals[a]

(a) Coral species	*n*	Results after incubation at 30 °C (days)[b]
Acropora sp. (1)	3	Healthy (60)
Acropora sp. (2)	6	Healthy (114)
Acropora sp. (3)	8	Healthy (84)
Fungia granulosa	1	Healthy (114)
Fungia echinata	1	Healthy (114)
Platygyra lamellina	7	Healthy (84)
Favia favus	6	Healthy (100)
Favites sp.	1	Healthy (114)
Goniastrea sp.	12	Healthy (100), 1/12 died
Acanthastrea sp.	2	Healthy (84)
Cyphastrea microphthalma	5	Healthy (84)
Stylophora pistillata	6	Lysis of tissue (24)

(b) Coral species	*n*	Results after incubation at 31 °C (days)[b]
Acropora sp. (1)	4	Bleached (53–60)
Acropora sp. (2)	4	Bleached (53–60)
Acropora sp. (3)	5	Bleached (40–45)
Fungia granulosa	1	Bleached (23)
Platygyra lamellina	5	Bleached (23–35)
Favia favus	5	Bleached (23–40)
Cyphastrea microphthalma	2	Bleached (40–45)
Goniastrea sp.	3	Bleached (23)
Acanthastrea sp.	1	Bleached (45)
Stylophora pistillata	6	Lysis of tissue (23)
Pocillopora damicornis	4	Healthy (30), lysis of tissue (32)

[a] Twelve different corals were collected from the Eilat reef (sea temperature 24 °C), and slowly adjusted to 30 and 31 °C. The corals were maintained constantly in two baths controlled at 30 and 31 °C during the experiment, in aerated 3-l aquaria containing artificial seawater, and illuminated by fluorescent lamps (4×10,000 K).
[b] Observations at the time noted, after a constant incubation at 30 °C (a) or 31 °C (b).

baths. The seawater temperature was increased by a maximum 0.5 °C every 2 days, starting from the temperature at which they were collected (24 °C). After reaching 30 °C, part of the corals was further increased to 31 °C, and observations were recorded every day. The corals were maintained in 3-l glass-covered aquaria, containing freshly prepared artificial seawater, which were changed every week. The corals were aerated and illuminated by fluorescent lamps (40,000 K).

Remarkably, all the corals maintained at 30 °C remained healthy for 2–4 months, except for *S. pistillata*, which remained healthy for ca. 3 weeks, but lysed after 24 days. Interestingly, similar corals that were originally collected from the same reef at the same time, but maintained at 31 °C, all bleached within 3–8 weeks, except for *P. damicornis* and *S. pistillata*, which lysed after 3–4 weeks. Bleaching occurred in two waves, i.e., part of the corals, such as *F. granulosa*, *F. favus*, and *Goniastrea*, bleached after 23 days at 31 °C. The rest of the corals bleached after ca. 45 days. Small differences occurred in the time a particular corals species bleached, but there were larger differences in the time of bleaching between different coral species (Table 17.7).

In order to examine whether certain bacteria were associated with the occurrence of bleaching at 31 °C, but not at 30 °C, three different corals that had been incubated at the two temperatures were crushed and the tissues were diluted and plated onto Marine and TCBS agar, for total and *Vibrio* sp. colony forming units (CFU), respectively. The total colony-forming bacteria in the bleached corals (31 °C) compared to the healthy corals (30 °C) ranged from $46–150\times10^5$ and from $0.5–18\times10^5$ per coral, respectively. Two of the bleached corals, *F. favus* and *Goniastrea* sp., contained ca. ten times more colony-forming *Vibrio* sp. bacteria in their tissue compared to the same healthy corals maintained at 30 °C. The third coral examined, *P. lamellina*, had similar *Vibrio* sp. concentrations in the bleached and healthy coral tissue.

17.11 General Discussion and Conclusions

The most significant findings in this study are:

1. *V. corallilyticus* is an etiological agent of the bleaching disease of *P. damicornis* at 24–25 °C.
2. *V. corallilyticus* is an etiological agent of coral tissue lysis at temperatures of 26–29 °C.
3. *V. corallilyticus* secrets a potent extracellular protease. Production of the protease increases greatly at the same temperature range (24–28 °C) as the transition from bleaching to tissue lysis of the corals.
4. *V. corallilyticus* is a new species that is widely distributed in marine environments throughout the world.
5. *V. coralliilyticus* is a *P. damicornis*-specific coral pathogen.

Koch's postulates were applied and fulfilled in order to demonstrate that *V. corallilyticus* was the etiological agent of bleaching and tissue lysing of the coral *P. damicornis*:

1. *V. corallilyticus* was present in diseased corals and absent from healthy tissue.
2. The strain was obtained in a pure culture.
3. Infection of healthy corals in controlled aquaria experiments caused the disease.
4. The pathogen was re-isolated from infected hosts.

This is the second demonstration of bacterial bleaching of corals, the first being the bleaching of the temperate coral *O. patagonica* by *V. shiloi* (Kushmaro et al. 1996, 1997). In each case, a specific and different *Vibrio* species was the responsible pathogen. These data support the bacterial hypothesis of coral bleaching.

In Chapter 25, Rosenberg suggests discriminating between "patchy-distributed bleaching" and "mass bleaching". We refer to the type of bleaching caused by *V. coralliilyticus* as patchy-distributed and spreading. It is patchy in the sense that bleached corals on the reef are surrounded by apparently healthy corals of the same, as well as other species. Several investigators have reported on the patchy spatial distribution and spreading nature of coral bleaching (Fisk and Done 1985; Oliver 1985; Jokiel and Coles 1990; Lang et al. 1992; Edmunds 1994). It has been argued that the random mosaic patterns of bleaching observed in coral colonies are difficult to attribute solely to environmental stress, since neighboring regions of the colony must be exposed to the same extrinsic conditions (Hayes and Bush 1990). Even more significantly, the spreading nature of patchy coral bleaching is highly symptomatic of an infectious disease. We suggest that the patchy pattern of bleaching of *P. damicornis* observed on the Eilat coral reef, which was shown to be an infectious disease, may be a model for patchy-distributed bleaching seen on other coral reefs.

The demonstration that bleaching and lysis of *P. damicornis* is the result of a bacterial infection does not diminish the importance of temperature stress on coral disease. Sea temperatures in the reef of Eilat, in the Red Sea, usually range from 20 °C in the winter to 26 °C in the summer. However, over the last few years, a warming trend has been observed, with temperatures reaching 28 °C, and considerably more diseased corals appeared than in former years. It is not clear however, at this time, whether the increased temperature causes *P. damicornis* to become more susceptible to infection or makes the bacterium more virulent. For *P. damicornis* to either bleach or lyse, in our experimental setup, both the pathogen and the permissive temperature must be present. In the case of bleaching of *O. patagonica*, in the Mediterranean Sea, the temperature stress was found to be primarily on the pathogen, causing the expression of a variety of virulence genes (Rosenberg and Ben-Haim 2002).

V. corallilyticus, a new pathogenic species, was found to be widely distributed. The type strain was isolated originally from the Indian Ocean (Zanzibar), and the five additional strains were present in the Red Sea (Eilat), Atlantic Ocean (Brazil), and Europe (Kent region). The fact that coral pathogenic strains of *V. corallilyticus* are widely distributed is important in understanding the source of coral disease and attempts to prevent their spread. *V. corallilyticus* infects and causes tissue damage to its host coral as a function of seawater temperature. It is likely that during the winter when temperatures are lower, the bacterium is present in different hosts, possibly bivalve larvae. It is presumed that the pathogen is transmitted to its coral host when appropriate environmental conditions occur. The mode of transmission is at present unknown.

Several marine *Vibrio* species are known as invertebrate pathogens (Hada et al. 1984; Egidus et al. 1986). The unusual feature of *Vibrio* bleaching of corals is that the target appears to be the intracellular zooxanthellae, rather than the animal tissue. *V. shiloi* does not infect corals that lack algae (Banin et al. 2001a), and the bacterium produces anti-algal toxins that inhibit photosynthesis and lyse the algae (Ben-Haim et al. 1999; Banin et al. 2001b). In the case of bleaching of *P. damicornis* by *V. coralliilyticus*, it also appears that the bacterium attacks the algae. The observed relatively higher level of photosynthetic pigments than intact-pigmented algae suggests that the intracellular algae were damaged, causing the release of their pigments. Based on the existing data, we suggest that bacterial bleaching of *P. damicornis* results from an attack on the algae, whereas bacteria-induced coral lysis and death is caused by bacterial extracellular proteases. The temperature-regulated production of protease by *V. coralliilyticus* supports this concept: the large increase in enzyme production occurs at the same temperature range (24–28 °C) as the transition from bleaching to lysis of the corals.

Jokiel and Coles (1990) reported that seawater temperatures slightly above the normal maximum cause bleaching of *P. damicornis*, while temperatures a few degrees higher cause lysis and death of Hawaiian reef corals. They showed that prolonged exposures of these corals to temperature elevations as small as 1–2 °C above normal summer ambient maxima, could produce bleaching, but often with low mortality rates. They suggested that bleaching was not induced by a "thermal shock" of rapidly fluctuating temperatures (i.e., increased temperature by 3–4 °C, for a short period of time), but rather was a response to increase in the prevailing mean temperature. These observations are similar to the data reported here for infected *P. damicornis* corals. Jokiel and Coles and others have not performed microbiological studies, so that it is not known if these corals were infected prior to elevating the temperature.

In general, studies examining the effect of temperature and light on bleaching of corals, have not taken into consideration the possibility that microorganisms present in their samples could influence the outcome of the experiment. For example, in testing the effect of elevated seawater temperatures on the bleaching patterns of Caribbean corals, Fitt and Warner (1995) used fresh unfiltered seawater that flowed from the sea into the chambers containing the corals. When these corals were exposed to 30–34 °C, bleaching and mortality of the corals resulted. Seawater could have introduced bacterial pathogens into the experimental system, which might have infected the corals. These pathogens could presumably turn virulent as the temperature increased. It is also possible that these corals could have been infected prior to being exposed to the increased water temperatures. Therefore, studies in the future should include microbiological aspects in the design and interpretation of the experiments.

The assumption that corals are living very close to their maximum temperatures and their bleaching threshold, and that a 1–2 °C increase in the average maximal temperature, would result in bleaching of the corals, might not be the case for the Eilat corals. Preliminary results showed that 12 different Eilat corals remained healthy for 2–4 months at 30 °C, which is ca. 4 °C above the average summer maximum in the Eilat coral reef. Interestingly, an increase of only 1 °C (from 30 to 31 °C) resulted in an apparent mass-bleaching event, which consisted of two main waves of bleaching. When plating coral tissue samples of the bleached (31 °C) and the healthy (30 °C) corals, the three bleached corals contained 10–100 times more total colony-forming bacteria, and ca. ten times more *Vibrio* sp. bacteria in two of the three corals examined. It is known that a combination of increased temperatures and increased solar irradiation, including ultraviolet irradiation, may enhance the incidence of bleaching (Harriott 1985; Brown and Suharsono 1990; Lesser et al. 1990; Williams and Bunkley-Williams 1990; Glynn et al. 1992; Dunne and Brown 2001; Jokiel, Chap. 23, this Vol.). The experiments reported here were performed under laboratory conditions using fluorescent lamps (40,000 K), with glass-covered aquaria. Exposing the experimental corals to solar irradiation and UV might have lowered their bleaching temperature threshold. In order to further study the effect of higher-than-normal temperatures on Eilat corals, additional experiments should be done in the future, preferably by combining increased temperature and solar irradiation effects. In addition, further investigations of the bacteria associated with bleaching of the corals, as well as direct contact experiments of bleached and healthy corals should be conducted in these experiments.

17.12 Unresolved Problems and Directions of Future Research

The discovery that bacterial bleaching of corals was, in fact, not restricted to the temperate coral, *O. patagonica*, in the Mediterranean Sea, but occurred also with the scattered reef-building coral, *P. damicornis*, in the Indian Ocean and the Red Sea, constitutes an important stepping-stone in studying the role of bacteria as coral pathogens. Although important experimental data have been obtained, several key questions remain unanswered and should be addressed in the future. First, the generality of the bacterial hypothesis of coral bleaching remains the key question. Many observed cases of bleaching throughout the world have been characterized by similar signs, e.g., spreading and patchy pattern, which hint of an infectious disease; yet, the evidence for bacterial bleaching in those cases is circumstantial and indirect. The following specific questions should be addressed:

1. Is patchy-distributed bleaching of additional species of reef-building corals caused also by bacterial infection? If so, are different bacterial pathogens responsible?
2. Are bacteria involved in mass bleaching of corals? Are there specific pathogens responsible for mass bleaching? If so, how do they overcome the host-specificity barrier?

In parallel to expanding the research on generalizing the bacterial hypothesis of coral bleaching, by studying other bleaching cases around the world, a deeper understanding of the *V. coralliilyticus/P. damicornis* pathogenic and defensive relationships should be addressed. Further investigation of the virulence mechanisms involved in bleaching and tissue lysis of *P. damicornis* by *V. coralliilyticus*, at both physiological and molecular levels, is required. We are currently attempting to obtain avirulent mutants of *V. coralliilyticus*, in order to identify and characterize specific virulence genes of the pathogen. This integrated knowledge could lead to possible future developments of technologies for early warning, and possibly for preventing or curing bacterial-induced bleaching of reef-building corals throughout the world.

References

Banin E, Ben-Haim Y, Israely T, Loya Y, Rosenberg E (2000) Effect of the environment on the bacterial bleaching of corals. Water Air Soil Pollut 123:337–352

Banin E, Israely T, Fine M, Loya Y, Rosenberg E (2001a) Role of endosymbiotic zooxanthellae and coral mucus in the adhesion of the coral-bleaching pathogen *Vibrio shiloi* to its host. FEMS Microbiol Lett 199:33–37

Banin E, Sanjay KH, Naider F, Rosenberg E (2001b) A proline-rich peptide from the coral pathogen *Vibrio shiloi* that inhibits photosynthesis of zooxanthellae. Appl Environ Microbiol 67:1536–1541

Banin E, Vassilakos D, Orr E, Martinez RJ, Rosenberg E (2003) Superoxide dismutase is a virulence factor produced by the coral bleaching pathogen *Vibrio shiloi*. Curr Microbiol 46(6):418–422

Ben-Haim Y, Rosenberg E (2002) A novel *Vibrio* sp. pathogen of the coral *Pocillopora damicornis*. Mar Biol 141:47–55

Ben-Haim Y, Banin E, Kushmaro A, Loya Y, Rosenberg E (1999) Inhibition of photosynthesis and bleaching of zooxanthellae by the coral pathogen *Vibrio shiloi*. Environ Microbiol 1:223–229

Ben-Haim Y, Thompson FL, Thompson CC, Cnockaert MC, Hoste B, Swings J, Rosenberg E (2003a) *Vibrio coralliilyticus* sp. nov., a temperature dependent pathogen of the coral *Pocillopora damicornis*. Int J Syst Evol Microbiol 53:309–315

Ben-Haim Y, Zicherman-Keren M, Rosenberg E (2003b) Temperature-regulated bleaching and lysis of the coral *Pocillopora damicornis* by the novel pathogen *Vibrio coralliilyticus*. Appl Environ Microbiol 69(7):4236–4242

Brown BE (1997) Coral bleaching: causes and consequences. Proc of 8th Int Coral Reef Symp 1:65–74

Brown BE, Suharsono (1990) Damage and recovery of coral reefs affected by El Niño related seawater warming in the Thousand Islands, Indonesia. Coral Reefs 8:163–170

Brown BE, Dunne RP, Scoffin RP, Le Tissier MDA (1994) Solar damage in intertidal corals. Mar Ecol Prog Ser 105:219–230

Brown BE, Dunne RP, Goodson MS, Douglas AE (2002) Experience shapes the susceptibility of a reef coral to bleaching. Coral Reefs 21:119–126

Coles SL, Jokiel PL (1977) Effects of temperature on photosynthesis and respiration in hermatypic corals. Mar Biol 43:209–216
Coles SL, Fadlallah YH (1991) Coral reef survival and mortality at low temperatures in the Arabian Gulf: new species-specific lower temperature limits. Coral Reefs 9:231–237
Coll JC (1992) The chemistry and chemical ecology of octocorals (Coelenterata, Anthozoa, Octocorallia). Chem Rev 92(4):613–631
Dunne RP, Brown BE (2001) The influence of solar radiation on bleaching of shallow water reef corals in the Andaman Sea, 1993–1998. Coral Reefs 20:201–210
Edmunds PJ (1994) Evidence that reef-wide patterns of coral bleaching may be the result of the distribution of bleaching-susceptible clones. Mar Biol 121:137–142
Egidus E, Wiik R, Anderson K, Hoff KA, Hjeltnes H (1986) *Vibrio salmonicida* sp. nov., a new fish pathogen. Int J Syst Bacteriol 36:518–520
Fang LS, Liao CW, Liu MC (1995) Pigment composition in different-colored scleractinian corals before and during the bleaching process. Zool Stud 34:10–17
Farmer JJ, Hickman-Brenner FW (1992) The genera *Vibrio* and *Photobacterium*. In: Balows A, Trüper HG, Dworkin M, Harder W, Schleifer KH (eds) The prokaryotes, 2nd edn. Springer, Berlin Heidelberg New York, pp 2952–3011
Fine M, Zibrowius H, Loya Y (2001) *Oculina patagonica*, a non lessepsian scleractinian coral invading the Mediterranean Sea. Mar Biol 138:1195–1203
Fine M, Banin E, Israely T, Rosenberg E, Loya Y (2002a) Ultraviolet (UV) radiation prevents bacterial bleaching of the Mediterranean coral *Oculina patagonica*. Mar Ecol Prog Ser 226:249–252
Fine M, Oren U, Loya Y (2002b) Bleaching effect on regeneration and resource translocation in the coral *Oculina patagonica*. Mar Ecol Prog Ser 234:119–125
Finkelstein RA, Hanne LF (1982) Purification and characterization of the soluble hemagglutinin (cholera lectin) produced by *Vibrio cholerae*. Infect Immun 36:1199–1208
Fisk DA, Done TJ (1985) Taxonomic and bathymetric patterns of bleaching in corals, Myrmidon Reef (Queensland). Proc 5th Int Coral Reef Symp 6:149–154
Fitt WK, Warner ME (1995) Bleaching patterns of four species of Caribbean reef corals. Biol Bull (Woods Hole) 187:298–307
Gates RD, Baghdasarian G, Muscatine L (1992) Temperature stress causes host cell detachment in symbiotic cnidarians: implications for coral bleaching. Biol Bull (Woods Hole) 182:324–332
Gleason DF, Wellington GM (1993) Ultraviolet radiation and coral bleaching. Nature 365:836–838
Glynn PW (1984) Widespread coral morality and the 1982–1983 El Niño warming event. Environ Conserv 11:133–146
Glynn PW (1991) Coral reef bleaching in the 1980s and possible connections with global warming. Trends Ecol Evol 6:175–179
Glynn PW (1993) Coral reef bleaching ecological perspectives. Coral Reefs 12:1–17
Glynn PW, D'Croz L (1990) Experimental evidence for high temperature stress as the cause of El-Niño-coincident coral mortality. Coral Reefs 8:181–191
Glynn PW, Imai R, Sakai K, Nakano Y, Yamazato K (1992) Experimental responses of Okinawan (Ryukyu Islands, Japan) reef corals to high sea temperature and UV radiation. Proc 7th Int Coral Reef Symp 1:27–37
Goreau TE (1964) Mass expulsion of zooxanthellae from Jamaican reef communities after hurricane Flora. Science 145:383–386
Hada HS, West PA, Lee JV, Stemmler J, Colwel RR (1984) *Vibrio tubiashii* sp. nov., a pathogen of bivalve mollusks. Int J Syst Bacteriol 34:1–4
Harrington DJ (1996) Bacterial collagenases and collagen-degrading enzymes and their potential role in human disease. Infect Immun 64:1885–1891
Harriott VJ (1985) Mortality-rates of scleractinian corals before and during a mass bleaching event. Mar Ecol Prog Ser 21:81–88
Harvell D, Kiho K, Quirlo C, Weir J, Smith G (2001) Coral bleaching and disease: contributors to 1998 mass mortality in *Briareum asbestinum* (Octocorallia, Gorgonacea). Hydrobiologia 460:97–104

Hayes RL, Bush PG (1990) Microscopic observations of recovery in the reef-building scleractinian coral, *Monastrea annularis*, after bleaching on a Cayman reef. Coral Reefs 8:203–209

Hayes RL, Goreau NI (1998) The significance of emerging diseases in the tropical coral reef ecosystem. Rev Biol Trop 46:173–185

Hoegh-Guldberg O (1999) Climate change coral bleaching and the future of the world's coral reefs. Mar Freshwater Res 50:839–866

Hoegh-Guldberg O, Smith GJ (1989) The effect of sudden changes in temperature, irradiance and salinity on the population density and export of zooxanthellae from the reef corals *Stylophora pistillata* (Esper 1797) and *Seriatopora hystrix* (Dana 1846). J Exp Mar Biol Ecol 129:279–303

Hoegh-Guldberg O, Salvat B (1995) Periodic mass-bleaching and elevated sea temperatures: bleaching of outer reef slope communities in Moorea, French Polynesia. Mar Ecol Prog Ser 121:181–190

Iglesias-Perieto R, Matta JL, Robins WA, Trench RK (1992) Photosynthetic response to elevated temperature in the symbiotic dinoflagellate *Symbiodinium microadriaticum* in culture. Proc Natl Acad Sci USA 89:302–305

Israely T, Banin E, Rosenberg E (2001) Growth, differentiation and death of *Vibrio shiloi* in coral tissue as a function of seawater temperature. Aquat Microb Ecol 24:1–8

Jensen PR, Harvell CD, Wirtz K, Fencial W (1996) Antimicrobial activity of extracts of Caribbean gorgonian corals. Mar Biol 125:411–419

Jokiel PL, Coles SL (1977) Effects of temperature on the mortality and growth of Hawaiian reef corals. Mar Biol 43:201–208

Jokiel PL, Guinther EB (1978) Effects of temperature on reproduction in the hermatypic coral *Pocillopora damicornis*. Bull Mar Sci 28(4):786–789

Jokiel PL, Morrissey JI (1986) Influence of size on primary production in the reef coral *Pocillopora damicornis* and the macroalga *Acanthophora spicifera*. Mar Biol 91:15–26

Jokiel PL, Coles SL (1990) Response of Hawaiian and other Indo Pacific reef corals to elevated temperatures. Coral Reefs 8:155–162

Kobluk DR, Lysenko MA (1994) Ring bleaching in Southern Caribbean *Agaricia agaricites* during rapid water cooling. Bull Mar Sci 54:142–150

Koh EGL (1997) Do scleractinian corals engage in chemical warfare against microbes? J Chem Ecol 23(2):379–398

Kushmaro A, Loya Y, Fine M, Rosenberg E (1996) Bacterial infection and coral bleaching. Nature 380:396

Kushmaro A, Rosenberg E, Fine M, Loya Y (1997) Bleaching of the coral *Oculina patagonica* by *Vibrio* AK-1. Mar Ecol Prog Ser 147:159–165

Kushmaro A, Rosenberg E, Fine M, Ben-Haim Y, Loya Y (1998) Effect of temperature on bleaching of the coral *Oculina patagonica* by *Vibrio shiloi* AK-1. Mar Ecol Prog Ser 171:131–137

Lang JC, Lasker HR, Gladfelter EH, Hallock P, Jaap WC, da Losa FJ, Muller RC (1992) Spatial and temporal variability during periods of 'recovery' after mass bleaching on Western Atlantic coral reefs. Am Zool 32:696–706

Lesser MP, Lewis S (1996) Action spectrum for the effects of UV radiation on photosynthesis in the hermatypic coral *Pocillopora damicornis*. Mar Ecol Prog Ser 134:171–177

Lesser MP, Stochaj WR, Tapley DW, Shick JM (1990) Bleaching in coral reef anthozoans: effects of irradiance, ultraviolet radiation, and temperature on the activities of protective enzymes against active oxygen. Coral Reefs 8:225–232

Loya Y, Sakai K, Yamazato K, Nakano Y, Sembali H, van Woesik R (2001) Coral bleaching: the winners and losers. Ecol Lett 4: 122–131

Lyons MM, Aas P, Pakulski JD, Van Waasbergen L, Miller RV, Mitchell DL, Jeffrey WH (1998) DNA damage induced by ultraviolet radiation in coral-reef microbial communities. Mar Biol 130:537–543

Muscatine L, Grossman D, Doino J (1991) Release of symbiotic algae by tropical sea-anemones and corals after cold shock. Mar Ecol Prog Ser 77:233–243

Oliver J (1985) Recurrent seasonal bleaching and mortality of corals on the Great Barrier Reef. Proc 5th Int Coral Reef Symp 4:201–206

Peters EC (1984) A survey of cellular reactions to environmental stress and disease in Caribbean scleractinian corals. Hegol Wiss Meeresunters 37:113–137
Richardson LL (1998) Coral diseases: what is really known? Trends Ecol Evol 13:438–443
Rosenberg E, Ben-Haim Y (2002) Microbial diseases of corals and global warming. Environ Microbiol 4:318–326
Sammarco PW, Coll JC (1988) The chemical ecology of alcyonarian corals (Coelenterata: Octocorallia). In: Scheuer PJ (ed) Bioorganic marine chemistry, vol 2. Springer, Berlin Heidelberg New York, pp 87–116
Sammarco PW, Coll JC (1992) Chemical adaptations in the Octocorallia: evolutionary considerations. Mar Ecol Prog Ser 88:93–104
Sebens PK (1994) Biodiversity of coral reefs: what are we losing and why? Am Zool 34:115–133
Stimson J (1997) The annual cycle of density of zooxanthellae in the tissues of field and laboratory-held *Pocillopora damicornis* (Linnaeus). J Exp Mar Biol Ecol 214:35–48
Szmant A, Gassman NJ (1990) The effects of prolonged bleaching on the tissue biomass and reproduction of the reef coral *Monastrea annularis*. Coral Reefs 8:217–224
Toren A, Landau L, Kushmaro A, Loya Y, Rosenberg E (1998) Effect of temperature on adhesion of *Vibrio* strain AK-1 to *Oculina patagonica* and on coral bleaching. Appl Environ Microbiol 64:1379–1384
Van Woesik R, de Vantier LM, Glazebrook JS (1995) Effects of the cyclone Joy on near shore coral communities of the Great Barrier Reef. Mar Ecol Prog Ser 128:261–270
Vernon JEN, Stafford-Smith H (2000) Corals of the world. Odyssey Publishing, Broomfield, CO, USA
Warner ME, Fitt WK, Scmidt GW (1996) The effects of elevated temperature on the photosynthetic efficiency of zooxanthellae in hospite from four different species of reef corals: a novel approach. Plant Cell Environ 19:291–299
Williams EH Jr, Bunkley-Williams L (1990) The world-wide coral reef bleaching cycle and related sources of coral mortality. Attol Res Bull 355:1–72
Wright AC, Morris JG Jr (1991) The extracellular cytolysin of *Vibrio vulnificus*: inactivation and relationship to virulence in mice. Infect Immun 59:192–197

18 Black Band Disease

Laurie L. Richardson

18.1 The First Coral Disease

Black band disease (BBD) was the first coral disease to be reported in the literature (Antonius 1973). It was first noted on reefs of Belize in the western Caribbean, and was described as a striking microbial assemblage that formed a band which moved across apparently healthy coral colonies, actively destroying coral tissue and leaving behind the bare coral skeleton (Fig. 18.1). The band appeared dark, which was the basis of the descriptive name. During this same decade, two additional coral diseases were reported, white plague (Dustan 1977) and white band (Gladfelter et al. 1977). BBD differed from the other two in that a very obvious microbial biomass was associated with the pathology, whereas no discernible microorganisms were observable for white plague or white band. Despite the distinct presence of a dense microbial community associated with active coral tissue lysis, identification of the BBD pathogen has proven to be challenging and is still the subject of ongoing investigation.

Fig. 18.1. Multiple black band disease infections on *Diploria strigosa*

18.2 Epizootiology

BBD typically progresses at rates of 3 mm/day across living coral colonies, although rates of up to 1 cm/day have been recorded. The disease affects both scleractinian (stony) and gorgonian (soft) corals. Band width is generally 0.5–1 cm, although bands as narrow as 1–2 mm (Fig. 18.2) and as wide as almost 7 cm have been documented. Band thickness is approximately 1 mm. The band is loosely attached to the coral surface, and can easily be dislodged by fanning the water adjacent to the band. While primarily recorded as present on reefs throughout the Caribbean in the 1970s and 1980s, it was found in the 1980s to also occur in the Indo-Pacific (Antonius 1985) and the Red Sea (Antonius 1988), and by the 1990s had spread to the Great Barrier Reef (Dinsdale 2002). It now exhibits a global distribution. A comprehensive list of coral species affected, as well as the global geographic distribution of BBD, is presented in detail in Green and Bruckner (2000).

On temperate reefs, BBD exhibits a distinct seasonal cycle in which the disease is present primarily during the warmest months of the year, when water temperatures rise above 28 °C (Rützler et al. 1983; Edmunds 1991; Kuta and Richardson 1996). Occasionally, it occurs year round and has been observed rarely at temperatures below 20 °C. The disease is most active when water temperatures rise above 30 °C, which is in agreement with laboratory results of BBD-associated microorganisms (discussed below). The most prominent BBD spatial distribution parameter is defined by water depth, with BB-diseased colonies usually limited to shallow waters of 10 m or less (Rützler et al. 1983; Taylor 1983; Antonius 1985).

Although it has been suggested that BBD is associated with elevated nutrient input and/or sewage outflows (Taylor 1983; Antonius 1988; Bruckner et al. 1997), most such reports are based on qualitative observations. Only one quan-

Fig. 18.2. Thin BBD on *Colpophyllia natans*. This coral colony is >3 m in diameter

titative study has been published to date aimed at assessing a relationship between water quality (including nutrients) and BBD activity (Kuta and Richardson 2002). In this study, we found a positive relationship between elevated nitrite and BBD incidence, but no positive correlation with nitrate, ammonium, or phosphate, the most readily assimilable (and nutrient-limiting) chemical species. BBD is routinely observed on pristine reefs far from anthropogenic perturbations. Any relationship between degraded water quality and BBD remains obscure.

BBD commonly exhibits very low prevalence values of <1% when active on reefs (Edmunds 1991; Kuta and Richardson 1996; Dinsdale 2002), although values over 6% have been documented (Bruckner and Bruckner 1997). Despite such low occurrence, BBD is considered to be an important factor negatively affecting coral reef health. Affected coral colonies are not observed to recover to any great extent, and the newly exposed substrate (dead coral) does not become colonized by coral recruits until after many years (Kuta and Richardson 1997; Edmunds 2000). Furthermore, it tends to affect large reef-building corals with extremely low growth rates of 1 cm/year or less (Hudson 1981).

18.3 The Search for the Black Band Disease Pathogen

The history of the search for a BBD pathogen has been ongoing and continual for the past 30 years. The subject of the search is surprisingly elusive. When freshly collected black band is viewed under the microscope, a diverse microbial community is always present, thus potential pathogens are not difficult to find. The diverse community is highly distinctive, and is consistently dominated by a population of gliding, filamentous, cyanobacteria. A multitude of small rods and cocci are also always present, as is a population of colorless gliding bacteria with internal sulfur granules.

Early work aimed at determining the identity of the BBD pathogen was focused on finding a single pathogen. In the 1970s, three different pathogens were proposed. These were the cyanobacterium, the gliding bacterium with sulfur inclusions, and sulfate-reducing bacteria. The basis for the proposal that the cyanobacterium was responsible for the disease was the fact that it is always present, and that it is the dominant member of the community in terms of biomass (Antonius 1981). Antonius identified the BBD cyanobacterium as *Oscillatoria submembranaceae*, a species that is quite common on various substrates found on coral reefs, such as sediments, corals, and macroalgae. This species was not known to be pathogenic. During the same decade, a different group (Garrett and Duckow 1975) proposed that the gliding, filamentous BBD microbe with obvious internal sulfur granules was involved. They identified this BBD member as within the genus *Beggiatoa* (species unknown), which, along with the cyanobacterium, was not known to be pathogenic. The basis for this suggestion was that *Beggiatoa* sp. is commonly found associated with many dying corals. The third study focused on a different group of sulfur cycle bacteria, the sulfate reducers. As sulfide oxidizers such as *Beggiatoa* are only found in environments with both oxygen and sulfide, and in marine environments the source of environmental sulfide is almost always sulfidogenic (i.e., sulfate-reducing) bacteria, the premise was that this group of organisms was essential to BBD pathology and thus a suspect primary pathogen. Thus, in 1975, Mitchell and Chet implicated *Desulfovibrio* spp. as the BBD pathogen. During this first decade of BBD research, it was also proposed that both the BBD cyanobacterium and *Beggiatoa* were required for active black band disease (Ducklow and Mitchell 1979). Experimental manipulations revealed that stressing corals could result in tissue necrosis, elevated bacterial activity, and development of a population of sulfur cycle bacteria, but that only when cyanobacteria were present would BBD develop.

Each of these early studies was based solely on microscopic observations, with no attempts made at isolation. Following this approach, in the early 1980s, another group suggested that BBD also involved a marine fungus (Ramos-Flores 1983).

18.3.1 *Phormidium corallyticum*

In the 1980s, a detailed observational study resulted in an assertion that the primary pathogen was unquestionably a BBD-specific cyanobacterium, and that it was a species other than *O. submembranaceae*. This group proposed that the BBD cyanobacterium was a new cyanobacterial species, and gave it the name of *Phormidium corallyticum* (Rützler and Santavy 1983). Their characterization was based on filament morphology using both light and transmission electron microscopy, and the observation that BBD consistently contained a cyanobacterium that was morphologically unique from other coral reef cyanobacteria. Specifically, the BBD cyanobacterium possessed one rounded and one pointed tip on the ends of each filament.

This group was unable to attain *P. corallyticum* in pure culture, which is not unusual when working with cyanobacterial isolations. Therefore, inoculation experiments aimed at demonstrating BBD induction were conducted using clumps of fresh BBD material (Rützler and Santavy 1983). Although filaments were repeatedly allowed to disperse and form clumps (a typical behavioral pattern of motile, filamentous cyanobacteria) in an effort to minimize associated microorganisms, the authors noted that an assortment of BBD bacteria were always present in their infectivity studies. Thus, Koch's postulates were not satisfied. Despite this, it became widely believed that *P. corallyticum* was, in fact, the primary pathogen of BBD. This belief was held by many throughout the 1980s and 1990s.

18.4 An Alternative View – Black Band Disease as a Pathogenic Microbial Consortium

In the 1990s, our group proposed that BBD is caused by a pathogenic microbial consortium, and that there is no primary pathogen (Carlton and Richardson 1995). The premise for this proposal was that the BBD microbial community consistently contained dominant populations of the same microorganisms. We noted that the BBD microbial community always contained populations of both sulfate-reducing and sulfide-oxidizing bacteria as well as phototrophic cyanobacteria. We also noted that such communities form the functional basis of microbial mats commonly found in many illuminated, sulfide-rich environments such as hot-spring outflows (see Jørgensen 1982).

Our belief that BBD is a true pathogenic microbial consortium was first based on observations of vertical migrations of BBD community members, evidence that the microbial community is both dynamic and organized (Richardson 1996). The community is normally dark ("black") due to the presence of the cyanobacterial light-harvesting photosynthetic accessory pigment phycoerythrin. Phycoerythrin is a red pigment, absorbing light maximally in a strong peak centered near 586 nm. When BBD is collected and dispersed, it appears red. In situ on the reef the community is organized into a dense matrix. The compacted biomass, plus the fact that water itself absorbs 90% of red light in the top 1 m, makes the band appear black. Yet one can often see white speckles and patches on the surface of BBD (Fig. 18.3) due to the highly refractile properties of elemental sulfur granules deposited by BBB-associated sulfide-oxidizing bacteria. Thus, BBD also appeared to resemble other microbial mats in terms of the presence of vertically migrating populations of both cyanobacterial and sulfur cycle microbes (Jørgensen 1982).

Fig. 18.3. Photomacrograph of BBD on *Diploria strigosa*. *Beggiatoa* sp. (*white specks*) can be seen on the dark *P. corallyticum* mat where incident light is highest (*top of ridges*). (Photo by G. McFall)

18.4.1 Chemical Microenvironments Within Black Band Disease

In the 1990s we conducted a study of intact BBD using oxygen- and sulfide-sensitive microelectrodes, and confirmed that BBD is structurally and functionally identical to other, well-characterized, microbial mats (Carlton and Richardson 1995; Richardson et al. 1997). We demonstrated that the top of the band is supersaturated with O_2 during daylight hours (the result of cyanobacterial oxygenic photosynthesis); that most of the band is anoxic due to the presence of sulfide generated within the band (by dissimilatory sulfate reduction); and that an interface of oxygen and sulfide is present that migrates vertically within the band dependent upon incident light intensity. Our results showed that the chemical microgradients present in BBD were the same as those of other microbial mats (Revsbech et al. 1983), both in terms of concentrations present and the dynamic, vertical migration of the oxygen/sulfide interface.

18.4.2 Characterization of Black Band Disease Functional Groups

The results of the microelectrode studies revealed that the members of the BBD community are exposed to chemical microenvironments that range from highly oxidizing to anaerobic and reducing. Such environments are of direct importance to the BBD microbial community members because they often dictate the energy-yielding metabolic pathway that may be employed (for example, aerobic vs. anaerobic respiration, sulfide oxidation, etc.) Thus, BBD can be considered as having a chemically dynamic internal environment that is both produced by the BBD microorganisms themselves, and that, once developed, will influence the physiology of the BBD members at any given time depending upon where they are in the band (i.e., at the highly oxygenated surface, in microaerobic zones above the oxygen/sulfide interface, at the oxygen/sulfide interface itself, or in an anaerobic, reducing environment below the interface). The results of our work on chemical microenvironments within BBD led us to believe that BBD is best described as a tightly organized (spatially and physiologically) synergistic community. We have approached this hypothesis by investigating the individual microbial members of the consortium in terms of the functional community as a whole.

18.4.2.1 Black Band Disease Cyanobacteria

In 1991, we isolated a cyanobacterium from active, in situ black band disease that we identified, based on morphology as described in Rützler and Santavy (1983), as *Phormidium corallyticum*. We have conducted substantial work on the physiology of the black band cyanobacterium in terms of photosynthetic capabilities and potential sulfide use (Richardson and Kuta 2003). We found that this cyanobacterium performs oxygenic photosynthesis in the presence of sulfide, and that it does not perform DCMU [(3-(3′,4′-dichlorophenyl)-1,1-dimethylurea – an inhibitor of electron transport in photosystem II)] forced anoxygenic photosynthesis with sulfide as electron donor. Thus, under illuminated conditions this member of the consortium continually produces oxygen in the presence of sulfide. This was an unusual finding because most cyanobacteria are sulfide-intolerant, due to the fact that exposure to environmental sulfide blocks electron transport in Photosystem II (Vismann 1991). While some cyanobacteria can use sulfide as an electron donor, we determined that the BBD cyanobacterium does not (Richardson and Kuta 2003). We also found that the BBD cyanobacterium cannot fix N_2 (both a laboratory culture and freshly collected black band material were used in these experiments), and thus are not performing a role of supplying fixed nitrogen to the BBD microbial community.

We experimentally determined photosynthetic and behavioral (tactic) responses to varying light intensity (Kuta 2000; Viehman 2002; Viehman and Richardson 2002; Richardson and Kuta 2003). We found that the motile cyanobacterial filaments exhibit a very strong negative phototaxis, which results in migration of the cyanobacterial population to a position underneath the population of *Beggiatoa* spp. during high light periods on the reef (Viehman and Richardson 2002). *P. corallyticum* appears to form the compact band matrix as a result of this negative phototaxis, which may allow for the development of high concentrations of sulfide at the base of the band. In our laboratory physiology experiments, we found that maximal photosynthetic activity occurred at very low light levels ($<30\ \mu E\ m^{-2}\ s^{-1}$), in agreement with the field behavioral studies, indicating the strong preference for low light (Richardson and Kuta 2003). In addition, growth temperature optima were realized at and above 30 °C (Richardson and Kuta 2003), which agrees with field observations of BBD activity and previous results by Taylor (1983), and coincides with the temperature at which the coral host becomes physiologically stressed (Fitt et al. 2001).

18.4.2.2
Black Band Disease Sulfate Reducers

In a series of manipulative experiments, we determined that anoxia plus sulfide at concentrations present at the base of active BBD is lethal to coral tissue (Richardson et al. 1997). These results formed the basis of our belief that sulfide produced by BBD sulfate reducers is integral to BBD pathogenicity. Thus, we concur with earlier investigators that sulfate reducers are important BBD microbes. In a suite of enrichment and culturing experiments using BBD samples collected from corals of reefs of Florida and Dominica, followed by PCR and 16S rRNA gene sequencing, we found 98% homologous matches to two strains of *Desulfovibrio* (*D. acrylicus* and *Desulfovibrio* sp. strain TBP-1) in each of ten BBD samples analyzed (Viehman 2002). For the enrichments, BBD was collected from six corals of four different species. Therefore, these results support the hypothesis that BBD sulfate-reducing bacteria are functionally specific to BBD pathogenicity, and suggest that they may be species-specific. The latter conclusion differs from those of two other groups (discussed below in Sect. 18.5) who found more variable BBD-associated sulfate reducers using similar molecular techniques.

18.4.2.3
Black Band Disease Sulfide Oxidizers

We conducted preliminary work on the BBD sulfide oxidizers, specifically the gliding population that migrates within the band. In addition to light manipulation experiments focused on understanding BBD microbial vertical migration patterns, we successfully grew sulfide-oxidizing bacteria from BBD that morphologically appear to be *Beggiatoa* (Nelson 1992). This was accomplished using two enrichment culture techniques: inoculation into sulfide-gradient tubes and growth under microoxic conditions on sulfide plates. While we were successful in enriching for and growing BBD-associated gliding, filamentous sulfide oxidizers, we were unable to attain pure cultures. At the present time, we believe that the role of sulfide oxidizers in BBD is to stabilize the oxygen/sulfide dynamics within the band.

18.5
Molecular Analysis of Black Band Disease-Associated Microorganisms

Recently, molecular studies by two groups have individually led to proposals that the BBD pathogen is not a sulfur cycle bacterium, cyanobacterium, fungus, or a pathogenic consortium, but may be a nonphotosynthetic eubacterial heterotroph (Cooney et al. 2002; Frias-Lopez et al. 2002). As noted above, all investigators had observed that the BBD microbial consortium contains a diverse assortment of bacteria. However, until these studies, no one had targeted the general heterotrophic microbial population as harboring the primary pathogen.

Both Cooney et al. (2002) and Frias-Lopez et al. (2002) conducted studies in which bacterial populations associated with BBD were subjected to PCR and analyzed by comparing 16S rRNA gene sequence data to sequences deposited in GenBank. In both studies bacterial populations associated with BBD and the healthy (i.e., not BBD infected) surface of the same colonies were assessed. One study (Frias-Lopez et al. 2002) also examined samples from the water column and the dead surface of BBD-infected colonies. Both studies were carried out using samples collected in the Caribbean. Cooney et al. analyzed samples from 11 individual coral colonies of three coral species, collected from Barbados and St. Croix. Frias-Lopez et al. collected and analyzed samples from three coral colonies in Curaçao, with one sample collected from each colony. Each study analyzed partial sequences of the 16S rRNA gene, with sequences in the range of 200–600 nucleotides.

In addition to new proposals of heterotrophs as BBD pathogens, the results of both studies are quite interesting when considered in terms of the previous work on potential BBD pathogens. As a result of the findings of both studies, there is now controversy in the literature as to the validity of *P. corallyticum* as the BBD cyanobacterium. Frias-Lopez et al. (2002) found a single black band-associated cyanobacterium that was most closely related to the planktonic species *Trichodesmium tenue* (93% match). They found this sequence in two of the three BBD-infected corals sampled, and found no cyanobacterial sequence in their molecular analysis of the third (despite the observation that the band was dominated by cyanobacteria). In the other study, Cooney et al. (2002) found two entirely different cyanobacteria in BBD. These were *Oscillatoria cf. corallinae* (92% sequence homology) and a planktonic freshwater strain (91% homology). None of these constitute a generally accepted identification (97.5% or higher sequence homology). Neither of the two studies found (partial) sequences that matched any *Phormidium* species or strain deposited in GenBank. Our laboratory strain (discussed above) of *P. corallyticum* was given to Cooney et al. and was partially sequenced (500 nucleotides) by this group and included in their data analysis. Cooney et al. found our laboratory culture to be most closely matched (97% homology) to an entirely different cyanobacterial genus, *Geitlerinema*. This strain was not found in the molecular analyses of BBD samples conducted by either Frias-Lopez et al. or Cooney et al.

Similarly, these studies have raised new questions about the presence of *Beggiatoa* in BBD. Neither group found a sequence from BBD samples that matched *Beggiatoa* in GenBank, and both groups questioned the morphology-based identification of this genus as a common BBD member.

The PCR results in terms of BBD sulfate-reducing bacteria were more consistent, with both groups finding sequences with higher homologies (94–99%) to different species and strains of sulfate reducers. Frias-Lopez et al. found sequence homologies corresponding to either *Desulfovibrio* or *Desulfobotulus* (another sulfate-reducer) in all three BBD samples. Cooney et al. found sequences implicating five sulfidogenic genera: *Desulfovibrio*, *Desulfocella*, *Dethiosulfovibrio*, *Desulfotomaculum*, and *Sulphurospirillum*. Both groups concluded that the presence of BBD sulfate-reducers was functionally important for the BBD microbial community, but that no sulfate-reducer (or sulfidogen) was the primary pathogen. As mentioned above, our group agrees with the conclusion that BBD sulfate reducers are functionally important, but our results differ in that we found 98% homology matches to two consistent *Desulfovibrio* strains (Viehman 2002).

18.5.1 Black Band Disease-Associated Heterotrophs as Primary Pathogens

The two molecular studies of Cooney et al. (2002) and Frias-Lopez et al. (2002) have individually supported the new proposal that the BBD pathogen may be a pathogenic heterotrophic bacterial species. Cooney et al. consistently found two (partial) 16S rRNA gene sequences in 11 BBD samples. These were one sequence with 93% homology to *Cytophaga* sp., and a second sequence 98% homologous to alpha proteobacteria isolated from Juvenile Oyster (JOD) disease. The GenBank deposit which yielded the 93% *Cytophaga* match was isolated from a deep-sea trench sediment and was not grouped with pathogenic species, although the authors note that some members of this genus are known pathogens of marine organisms. The sequence found in all BBD samples that corresponded to JOD alpha proteobacteria exhibited 98% homologies to two separate strains isolated from JOD in cultured oysters. The only other sequence common to all BBD samples in the Cooney et al. study was a cyanobacterial sequence with 92% homology to *Oscillatoria corallinae* (also found on non-BBD samples). These authors have concluded that further investigations of BBD pathogenesis should target these three species.

Frias-Lopez et al. (2002), who also used culture-independent molecular analysis of BBD samples following PCR, did not find any sequences consistently present in all of their BBD samples (n=3). Rather, they found that two of their BBD samples yielded sequences with moderate homologies (range 87–94%) to *Cytophaga fermentens*. (This sequence, however, was also detected in a sample collected from dead coral.) The only other common (again two of three) sequence match in this study was to the ubiquitous (in sub-tropical marine waters) planktonic cyanobacterium *T. tenue* discussed previously. Thus, the two studies did not find the same BBD specific sequences. Both groups emphasized that non-cultivative molecular techniques should be used in future BBD investigations.

18.6 Histopathology of Black Band Disease-Affected Coral Tissue

By far the most work conducted to date on BBD has been in the areas of microbiology and epizootiology. Only two studies have provided histology-based results that reveal the effect of BBD on coral tissue. Peters (1993) showed that necrotic tissue associated with BBD had a predominance of cyanobacterial filaments in the tissue. Similarly, Bythell et al. (2002) found cyanobacterial filaments within BBD degraded tissue, and also showed the presence of coccoid bacterial cells.

18.7 Some Unresolved Questions and Future Directions

There are many unresolved questions concerning the etiology of black band disease, and a multitude of promising future research directions. Perhaps most important is the unresolved question of the identity of the BBD pathogen. In addition, we do not know the mode of transmission of BBD infections on the reef, nor do we know the mechanism by which the disease has spread to infect corals worldwide. We do not know the relationships between disease incidence and/or prevalence and (with the exception of elevated water temperature) environmental factors. We also do not know if BBD is promoted by environmental degradation, and we do not know why it appears on pristine reefs.

Perhaps the most promising future direction for BBD research is implementation of an integrative approach that combines molecular techniques with field- and laboratory-based programs on BBD etiology. This approach has been suggested by BBD researchers in the past (Richardson et al. 2001; Cooney et al. 2002). Such studies should include continued efforts to culture BBD microorganisms with the goal of understanding the role of each BBD microorganism in the context of the disease process. An encouraging approach is to construct species and strain-specific molecular probes labeled with fluorochromes or other tracers. Such probes, when used in conjunction with fluorescent in situ hybridization, may yield information on both BBD pathogenesis and interactions of disease processes (transmission, etc.) in the reef environment.

References

Antonius A (1973) New observations on coral destruction in reefs. 10th Meeting Assoc Isl Mar Lab Carib 10:3

Antonius A (1981) Coral reef pathology: a review. Proc 4th Int Coral Reef Symp 2:3–6

Antonius A (1985) Coral diseases in the Indo-Pacific: a first record. PSZNI Mar Ecol 6:197–218

Antonius A (1988) Distribution and dynamics of coral diseases in the eastern Red Sea. Proc 6th Int Coral Reef Symp 3:145–150

Bruckner A, Bruckner R (1997) The persistence of black band disease in Jamaica: Impact on community structure. Proc 8th Int Coral Reef Symp 1:601–606

Bruckner AW, Bruckner RJ, Williams EH (1997) Spread of a black-band disease epizootic through the coral reef system in St Ann's Bay, Jamaica. Bull Mar Sci 61:919–928

Bythell J, Barer M, Cooney R, Guest J, O'Donnell A, Pantos O, Le Tissier M (2002) Histopathological methods for the investigation of microbial communities associated with disease lesions in reef corals. Lett Appl Microbiol 34:111–116

Carlton R, Richardson LL (1995) Oxygen and sulfide dynamics in a horizontally migrating cyanobacterial mat: black band disease of corals. FEMS Microbiol Ecol 18:155–162

Cooney R, Pantos O, Le Tissier M, Barer M, O'Donnell A, Bythell J (2002) Characterization of the bacterial consortium associated with black band disease in coral using molecular microbiological techniques. Environ Microbiol 4:401–413

Dinsdale E (2002) Abundance of black band disease on corals from one location on the Great Barrier Reef: a comparison with abundance in the Caribbean region. Proc 9th Int Coral Reef Symp 2:1239–1244

Ducklow H, Mitchell R (1979) Observations on naturally and artificially diseased tropical corals: a scanning electron microscope study. Microb Ecol 5:215–223

Dustan P (1977) Vitality of reef coral populations off Key Largo, Florida: recruitment and mortality. Environ Geol 2:51–58

Edmunds PJ (1991) Extent and effect of black band disease on Caribbean reefs. Coral Reefs 10:161–165

Edmunds P (2000) Recruitment of scleractinians onto the skeletons of corals killed by black band disease. Coral Reefs 19:69–74

Fitt WK, Brown BE, Warner ME, Dunne RP (2001) Coral bleaching: interpretation of thermal tolerance limits and thermal thresholds in tropical corals. Coral Reefs 20:51–65

Frias-Lopez J, Zerkle A, Bonheyo G, Fouke B (2002) Partitioning of bacterial communities between seawater and healthy, black band diseased, and dead coral surfaces. Appl Environ Microbiol 68:2214–2228

Garrett P, Ducklow H (1975) Coral disease in Bermuda. Nature 253:349–350

Gladfelter WB, Gladfelter EH, Monahan RK, Ogden JC, Dill RF (1977) Environmental studies of Buck Island Reef National Monument, St Croix US Virgin Islands. Spec Rep Nat Park Serv US Dept Int 173

Green EP, Bruckner AW (2000) The significance of coral disease epizootiology for coral reef conservation. Biol Conserv 96:347–361

Hudson J (1981) Growth rates of *Montastraea annularis*: a record of environmental change in the Key Largo National Marine Sanctuary, Florida. Bull Mar Sci 31:444–459

Jørgensen B (1982) Ecology of the bacteria of the sulphur cycle with special reference to anoxic-oxic interface environments. Philos Trans R Soc Lond 298:543–561

Kuta KG (2000) Black band disease of corals: ecology and physiology of *Phormidium corallyticum*. Florida International University, Miami, 79 pp

Kuta KG, Richardson L (1996) Abundance and distribution of black band disease of corals in the northern Florida Keys. Coral Reefs 15:219–223

Kuta KG, Richardson LL (1997) Black band disease and the fate of diseased coral colonies in the Florida Keys. Proc 8th Int Coral Reef Symp 1:575–578

Kuta KG, Richardson LL (2002) Ecological aspects of black band disease of corals: relationships between disease incidence and environmental factors. Coral Reefs 21:393–398

Mitchell R, Chet I (1975) Bacterial attack of corals in polluted seawater. Microb Ecol 2:227–233

Nelson D (1992) The genus *Beggiatoa*. In: Balows A, Truper H, Dworkin M, Harder W, Schleifer K (eds) The prokaryotes, 2nd edn. Springer, Berlin Heidelberg New York, pp 3171–3180

Peters E (1993) Diseases of other invertebrate phyla: Porifera, Cnidaria, Ctenophora, Annelida, Echinodermata. In: Couch JA, Fournie JW (eds) Advances in fisheries science, Pathobiology of marine and estuarine organisms. CRC Press, Boca Raton, pp 393–449

Ramos-Flores T (1983) Lower marine fungus associated with black line disease in star corals (*Montastraea annularis*). Biol Bull 165:429–435

Revsbech NP, Jørgensen BB, Blackburn TH, Cohen Y (1983) Microelectrode studies of the photosynthesis and O_2, H_2S and pH profiles of a microbial mat. Limnol Oceanogr 28:1062–1074

Richardson LL (1996) Motility patterns of *Phormidium corallyticum* and *Beggiatoa* spp. associated with black band disease of corals. Microb Ecol 32:323–335

Richardson LL (1997) Occurrence of the black band disease cyanobacterium on healthy corals of the Florida Keys. Bull Mar Sci 61:485–490

Richardson LL, Kuta KG (2003) Ecological physiology of the black band disease cyanobacterium *Phormidium corallyticum*. FEMS Microbiol Ecol 43:287–298

Richardson LL, Kuta KG, Schnell S, Carlton RG (1997) Ecology of the black band disease microbial consortium. Proc 8th Int Coral Reef Symp 1:597–600

Richardson LL, Smith GW, Ritchie KB, Carlton RG (2001) Integrating microbiological, microsensor, physiologic and molecular techniques in the study of coral disease pathogenesis. Hydrobiologia 460:71–89

Rützler K, Santavy D (1983) The black band disease of Atlantic reef corals. I. Description of the cyanophyte pathogen. PSZNI Mar Ecol 4:301–319

Rützler K, Santavy D, Antonius A (1983) The black band disease of Atlantic reef corals. III. Distribution, ecology and development. PSZNI Mar Ecol 4:329–358

Taylor D (1983) The black band disease of Atlantic reef corals. II. Isolation, cultivation, and growth of *Phormidium corallyticum*. PSZNI Mar Ecol 4:320–328

Viehman T (2002) Black band disease of corals: ecology and physiology of the sulfur bacteria *Desulfovibrio* spp. and *Beggiatoa* spp. Biology, Florida International University, Miami, 100 pp

Viehman TS, Richardson LL (2002) Motility patterns of *Beggiatoa* and *Phormidium corallyticum* in black band disease. Proc 9th Int Coral Reef Symp 1:1251–1256

Vismann B (1991) Sulfide tolerance: physiological mechanisms and ecological implications. Ophelia 34:1–27

19 Dark Spots Disease and Yellow Band Disease, Two Poorly Known Coral Diseases with High Incidence in Caribbean Reefs

Diego L. Gil-Agudelo, Garriet W. Smith,
Jaime Garzón-Ferreira, Ernesto Weil, and Dirk Petersen

19.1 Introduction

Coral reefs have been described as very stable systems in delicate balance with nature. They are also the most sensitive and vulnerable ecosystems in the sea (Grigg and Dollar 1990). Their exposure to some of the most common natural and anthropogenic perturbations have caused considerable decline of their health during the last three decades (Grigg and Dollar 1990; Ginsburg 1994; Birkeland 1997).

Like many other marine taxa, corals have been affected by new diseases over the last 30 years (Harvell et al. 1999). Some of these diseases are relatively well known, such as black band disease in hard corals (Santavy and Peters 1997) and aspergillosis in sea fans (see Chap. 15, Smith and Weil, this Vol.). Other diseases have been observed for many years but their etiologies are still not clear (Santavy and Peters 1997).

Dark spots disease and yellow band disease are relatively new diseases affecting Caribbean corals. These diseases were first reported in the 1990s, although both of them may have been present for longer time periods in these reefs. In some regions, these diseases are distributed sparsely, but in other areas these diseases are distressing and killing coral populations to a significant degree, including key species in the region.

These diseases are a challenge to study from the point of view of their pathology. In the case of yellow band, prior studies suggested that this syndrome might be the result of a zooxanthellae disease and not a coral disease (Cervino et al. 2001). In the case of dark spots disease, which affects numerous coral species, it is not clear if disease signs present in all species are the result of the same pathogen or whether different diseases exhibit similar signs in different coral species.

Our goal in this chapter is to present an update of the current knowledge on the pathology (pathogen identification and characterization) and epizootiology (distribution, host ranges, prevalence, and impact) of these two diseases, as a starting point for future studies. Due to the recent discovery of these two diseases, little information is available to date; therefore, most of the information included in this chapter is recent and has not been published or is in the process of being published.

19.2 Dark Spots Disease

19.2.1 Original Description

Dark spots disease (DSD) was first observed in the early 1990s during a bleaching event in Nuestra Señora de las Islas del Rosario Archipelago (Colombian Caribbean). At the time, it was reported as a type of bleaching disease (Solano et al. 1993).

Later, during a 1994 expedition to the San Andrés and Providencia archipelago (Colombian Caribbean), Garzón-Ferreira and other researchers found this disease impacting heads of the *Montastraea annularis* complex, mainly the columnar form. These researchers recognized it as a nonbleaching disease for the first time and named it dark spots disease. Later, similar signs were observed by the same group of researchers on *Siderastrea siderea*, living in the coral for-

Fig. 19.1a–d. Examples of coral species showing signs of dark spots disease. **a** *S. siderea* in Puerto Rico (photo by Ernesto Weil). **b** *M. annularis* in Santa Marta, Colombia (photo by Diego L. Gil-Agudelo). **c** *S. intersepta* in San Andrés Islands, Colombia (photo by Jaime Garzón-Ferreira). **d** *A. agaricites* in San Andrés Islands, Colombia (photo by Jaime Garzón-Ferreira)

mations of the Urabá Gulf (Colombian Caribbean). During these expeditions, the disease was described as small, round, dark spots that apparently grow in size over time. Some can be associated with a depression of the coral surface and others expand into a ring surrounding dead coral (Garzón-Ferreira and Gil 1998; Fig. 19.1).

The first in-depth study of this disease was performed in 1997. This study showed that, in fact, the species most affected (at least in the study site) were *M. annularis* and *S. siderea* (Gil-Agudelo 1998; Gil-Agudelo and Garzón-Ferreira 2001). Nevertheless, other species were also affected by this disease, although in lower proportions. These included *Stephanocoenia intersepta* and other species of the *M. annularis* complex (*M. faveolata* and *M. franksi*).

In other parts of the Caribbean (i.e., Virgin Islands and Puerto Rico), *S. siderea* has been observed with a band of tissue with coloration resembling the one present in DSD, but killing the coral from the outside border to the inside (Fig. 19.1). Some researchers have called this syndrome "ring disease" (Ernesto Weil, University of Puerto Rico), but we consider it to be the same DSD.

19.2.2 Species Affected by the Disease

DSD is mainly found on *S. siderea* and the *M. annularis* complex (*M. annularis*, *M. faveolata*, and *M. franksi*.) Nevertheless, similar signs have been observed in other corals species (Garzón-Ferreira et al. 2001). Monitoring efforts have only been performed so far on *M. annularis* and *S. siderea*, but direct observations have shown extensive mortality in *Stephanocenia intersepta*. Furthermore, during a visit to a sampling site in San Andrés Island (Colombian Caribbean) in 1998, signs of a similar syndrome were found in colonies of *Agaricia agaricites* in shallow waters (less than 5-m depth). A year later, extensive mortality was observed on the same coral colonies. Although no disease data were collected during these visits, our observations suggested that the corals were killed by DSD (Fig. 19.1). Other species such as *Colpophyllia natans*, *M. cavernosa*, and *Diploria stigosa* have also been seen with signs resembling DSD.

19.2.3 Geographic Range and Prevalence of the Disease

To date, the only study quantitatively determining the incidence of this disease was performed in Colombia. In this study, more than 16% of the coral colonies analyzed (9398) were found to be affected by DSD. Almost 28% of *S. siderea* (*n*=1875), 20% of *M. faveolata* (*n*=347), and more than 17% of M. annularis (*n*=5694) were found with signs of this disease (Gil-Agudelo 1998).

In other parts of the Caribbean (such as the San Andrés islands, Colombia), species such as *A. agaricetes* and *S. intersepta* have been found with a high incidence (although the precise incidence has not been determined) of the disease. Goreau et al. (1998) found a high incidence of DSD in corals in Bonaire and

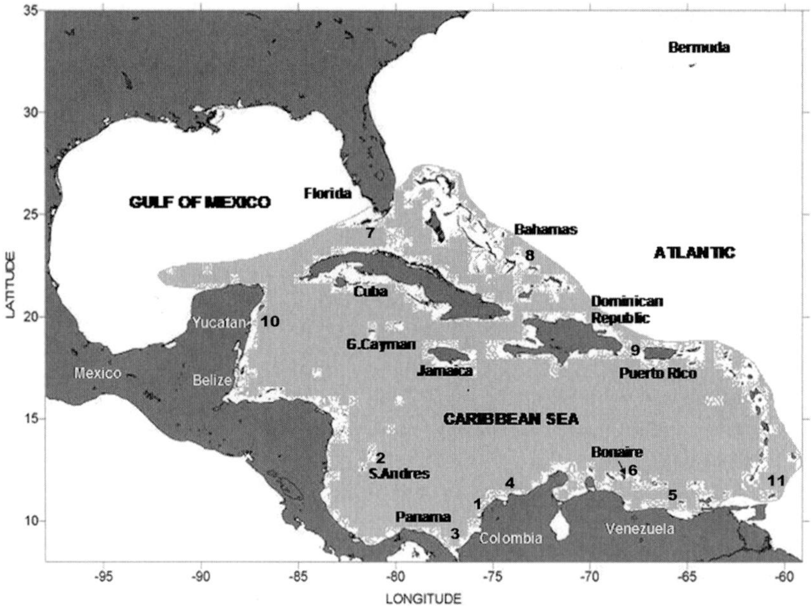

Fig. 19.2. Observations of dark spot disease in the Caribbean. *1* Islas del Rosario (Solano et al. 1993), *2* San Andrés and Providencia Archipelago (Garzón-Ferreira et al. 2001), *3* Urabá Gulf (Garzón-Ferreira et al. 2001), *4* Santa Marta (Gil-Agudelo and Garzón-Ferreira 2001), *5* Venezuela (Ernesto Weil, pers. observ.), *6* Curaçao (James Cervino, Univerity of South Carolina, pers. comm.), *7* Florida Keys (Diego L. Gil-Agudelo, pers. observ.), *8* San Salvador, Bahamas (Diego L. Gil-Agudelo, pers. observ.), *9* Puerto Rico (Ernesto Weil, pers. observ.), *10* Puerto Morelos, Mexico (Diego L. Gil-Agudelo, pers. observ.), *11* Grenada (Ernesto Weil, pers. observ.)

Curaçao (Netherlands Antilles). Recently, a relatively high incidence of DSD in coral reefs in Puerto Rico, Grenada, Mexico, and Grand Cayman Islands has been reported.

Further north, the incidence of DSD appears to diminish. In places such as Florida Keys (USA), San Salvador (Bahamas), Barbados, and Puerto Morelos (Mexico), a lower incidence of the disease has been observed but has not been quantified (Fig. 19.2).

19.2.4
Environmental Parameters Affecting Dark Spots Disease

Gil-Agudelo and Garzón-Ferreira (2001) showed that the incidence of this syndrome appeared to be influenced by water depth and temperature, being more prevalent during periods of warm water (over 28 °C), and at shallow depths (less than 10 m). They also confirmed DSD can affect corals over long periods of time, resulting in the creation of the depressions on the coral surface, typical of this disease. The origin of the depressions is still unknown, but we suspect that they originate from a local decrease in the calcification process.

Healthy polyps continue their normal calcification process resulting in increased calcification around the borders of the spot, thus, creating the depression.

19.2.5 Growth Rate of Dark Spots Disease

Garcés-Baquero (2000) quantified the growth rate of the DSD in two coral species (*M. annularis* and *S. siderea*) in the Colombian Caribbean. During this study, in both species it was found that the spreading rate of the disease was very slow. The average loss of tissue due to the action of DSD was 0.51 cm^2/month (SE=0.19) for *S. siderea* and 1.33 cm^2/month (SE=0.23) for *M. annularis*.

Monitoring of DSD showed also that this disease might decrease the regeneration capability of corals. In fact, lesions as small as 1 cm^2 showed no regeneration of tissue during a period of several months (Garcés-Baquero 2000). Continuous monitoring also showed that signs of this syndrome could be present in some corals for several years, showing little change. Nevertheless, DSD in other species shows higher rates of progression. Although not quantitatively determined, *S. intersepta* appeared to have a much faster progression rate of DSD, resulting in much larger lesions.

Due to the small lesions of DSD and its slow growth rate, it might appear that the effect of this disease in corals is very limited and perhaps almost insignificant for the reef ecosystem. However, the high incidence of DSD in some reefs and the fact that it mainly affects two of the most important reef builder Caribbean corals (*M. annularis* and *S. siderea*), indicates that it can be important for the general health of the reef.

19.2.6 Pathological Studies

To date, only two species have been studied in an attempt to identify the pathogen (*S. siderea* and *M. annularis*). Microbiological studies of the mucus layers of affected colonies showed small changes in the composition of the microbial communities living in association with healthy and disease corals.

Moreover, a bacterial strain metabolically related to *Vibrio carchariae* was found in samples of disease corals from both species (Table 19.1). These bacterial strains were tested in the field by inoculating healthy corals, but no signs of the disease were observed after 1 month of observations.

19.2.7 Other Studies and Observations

Recently, the presence of a "pigmented spot" with a diameter of about 3–5 cm on a colony of *M. annularis* has been noted in an experimental closed system aquarium. In total, 50 specimens of this species were collected for studies on

Table 19.1. Differences in the structure of the microbial community from mucus from healthy corals (*M. annularis* and *S. siderea*) and corals showing signs of dark spots disease based on the metabolic profile of culturable bacteria. Note that group VII is present only in mucus from disease corals

Group no.	Associated genus/species	Diseased		Healthy		Water	
		n	(%)	*n*	(%)	*n*	(%)
I	Enterobacter	11	11.83	8	9.64	3	4.48
II	*Salmonella/Serraita*	8	8.60	6	7.23	4	5.97
III	*Klebsiella/Cedecea*	4	4.30	7	8.43	2	2.99
IV	*Deleya*	6	6.45	3	3.61	4	5.97
V	*V. natriegens/Kluyvena*	5	5.38	5	6.02	2	2.99
VI	*Serratia/V. metschnikovii*	5	5.38	5	6.02	2	2.99
VII	*V. carchariae*	7	7.53	0	0.00	0	0.00
VIII	*K. planticola*	10	10.75	3	3.61	3	4.48
IX	*V. metschnikovii/carchariae*	9	9.68	11	13.25	6	8.96
X	*Yersinia*	8	8.60	8	9.64	4	5.97
XI	*Aeromonas*	1	1.08	2	2.41	10	14.93
XII	*V. pelagius/splendidus*	3	3.23	7	8.43	7	10.45
XIII	*V. harveyii/anguillarum*	4	4.30	1	1.20	4	5.97
XIV	*Kingella/Brucella*	2	2.15	5	6.02	6	8.96
XV	*Xanthomonas*	3	3.23	5	6.02	4	5.97
XVI	*Alteromonas*	6	6.45	6	7.23	5	7.46
XVII	*Agrobacterium/Rhizobium*	1	1.08	1	1.20	1	1.49
Total		93		83		67	

coral reproduction in Curaçao (Netherlands Antilles) in November 2001 and February 2002. Although signs of DSD were observed in some colonies at the collection site, none of the collected colonies showed any abnormal pigmentation at this time. The water temperature of the aquarium setup was around 28 °C when a spot was observed for the first time. The water temperature had risen gradually from 26.5 °C in May to a maximum of 28.1 °C in September, at which time the spot was detected. The progress of the spot was not monitored during September and October, but during the first days of November, a rapid spread of these dark spots was noted (water temperature decreased to 27.0 °C by the end of November). Meanwhile, other tanks containing similar corals with constant temperatures around 26 °C did not present signs of this syndrome. These observations are consistent with prior studies that showed the increase of incidence and progression rate of DSD when water temperature is around 28 °C (Garcés-Baquero 2000; Gil-Agudelo and Garzón-Ferreira 2001).

By the end of November, treatments were initiated to try to stop the spread of the disease. Colonies with signs of DSD were treated with Baytril. The active substance of this antimicrobial is Enrofloxacin, which inhibits the DNA-gyrase in a broad spectrum of Gram-negative and Gram-positive bacteria (http://www.

compasnac.com/cvp/10/1040/1040008.htm). The concentration used was 0.2 ml of Baytril 10% in a 15-l aquarium. Experiments were carried out at 25.4 °C with 26-W fluorescent daylight lamps for 8 days. Daily exchanges of 50% water were performed, each time adding 0.1 ml of the antibiotic. After the treatment, the coral colonies showed increases in the necrotic area and some died within 2 days after finishing the treatment.

Treatments with Advocin, another DNA-gyrase and topoisomerase IV inhibitor with Danofloxacin as active ingredient, (http://www.hevra.org/vmri_spc/spc.asp?Product_Identifier=FR/V/0126/001) were also performed, resulting in a heavy increase in the necrotic area and no reduction of disease signs. These observations might show that the pathogen is not sensitive to these antibiotics, or that the change of the normal microbiota due to the action of the antibiotics allowed the rapid development of the disease. Additional stress influences, for instance, transfer from holding tanks to treatment aquariums or changed light regime, might have enhanced the necrotic process.

At the same time, corals in the tanks that did not show disease signs began presenting some dark spots. This outbreak may have been the result of transmission of the disease from the previously affected tanks due to nonaseptic techniques. Rapid action was taken to control the spread of the disease, moving newly infected corals to other tanks. Also, the lowering of water temperature (around 26 °C) may have arrested the outbreak in these tanks. Due to this action, one third of the *Montastrea* colonies remained unaffected by the disease.

Another factor that possibly contributed to this outbreak was the relatively low pH of the water, which was 8.04 between March and August of 2002. The effect of water pH in coral diseases and its interaction with other variables has not been studied in depth, but might play an important role in these phenomena.

19.3 Yellow Band and/or Blotch Disease

19.3.1 Original Description

This disease was first reported in 1994 by Craig Quirolo (Reef Relief, Key West, Florida, USA) from corals in the lower Florida Keys (Santavy and Peters 1997). Preliminary data from histological studies revealed a degeneration of the tissue in the affected areas, as well as the presence of crystalline-like material in the gastric cavity (Santavy and Peters 1997). By 1996, a major outbreak affecting corals of the *M. annualis* species off the shore of Panama was reported (Santavy et al. 1999). Another outbreak was reported in the Netherlands Antilles in 1997 (Cervino and Smith 1997). Jim Porter (University of Georgia, Athens) described this syndrome as "yellow blotch disease" in 1997, due to the color of patches of coral tissue, which were yellowish. Cervino and Smith (1997) also described this syndrome as "round yellow rings on the sides of the coral." They also reported that during the early stages of this disease, bands or rings are noticed, which change to yellow blotches surrounding dead patches of coral. Korrubel and Riegl (1998) reported similar signs in corals of the southern Arabian Gulf. Although it is not certain that these syndromes are the same, in this paper, we consider them to be related.

Although this disease is considered new, Dustan described similar signs in the 1970s as "ring bleaching".

19.3.2 Species Affected by the Disease

Initially, this syndrome was shown to attack the *M. annularis* complex, mainly *M. faveolata* in the Caribbean. Garzón-Ferreira and collaborators (2001) found signs of yellow band disease in several other coral species in the Colombian Caribbean. These included *A. agaricites*, *Diploria labyrinthiformis*, *D. strigosa*, *Favia fragum*, *Porites astreoides*, *C. natans*, and the *M. annularis* complex (Fig. 19.3).

Fig. 19.3a–d. Corals affected by yellow band disease. **a** *M. faveolata* in Puerto Rico (photo by Ernesto Weil). **b** *C. natans* in San Andrés Island, Colombia (photo by Jaime Garzón-Ferreira) **c** *D. strigosa* in San Andrés Island, Colombia (photo by Garzón-Ferreira). **d** *Alveopora viridis* in the Gulf of Aqaba, Israel (photo by Esther Winters)

Kurubel and Riegl (1998) found signs of a disease similar to yellow band disease (YBD) in the southern Arabian Gulf affecting *Acropora clahrata*, *A. pharaonis*, *A. tenuis*, *A. valida*, *A. florida*, *Porites lutea*, *P. lichen*, *P. nodifera*, *Turbidina reniformis*, and *Cyphastrea microphthalma*.

19.3.3 Geographic Range and Prevalence of the Disease

Although first seen in the Florida Keys, this syndrome was soon detected in the southwestern Caribbean, off the coasts of Panama (Cervino and Smith 1997). Presently, yellow band disease is commonly observed on almost all reefs in the Caribbean from Colombia and Venezuela to the Bahamas and Bermuda (Fig. 19.4), generally affecting the *M. annularis* complex. In some Caribbean reefs, YBD is the most prevalent disease, and it has been observed that this disease is more prevalent in reefs with low anthropogenic impacts than those located in close proximity to cities and coastal developments (E. Weil, unpubl. data).

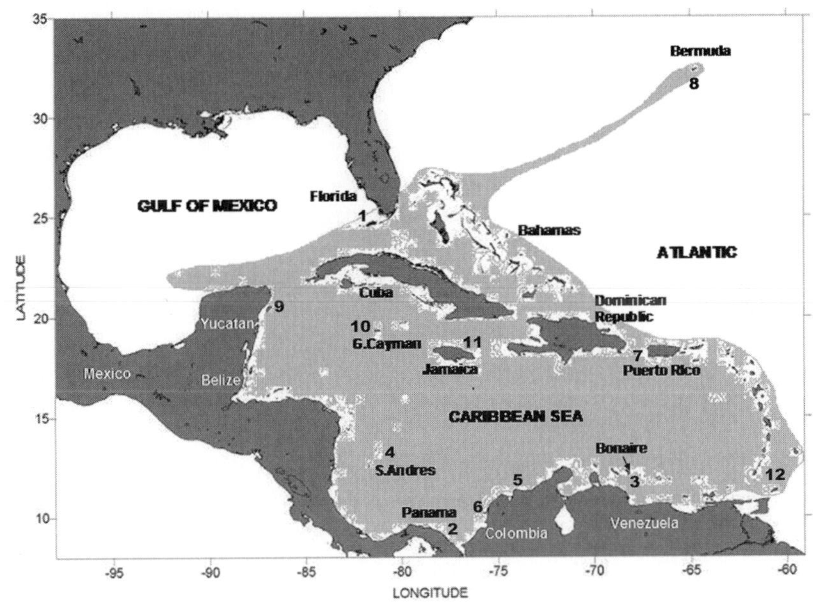

Fig. 19.4. Observations of yellow band disease in the Caribbean. *1* Florida Keys (observation by Craig Quirolo in 1994), *2* Panama (Santavy et al. 1999), *3* Netherlands Antilles (Cervino and Smith 1997), *4* San Andrés and Providencia Archipelago (Garzón-Ferreira et al. 2001), *5* Santa Marta (Garzón-Ferreira et al. 2001), *6* San Bernardo (Garzón-Ferreira et al. 2001), *7* Puerto Rico (Ernesto Weil, pers. observ.), *8* Bermuda (Ernesto Weil and Garriet W. Smith, pers. observ.), *9* Puerto Morelos, Mexico (Diego L. Gil-Agudelo, pers. observ.), *10* Grand Cayman Islands (Ernesto Weil, pers. observ.), *11* Jamaica (Ernesto Weil, pers. observ.)

The Colombian Coral Reef Monitoring System (SIMAC) showed an increase in the incidence of YBD during 1999, although none was observed in 1998. As much as 12.5% of the colonies of *M. annularis* was affected by this disease in Chengue Bay (Santa Marta) and 7.76% of the colonies of *D. strigosa* in San Andrés Island. These high values may be an overestimate of the real incidence of the disease due to the small number of colonies of these species in the evaluated transects (only eight colonies of *M. annularis*). Furthermore, during 1998, a bleaching event occurred in San Andrés, which may have contributed to the misidentification of the disease during this year (Garzón-Ferreira et al. 2001).

Later, in 2000, no signs of yellow band disease were reported at any of the stations monitored in Colombia. Species suffering from YBD other than the *M. annularis* complex (except for those found in the Red Sea) have been reported only in Colombia (Garzón-Ferreira et al. 2001).

19.3.4 Growth Rate of Yellow Band Disease

In the Caribbean, the growth rate of YBD was measured at 0.70 cm/month (SD=0.2) for *M. faveolata*. These linear measurements do not represent the total amount of tissue lost by the disease, because a small lesion surrounded by YBD loses less tissue than a larger area with the band progressing at the same rate. Nevertheless, it gives a good idea of the slow progression of the disease, compared to other diseases such as black band disease and white band disease that can grow several centimeters per month (Ritchie and Smith 1998).

Similar measurements of the progression of YBD in *M. franksi* (less than 0.1 cm/month) showed that the disease progression rate might be determined by the coral species that is being attacked. The progression rate of YBD in other coral species has not yet been determined.

Similar to other diseases, YBD progresses faster during warm-water than during cold-water seasons.

19.3.5 Pathological Studies

Comparisons of the bacterial communities living on healthy *M. annularis* and the ones with signs of YBD found differences between the microbial communities, similar to that seen in other diseases. Also, several bacterial strains metabolically related to the genus *Vibrio* were found only in the mucus of diseased corals, which make them a suspected cause of YBD. These strains are now being tested for their possible role in this disease.

19.3.6
Other Studies and Observations

Cervino et al. (2001) proposed that disease signs of YBD could actually be a zooxanthellae disease, which can kill the coral by damage produced by its symbiont. They base this conclusion mainly on their finding of a reduction of 41–97% of the number of zooxanthellae in corals with yellow band, and a reduction of the mitotic index of zooxanthellae from 2.5% to nearly 0%.

This hypothesis has led Dr. Roberto Iglesias (Universidad Autónoma de México) to investigate the photosynthetic characteristics of this syndrome. Preliminary results suggest that, in fact, some characteristics such as the quantum yield of the algae are decreased in parts of the coral affected by YBD, also suggesting damage in the coral symbionts.

19.4
Unresolved Questions and Future Research

The knowledge of both diseases is very limited at this time. A lot of work has to be done in many areas, from field determinations of the incidence of the disease and its relationship to environmental parameters, to detail studies of the characteristics of the disease and determination of the pathogenic agent.

19.4.1
Field Determinations

Probably the most immediate need is the determination of the real extent of these diseases in the Caribbean. For this, determination of the incidence of the disease with standard techniques must be performed in different Caribbean reefs. Comparisons between the incidence of the diseases and environmental parameters from different places (short and long term) in the Caribbean must be done in order to elucidate possible links between diseases and changes in the environment.

Work is also needed to determine the rate of tissue loss (especially yellow band disease) in different coral species and its changes due to environmental conditions. These studies will be important to determine the velocity of tissue loss due to the diseases and to compare the tissue lost by these diseases with new tissue formation. This will tell us if the reef is dying or recovering.

In this way, the real incidence of the diseases on coral communities can be determined.

19.4.2
Laboratory Determinations

The most important need at this time is to fulfill the Koch postulates for an adequate identification of the pathogenic agent(s) causing these diseases. This will allow studies to be designed to determine if the variable signs of these diseases is the product of natural variation, or due to the action of different pathogenic agents (different disease). Moreover, it will help to determine if the presence of similar signs in different coral species is due to the same disease.

An important aspect in the study of coral diseases is the determination of their origin. Once the pathogenic agent is determined, studies need to be carried out to determine the sources and vectors of the pathogenic agents involved in diseases. These kinds of studies will be crucial from a management point of view to try to control their spread by controlling the possible sources.

From the pathological point of view, it is very important to determine the mechanisms of the infection. This will also help to clarify of the role of some environmental factors in the disease process.

Acknowledgments. The authors would like to thank everyone directly and indirectly involved in the research of these diseases throughout the years, too many to be enumerated individually. We also want to thank Dr. Eugene Rosenberg and the Israeli Center for the Study of Emerging Diseases for their invitation to write this chapter. The following institutions helped support the study of these diseases: INVEMAR, CORALINA, UAESPNN, University of South Carolina at Aiken, University of Puerto Rico, Rotterdam Zoo. Portions of this work have been supported by the following institutions: COLCIENCIAS-BID (#2105-09-327-97), MinAmbiente-FONAM (#BID-774-OC/CO) UNEP-UCR/CAR, NOAA-CRES (#0648-0384), and World Bank-GEF.

References

Birkeland C (1997) Life and death of coral reefs. Chapman and Hall, New York

Cervino JM, Smith GW (1997) Corals in peril. Ocean realm. Summer 1997, 42:33–35

Cervino J, Goreau TJ, Nagelkerken I, Smith GW, Hayes, R (2001) Yellow band and dark spot syndromes in Caribbean corals: distribution, rate of spread, cytology, and effects on abundance and division rate of zooxanthellae. Hydrobiologia 460:53–63

Garcés-Baquero AM (2000) Desarrollo de la Enfermedad de Lunares Oscuros en los corales pétreos *Montastraea annularis* y *Siderastrea siderea* en la Bahía de Gayraca, Caribe colombiano. Thesis (Marine Biologist), Fundación Univ Bogotá Jorge Tadeo Lozano

Garzón-Ferreira J, Gil DL (1998) Another unknown Caribbean coral phenomenon? Reef Encounter 24, Dec 1998, 10 pp

Garzón-Ferreira J, Gil-Agudelo DL, Barrios LM, Zea S (2001) Stony coral diseases observed in southwestern Caribbean reefs. Hydrobiologia 460:65–69

Gil-Agudelo DL (1998) Características, incidencia y distribución de la Enfermedad de "Lunares Oscuros" en corales pétreos del Parque Nacional Natural Tayrona, Caribe colombiano. Thesis (Marine Biologist), Fundación Univ Bogotá Jorge Tadeo Lozano

Gil-Agudelo DL, Garzón-Ferreira J (2001) Spatial and seasonal variation of the Dark Spots Disease in coral communities of the Santa Marta area (Colombian Caribbean). Bull Mar Sci 69(2):619–629

Ginsburg RN (1994) Proceedings of the colloquium on global aspects of coral reefs. Health, hazards, and history. Rosenstiel School of Marine and Atmospheric Science, University of Miami

Goreau TJ, Cervino J, Goreau M, Hayes R, Hayes M, Richarson L, Smith G, DeMeyer K, Nagelkerken I, Garzón-Ferreira J, Gil D, Garrison G, Williams EH, Bunkley-Williams L, Quirolo C, Petterson K, Porter J, Porter K (1998) Rapid spread of diseases in Caribbean coral reefs. Rev Biol Trop 46(5):157–171

Grigg RW, Dollar SJ (1990) Natural and anthropogenic disturbances on coral reefs. In: Dubinsky Z (ed) Coral reefs. Elsevier, Amsterdam, pp 439–452

Harvell CD, Kim K, Burkholder JM, Colwell RR, Epstein PR, Grimes DJ, Hofmann EE, Lipp EK, Osterhaus ADME, Overstreet RM, Porter JW, Smith GW, Vasta GR (1999) Emerging marine diseases - climate links and anthropogenic factors. Science 285:1505–1510

Korrubel JL, Riegl BR (1998) A new disease from the southern Arabian Gulf. Coral Reefs 17:22

Ritchie KB, Smith GW (1998) Type II white-band disease. Rev Biol Trop 46(5):199–203

Santavy DL, Peters EC (1997) Microbial pests: coral diseases in the Western Atlantic. Proc 8th Int Coral Reef Symp 1:607–612

Santavy DL, Peters EC, Quirolo C, Porter JW, Bianchi CN (1999) Yellow-blotch disease outbreak on reefs of the San Blas Islands, Panama. Coral Reefs 18:97

Solano OD, Navas-Suarez G, Moreno-Forero SK (1993) Blanqueamiento coralino de 1990 en el Parque Nacional Natural Corales del Rosario (Caribe colombiano). An Inst Invest Mar Punta Betín 22:97–111

20 White Plague, White Band, and Other "White" Diseases

John Bythell, Olga Pantos, and Laurie Richardson

20.1 Introduction

Although commonly reported among disease occurrences of reef corals (Weil et al. 2002), the "white" diseases are probably the most enigmatic. There are a number of these diseases, or syndromes, and confusion arises because of the very similar disease signs that are present in association with each of them. The 'white syndromes' include white band disease (Antonius 1981b; Gladfelter 1982), more recently divided into two groups, white band type I and white band type II (Ritchie and Smith 1998); white plague (or simply plague; Dustan 1977), again recently divided into white plague types (WP) I, II and III (Richardson et al. 1998a, b, 2001); and 'stress-related necrosis', "patchy necrosis", or 'shut-down reaction' (Antonius 1981a; Peters 1983, 1984). Confusion between white plague and white band disease is problematic because the visible signs of the diseases are virtually identical in that there is a sharp, distinct line between apparently healthy coral tissue and freshly exposed coral skeleton with no obvious microbial material present. At times, some types of both white band and white plague may have a zone of bleached coral tissue at the disease boundary, but in both cases this is usually transient and variable. The distinction has been based primarily on the host species affected, with *Acropora* species in the Caribbean commonly reported to be affected by white band disease and other scleractinian species, including massive species, affected by white plague. Early studies (e.g. Antonius 1981a, b) used the terms interchangeably and any historical analysis must therefore account for both diseases. Both Antonius (1981b) and Peters (1984) reported the presence of white band disease on a number of massive corals as well as the branching acroporids. More recently, investigators have been studying the etiology of a coral disease associated with white lesions (Bythell et al. 2002), which appears similar to WPI as well as patchy necrosis.

This review specifically explores what is known about the different "white" disease types, including an assessment of their comparative prevalence and impacts on reef coral populations. Recent results of culture-independent studies of bacterial 16S rRNA gene diversity of white plague disease are discussed.

20.2 White Plague – One Disease or Many?

Dustan (1977) originally proposed that "Plague appears to be a suite of diseases that result in the death of the coral tissue...", but offered little evidence to support the statement. However, this does indicate that different types of disease signs were observed by Dustan (1977) during the surveys. The case for different types of white plague disease is based primarily on three factors: (1) rate of progression of lesions on affected coral colonies (and therefore the time to total colony mortality, depending on colony size), (2) prevalence rates within the population, and (3) the species of coral that are affected (Richardson et al. 2001). Another indicator is whether the lesion starts at the base or on other parts of the colony, which requires in situ monitoring and tracking of newly diseased colonies. Observations of differences in these variables may be due to the involvement of different causal agents, but could equally be explained by different susceptibilities of the host coral. Currently, the link between coral diseases and environmental stress is poorly understood (Richardson 1998), although it is highly likely that the physiological status of the coral will affect disease resistance.

White plague is the most well characterised of the white diseases (with the exception of white pox, which is addressed in a separate chapter). White plague was first reported by Dustan (1977) as affecting six species of massive and encrusting corals in the Florida Keys. The disease was characterised by lesions on the coral colony that expanded at a rate of a few millimetres per day and often resulted in whole-colony mortality. The edges of the lesions showed a sharp boundary between apparently healthy tissue and freshly exposed skeleton, with no buildup of micro-organisms or necrotic tissues visible to the naked eye. The disease was shown to be transmissible between adjacent colonies, with the implication that a microbial pathogen was involved. In the summer of 1995, a disease with similar visible signs, but with a more rapid rate of lesion progression (up to 2 cm/ day) was observed on 17 species of hard coral on the same reefs of the northern Florida Keys. The new outbreak was designated white plague type II because of the similar disease signs to those reported by Dustan (1977), but a faster rate of progression (Richardson et al. 1998a, b). The disease was most prevalent on the elliptical star coral, *Dichocoenia stokesii*. It was observed to affect small colonies (typically 10 cm diameter or less) of this and the 16 other coral species. In addition to these indications of greater virulence, WP type II was invariably observed to start from the base of the colony, whereas WP type I lesions were observed to occur more variably across the colony. A third and even more aggressive type of white plague (WP type III) was reported from the northern Florida Keys in 1999 (Richardson et al. 2001). WP type III is characterised by extremely rapid rates of lesion progression (dm/day) and was found on the largest colonies of *Colpophyllia natans* and *Montastraea annularis*. Both WPII and WPIII have been observed to have spread throughout the wider Caribbean in the last 3 years (Borger 2004; Miller et al. 2001; Nugues 2002; Weil et al. 2002).

White plague produces a more or less uniform arc separating apparently healthy tissue from bare skeleton. One form, WP type II, is distinguished from the others in that tissue necrosis progresses in a very specific pattern, i.e. from the entire base of the colony to the apex. At times it appears to radiate from a single point of infection, but this point is always at the colony base. Thus, it tends to affect the entire circumference of a living colony, progressing in this manner towards the apex (Fig. 20.1d–f). The rate of band progression is faster than that of white band disease (discussed below), and so the secondary succession of an algal community succeeding the band does not show a clear successional gradation. One observation, for which there is no explanation to date, is that at times the progression of WPII halts. It may start again after a few hours or days, or even weeks, or may not ever manifest, resulting in a colony that is "mushroom-shaped" with a "stalk" of dead skeleton with a cap of living tissue. WPIII starts in the centre of colonies, with the most striking characteristic being the extremely rapid rate of tissue degradation.

An additional white syndrome has also been termed white plague in the literature (Bythell et al. 2002; Pantos et al. 2003). In this case, the tissue degradation does not form a uniform arc at the lesion boundary, but is characterised by more random patches of denuded skeleton that extend sporadically and also do not give rise to a graded algal community (Fig. 20.1g–i). Visible signs of this latter syndrome are in some cases indistinguishable from predation by corallivorous snails (*Coralliophila* sp.) and fireworms (*Hermodice* sp.), however, there is enough evidence of the occurrence of this syndrome in the absence of these predators and the rare observation of tissues sloughing from the skeleton to indicate that it may also be caused by an infectious disease (Bythell et al. 1993; J.C. Bythell, pers. observ.).

Of all the types of white plague, the causal agent of only one, WP type II, is known. The WPII pathogen was originally identified as an α-proteobacterium most closely related (86% homology of a partial 16S rRNA gene sequence) to a *Sphingomonas* sp. (Richardson et al. 1998a, b). It has since been characterised in more detail, and recently published as a novel genus (*Aurantimonas coralicida, gen nov.* sp. *nov.*; Denner et al. 2003). In these studies, plating of fresh samples of diseased tissue onto Marine agar plates followed by microscope examination of the most abundant bacteria that grew revealed a predominance of gram-negative bacteria. Colonies typically were translucent for 2–3 days, and then developed a distinctive golden pigmentation. A culture was obtained from a single colony, determined to be pure, and shown to instigate the disease, thus fulfilling Koch's postulates (Richardson et al. 1998a, b). Although in this case a pathogen has been isolated and characterised, further work is needed to understand the pathology and etiology of white plague disease and, critically, to determine the relationships between the different types of the disease and white band disease. For example, the mechanism of coral tissue death is not known, nor are the disease reservoir or mode of transmission between colonies.

a
b
c
d
e
f
g
h
i

20.3
White Band Disease – A Different Pathology?

White band disease (WBD) was also first noted in the 1970s and also in the Greater Caribbean (Gladfelter et al. 1977). This original outbreak was reported on Caribbean acroporids (*Acropora cervicornis* and *A. palmata*), and is believed to have completely restructured some Caribbean reefs as a result of epizootics (Aronson and Precht 1997, 2001). WBD was reported as typically starting at the base of acroporid branches, and moving outwards towards the growth tips of the colony. Tissue destruction was rapid and the death of entire colonies was common.

Classic experiments by Antonius and others in the 1980s (Antonius 1981a, b, 1988) showed that corals that had been affected by white band disease for days or weeks would die overnight if stressed by being transferred to laboratory aquaria, conditions in which healthy corals would survive. The signs of white band appear to be virtually indistinguishable from those of white plague (Table 20.1), with the exception of the coral genera affected.

While the early studies by Dustan (1977) and Antonius (1981b) suggested that white plague was readily transferred between adjacent colonies, similar studies conducted for white band disease did not. Because of this, Antonius (1981b) suggested that white band disease may be a form of 'stress-related necrosis' or, as he termed it, 'shut-down reaction' caused by environmental stress rather than a pathogen. However, Gladfelter (1982) clearly showed progression of white band disease along the reef in Tague Bay, St Croix, with the disease signs visible from aerial photography. This transmission along the reef indicates an infectious agent. The disease spread eastwards, against the prevailing currents, suggesting that the vector may have been another organism rather than being water-borne. There is no environmental gradient along this reef that would explain the pattern of spread of the disease as stress-related necrosis. By contrast, Peters (1984) examined the histopathology of several diseases and found bacteria associated with white band disease, but not white plague, which she referred to as 'stress-related necrosis'. Antonius (1981b) reported an intriguing seasonal shift from white to black band disease, implying that black band disease may develop on existing white bands. More recently published studies, however, have not seen this pattern. There are some field observations that suggest that one disease may develop into the other (Fig. 20.2), but such observations are too rare to be able to draw any general conclusions. In addi-

◀ **Fig. 20.1.** Examples of 'white syndromes' on Caribbean reef corals. We distinguish here between white disease that radiates evenly from a point of infection (**a–c**), white plague type II that progresses evenly upwards from the edge of the tissues at the base towards the apex of the colony (**d–f**), and white plague/patchy necrosis that forms more random patches of denuded skeleton and an uneven lesion boundary (**g–i**). Coral host species are *Siderastrea siderea* (**a**, **g**); *Colpophyllia natans* (**b**); *Acropora palmata* (**c**); *Montastraea annularis* (**d**, **f**, **h**, **i**); and *Diploria strigosa* (**e**)

Table 20.1. Visible signs of plague and related diseases reported in the literature

Disease/ syndrome	Rate of lesion progress (mm/day)	Lesion starts at base?	Bleached band precedes lesion?	Species affected	Reference
WP (type 1)	Up to 3	Often	No	Several massive and encrusting species	Dustan (1977)
WP (type 2)	Up to 20	Yes	Sometimes	37 Species	Richardson et al. (1998a, b); Weil and Smith (Chap. 2, this Vol.)
WP (type 3)	20+	No	No	*Montastraea annularis* and *Colpophyllia natans*	Richardson et al. (2001)
WBD (type 1)	0.8–20	Yes[a]	No	*Acropora* sp.	Antonius (1981b); Gladfelter (1982)
WBD/WP?	1–3	?	No	5 Massive/ encrusting species	Antonius (1981b)
Shut down reaction				*Acropora* sp.	Weil and Smith (Chap. 2, this Vol.)
Patchy necrosis WBD (type 2)	0.5	Often	Yes	*Acropora* sp.	Ritchie and Smith (1998)
White pox	Up to 10.5 cm^2/ day	No	No	*Acropora* sp.	Patterson et al. (2002)

[a] At the base of the branch.

Fig. 20.2. *Diploria strigosa* colony showing apparent progression of black band and white band/white plague disease in separate parts of the same lesion

tion, it is known that some coral diseases can occur simultaneously on the same coral colony (Weil et al. 2002), thus the question of the relative roles of onset of each disease is further complicated.

White band disease is characterised by a progressive lesion which, starting from a point of infection, radiates outwards at an extremely uniform rate along the length of the lesion boundary. The lesion boundary thus forms a uniform arc which progresses up separate branches of branched species at an equal rate. A succession of algae succeeds the white band of exposed skeleton, forming a visibly graded algal community on the recently denuded skeleton. Earlier workers (Antonius 1981a, b, 1988; Peters 1984) reported these consistent disease signs as white band disease on a variety of coral species, including massive species, in both the Indo-Pacific and Caribbean. Since then, it has become accepted practice that this disease is confined to the branching *Acropora* species (Fig. 20.1c) and that where these signs appear on massive species (Fig. 20.1a, b), it is termed white plague. However, it is not clear where this distinction arose. Two types of white band disease have been reported. In type II white band disease, the lesion boundary is often preceded by a zone of bleached tissues (Ritchie and Smith 1998), whereas this bleached zone was not observed in the original (Type I) outbreaks of the late 1970s (Gladfelter 1982). The advancing lesion often 'catches up' with the advancing bleaching zone in WBD type II (Ritchie and Smith 1998), meaning that extended surveillance of affected coral colonies must be carried out to distinguish between them. Type II white band disease is consistently associated with a bacterium related to *Vibrio carchariae* (Ritchie and Smith 1995a, b, 1998), but re-infectivity studies have so far not proved it to be the causal agent.

The WBD pathogen has remained elusive despite two decades of investigation. Using a histological approach, Peters (1983) found aggregates of Gram-negative bacteria associated with WB-diseased tissue. More recently, however, this same group reported that these suspect aggregates were, at times, found in healthy tissues of acroporids (Santavy and Peters 1997). In the same report, it was also noted that the aggregates were absent from WP-diseased tissue samples (Santavy and Peters 1987). No isolation or culture of bacteria from the aggregates has been reported.

Using a different approach, Ritchie and Smith (1995a, b, 1998) investigated the etiology of WBII by comparing the microbial community structure of diseased and non-diseased acroporids. Specifically, they studied the microbial community associated with the surface mucopolysaccharide layer (SML) of healthy and diseased corals. Using carbon source utilisation pattern analyses (BioLog), they found that the WB community had a distinctly different composition to that of healthy tissues, and was less diverse (Ritchie and Smith 1995a). They found that although, in general, the members of the bacterial community remained the same, in WB-diseased samples the relative proportion of vibrios increased, and the relative proportion of pseudomonads decreased (Ritchie and Smith 1995b). They found the WB-associated vibrios to cluster most closely with the species *Vibrio charcharia* (Ritchie and Smith 1995b). In one study (Ritchie and Smith 1998), they found that this species was approximately

18% of the entire heterotrophic bacterial community in tissue samples from WBD type II diseased colonies of *A. cervicornis*. Although strong evidence thus points to this species as the WBII pathogen, and isolates are readily culturable, studies to determine the role of *V. charcharia* as the WB pathogen have been hindered by the difficulty in maintaining healthy colonies of Caribbean acroporids in laboratory aquaria (G. Smith, pers. comm.).

20.4 Other "White" Diseases

A review of disease signs from corals affected by various 'white syndromes' suggests that there are at least four general types that can be distinguished using visible signs alone. White plague, white band, and white plague-like lesions have been described above and are illustrated in Fig. 20.1. A fourth 'white' disease, white pox, was first reported affecting *Acropora palmata* populations in the Florida Keys in 1996 (see Chap. 16, Patterson Sutherland and Ritchie, this Vol.). The causal agent has recently been identified as the ubiquitous bacterium *Serratia marcescens* (Patterson et al. 2002). It is present as round lesions on the surface of colonies.

Another proposed disease, 'patchy necrosis' (Bruckner and Bruckner 1997; Rodriguez-Martinez et al. 2001), also manifests as variably sized, rounded patches that appear simultaneously across the colony. Patches can expand rapidly (up to 10.5 cm^2/day, average 2.5 cm^2/day; Patterson et al. 2002), and may coalesce to eventually kill the entire colony. The disease is epizootic in the Florida Keys and has recently been reported in Mexico and Puerto Rico. The signs of this disease appear to be indistinguishable from the patchy lesions created by white plague type I on massive coral species (Fig. 20.1g–i). Comparative bacteriology and pathology studies have so far not been carried out. The distinction between these diseases remains unclear.

20.5 Outbreaks of the White Diseases: Relative Severity and Prevalence

Because of the difficulty in distinguishing between most of the white diseases, and also distinguishing between certain white plague lesions and predation, it is difficult to determine whether prevalence rates have changed over time. Certainly there have been more frequent reports of significant outbreaks of the disease, but it is virtually impossible to tell whether this is due to changes in prevalence or to an increase in the observation and monitoring of coral reefs (Richardson 1998). The picture is further complicated by the fact that there appear to be different periodicities of disease outbreak depending on location. In the Florida Keys, significant outbreaks have typically been reported to occur in summer (Dustan 1977; Richardson et al. 1998a, 2001), whereas a white plague type II outbreak was first reported in St John, US Virgin Islands, in 1997 and has continued for several years with disease incidence being sustained throughout the year (Miller et al. 2001). Studies of individually tagged corals in a marine protected area since the late 1980s (Bythell et al. 1993) showed that tissue mortality with white plague-like signs caused between 0.1 and 5% tissue loss, depending on species, over a 26-month period. There was an increased incidence and prevalence rate of this type of mortality following a major hurricane impact, but prevalence rates have since returned to levels seen in the late 1980s (Bythell et al. 2000). In St Lucia, Nugues (2002) reported a major outbreak of white plague type II in 1997 that caused 7% mortality at one site within 8 months. Overall, it appears that outbreaks of white plague have been sporadic, but sustained within the Caribbean region since the 1990s. The high prevalence rates and likelihood of mortality, coupled with its propensity to affect dominant reef-building species such as *Montastraea annularis* (complex) and *Colpophyllia natans*, means that this disease represents a significant and continuing threat to Caribbean coral reefs. Although no long-term studies of the effect of white plague on reef communities have yet been published, Aronson and Precht (1997, 2001) reported that an outbreak of white band in the 1980s restructured the coral community composition of an entire reef in Belize, with the virtual elimination of *Acropora cervicornis* from reefs in the central shelf lagoon of Belize. *Agaricia tenuifolia* replaced *Acropora cervicornis* in the early 1990s, but anomalously high water temperatures in 1998 caused severe bleaching and catastrophic mortality of the emergent *A. tenuifolia* populations. These dramatic shifts in community structure left a clear signal in the fossil record and indicated that neither transition had occurred in the previous 3000+ years (Aronson et al. 2002). More recently, new outbreaks of patchy necrosis have been reported in the Caribbean (see Chap. 2 Weil and Smith, this Vol.).

20.6 Bacterial 16S rDNA Diversity of White Plague Affected Corals

Pantos et al. (2003) have recently examined the bacterial 16S rDNA diversity associated with corals affected with white plague (type 1) or possibly patchy necrosis lesions (Fig. 20.1g–i). An important finding of this study was that there was a difference in the 16S rRNA gene profiles, which can estimate bacterial community composition between the apparently healthy tissues of diseased corals and the tissues of non-diseased corals (Fig. 20.3). Although the tissues of white plague-affected colonies may appear normal immediately adjacent to the lesion, they support a distinct bacterial community. Whether this is due to pathogenic bacteria invading the tissues in advance of development of visible signs or a change in bacterial flora due to stress remains to be determined. This study clearly shows the development of a bacterial community at the lesion boundary that is distinct from that of healthy corals, with the loss of some bacteria normally seen in healthy tissues and the appearance of other bacteria that occur consistently in all diseased samples (Fig. 20.3). The same general pattern has been seen in white band type II (Ritchie and Smith 1995a, b,

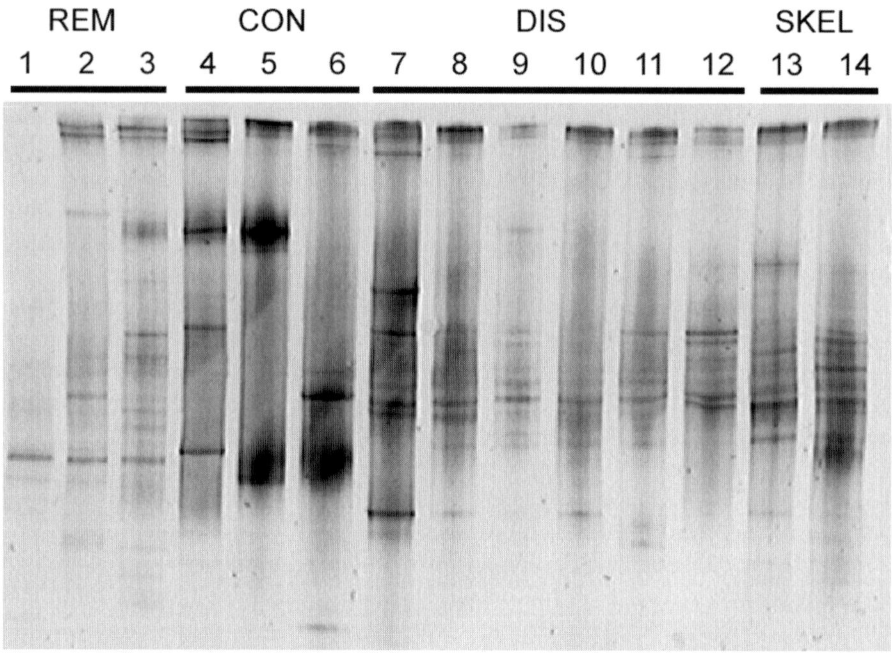

Fig. 20.3. DGGE profile of bacterial 16S rDNA PCR products from non-diseased *Montastraea annularis* colonies (*REM*; *lanes 1–3*); apparently healthy coral tissue from diseased colonies (*CON*; *lanes 4–6*); diseased coral tissue (*DIS*; *lanes 7–12*); and, exposed skeleton (*SKEL*; *lanes 13–14*). Independent samples were taken from separate coral colonies. Source: Pantos et al. (2003)

1998) comparing patterns of carbon source utilisation by surface bacteria associated with diseased and healthy corals, as discussed above. Sequence analysis from clone libraries generated from samples of white plague lesions (Pantos et al. 2003) consistently showed the presence of an α-proteobacterium from the *Roseobacter/Rhodobacter* group that is closely related (98%) to the causal agent of juvenile oyster disease (JOD) in the cultured eastern oyster *Crassostrea virginica* (Boettcher et al. 1999, 2000). This ribotype was absent from healthy coral samples, indicating that it certainly plays a key role in the disease and should be targeted for culture and testing of Koch's postulates as a potential pathogen. Pantos et al. (2003) did not find *Aurantimonas coralicida* (Denner et al. 2003) or related bacteria in the diseased lesions, and concluded that this particular disease (Fig. 20.1g–i) was distinct from white plague type II (Richardson et al. 1998a, b, 2001). This finding agrees with the difference in overall disease etiology.

20.7 Histopathology

Peters (1984) examined the histology of coral tissues affected by various diseases and syndromes and found little evidence of microbial infection at the lesion boundary except in certain cases of white band disease. In this case (i.e. observation of aggregates of Gram-negative bacteria associated with white band diseased tissue) interpretation was complicated in that the bacteria could not be cultured, were not assessed using non-cultivative techniques, and were also found to be present in healthy coral tissue of the same coral species (Santavy and Peters 1997). More recently, Bythell et al. (2002) have used an agarose pre-embedding technique to preserve the fragile necrotic tissues at the coral lesion boundary in a study targeting several coral diseases. Their results showed that even at a microscopic scale, the lesion boundary of white plague and white band disease forms a sharp break between apparently healthy tissues and denuded skeleton. The pathological process of white band disease involves the detachment of these apparently healthy tissues from the skeleton (Fig. 20.4a). Gladfelter (1982) reported macroscopic observations of these de-

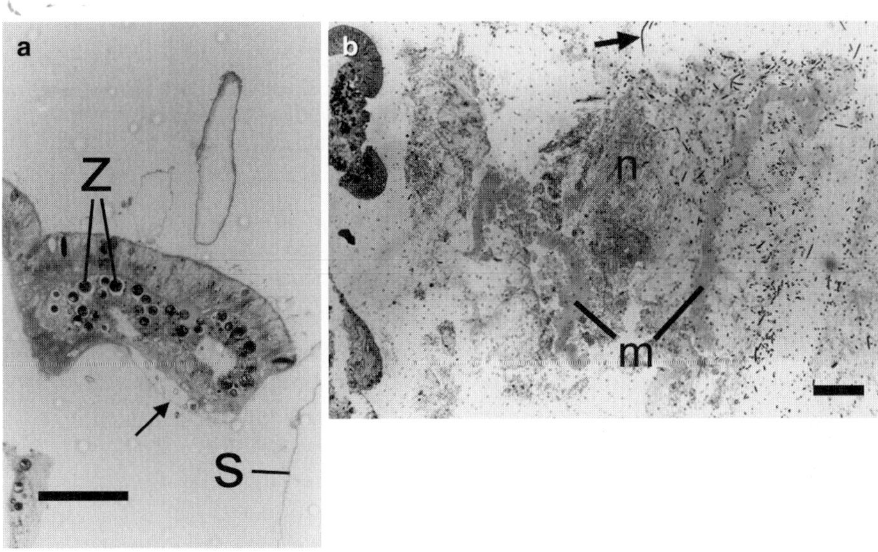

Fig. 20.4. Histopathology of white band disease (type I) on *Acropora palmata* and black band disease on *Diploria strigosa.* The lesion boundary of WBD type I (**a**) is characterised by a clean break between denuded skeleton (S) and apparently healthy tissues. The tissues can be seen in the process of detaching from the skeleton (*arrow*). Tissues have a full complement of zooxanthellae (*z*) up to the lesion boundary. The lesion boundary of BBD (**b**) consists of a wide area (several mm) of necrotic tissues (*n*) with the acellular mesoglea (*m*) clearly visible within the necrotic tissues. Darkly staining, apparently healthy tissues can be seen at *left* for comparison. Numerous filamentous cyanobacteria are present in the necrotic tissues (*arrow*). *Scale bars* 0.5 mm

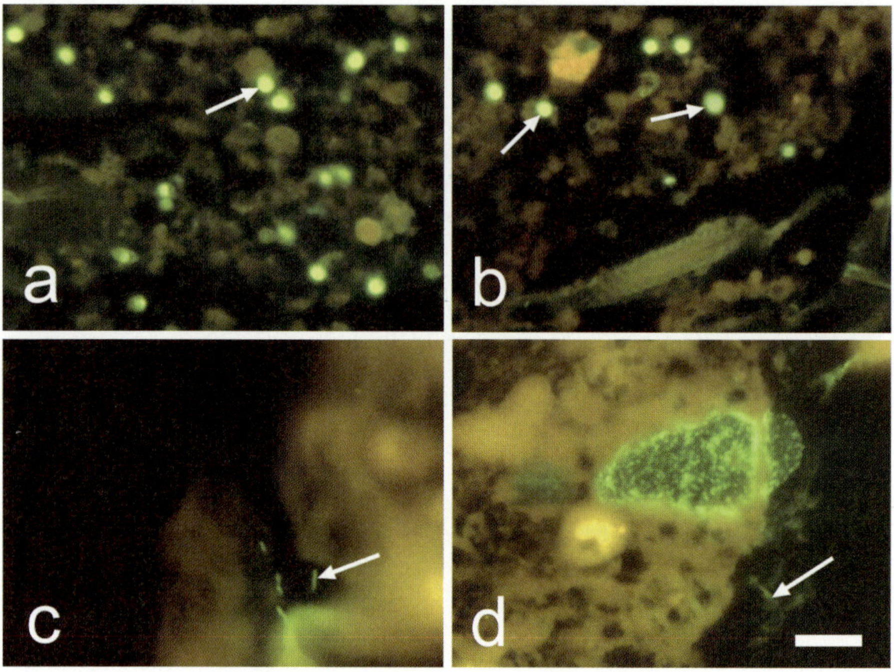

Fig. 20.5. Fluorescence in situ hybridisation (FISH) of coral tissue sections of samples from the lesion boundary of white plague affected (**a**, **b**) and from non-diseased (**c** and **d**) *Montastraea annularis*. Sections are fluorescently labelled using tyramide signal amplification of horseradish peroxidase-labelled EUB338 oligonucleotide probes which hybridise to a wide range of bacteria. Concentrations of coccoid bacteria ca. 1 μm diameter are visible (*arrows*) in diseased tissues. Bacteria were present in healthy coral tissues, but were typically present on the surface and concentrated around mucus glands (*arrows*, **c** and **d**), or as isolated cells within the tissues. The extremely small (ca. 0.1 μm) particles labelled in **d** are likely to be non-specific staining of mucus. Nematocycts also show some non-specific staining (**b**, *lower centre*). *Scale bar* 5 μm

tached tissues forming small bodies (ca. 1 mm diameter) and floating away from the lesion boundary, aided by their own ciliary action. This contrasts markedly with black band disease, for example, where the transition zone from healthy tissues to denuded skeleton extends over several millimetres and consists of necrotic tissues and a clearly visible microbial mat dominated by cyanobacterial filaments (Fig. 20.4b). Using a eubacterial 16S rRNA oligonucleotide probe, Bythell et al. (2002) were able to show a buildup of coccoid bacteria in small fragments of necrotic tissue at the lesion boundary of white plague disease in *Montastraea annularis* (Fig. 20.5a, b). Bacteria were extremely scarce in non-diseased coral samples as well as the apparently healthy tissues adjacent to the lesion of affected colonies (Fig. 20.5c, d). Further work is now needed to specifically label potential pathogens at the lesion boundary. Ideally this should be conducted for all of the white plague diseases.

20.8 Some Unresolved Questions and Future Directions

The various presentations of white coral diseases and syndromes make them among the most difficult coral diseases to define and characterize, both in the laboratory and in the field. The term 'white plague' was originally used to collectively describe a suite of syndromes causing tissue lesions with an abrupt boundary between denuded skeleton and apparently healthy tissues (Dustan 1977). As such it would include white band disease and white pox. The literature has been further confused as earlier researchers apparently used the term white plague synonymously with white band disease (Antonius 1981a, b, 1984; Peters 1984) and these workers reported white band disease on massive species as well as branching acroporids.

So what do we know, at this point, about the white diseases? At one end of the spectrum, white plague type II (WPII) has been well characterised, with the causal agent identified as a novel bacterium *Aurantimonas coralicida* (Denner et al. 2003). This disease causes consistent, characteristic signs and has been identified as the cause of significant outbreaks of coral mortality in the Florida Keys and elsewhere in the Caribbean since the mid-1990s (Richardson et al. 1998a, b; Nugues 2002). Since the term white plague was originally used to collectively describe a suite of syndromes (Dustan 1977), perhaps a more specific disease name is warranted for WPII, but it is clearly a well-defined, identifiable disease.

White pox has also recently been characterised and a pathogen identified for diseased *Acropora palmata* in the Florida Keys (Patterson et al. 2002). It is not clear, however, whether this disease is confined to the acroporids. The visible signs of white pox appear to be virtually identical to those described for some forms of white plague on massive corals (Fig. 20.1g–i), although recent studies of this form of white plague have not detected either the white pox pathogen *Serratia marcescens* or the WPII pathogen *Aurantimonas coralicida*, suggesting that it may be a separate disease. Thus, although white pox is well described in one host species and location, further work is needed to define the host extent and range.

It appears that the visible signs of white band disease are well defined and distinct from those of WPII and white pox, indicating that there are at least three separate "white" diseases. However, the distinction between WB, WPI, WPIII, patchy necrosis, rapid tissue necrosis and shut down reaction is unclear. No causal agent has been proven for these latter diseases, and until microbiological evidence is found to the contrary, we argue that the term 'white disease' should perhaps be used to collectively describe this set of signs on both massive corals and branched acroporid species.

We argue here that differentiation between the various white plague diseases, as well as white band types I and II, white pox, patchy and rapid tissue necroses, as well as shut down reaction, should be based on a more complete survey of the characteristics of associated microbial pathogens, using a combination of cultivative and non-cultivative (molecular) approaches. More specifically, we recommend that as each pathogen is physically isolated (cultured and tested using Koch's postulates), or strongly suspected based on non-cultivative (molecular) approaches, that specific molecular probes are constructed. These probes can then be used to test for the presence of "known" pathogens in the different disease manifestations.

References

Antonius A (1981a) Coral reef pathology: a review. Proceedings of the 4th International Coral Reef Symposium, Manila 2:3–6

Antonius A (1981b) The "band" disease in coral reefs. Proceedings of the 4th International Coral Reef Symposium, Manila, 2, pp 7–14

Antonius A (1984) Coral diseases. Umschau 84:706–708

Antonius A (1988) Distribution and dynamics of coral disease in the Eastern Red Sea. Proceedings of the 6th International Coral Reef Symposium, Australia, 2, pp 293–298

Aronson RB, Precht WF (1997) Stasis, biological disturbance, and community structure of a Holocene coral reef. Paleobiology 23:326–346

Aronson RB, Precht WF (2001) White-band disease and the changing face of Caribbean coral reefs. Hydrobiologia 460:25–38

Aronson RB, Macintyre IG, Precht WF, Murdoch TJT, Wapnick CM (2002) The expanding scale of species turnover events on coral reefs in Belize. Ecol Monogr 72:233–249

Boettcher KJ, Barber BJ, Singer JT (1999) Use of antibacterial agents to elucidate the etiology of juvenile oyster disease (JOD) in *Crassostera virginica* and numerical dominance of an alpha-proteobacterium in JOD-affected animals. Appl Environ Microbiol 65:2534–2539

Boettcher KJ, Barber BJ, Singer JT (2000) Additional evidence that juvenile oyster disease is caused by a member of the Roseobacter group and colonization of nonaffected animals by *Stappia stellulata*-like strains. Appl Environ Microbiol 66:3924–3930

Borger JL (2003) Three scleractinian coral diseases in Dominia, West Indies: distribution, infection patterns, and contribution to coral tissue mortality. Rev Biol Trop 51 (Supl. 4):25–38

Bruckner AW, Bruckner RJ (1997) Outbreak of coral disease in Puerto Rico. Coral Reefs 16:260

Bythell JC, Gladfelter EH, Bythell M (1993) Chronic and catastrophic natural mortality of three common Caribbean reef corals. Coral Reefs 12:143–152

Bythell JC, Hillis-Starr ZM, Rogers CS (2000) Local variability but landscape stability in coral reef communities following repeated hurricane impacts. Mar Ecol Prog Ser 204:93–100

Bythell JC, Barer MR, Cooney RP, Guest JR, O'Donnell AG, Pantos O, Le Tissier MDA (2002) Histopathological methods for the investigation of microbial communities associated with disease lesions in reef corals. Lett Appl Microbiol 34:359–364

Denner EBM, Smith G, Busse H-J, Schumann P, Narzt T, Polson SW et al. (2003) *Aurantimonas coralicida* gen. nov., sp. nov., the causative agent of white plague type II on Caribbean scleractinian corals. Int J Syst Evol Microbiol 53:1253–1260

Dustan P (1977) Vitality of reef coral populations off Key Largo, Florida: recruitment and mortality. Environ Geol 2:51–58

Gladfelter WB (1982) White-band disease in *Acropora palmata* – implications for the structure and growth of shallow reefs. Bull Mar Sci 32:639–643

Gladfelter WB, Gladfelter EH, Monahan RK, Ogden JC, Dill RF (1977) Environmental studies of Buck Island Reef National Monument. US National Park Service Report, 144 pp

Miller J, Rogers C, Waara R (2003) Monitoring the coral disease, plague type II, on coral reefs in St. John, US Virgin Islands. Rev Biol Trop 51 (Supl. 4):47–55

Nugues MM (2002) Impact of a coral disease outbreak on coral communities in St Lucia: what and how much has been lost? Mar Ecol Prog Ser 229:61–71

Pantos O, Cooney RP, Le Tissier MDA, Barer MR, O'Donnell AG, Bythell JC (2003) the bacterial ecology of a plague-like disease affecting the Caribbean coral *Montastrea annularis*. Environ Microbiol 5:370–382

Patterson KL, Porter JW, Ritchie KB, Polson SW, Mueller E, Peters EC, Santavy DL, Smith GW (2002) The etiology of white pox, a lethal disease of the Caribbean elkhorn coral, *Acropora palmata*. Proc Natl Acad Sci USA 99:8725–8730

Peters EC (1983) The role of environmental-stress in the development of coral diseases and microparasite infestations. Am Zool 23:960

Peters EC (1984) A survey of cellular reactions to environmental-stress and disease in Caribbean scleractinian corals. Helgol Meeresunters 37:113–137

Richardson LL (1998) Coral diseases: what is really known? Trends Ecol Evol 13:438–443

Richardson LL, Goldberg WM, Carlton RG, Halas JC (1998a) Coral disease outbreak in the Florida keys: plague type II. Rev Biol Trop 46 (Supl. 5):187–198

Richardson LL, Goldberg WM, Kuta KG, Aronson RB, Smith GW, Ritchie KB et al. (1998b) Florida's mystery coral-killer identified. Nature 392:557–558

Richardson LL, Smith GW, Ritchie KB, Carlton RG (2001) Integrating microbiological, microsensor, molecular, and physiologic techniques in the study of coral disease pathogenesis. Hydrobiologia 460:71–89

Ritchie KB, Smith GW (1995a) Carbon-source utilization patterns of coral-associated marine heterotrophs. J Mar Biotechnol 3:107–109

Ritchie KB, Smith GW (1995b) Preferential carbon utilization by surface bacterial communities from water mass, normal and white-band diseased *Acropora cervicornis*. Mol Mar Biol Biotechnol 4:345–352

Ritchie KB, Smith GW (1998) Type II white-band disease. Rev Biol Trop 46 (Supl. 5):199–203

Rodriguez-Martinez RE, Banaszak AT, Jordan-Dahlgren E (2001) Necrotic patches affect *Acropora palmata* (Scleractinia : acroporidae) in the Mexican Caribbean. Dis Aquat Org 47(3):229–234

Santavy DL, Peters EC (1997) Microbial pests: coral disease in the Western Atlantic. Proc 8th Int Coral Reef Symp 1:607–612

Weil E, Urreiztieta I, Garzon-Fereira J (2002) Geographic variability in the incidence of coral and octocoral diseases in the wider Caribbean. Proc 9th Int Coral Reef Symp 2:1231–1237

21 Monitoring the Health of Coral Reef Ecosystems Using Community Metabolism

Jack Silverman, Boaz Lazar, and Jonathan Erez

21.1 Introduction

In the last few decades, anthropogenic influences on coral reefs have become a major threat to their existence (Hoegh-Guldberg 1999). Global changes, such as atmospheric CO_2 increase, global warming and increased UV radiation associated with stratospheric ozone depletion, can influence the well-being of coral reefs (Smith and Buddemeir 1992; Kleypas et al. 1999; Wilkinson 1999; Wellington et al. 2001). On a local scale, human economical exploitation and various types of pollution, mainly eutrophication and siltation have exposed many fringing and offshore reefs to severe stresses (Brown 1997). In this chapter, we discuss the use of community metabolism to follow the activity of benthic primary producers and growth of frame builders (mainly corals) in response to local environmental changes. In particular, we concentrate on the metabolic response of a coral reef in the Gulf of Eilat to anthropogenic eutrophication associated with fish farming.

21.2 Coral Reef Metabolism

Calcification, photosynthesis and respiration are the three major metabolic processes dominating the community metabolism of coral reef ecosystems (Hatcher 1997). By following changes in the dissolved oxygen (O_2) and total alkalinity (A_T) of reef water, it is possible to evaluate the intensity of these processes at the community level (Smith 1978). Equations (21.1)–(21.3) describe the stoichiometry of these processes and convey their interdependence.

Gross Production - P_g:
$$106[CO_2] + 16[HNO_3] + [H_3PO_4] + 122[H_2O] + \text{light} \rightarrow [(CH_2O)_{106}\_(NH_3)_{16}\_(H_3PO_4)] + 138[O_2] \quad [21.1]$$

Respiration - R:
$$[(CH_2O)_{106}\_(NH_3)_{16}\_(H_3PO_4)] + 138[O_2] \rightarrow 106[CO_2] + 16[HNO_3] + [H_3PO_4] + 122[H_2O] \quad [21.2]$$

Calcification - G:
$$[Ca^{+2}] + 2[HCO_3^{-1}] \rightarrow [CaCO_3] + [CO_2] + [H_2O] \quad [21.3]$$

Gross Production (P_g; Eq. 21.1) is the rate of photosynthetic carbon and nutrient fixation into organic matter resulting in O_2 evolution. The nutrients are derived from the dissolved inorganic pool in the surrounding sea, or by recycling of organic matter by reef heterotrophs (Erez 1990; Hatcher 1997). Community respiration (R; Eq. 21.2) is the rate of aerobic decomposition of organic matter, which supplies energy to reef heterotrophs causing inorganic carbon (C_T) increase and O_2 decrease. Community calcification (G; Eq. 21.3) is the rate of $CaCO_3$ deposition, where Ca^{2+} and HCO_3^- ions combine to precipitate $CaCO_3$ and release CO_2. Calcification rates during daytime are higher than nighttime for most calcifying organisms in the reef. This phenomenon is referred to as light-enhanced calcification (Barnes and Chalker 1990 and references therein), and is often attributed to the mutual dependence between photosynthesis and calcification (Eqs. 21.1, 21.3; Barnes and Chalker 1990).

The difference between P_g and R is termed net production (P_n). In coral reef ecosystems, P_g and R are very high and almost equal, hence P_n is often close to zero (Kinsey 1983, 1985; Pichon 1995, 1997). It is also common to calculate the ratio P_g:R, which is close to unity when the reef is in trophic balance. When P_g:R >1, the reef is autotrophic and when P_g:R <1 the reef is heterotrophic (Kinsey 1983, 1985; Pichon 1995, 1997). Autotrophic reefs require an increased external supply of dissolved nutrients, while heterotrophic reefs require an increased supply of particulate organic matter (plankton) from the open sea to support its growth (Marsh and Smith 1978; Erez 1990).

The metabolic rates are expressed as carbon fluxes (Smith 1978; Hatcher 1997). Estimates of P_g and R are usually based on oxygen budgets and G is estimated from total alkalinity (A_T). They are estimated in a number of different ways:

1. Flowing water respirometry, measurements of chemical changes (mainly O_2 and pH) in water parcel as it flows over the reef (Odum 1956; Barnes 1983).
2. Temporal changes in water chemistry of isolated water bodies such as slack water tidal pools (Kinsey and Davies 1979), selected micro-environments (Hoegh-Guldberg et al. 1997) and in situ respirometry conducted on artificially isolated coral knolls (Gattuso et al. 1997).
3. The diurnal mass-balance differences between reef water and open-seawater of O_2, C_T, A_T and pCO_2, multiplied by the rate of water exchange between these two reservoirs (Smith 1978; Kayanne et al. 2000).

Based on the measurements available until 1983, Kinsey (1983, 1985) proposed a "standard" metabolic performance for coral reefs, which was defined as follows: P_g=600±50 mmol C m^{-2} day^{-1}, P_g:R=1±0.1 and G=165±20 mmol C m^{-2} day^{-1}. However, data collected during the 1980s and the 1990s revealed a rather wide range of metabolic performance values, which challenged the concept of a "standard" performance for a coral reef ecosystem (e.g., Pichon 1997). For example, metabolic rates and ratios of several reefs in the Indo-Pacific and Red Sea varied between: P_g=220–1630 mmole C m^{-2} day^{-1}, G=40–340 mmole C m^{-2} day^{-1},

P_g:R=0.8–1.8 (Op. cit.). This range was attributed to differences in community structure and live cover, as well as seasonal changes in environmental parameters such as temperature, nutrients and light (Smith 1981; Kinsey 1985; Pichon 1997).

21.3 The Response of Coral Reefs to Eutrophication

Reefs that are exposed to substantial nutrient supply show an increase in their primary productivity, mainly due to development of macro-algae (Smith et al. 1981; Hatcher et al. 1989; Bell 1992; Done 1992; Hughes 1994; Lapointe 1997; Wachenfield 1998). In extreme cases, macro-algae overgrow, smother and kill the underlying corals (Smith et al. 1981; Genin et al. 1995). Reef eutrophication causes marked reduction in coral recruitment, possibly because of competition with algae, which rapidly occupy the hard substrate (Richmond 1997; Epstein et al. 1999; Abelson et al. 2000). Furthermore, laboratory and field experiments have shown that nutrient loading of reef building corals significantly suppressed their rates of calcification (Kinsey and Davies 1979; Marubini and Davies 1996; Ferrier-Pages et al. 2000) and reproduction (Koop et al. 2001 and literature therein). It is most likely that eutrophication will gradually increase the benthic autotroph population relative to frame builders ("trophic phase shift", according to Lapointe 1997). This change may be accompanied by higher grazing pressure and its associated bioerosion (Hallock 1988 and literature therein), thus upsetting the delicate balance between growth and erosion of the reef framework. The trophic phase shift is reflected in the reef's metabolic performance by an increase in community P_n and P_g:R and decrease in G (Silverman et al. 2002, 2003).

21.4 Case Studies in the Red Sea System

We conducted metabolic performance studies on three reefs in the Red Sea system:

1. Bullisar Island Reef (BIR) in the Dahlak Archipelago (15°22. 06′N, 40°27. 96′E), southern Red Sea. The reef fringes a low coral island roughly 1 km^2, on its northern and southern sides at a depth of 2–3 m, with close to 100% coral cover. It is a pristine reef thriving in waters that are in the upper temperature range for the global distribution of coral reefs (25–34 °C). The reef was studied during May 1995.
2. Ras Abu-Galum (RAG), east coast of Sinai Peninsula, Gulf of Aqaba, Red Sea, Egypt (28°38.4′N, 34°35.0′E). This is a well-developed fringing reef, densely populated mainly by a variety of branching corals and dominated by large colonies of *Acropora*. It is a pristine reef located in relatively cold water (21–27 °C; Reiss and Hottinger 1984). The reef was studied during July 1996.

3. The Nature Reserve Reef (NRR), Eilat, Israel, Gulf of Eilat (Aqaba) (29°30.23′N, 34°55.14′E). This is a fringing reef with ~30% live coral cover. Currently the reef is stressed by increased eutrophication associated with fish farming in cages at the northern tip of the Gulf. In the early 1990s, despite chronic eutrophication from municipal sewage and oil pollution, this reef had a live cover of 70% (Zakai 2000; Erez et al. 2003) and normal metabolic rates (see below). The reef was studied occasionally during the early and late 1990s and almost monthly from 2000 to 2002.

The results of community metabolism studies conducted at the three sites are presented in Figs. 21.1 and 21.2. P_n and G were estimated from dissolved oxygen and A_T, respectively, using methods 1 and 3 above. R was calculated from nighttime oxygen depletions relative to the open sea using the above methods, considering a photosynthetic and respiratory quotient of unity, typical for the Eilat reef (Barnes and Lazar 1993). P_g was calculated from the sum P_n+R (Marsh and Smith 1978). In addition, we define an environmental parameter (*EP*) that is the ratio $G/(P_g{:}R)$. This parameter represents the net $CaCO_3$ addition by frame builders (mainly hermatypic corals), relative to the net growth of macro algae. High *EP* values represent a healthy reef under oligotrophic conditions, where frame builders calcify normally, and $P_g{:}R$ is low

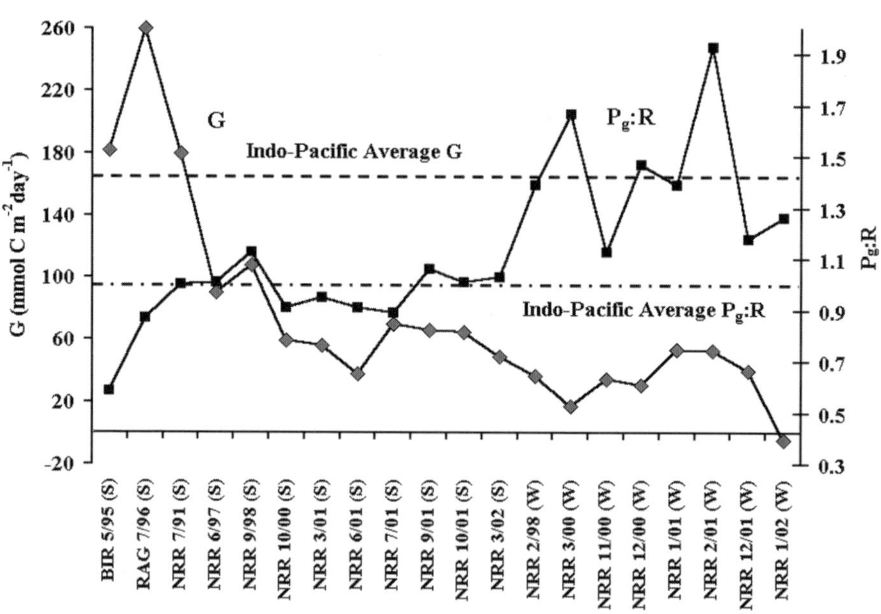

Fig. 21.1. Community calcification (*grey diamonds*) in Red Sea reefs ordered chronologically and according to season and photosynthesis to respiration ratio (*black squares*). The *dashed lines* represent the "standard" community calcification (*dashes*) and $P_g{:}R$ (*dot–dash*) according to Kinsey (1985). *S* Summer, *W* winter

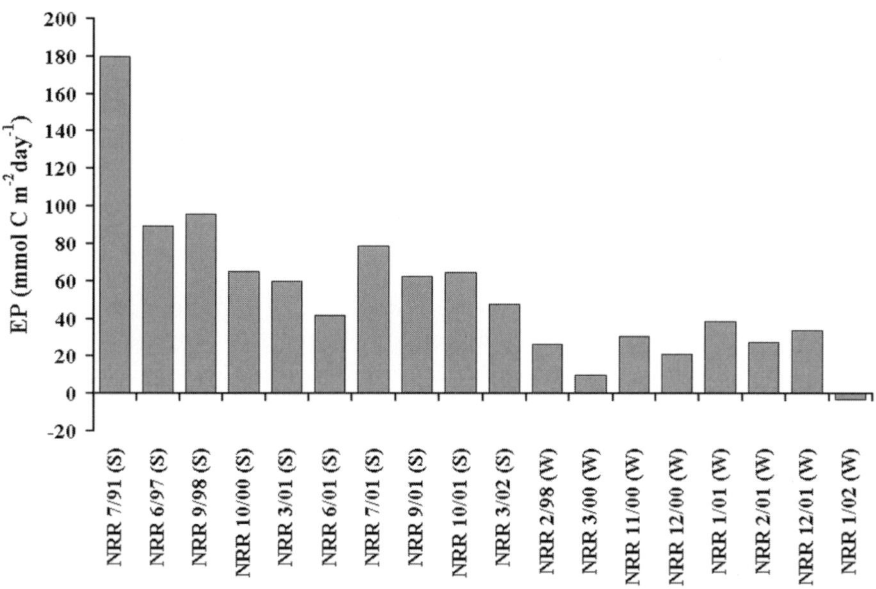

Fig. 21.2. Deterioration in environmental parameter (*EP*) ordered chronologically and according to season of the NRR since the early 1990s up till 2002. *S* Summer, *W* winter

(usually P_g:R=1), indicating considerable heterotrophic consumption of plankton from the open sea. Low *EP* values represent stressed reefs, infested with macro-algae due to high nutrient levels, and high P_g, and/or reduction in R, which may indicate less heterotrophy (lower plankton consumption), both causing P_g:R>1. At the same time, G is lower representing the reduction in the activity of frame builders either because of direct effect of nutrients on calcification or decrease in live cover.

The results (Fig. 21.1) suggests a general trend of increase in P_g:R and decrease in G. This trend is attributed to the continued eutrophication process in the northern Gulf of Eilat, as will be discussed below. The pristine Red Sea reefs (RAG, BIR) behave in a similar way to the Eilat reef, NRR, in 1991, showing higher G and *EP* (Fig. 21.2) compared to the "standard performance" (Kinsey 1985). This is in spite of their location at the thermal margins (high and low) of the global distribution of reefs. The negative trend in *EP* and G at NRR with time (Fig. 21.2) represents both the decrease in live cover (>50%) of hermatypic corals (mentioned above) and the increase in P_g:R (Fig. 21.1) associated with the ongoing eutrophication process in the northern Gulf of Eilat.

The average metabolic performance at NRR for the period 1997–2002 shows a clear seasonal variability (Table 21.1).

G values in the winter are half of those in the summer. The P_g:R ratios are higher by 50% during the winter than during the summer, hence, *EP* is lower by a factor of 3 during winter than summer. Similar differences were observed also between the winter and summer of 1990/1991; however, the absolute values of G

Table 21.1. Seasonal differences in metabolic performance of the Nature Reserve Reef (NRR), Eilat (1997–2002)

Season	Average *G*	Average P_g:*R*	Average *EP*
	(mmol C m^{-2} day^{-1}		mmol C m^{-2} day^{-1}
Summer	67	1.0	67
Winter	33	1.4	23

and *EP* were higher by a factor of 2–3 (*G*=180 and 165 mmol C m^{-2} day^{-1}; *Pg*:*R*=1 and 1.65; *EP*=180 and 100 mmol C m^{-2} day^{-1}, for summer and winter respectively (Korpal 1991; Barnes and Lazar 1993). These seasonal changes are associated with the natural oceanographic regime of the Gulf of Eilat (Reiss and Hottinger 1984). The stratified summer period is characterized by highly oligotrophic conditions (nitrate and phosphate are below 0.07 and 0.03 μM, respectively). During the winter, vertical mixing brings nutrient-enriched deep water to the photic zone and nitrate and phosphate values can rise up as high as 1 and 0.3 μM, respectively (depending on the depth of mixing). Thus, as expected during the winter mixing, increased nutrients support the rapid development of macro-algae population and cause an increase in P_g:*R* relative to the stratified summer. Furthermore, *G* may be suppressed either by decreased heterotrophy of frame work builders (increase in P_g:*R*) and/or by direct effect of nutrients on calcification (see above). In addition, part of the variability in *G* may be attributed to temperature (Shinn 1966; Lough and Barnes 2000, and literature therein), which affects the saturation level of aragonite (Gatuso et al. 1998; Lecqlerc et al. 2000; Langdon et al. 2000) and possibly the physiology of coral growth.

The seasonal variability is superimposed on the long-term trend in *G* and *EP* discussed above. The long-term increase in nutrient levels is associated with anthropogenic inputs that have increased dramatically during the last 6 years (Erez et al. 2000, 2003; Lazar et al. 2000). During the early 1990s, the sewage of Eilat was the main source of pollution. In the last year before it was diverted to inland oxidation ponds (1995–1996), it added a maximum of 140 tons of N per year (Shlesinger, pers. comm.). In parallel and during the following years, production of fish in cages located in the North Beach increased and were responsible for the addition of similar and increasingly larger amounts of N reaching a maximum of 320 tons N per year since 1999. As a result, nutrient levels have increased in both the coastal waters and in the deep basin below the thermocline. The integrated levels of nitrate in the water column have increased by a factor of 2.5 since 1997, and the oxygen levels have decreased accordingly. The most dramatic change has been observed in phytoplankton productivity in the open sea, which has increased, by a factor of 3 (Erez et al. 2003).

The time coincidence between the increase in the fish farming and its associated eutrophication and the deterioration in the Eilat reef, in both metabolic performance and in other ecological indicators (Abelson et al. 2000; Zakai 2000; Erez et al. 2003; Chap. 1, Loya , this Vol.), suggests a cause and effect relation. Indeed, the literature has documented numerous cases in which eutrophication caused deterioration of coral reefs (Smith et al. 1981; Pastorok and Bilyard 1985; Wittenberg and Hunt 1992; Hunter and Evans 1995; Lapointe 1997; Roberts et al. 2002, and literature therein).

In conclusion, we have shown that community metabolism is a sensitive monitoring tool for assessing the health of a coral reef. It can detect both seasonal variability and long-term trends that are associated with nutrient pollution, as well as other environmental changes.

21.5 Open Questions and Future Research

Monitoring the environmental parameter (*EP*) and other similar parameters which, if properly developed, can be used as a diagnostic tool for coral reef health on a frequent basis (perhaps even daily). The changes in the metabolic performance may be analogous to the medical monitoring of a human patient's vital signs in hospitals. An on-going project is the development of algorithms that will enable real-time automatic calculations of P_g, *R*, *G* and *EP*. The algorithms will employ the values of pH, O_2, temperature and tide in the reef, all of which are continuously measured by a submersible data logger. The easy (internet) real-time accessibility of natural resources management authorities to the processed data will greatly enhance their response time.

Gathering a long-term continuous record of community metabolic data will allow us to evaluate the effect of anomalous environmental conditions such as storms, extreme temperatures (high and low), abnormal atmospheric conditions (cloud cover, volcanic eruptions, dust storms or rain storms), abnormal tides, earthquakes and coral disease (e.g., bacterial infection) on reef performance. Characterizing the reef response to natural anomalous conditions provides a clue to identify anthropogenically induced stress and to take measures to reduce its cause to a sustainable level. During our ongoing research program in the NRR, we are planning to follow the *EP* in order to record the reef response to future relocation of the fish farms to land.

References

Abelson A, Olinki R, Shimla A, Bresler V (2000) Coral recruitment to the reefs in Eilat: current status, influencing factors and their effects on the reef's status. Ecology and environment (Hebrew publication), July 2000, vol. 6, pp 116–117

Barnes DJ (1983) Profiling coral reef productivity and calcification using pH and oxygen electrodes. J Exp Mar Biol Ecol 65:149–161

Barnes DJ, Chalker BE (1990) Calcification and photosynthesis in reef-building corals and algae. In: Dubinsky Z (ed) Ecosystems of the world, vol 25. Coral reefs. Elsevier, Amsterdam, pp 109–125

Barnes DJ, Lazar B (1993) Metabolic performance of a shallow reef patch near Eilat on the Red-Sea. J Exp Mar Biol Ecol 174:1–13

Bell PRF (1992) Eutrophication and coral reefs – some examples in the Great Barrier Reef Lagoon. Water Resour 5:553–568
Brown BE (1997) Disturbances to reefs in recent times. In: Birkeland C (ed) Life and death of coral reefs. Chapman and Hall, New York, pp 354–378
Done TJ (1992) Phase shifts in coral reef communities and their ecological significance. Hydrobiologia 247:121–132
Epstein N, Bak RPM, Rinkevich B (1999) Implementation of small-scale "no-use zone" policy in a reef ecosystem: Eilat's reef-lagoon six years later. Coral Reefs 18:327–332
Erez J (1990) On the importance of food sources in coral-reef ecosystems. In: Dubinsky Z (ed) Ecosystems of the world, vol 25. Coral reefs. Elsevier, Amsterdam, pp 411–415
Erez J, Silverman J, Rivlin T, Dray M, Nehorai A, Lazar B (2000) Increase in nutrient levels in the Gulf of Eilat and their negative impact on the Nature Reserve Reef in Eilat. Ecology and environment (Hebrew publication), July 2000, vol. 6, pp 118–119
Erez J, Iluz D, Zakai D, Silverman J, Lazar B (2003) Eutrophication processes in the Gulf of Eilat (Aqaba), Red-Sea, and their effects on the coral reef ecosystem. Geophys Res Abstr 5:11772
Ferrier-Pages C, Gattuso JP, Dallot S, Jaubert J (2000) Effect of nutrient enrichment on growth and photosynthesis of the zooxanthellate coral *Stylophora pistillata*. Coral Reefs 19:103–113
Gattuso JP, Payri CE, Pichon M, Delesalle B, Frankignoulle M (1997) Primary production, calcification, and air-sea CO_2 fluxes of a macroalgal-dominated coral reef community (Moorea, French Polynesia). J Phycol 33:729–738
Gattuso JP, Frankignoulle M, Bourge I, Romaine S, Buddemeier RW (1998) Effect of calcium carbonate saturation of seawater on coral calcification. Global Planet Change 18:37–46
Genin A, Lazar B, Brenner S (1995) Vertical mixing and coral death in the Red-Sea following the eruption of Mount Pinatubo. Nature 377:507–110
Hallock P (1988) The role of nutrient availability in bioerosion: consequences to carbonate buildups. Palaeogeogr Palaeoclimatol Palaeoecol 63:275–291
Hatcher BG (1997) Organic production and decomposition. In: Birkeland C (ed) Life and death of coral reefs. Chapman and Hall, New York, pp 140–174
Hatcher BG, Johannes RE, Robertson AI (1989) Review of research relevant to the conservation of shallow tropical marine ecosystems. Oceanogr Mar Biol Ann Rev 27:337–414
Hoegh-Guldberg O (1999) Climate change, coral bleaching and the future of the world's coral reefs. Mar Freshwater Res 50:839–866
Hoegh-Guldberg O, Takabayashi M, Moreno G (1997) The impact of long term nutrient enrichment on coral reef calcification and growth. Proc 8th Int Coral Reef Symp 1:861–866
Hughes TP (1994) Catastrophes, phase shifts and large-scale degradation of a Caribbean coral reef. Science 265:1547–1551
Hunter CL, Evans CW (1995) Coral reefs in Kaneohe Bay, Hawaii: two centuries of western influence and two decades of data. Bull Mar Sci 57:501–515
Kayanne H, Kudo S, Hata H, Yamano H, Nozaki K, Kato K, Negishi A, Saito H, Akimoto F, Kimoto F (2000) Integrated monitoring system for coral reef water pCO_2, carbonate system and physical parameters. 9th Int Coral Reef Symp, Bali, Indonesia, Book of Abstracts, p 249
Kinsey DW (1983) Standards of performance in coral reef primary production and carbon turnover. Perspective on Coral Reefs. Australian Institute of Marine Science, Townsville, Australia, pp 209–218
Kinsey DW (1985) Metabolism, calcification and carbon production I: system level studies. Proc 5th Int Coral Reef Symp, Tahiti 4:505–526
Kinsey DW, Davies PJ (1979) Effects of elevated nitrogen and phosphorus on coral reef growth. Limnol Oceanogr 24:935–940
Kleypas JA, Buddemeier RW, Archer D, Gattuso JP, Langdon C, Opdyke BN (1999) Geochemical consequences of increased atmospheric carbon dioxide on coral reefs. Science 284:118–120
Koop K, Booth D, Broadbent A, Brodie D, Bucher D, Capone D, Coll J, Dennison W, Erdmann M, Harrison P, Hoegh-Guldberg O, Hutchings P, Jones GB, Larkum AWD, O'Neil J, Steven A, Tentori E, Ward S, Williamson J, Yellowlees D (2001) ENCORE: the effect of nutrient enrichment on coral reefs. Synthesis of results and conclusions. Mar Pollut Bull 42(2):91–120

Korpal T (1991) Dissolved nutrients in the reef of the Northern Gulf of Eilat – Sources and fluxes. Msc Thesis, Hebrew University of Jerusalem, H Steinitz Marine Laboratory in Eilat, Israel
Langdon C, Takahashi T, Marubini F, Atkinson M, Sweeney C, Aceves H, Barnett H, Chipman D, Goddard J (2000) Effect of calcium carbonate saturation state on the calcification rate of an experimental coral reef. Global Biogeochem Cycles 14:639–654
Lapointe BE (1997) Nutrient threshold for bottom-up control of macro-algal blooms on coral reefs in Jamaica and Southeast Florida. Limnol Oceanogr 42:1119–1131
Lazar B, Erez J, Iluz D, Al-Hatib M, Rivlin T, Dray M, Nehorai A, Stambler N, Dubinsky Z (2000) The effect of eutrophication process in the Northern Gulf of Eilat on its ecological systems. Ecology and environment (Hebrew publication), July 2000, vol. 6, pp 116–117
Leclercq N, Gattuso JP, Jaubert J (2000) CO_2 partial pressure controls the calcification rate of a coral community. Global Change Biol 6:1–6
Lough JM, Barnes DJ (2000) Environmental controls on growth of the massive coral *Porites*. J Exp Mar Biol Ecol 245:225–243
Loya Y (1972) Community structure and species diversity of hermatypic corals at Eilat, Red-Sea. Mar Biol 13:100–123
Marsh JA Jr, Smith SV (1978) Productivity measurements of coral reefs in flowing waters. In: Stoddart DR, Dohanner RE, Rakat M, Johannes RE (eds) Coral reef research methods. Monographs on oceanographic methodology 5. UNESCO, Paris, pp 361–377
Marubini F, Davies PS (1996) Nitrate increases zooxanthellae population density and reduces skeletogenesis in corals. Mar Biol 127:319–328
Odum TH (1956) Primary production in flowing waters. Limnol Oceanogr 1:102–117
Pastorok RA, Bilyard GR (1985) Effects of sewage pollution on coral reef communities. Mar Ecol Prog Ser 21:175–189
Pichon M (1995) Coral reef ecosystems. Encycl Environ Biol 1:425–443
Pichon M (1997) Coral reef metabolism in the Indo-Pacific: the broader picture. Proc 8th Int Coral Reef Symp, Panama 1:977–980
Reiss Z, Hottinger L (1984) The Gulf of Aqaba. Ecological micropaleontology. In: Billings WD, Golley F, Lange OL, Olson JS, Remert H (eds) Ecological studies, vol 50. Springer, Berlin Heidelberg New York, pp 48–56
Richmond RH (1997) Reproduction and recruitment in corals: critical links in the persistence of reefs. In: Birkeland C (ed) Life and death of coral reefs. Chapman and Hall, New York, pp 175–197
Roberts CM, McClean CJ, Veron JEN, Hawkins JP et al. (2002) Marine biodiversity hotspots and conservation priorities for tropical reefs. Science 295:1280–1284
Shinn EA (1966) Coral growth-rate, an environmental indicator. J Paleontol 40:233–240
Silverman J, Lazar B, Erez J (2002) The metabolic response of a coral reef to eutrophication. 2002 Ocean Sciences Meeting in Hawaii, Book of Abstracts
Silverman J, Lazar B, Erez J (2003) Sorting out the effects of temperature, carbonate chemistry and nutrients on the metabolic performance of a coral reef. Geophys Res Abstr 5:11901
Smith SV (1978) Alkalinity depletion to estimate the calcification of coral reefs in flowing waters. In: Stoddart DR, Dohanner RE, Rakat M, Johannes RE (eds) Coral reefs: research methods. Monographs on oceanographic methodology 5. UNESCO, Paris, pp 397–404
Smith SV (1981) The Houtman Abrolhos Islands: carbon metabolism of coral reefs at high latitude. Limnol Oceanogr 16:612–621
Smith SV, Buddemier RW (1992) Global change and coral reef ecosystems. Annu Rev Ecol Syst 23:89–118
Smith SV, Kimmerer WJ, Laws EA, Brock RE, Walsh TW (1981) Kaneohe Bay sewage diversion experiment: perspectives on ecosystem responses to nutritional perturbation. Pac Sci 35:279–385
Stimson J, Larned S, McDermaid K (1996) Seasonal growth of coral reef macroalga *Dictyosphaeria cavernosa* and the affects of nutrient availability, temperature and herbivory on growth rate. J Exp Mar Biol 196:53–77
Wachenfield D (1998) State of the Great Barrier Reef World Heritage Area 1998. GBRMPA, Townsville, p 139

Wellington GM, Glynn PW, Strong AE, Navarette SE, Wieters E, Hubbard D (2001) Crisis on coral reefs linked to climate change. EOS Transact AGU 82(1):1–3
Wilkinson CR (1999) Global and local threats to coral reef functioning and existence: review and predictions. Mar Freshwater Res 50:867–878
Wittenburg M, Hunte W (1992) Effects of eutrophication and sedimentation on juvenile corals I. Abundance, mortality and community structure. Mar Biol 112:131–138
Zakai D (2000) Planulae recruitment and live coral cover as indicators of reef status in the Nature Reserve Reef, Eilat. Ecology and Environment (Hebrew publication), July 2000, vol. 6, pp 115–116

22 Coral Resistance to Disease

KERRI M. MULLEN, ESTHER C. PETERS, and C. DREW HARVELL

22.1 Theoretical Framework

Understanding the dynamics of resistance is particularly important for understanding the impacts of disease and predicting evolutionary outcomes for diseases. Predictive epidemiological models include not only terms for transmission of infectious microorganisms, but also terms for host resistance. In susceptible-infected-resistant (SIR) epidemiological models, timing and degree of resistance can determine the spread rate and impact of disease (Anderson and May 1979, 1991). Resistance is defined as "the natural or acquired ability of an organism to maintain its immunity to or to resist the effects of an antagonistic agent, e.g., pathogenic microorganism, toxin, drug (Stedman 1995)." An organism that is immune to an infectious disease will not acquire it because it has a particular suite of complex structural and functional features. These features prevent the pathogenic microorganism from entering, surviving in, or multiplying within its body and causing disease by disrupting key cellular metabolic processes through the release of toxins or enzymes or by altering its structure (e.g., tissue damage through scarring), or causing cell death. Many factors can affect the condition of this system and the response to a pathogen that an individual host is capable of generating at a particular time. The interaction of host and pathogen, and how they are affected by changing environmental conditions, can affect the populations of both organisms (Garnett and Holmes 1996).

Understanding the mechanisms of coral resistance to disease is of particular importance because in warming oceans, corals are demonstrably stressed by high summer temperatures. Stress in corals can be identified by an increased rate of bleaching (Hoegh-Guldberg 1999; Bruno et al. 2001; see other chapters in this Vol.), which may be linked to the appearance of some diseases (Kushmaro et al. 1997; Harvell et al. 2001; Porter et al. 2001), suggesting a role for compromised resistance. In some cases, bleaching itself is an infectious disease (Kushmaro et al. 1997; Ben-Haim et al. 1999; Ben-Haim and Rosenberg 2002). The rates of coral bleaching have increased in the last three decades and impacts of coral disease also appear to have increased (Santavy and Peters 1997; Hoegh-Guldberg 1999; Porter et al. 2001; Bruckner 2002; Harvell et al. 2002; Ward and Lafferty 2004).

22.2 Known Mechanisms of Coral Resistance

Although current rates and impacts of disease in corals are high (Aronson and Precht 1997; Richardson 1998; Richardson et al. 1998; Harvell et al. 1999; Weil et al. 2000, 2001; Porter et al. 2001), little is known about the resistance of coral to infectious disease. Bigger and Hildemann (1982) reviewed cellular defense systems of the Cnidaria, including pathogen defense, wound healing and inflammation, and response to foreign tissue. There is no previous work on coral resistance to pathogen infections, except for reviews of generalized coral responses to stress and injury (Peters 1984b; Hayes and Goreau 1998; Olano and Bigger 2000). Recent experiments and histological observations of scleractinian (Hexacorallia) and gorgonian (Octocorallia) corals provide insights into how resistant these organisms might be to pathogenic microorganisms.

22.2.1 Structure and Function of Coral Cells

The anatomy and histology of corals have been described by Hyman (1940), Bayer (1974), Chapman (1974), Peters (1984a), Fautin and Mariscal (1991), and others. The basic structure in each group is the polyp, a hollow cylindrical blind-ended sac like a sea anemone, often connected to other polyps by gastrovascular tissue, forming a colony. The polyp has a mouth, surrounded by a ring of hollow retractable tentacles, and is connected to the gastric cavity by a pharynx. The internal gastric cavity is divided by partitions called mesenteries. The mesenteries connect to the pharynx; within the gastric cavity the free edges of the mesenteries form mesenterial filaments.

Colony formation differs between the groups. For scleractinia, the bases of the polyp sacs are embedded in the aragonite exoskeleton produced by the calicoblastic epithelium of the polyps, which lines the skeleton everywhere. In the octocorals, the bases of the polyp sacs are embedded in a thick layer of the primitive connective tissue known as mesoglea. Scleroblasts, modified epithelial cells within the mesoglea, form calcium carbonate sclerites varying in morphology from thin, spindle-shaped to thick, polymorphic, with variable surface projections to support and protect the tissue from predators. The horny corals or gorgonians are further supported by a proteinaceous rod produced by the axis epithelium. Polyps are connected to one another by cell-lined tubes known as gastrovascular canals in the scleractinia and solenia in the octocorals. The polyps are attached to their supporting exoskeletons or axial rods by cells called desmocytes (Bayer 1974; Muscatine et al. 1997).

In both groups, a simple columnar or pseudostratified columnar epithelium, the epidermis, covers the external surfaces of the polyps and interpolypal tissue or coenosarc (coenenchyme). This epithelium covers the layer of mesoglea. Internally, the gastric cavity and canals that connect the polyps are lined by a generally cuboidal epithelium, the gastrodermis, also covering the mesoglea. The mesenteries and their filaments within the gastric cavity are lined on both surfaces by gastrodermis with mesoglea between.

The external and internal epithelial layers consist of several types of cells which provide protection or enable the polyps to capture and digest food, and support dinoflagellate algal cells that have a symbiotic relationship with their host coral (mainly in tropical shallow-water species). These algal cells or zooxanthellae are phagocytosed into vacuoles within the gastrodermal cells, but are not digested. They undergo photosynthesis in light and exchange nutrients and waste molecules with the polyp cells. Epitheliomuscular cells or myonemes and the subepidermal nerve net provide the polyps with the ability to expand or contract their bodies and tentacles, detect changes in the environment, and communicate with other polyps in the colony.

Although the nematocysts and spirocysts are important in capturing zooplankton prey and protecting the coral from predators, they probably do not provide much protection against pathogenic microorganisms (viruses, bacteria, fungi, protozoa). The surface epidermis of scleractinia contains unicellular secretory or gland cells and ciliated supporting cells. These cell types can be reduced in size or fewer in number in the gorgonian epidermis. The gastrodermis also contains supporting and gland cells. The mesoglea binding the two layers of epithelia together throughout the colony consists of a gelatinous substance, collagen fibers, and cells. Although generally referred to as mesogleal cells, they represent different cell populations. Some appear to be fibroblasts and secrete the matrix and collagen fibers; others, called amoebocytes, can be granular or agranular and function as phagocytes (Bigger 1984; Olano and Bigger 2000). Some of these cells have also been identified as pluripotential stem cells, capable of dividing and differentiating into various cell types as needed, such as cnidoblasts, scleroblasts, or germ cells. The latter two groups are capable of migrating through the mesoglea to distant locations when needed in the epithelia.

22.2.2 Innate Immune Response

Like other invertebrates, corals possess innate or natural immunity, a nonspecific ability to react to many potentially pathogenic organisms that is not altered with subsequent exposure. Basic host defenses include mechanical or physical barriers (e.g., epidermis), the ability to move, to shed or expel pathogens, secretion of chemicals (e.g., acid) or production of bioactive compounds (e.g., antimicrobial peptides), and phagocytic cells that can engulf and destroy microorganisms on contact (Cotran et al. 1999).

The cellular response consists of fixed or circulating amoeboid phagocytes that ingest microscopic organisms and kill them by exposure to proteolytic enzymes and free oxygen radicals. These cells go by different names in different phyla, e.g., leukocytes (macrophages) in vertebrates, hemocytes in mollusks, coelomocytes in echinoderms. For larger tissue-invading organisms, the amoebocytes can surround the foreign form to encapsulate or wall it off, or form a nodule, an aggregation of amoebocytes and bacteria or other pathogenic microorganisms; these structures can be accompanied by the deposition of a layer of melanin. The humoral response in innate immunity consists of secreted antimicrobial peptides, macrokines (similar to cytokines), and lectins (to agglutinate microorganisms to make them easier to phagocytose). Acquired or adaptive immunity, cell-mediated and humoral, involves the production of specific antibodies and T lymphocytes to eliminate the invading microorganisms through the operation of the major histocompatibility complex restriction that protects normal cells from attack (Clancy 1998). Adaptive immunity against pathogenic microorganisms has not been demonstrated in invertebrates.

Corals are animals, but because of their sessile nature and symbiosis with carbon-fixing algae, they have many plant-like physiological qualities. Therefore, in mapping out components of coral resistance to disease, it is useful to consider both plant and animal models. Plant inducible responses to fungi include constitutive and inducible components (Levin 1976; Agrawal et al. 1999; Berenbaum and Zangerl 1999). The main components of pathogen resistance are inducible and were classified by Kombrink and Somssich (1995), depending on speed of response and localization. Immediate early responses involve recognition and signaling processes, followed by locally initiated mechanisms such as phenylproponoid pathways, peroxidases and intracellular pathogenesis proteins. Finally, broad-spectrum systemic responses begin, such as production of chitinase and 1,2 beta-glucanases. Plant inducible responses to pathogens appear to diverge from responses to herbivores in using a salicylic acid pathway (Thaler et al. 2002a).

Invertebrate defenses against microbial infections are diverse, as noted above, including largely inducible components such as encapsulation via prophenoloxidase (PPO)-catalyzed melanization (Aspan and Soderhall 1995), direct production of antimicrobial peptides, and multistep processes such as opsonization and phagocytosis initiated by lectin recognition. What is common to both plants and animals is the inducibility of the dominant mechanisms, rendering detection and timing of resistance components in corals a high priority. Many microorganisms have, however, developed their own protection against one or more of these defenses, with the result that infections and disease are present in host populations (Clancy 1998). Alternatively, anything that adversely affects the integrity of the coral cells or their ability to produce defense compounds by induction of key processes can permit infection by microorganisms and initiation of disease.

22.2.3 Coral Immune System

Several studies have provided insights into how corals resist infection. For the sedentary scleractinian corals, the mucociliary system of the epidermis plays an important role in contrast to gorgonians. Mucous secretory cells are usually abundant in the epidermis and copious quantities of mucus are released (Bigger and Hildemann 1982); the composition and structure of the mucus varies among species (Meikle et al. 1988). The acidic mucopolysaccharides can trap or repel bacteria; in other cases bacteria might use the compounds in the mucus as substrates (Rublee et al. 1980; Paul et al. 1986). Santavy (1995) noted that scleractinian corals infected by black-band disease had higher surface microbial productivity than healthy or otherwise compromised corals. Apical cilia on the supporting cells wave constantly, producing water flows to sweep mucus and trapped particles (e.g., bacteria, sediment) off the surface of the colony to fall to the base of the colony or be disbursed by reef currents. The production of mucus and ciliary beating require expenditure of much cell energy. Peters (1984b) found that the epidermis at the base or sediment margin of massive corals lacked mucous secretory cells, perhaps due to the constant work involved in trying to keep sediment off the coral. In a laboratory study, constant exposure to sedimentation for 3 months caused a reduction in the number of mucous secretory cells and changes in the pH of the mucus (Peters and Pilson 1985). Tissue loss due to sedimentation has been shown to be preventable in the laboratory when antibiotics are present (Hodgson 1990). Bacterial diseases such as white plague and black-band disease typically start at tissue margins (Antonius 1985; Richardson et al. 1998) where this defense could be weakened or nonexistent.

Gorgonians, however, generally have fewer mucous secretory cells, although this depends on the species. Morphology of the colony, including vertical cylindrical growth to enable the polyps to extend into currents for food capture and maximum light exposure, also reduces the need for mucus. Cilia are present on cells of the epidermis, cnidoglandular tract of the mesenterial filaments, and pharynx to produce currents within the polyp to remove wastes.

Phagocytosis is the dominant mechanism of defense in invertebrates. In Cnidaria, phagocytosis is accomplished by amoebocytes, motile phagocytic cells that take part in wound healing and tissue reorganization (Chapman 1974; Mattson 1976; Bigger and Hildemann 1982), as well as cells of the gastrodermis and epidermis when the host is traumatized (Olano and Bigger 2000). The amoebocytes can be agranular or contain numerous neutral or acidophilic granules under the light microscope. The density of the cells and their appearance varies between taxa as well as within colonies (Figs. 22.1, 22.2).

Amoebocytes in the scleractinia are few and scattered within the mesoglea; they are best viewed in tissue sections of the fleshy species with larger polyps and thicker mesoglea. It is difficult to detect them in areas of thin mesoglea. In the mesoglea, they appear to be round to spindle-shaped, sometimes surrounded by a lacuna or space. The acidophilic granules have been considered to be lysosomes or peroxisomes (Olano and Bigger 2000).

In the Gorgonia, amoebocytes occur in dense clusters throughout the thicker mesoglea. They can form a layer beneath the epidermis or be present between epidermal cells or on the surface of the epidermis, perhaps a first line of defense against bacteria through phagocytosis and within-cell destruction by enzymes.

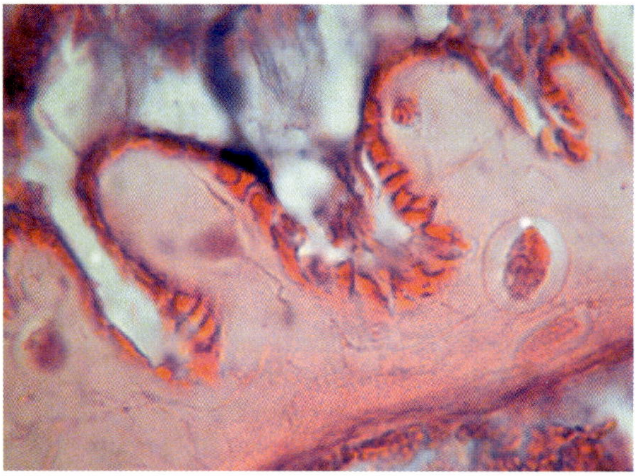

Fig. 22.1. Light microscopic view using oil immersion to show cells present in the mesoglea of a brain coral, *Diploria strigosa*. From *left* to *right*, the cells appear to be an agranular amoebocyte, a fibroblast, a stem cell, and a granular amoebocyte surrounded by a space

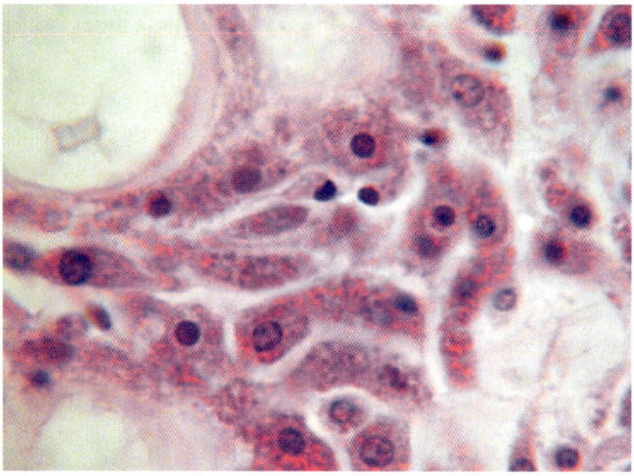

Fig. 22.2. Light microscopic view using oil immersion to show cells present in the mesoglea of a sea fan, *Gorgonia ventalina*. In the *center* is a fibroblast, surrounded by acidophilic granular amoebocytes, much more numerous and larger than their scleractinian counterparts

The inflammatory process in which these cells participate is less well understood in invertebrates than vertebrates (Sparks 1972). Infiltration of phagocytic cells (macrophages) is one of the characteristics of inflammation. The roles of the different kinds of cnidarian amoebocytes have been postulated to include production of collagen fibers within the mesoglea (like fibroblasts); stem cells (sometimes referred to as interstitial cells in the literature) to differentiate into scleroblasts, germ cells, or other cell types; or assisting in wound repair as phagocytes (increasing in numbers at wound sites as a result of migration), or differentiating into epidermal cells.

In the Anthozoa, studies on phagocytosis and wound healing have been limited to sea anemones and gorgonians. Under normal conditions, the connective tissue of anemones contains a homogenous population of amoebocytes, but following wounding, cell density increased significantly in a circular pattern around the region of damage (Patterson and Landolt 1979). The amoebocytes had secondary lysosomes and were observed to behave as phagocytes, cleaning up damaged cells. Within the repair zone in the mesoglea, swelling of the mesoglea was found, along with diapedesis of phagocytes through mesoglea and epidermal cells to discharge debris at the surface, like that reported for mollusks. Phagocytes derived from amoebocytes infiltrated the mesoglea by migration from other sites (mitotic activity was not observed in these cells). The atypical cells found in the zone appeared also to be morphologically suited for the production and secretion of unknown substances. Finally, cells infiltrated the lesion from the surrounding epithelium. The authors noted that this was more than a simple phagocyte response and that a distinct series of cellular events followed this injury. They concluded that the anemone has a "functional inflammatory response that predates the origin of a circulatory system or specialized organs."

The inflammatory response in the gorgonian *Plexaurella fusifera* is also caused by amoebocyte accumulation at the wound site, an effect of cells migrating from adjacent uninjured tissue (Meszaros and Bigger 1999). The migration of amoebocytes into a wound region to isolate the damaged region, prevent secondary infection, and initiate tissue repair by producing mesogleal fibers is further evidence of an organized reaction to injury and infection (Meszaros and Bigger 1999). Despite numerous histological examinations of scleractinian corals affected by various lesions (wounding, tissue infiltration by algae, bleaching, and diseases such as black band disease and white band disease), inflammatory responses characterized by infiltration of numerous amoebocytes have not been detected.

Both scleractinia and gorgonia are also capable of reacting to invading microorganisms by actively producing barriers to their penetration. For example, fungi that bore into the exoskeleton of scleractinians (Le Campion-Alsumard et al. 1995) induce activity by the calicoblasts, which lay down more skeleton. In histological preparations, the normally squamous calicoblastic epidermis becomes columnar with a more acidophilic staining cytoplasm adjacent to the fungal filaments. Layers of skeleton and organic material can be deposited to form a pearl. The axis epithelium and other cells of gorgonians can also be induced to begin more rapid production of gorgonin, with the deposition of melanin to wall off infiltrating fungi and algae (see below) and the formation of nodules (Morse et al. 1977).

In addition to cell-mediated immune functions, corals produce antibacterial, antifungal, and predator-deterrent compounds (Jensen et al. 1996; Kim et al. 2000a, b). For example, the anemone *Anthopleura elegantissima* mucus contains an enzyme that closely resembles lysozyme in its ability to lyse the bacterium *Micrococcus lysodeikticus* (Phillips 1963). Koh (1997) demonstrated that extracts from 100 coral species inhibited the growth of a marine cyanobacterium and extracts from eight of the species inhibited the growth of marine bacteria. Those eight species also had the fewest bacteria on their surfaces compared to corals lacking the antimicrobial compounds. Production of resistance compounds is also possible from associated surface bacteria. Twenty-nine percent of bacteria isolated from corals had antibacterial properties (Castillo et al. 2001). In other marine invertebrates, bacteria also appear as a source of antimicrobial compounds. Gil-Turnes et al. (1989) demonstrated that antifungal compounds that protect crustacean embryos from the fungal pathogen *Lagenidium callinectes* are produced by surface bacterial symbionts. The structural similarity between bryostatins of the bryozoan *Bugula neritina* and the bacterial symbiont *Candidatus* in *Endobugula sertula* suggests that the surface-associated bacteria produce the defensive compounds (Anthoni et al. 1990; Davidson and Haygood 1999).

Among cnidarians, gorgonians display some of the most potent antimicrobial activities (Burkholder and Burkholder 1958; Burkholder 1973; Bigger and Hildemann 1982; Jensen et al. 1996; Kim et al. 2000a, b). Crassin acetate, found in the gorgonians *Pseudoplexaura crassa* and *P. wagenaari* and in the endosymbiotic zooxanthellae, has antimicrobial and antiprotozoan activity and deters parrotfish. The hydroquinones of *Pseudopterogorgia rigida* and *P. acerosa* have antiviral and antibacterial activity and deter predatory fish (Harvell et al. 1988). Immunoglobulin A was reported to be secreted by cnidarian mucous secretory cells (Tomasi and Grey 1972, cited in Hayes and Goreau 1998), but this has not been confirmed by others (see also Chap. 12, Kelman, this Vol. for antimicrobial compounds in corals.)

The combination of cellular and humoral factors that make up the immune system varies from one individual to another; within the corals, it is clear that mucociliary activity, amoebocyte response, and production of antimicrobial compounds vary greatly among families, genera, and species. These genetically mediated differences might enable one group or one individual to have an advantage over others in resisting invasion by pathogens and reducing its susceptibility to disease. In addition, the age of the organism, its gender, reproductive state, and nutritional status can affect the immune system. For example, bleaching of tropical scleractinia or gorgonia for an extended period (weeks) removes a principal dietary resource, leading to atrophy and necrosis of the tissues (Lasker et al. 1984; Glynn et al. 1986). With loss of nutrients, mucus secretion, and ciliary beating, amoebocyte numbers are reduced, leaving polyps more susceptible to penetration by pathogenic microorganisms. Even if the polyps survive and recover their algal populations, reproduction and calcification can be inhibited for more than a year following the bleaching event, and other cellular processes might also be limited during this time (e.g., Szmant and Gassman 1990; Michalek-Wagner and Willis 2001; see also chapters in this Vol.). The line between reversible cellular changes and irreversible changes can be difficult to distinguish, including those changes that affect the cells of the immune system in corals. Tissue loss during bleaching events might be due to host cell necrosis, or it might be due to pathogenic microorganisms that have easily evaded the weakened defenses. A priority is understanding what deficits occur during bleaching that might directly affect coral immunity and susceptibility to infection.

Recent studies of aquatic organisms have sought to identify biomarkers, physiological, biochemical, or histological indicators, to show how well an organism's immune system is functioning under different environmental conditions, or when exposed to pathogens. Hawkridge et al. (2000) identified several antioxidant enzymes mainly in intracellular granules, as well as in accumulation bodies of the zooxanthellae and in different types of cnidae, in the sea anemone *Anemonia viridis* and the scleractinian coral *Goniopora stokesi.* Downs et al. (2000) reported development of biomarkers in *Montastrea faveolata* to detect coral responses to thermal stress. These include molecular chaperones of temperature-sensitive pathways (heat shock proteins 60 and 70, chloroplast small heat shock protein), indicators of cell integrity (lipid peroxide, alpha beta crystalline, glutathione, and ubiquitin), and antioxidant enzymes indicative of oxidative stress (manganese superoxide dismutase, copper/zinc superoxide dismutase). These markers represent both zooxanthellae and coral stress proteins and respond to changes in temperature and light level. Downs et al. (2002) showed significant variation in these biomarkers for corals from different depths during a bleaching event, supporting the hypothesis that bleaching is driven by oxidative stress. Banin et al. (2000) detected toxin P as a virulence factor of *Vibrio shiloi* that inhibits photosynthesis of zooxanthellae. The presence of virulence factors that operate differentially on zooxanthellae and the coral host indicates that origins of resistance from both coral and zooxanthellae should be considered.

In another experimental study of the basis of self-/nonself-recognition in the gorgonian *Swiftia exserta*, Salter-Cid and Bigger (1991) observed that histocompatibility reactions during tissue grafting met the minimal functional criteria of cytotoxicity, specificity, and altered secondary response (memory) that characterize an adaptive immune response. Autografts (host tissue applied to the same host) resulted in the fusion of the tissues. However, allografts (different donor tissue from the same species) resulted in rapid loss of tissue in the immediate contact area in 7–9 days. When another allograft was applied to the same host after a resting period, the same reaction occurred in only 3–4 days. Cell death was limited to the graft tissue interface, suggesting that this response was mediated by a contact or short-range cytotoxic molecule, rather than by a diffusible, long-range molecule (Salter-Cid and Bigger 1991). Additional studies are needed to confirm these observations.

In summary, the immune system of corals shares similarities with other invertebrates, but is so poorly known that important differences might yet surface. The least understood components of coral immunity involve any possible collaboration between coral and algal cells and the role of the symbiosis in im-

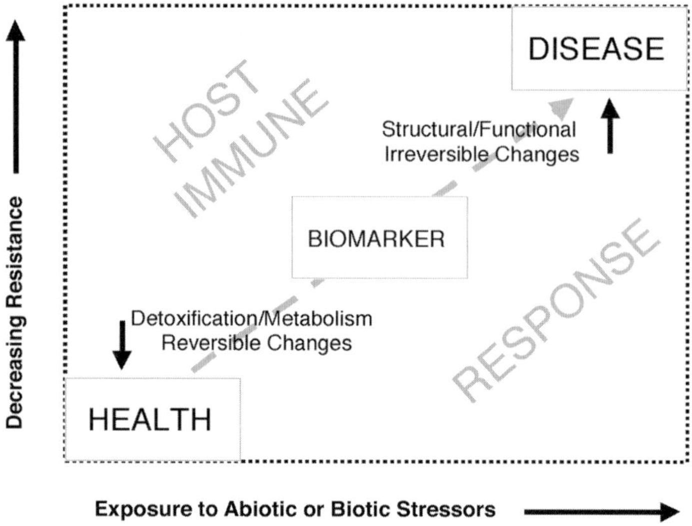

Fig. 22.3. Diagram of basic cellular changes occurring that adversely affect the host's resistance as exposure to stressors increases

munity. With increasing exposure to environmental stressors outside the normal range to which an individual is accustomed (e.g., increases or decreases in salinity, oxygen, light; chemical contaminants), or to pathogenic microorganisms, the host's immune system cells respond by undergoing detoxification or other metabolic reactions to try to reverse cellular changes and maintain the host organism's homeostasis. These reactions can produce biomarkers, which can be measured to provide an indication of the functioning of the organism and its immune system. As the stressors continue to exert their effects on the cells, irreversible changes in the nucleus, organelles, and membranes can occur, signaling impairment of vital functions or systems (disease). Although the host immune response in invertebrates is simpler in concept than in vertebrates, we have much to learn about how the cells function and interact to provide resistance to diseases in corals (Fig. 22.3).

22.3 Gorgonians: the Sea Fan as a Model System

In recent coral disease workshops (National Oceanic and Atmospheric Administration (NOAA) – Interagency Coral Disease and Health Consortium (CDHC), Charleston, SC, and World Bank, Akumal, Mexico), developing model systems for the study of coral resistance emerged as a research priority for future management and sustainability of reef habitats. A goal in our lab is to develop sea fans into such a model system to investigate chemical, cellular, and structural mechanisms of resistance. Critical priorities are to understand:

1. Cellular mechanisms of coral resistance;
2. Chemical mechanisms of coral resistance;
3. Relative contributions of zooxanthellae and corals to resistance; and
4. Genetic variation within and among colonies in resistance.

The focus of this review will be to identify what we know about each of these critical areas and then suggest future directions in research.

Aspergillosis is a disease of sea fan corals, first reported by Nagelkerken et al. (1996) and Smith et al. (1996). Because fungi in the genus *Aspergillus* are opportunistic pathogens in immune-compromised humans and other animals, the interaction between disease and resistance is of particular interest in this new outbreak in sea fans. We have shown that sea fans and other gorgonian coral species employ a battery of general antifungal and antibacterial compounds (i.e., secondary chemistry) for disease resistance (Jensen et al. 1996; Kim et al. 2000a, b), and have identified chitinase as a component of resistance extracts. In both plants and animals, systemic responses include hydrolytic enzymes such as chitinase (Tuzun and Bent 1999), a class of enzymes that hydrolyze chitin. Chitinases defend against fungal pathogens by destroying chitin-containing cell walls (Jolles and Muzzarelli 1999). Chitinolytic proteins are prominent, inducible components of antifungal resistance against *Aspergillus fumigatus* in guinea pigs (Overdijk et al. 1996) and humans (Tjoelker et al. 2000).

Field and laboratory studies have shown variability among fans in host resistance and aggregation of diseased individuals. Dube et al. (2002) detected significant differences in mean and variance of antifungal activity (AFA) for sea fan populations at different locations in the Florida Keys. They also detected a correlation between disease pressure and variance in antifungal activity that is consistent with selection acting on antifungal activity. Jolles et al. (2002) mapped all fans within three replicate 10×10 m grids to investigate spatial distribution of infected fans. Using geostatistical analyses to separate aggregation of diseased from possible underlying aggregation of all fans, they detected significant aggregation of diseased fans. This aggregation could be caused by either secondary transmission among neighbors or variation in resistance. Because the degree of aggregation increases with increasing disease severity in this data set, it seems more likely that aggregation is caused by factors affecting resistance. However, it is still not possible to rule out increased transmission in more aggregated locations as a cause of more severe disease.

To understand the relationship between disease outbreak and resistance response requires an experimental approach. Because *Aspergillus sydowii* can be readily cultured, this patho-system allows development of challenge inoculation experiments. The protocol we have developed involves growing *A. sydowii* on PYG agar (0.2% peptone, 0.2% yeast extract, 0.5% glucose, 3.6% bactoagar, 0.005% tetracycline) into which sterile cotton wicks are embedded. The wicks can then be applied to sea fans (and other gorgonians) in the lab and field to test response to infection. For field experiments, we were cautious in applying pure isolates of *A. sydowii* isolated from those same reefs. Using these inoculation protocols, we inoculated clonally replicated arrays of sea fans and showed that the level of AFA increased in inoculated fans and was higher in some clones (all the pieces from the same fan) than others (Harvell et al., unpubl.). This is the first experimental evidence for inducible AFA and for variation in levels of resistance among sea fans. Because corals are sessile like plants, there is considerable insight to be gained from plant studies about the importance of genetic neighborhoods and resistance structure of hosts under disease pressure. Studies on the anther smut disease *Usatilago violacea* and the dioecious perennial *Silene alba*, have shown the importance of fungal pathogen and host genetic neighborhoods and frequency-dependent selection (Antonovics and Thrall 1994; Thrall and Burdon 2003). Studies of disease spread in experimental populations of *S. alba*, where transmission rates were manipulated by varying genetically based host resistance, have confirmed the importance of frequency-dependent selection in this system (Thrall and Jarosz 1994). Host genetic structure was manipulated by establishing relatively resistant and susceptible host families. The progeny of susceptible families had higher infection levels than those from resistant families, and both frequency and density of hosts affected disease spread. More experimental field studies of coral resistance are needed to fill in this type of spatial detail for corals.

In our studies of resistance to fungal disease in gorgonians, we have identified several components

22.3.1 Generalized Antifungal Activity

Minimum inhibitory concentration (MIC) assays showed that of the 20 common gorgonian species in the Florida Keys, extracts from 15 species had MICs <15 mg /ml against *Aspergillus sydowii*, the fungus pathogenic to sea fans. Extracts from several species in two gorgonian genera (*Pseudoplexaura* and *Pseudopterogorgia*) were among the most active with MICs <10 mg/ml. *Gorgonia ventalina* L., one of two sea fan species known to be hosts to *A. sydowii* in the field, had an MIC <10 mg/ml, suggesting that complete disease resistance requires more active extracts (Kim et al. 2000b). Preliminary experiments show increasing levels of general antifungal activity 7 days post-inoculation (*t*-test, P=0.0025) and clone-specific profiles of resistance (Harvell et al., unpubl.). Previous surveys suggest highly localized (Kim et al. 2000a) antifungal activity adjacent to lesions. In addition to understanding mechanisms of resistance, our work will be guided by theoretical considerations for examining and modeling the evolution of a phenotypically plastic inducible response (Karban and Baldwin 1997; Harvell and Tollrian 1999): *time course* of induction and relaxation, *lag time* in response, and *norm of reaction profile* of colonies within different gorgonian populations (Schlichting and Pigliucci 1998). Adolph and Padilla (1996) highlight long lag times as a significant constraint in the evolution of inducible resistance. Preliminary work indicates that structural mechanisms (i.e., production of gorgonin and melanin, and changes in sclerite composition), as well as chemical response, have a time lag of at least 7 days, and nodule formation appears to be even slower.

22.3.2 Chitinase

Chitinases are widely distributed in marine invertebrates (Elyakova 1972), and could be an important source of induced antifungal resistance similar to that found in plants, insects and mammals. Chitinase is widely assayed in plant antifungal studies, and with other hydrolytic enzymes is an important tool in engineering plants resistant to fungal disease (Tuzun and Bent 1999). Recent studies have detected increased chitinolytic activity following systemic infection with *Aspergillus fumigatus* in humans and guinea pigs, providing evidence for a generalized and conservative inducible antifungal response (Overdijk et al. 1996; Tjoelker et al. 2000). Our preliminary studies show that chitinases are present in sea fans, with higher endochitinase than exochitinase levels (Mullen et al., unpubl.). Endochitinases can cleave any portion of a chitin polymer and is effective in cleaving the fungal cell wall, whereas exochitinases can only cleave N-terminal ends. Higher endochitinase activity is indicative of an induced antifungal response (Roberts and Selitrennikoff 1988). Work is underway to quantify constitutive and induced levels of chitinase in sea fans, and the degree of suppression of growth of *Aspergillus sydowii* in the presence of chitinase isolated from sea fans. We adapted and modified the rapid chitinase fluorogenic assay from Tronsmo and Harman (1993) to isolate, quantify and identify chitinolytic proteins and their relative activity. Before investing time in mapping genes for chitinase production, and before determining whether it is the coral host or the algal symbiont producing chitinases, it is critical to show that natural levels are inhibitory to the pathogen *A. sydowii.*

22.3.3 Melanin

In addition to chitinase, we identified substantial melanin deposits, a mechanism of fungal resistance, in the sea fan axial skeleton adjacent to areas with fungal hyphae. Sea fans were decalcified and prepared as histological slides at the Cornell Veterinary School, and we verified histochemically that the purpling response of diseased sea fans is associated with localized deposition of melanin in coenenchyme adjacent to fungal hyphae (Petes et al. 2003). Since quantification of melanin is technically challenging, we propose to assay prophenyloxidase (PPO), a melanin precursor, as a proxy for melanin production. This will link our sea fan resistance work with what is known of melanization (Leonard et al. 1985) and PPO activation as a common defense against fungi in other invertebrates.

Two distinct hyphal invasions have been observed in the axial skeleton of sea fans: sparse, thick hyphae (Fig. 22.4a) and dense, thin hyphae (Fig. 22.4b). We

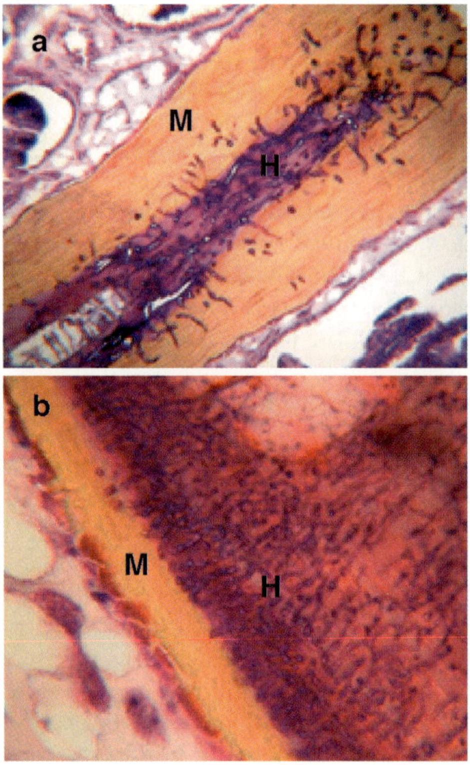

Fig. 22.4. Light microscopic view showing hyphal invasion of the axial skeleton of the common sea fan, *Gorgonia ventalina*. Hyphae (*H*) are dark purple filaments extending from the medulla (*middle*) of the axial rod into the cortex (outer layer of gorgonin), where a thick yellow melanin (*M*) layer is apparent. Two different fungi are shown, characterized by **a** thick, sparse hyphae or **b** thin, dense hyphae. Stain is H&E

hypothesize that the dense, thin hyphal invasions are *Aspergillus sydowii*. Invasion appears to begin where the axial skeleton has been denuded of tissue. The hyphae appear to migrate along the gorgonin axis, sending additional webs of fungi into the axis. Where the tissue remains covering the axis, sea fans respond to infection by sequestering invading hyphae with a thick melanin layer in the cortex (outer horny layer) of the axial skeleton. Since hyphae are rarely observed in the sea fan tissue, this melanized layer may deter infection from spreading into the coenenchyme. There is often a hypertrophied axis epithelial layer adjacent to the melanized gorgonin. Axis epithelial cells secrete layers of gorgonin and melanin pigment, and desmocytes (dark pink to purple cells along the outer axial layer) attach the axis epithelium to the axis. Desmocytes are easily recognized by striations that look like outstretched fingers reaching into the gorgonin. Even at locations distal to an aspergillosis lesion, fungal hyphae can be found sequestered in the medulla. Occasionally, hyphae have been observed radiating out of the medulla and penetrating the cortex.

In histological sections, we detected a new parasite that causes well-circumscribed purple spots on the surface of the sea fan. This parasite is characterized by purple to blue bodies associated with a densely staining blue mucus (Fig. 22.5a), and sometimes a matrix (Fig. 22.5b), when stained with haema-

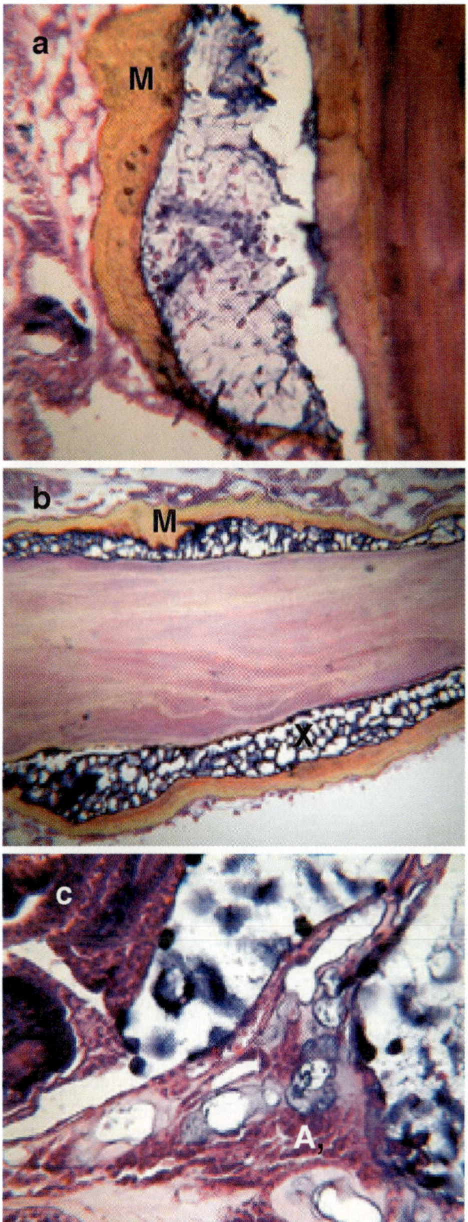

Fig. 22.5. Light microscopic view showing an unknown sea fan parasite characterized by the outward sign of dark purple spots and microscopic purple staining ovoid bodies surrounded by a blue mucous and often a matrix (*M*) associated with the axial skeleton. A melanin response in the cortex of the axial skeleton is visible (**a, b**), as well as an amoebocytic response (*A*), made evident by dense acidophilic granular cells (**c**), when the parasite invades the tissue of the host, *Gorgonia ventalina*. Stain is H&E

toxylin and eosin (H&E). This parasite is most often found in the axial skeleton of the sea fan, but occasionally has been observed invading a polyp and the surrounding tissue (Fig. 22.5c), and is associated with the gross sign of small purple spots. Sea fan response to the parasite is primarily a melanization response in the outer layer of the axial skeleton. Melanized gorgonin is visible as a thick bright yellow band, and appears to be a method of encapsulating the parasite to control spread into the adjacent coenenchyme. When this parasite invades a polyp, the blue bodies and mucus invade the gastric cavity. Darker purple staining of the polyp tissue suggests retraction. There appears to be an amoebocytic response in infected tissue adjacent to the polyp. Amoeboid cells with acidophilic granules (grainy deep pink cells visibly clustered in the coenenchyme) are more numerous in the area of infection.

22.3.4 Amoebocyte Recruitment

Tissue repair following a wound or invasion not only eliminates a pathogen or inhibits further spread, but significantly reduces the possibility of a secondary infection (Sparks 1972; Bigger and Hildemann 1982; Metchnikoff 1982). Evidence that sea fans can successfully defend against parasitic invasion and regenerate lost tissue is shown in Fig. 22.6. A layer of melanized gorgonin extending from the axial skeleton into the mesoglea suggests that a parasite was present (Fig. 22.6a, b), but it appears that encapsulation and phagocytosis have

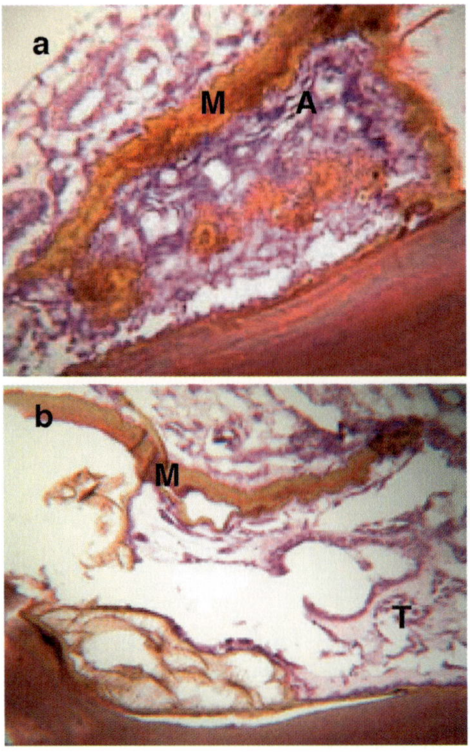

Fig. 22.6. Light microscopic view showing evidence that sea fans can successfully defend against parasitic invasion and regenerate lost tissue. A layer of melanized gorgonin (*M*) extends from the axial skeleton into the mesoglea suggesting that a parasite was present (**a, b**), but it appears that encapsulation and phagocytosis have not only prevented further infection, but abolished the parasite from the area. Acidophilic granular cells are observed in high density where the parasite was (**a**: contrast to the coenenchyme outside the area of encapsulation). The host is able to regenerate tissue (*T*) in the previously infected areas (**b**: observe coenenchyme moving into area of previous encapsulation). Stain is H&E

not only prevented further infection, but abolished the parasite from the area. Acidophilic granular cells are observed in high density where the parasite was (Fig. 22.6a: contrast to the coenenchyme outside the area of encapsulation). Once the host defends itself against invasion, it is able to regenerate tissue in the previously infected areas (Fig. 22.6b: observe coenenchyme moving into area of previous encapsulation).

22.3.5 Encapsulation with Gorgonin

Gorgonian corals may react to invasion by separating the parasite from the host tissue with a proteinaceous capsule (Goldberg et al. 1984). *Gorgonia ventalina* responds to infiltrating filamentous algae by the formation of a grossly visible nodule at that site, which upon microscopic examination reveals host tissue (coenenchyme) invaded with algal filaments, each surrounded by a tube of gorgonin, the same material that is secreted by the sea fan for its axial skeleton (Morse et al. 1977). Also in the region of infection were abnormally high numbers of amoebocytes. Goldberg et al. (1984) detected accumulation of granular

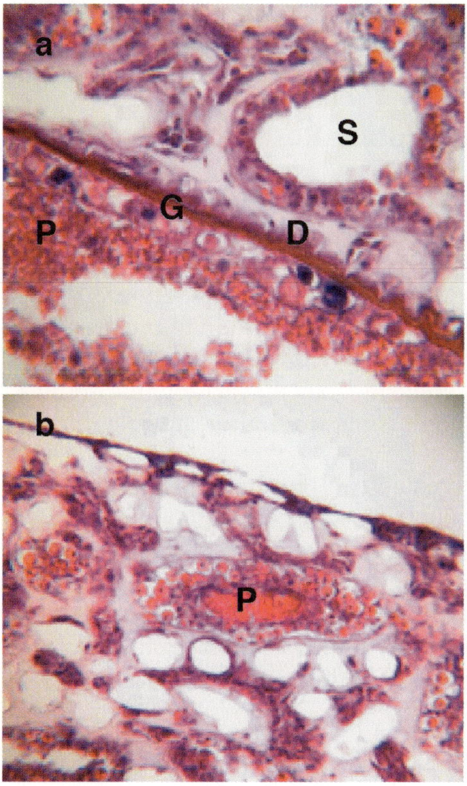

Fig. 22.7. Light microscopic view showing the production of a gorgonin (*G*) wall in response to an unknown parasite (*P*). Along the gorgonin layer is a hypertrophied axial epithelial layer of cells and what appears to be an inflammatory response (**a**). Desmocytes (*D*) are visible as purple finger-like extensions from the tissue into the gorgonin (**a**). A reaction in the adjacent tissue shows acidophilic granular cells in high density: amoebocytes appear in strands indicating migration through the mesoglea (**a**). The presence of parasite granules (*P*) in the sea fan tissue suggests that gastrodermal cells lining the gastrodermal canals are phagocytosing the parasite, evident by the dark pink granules present in the sea fan coenenchyme. Amoebocytes are lined up along the solenia (*S*), which is full of dark pink parasite granules (**b**). Stain is H&E

amoebocytic cells when the marine microalga *Entocladia endozoica* was present in *Pseudoplexaura* spp. When algal filaments extend beyond the gorgonin capsule into the mesoglea, the amoebocytes release vesicles in a process that appears to involve cell lysis, and the filaments are subsequently encapsulated by a skeletogenic epithelium (Goldberg et al. 1984). We found similar reactions to a dark pink staining unidentified organism present in the tissue of several sea fans, including encapsulation with gorgonin and an amoebocytic response (Fig. 22.7a, b). The production of a gorgonin wall in response to this unknown parasite is different from the melanization response associated with hyphal and purple spot "blue body" infections: the gorgonin layer is much thinner, and the yellow melanin layer is barely discernible (Fig. 22.7a, b). Along the gorgonin layer is a hypertrophied axial epithelial layer of cells and what appears to be an inflammatory response (Fig. 22.7b). Desmocytes are visible as purple finger-like extensions from the tissue into the gorgonin (Fig. 22.7b). A reaction in the adjacent tissue is evident by the presence of acidophilic granular cells in high density (Fig. 22.7a: compare to upper left corner where there is no visible amoebocytic response in the coenenchyme, i.e., very few dark purple staining granular cells). Amoebocytes migrate through the mesoglea and appear in strands. The presence of parasite granules in the sea fan tissue suggests that gastrodermal cells lining the gastrodermal canals are phagocytosing the parasite. Amoebocytes are lined up along the solenia, which is full of dark pink parasite granules (Fig. 22.7b).

22.4 Some Unresolved Questions and Future Research

In the study of coral resistance to disease, there are many unresolved questions. No research has been done on the resistance of Scleractinia to identified pathogen infections, although some patterns in species-specificity (e.g., black band disease most commonly affects faviids) suggest that genetic traits, perhaps expressed as variations in the quantity or composition of mucus, secretion of antimicrobial compounds, or other factors, control resistance (e.g., Weil et al. 2000). Limited studies on the Gorgonia indicate that acidophilic granular amoebocytes and antimicrobial compounds play an important role in resistance.

At the Bivalve Biomarker Workshop held in 1998, in Charleston, South Carolina, one of the working groups of scientists discussed immune function and disease responses in the diverse species of bivalves and how the immune system might be affected by exposure to toxicants (Ringwood et al. 1999). Compared to corals, the immune system of bivalves has been extensively explored and offers some models for approach. Some questions from that workshop that pertain to corals and others raised in this review include:

1. Can different subclasses of amoebocytes be separated on the basis of histochemical and immunohistochemical characteristics?
2. What traits of mucus repel or attract bacteria?
3. How does exposure to different stressors change mucus composition and microbial flora on the coral surface?
4. In what cells are antimicrobial peptides and antioxidant compounds in the coral immune system produced?
5. Can we apply procedures used to measure immune system parameters in other organisms to obtain quantitative indicators of immune function in corals (e.g., amoebocyte number, differential cell counts, killing index, phagocytic index, chemiluminescence, agglutinins)?
6. What biomarkers are useful for characterizing the condition of the immune system of corals and quantifying resistance?
7. What are "normal ranges" of biomarkers and other parameters that demonstrate the coral is healthy and the immune system functioning properly?
8. What is the relationship between immune function biomarkers and other endpoints of population condition (growth, reproduction, gamete viability)?
9. What are the primary mechanisms and time courses of resistance to bacteria, fungi and protozoans?
10. How does environmental stress and warming affect coral immunity?

This is only a starting point. The field of coral immunology is open to extensive exploration. The results of future studies should lead to insights on coral resistance to pathogenic microorganisms and direct us to approaches to mitigate the effects of disease on coral reefs.

References

Adolph S, Padilla D (1996) Plastic inducible morphologies are not always adaptive: the importance of time delays in a stochastic environment. Evol Ecol 10:105–117

Agrawal AA, Tuzun S, Bent E (1999) Induced plant defenses against pathogens and herbivores: biochemistry, ecology, and agriculture. APS Press, St Paul, MN

Anderson RM, May RM (1979) Population biology of infectious diseases I. Nature 280:367

Anderson RM, May RM (1991) Infectious diseases of humans: dynamics and control. Oxford Univ Press, Oxford

Anthoni U, Nielson PH, Perieira M, Christopherson C (1990) Bryozoan secondary metabolites: a chemotaxonomical challenge. Comp Biochem Physiol 96B:431–437

Antonius A (1985) Black band disease infection experiments on hexacorals and octocorals. Proc 5th Int Coral Reef Cong, Tahiti 6:155–160

Antonovics J, Thrall PH (1994) The cost of resistance and the maintenance of genetic polymorphism in host-pathogen systems. Proc R Soc Lond Ser B 257:105–110

Aronson RB, Precht WE (1997) Stasis, biological disturbance, and community structure of a Holocene reef. Paleobiology 23(3):326–346

Aspan AK, Soderhall AP (1995) The prophenoloxidase activating system in invertebrates: assays of the prophenoloxidase activating enzyme (a serine proteinase) and phenoloxidase. In: Stolen JS, Fletcher TC, Anderson DP, Roberson BS, van Muiswinkel WB (eds) Techniques in fish immunology, vol 4. SOS Publ, Fair Haven, NJ, pp 161–171

Banin E, Ben-Haim Y, Israely T, Loya Y, Rosenberg E (2000) Effect of the environment on the bacterial bleaching of corals. Water Air Soil Pollut 123:337–352
Bayer (1974) *Plexaura homomalla*: Brief historical background. In: Bayer FM, Weinheimer AJ (eds) Prostaglandins from *Plexaura homomalla*: ecology, utilization and conservation of a major medical marine resource. Univ Miami Press, Coral Gables, pp 1–8
Ben-Haim Y, Rosenberg E (2002) A novel *Vibrio* sp. pathogen of the coral *Pocillopora damicornis*. Mar Biol 141:47–55
Ben-Haim Y, Banin E, Kushmaro A, Loya Y, Rosenberg E (1999) Inhibition of photosynthesis and bleaching of zooxanthellae by the coral pathogen *Vibrio shiloi*. Environ Microbiol 1:223–229
Berenbaum MR, Zangerl AR (1999) Coping with life as a menu option: inducible defenses of the wild parsnip. In: Tollrian R, Harvell CD (eds) The ecology and evolution of inducible defenses. Princeton Univ Press, Princeton, NJ, pp 10–32
Bigger CH (1984) Immunorecognition among invertebrates. Dev Comp Immunol 3:29–34
Bigger CH, Hildemann WH (1982) Cellular defense systems of the coelenterata. In: Cohen N, Sigel MM (eds) The reticuloendothelial system. Plenum Press, New York, pp 59–87
Bruckner AW (2002) Priorities for effective management of coral diseases. NOAA Tech Mem NMFS-OFR-22. US Department of Commerce, National Oceanic and Atmospheric Administration, National Marine Fisheries Service, Silver Spring, MD
Bruno JF, Siddon CE, Witman JD, Colin PL (2001) El Niño related coral bleaching in Palau, western Caroline Islands. Coral Reefs 20:127–136
Burkholder PR (1973) The ecology of marine antibiotics and coral reefs. In: Jones OA, Endean R (eds) Biology and geology of coral reefs, vol II. Biology 1. Academic Press, New York, pp 117–182
Burkholder PR, Burkholder LM (1958) Antimicrobial activity of horny corals. Science 127:1174
Castillo I, Lodeiros C, Nunez M, Campos I (2001) In vitro study of antibacterial substances produced by bacteria associated with various marine organisms. Rev Biol Trop 49:1213–1222
Chapman D (1974) Cnidarian histology. In: Muscatine L, Lenhoff HM (eds) Coelenterate biology: reviews and new perspectives. Academic Press, New York, pp 93–128
Clancy J Jr (1998) Basic concepts in immunology: a student's survival guide. McGraw-Hill, New York
Cotran RS, Kumar V, Collins T (1999) Robbins pathologic basis of disease, 6th edn. Saunders, Philadelphia
Davidson SE, Haygood MG (1999) Identification of sibling species of the bryozoan *Bugula neritina* that produce different anticancer bryostatins and harbor distinct strains of the bacterial symbiont "*Candidatus* Endobugula sertula". Biol Bull 196:273–280
Downs CA, Mueller E, Phillips S, Fauth JE, Woodley CM (2000) A molecular biomarker system for assessing the health of coral (*Montastrea faveolata*) during heat stress. Mar Biotechnol 2:533–544
Downs CA, Fauth JE, Halas JC, Dustan P, Bemiss J, Woodley CM (2002) Oxidative stress and seasonal coral bleaching. Free Radical Biol Med 33(4):533–543
Dube D, Kim K, Alker AP, Harvell CD (2002) Size structure and geographic variation in chemical resistance of sea fan corals (*Gorgonia ventalina*) against a fungal pathogen. Mar Ecol Prog Ser 231:139–150
Elyakova LA (1972) Distribution of chitinases and cellulases in marine invertebrates. Comp Biochem Physiol B 43:67–70
Fautin DG, Mariscal RN (1991) Cnidaria: anthozoa. In: Hyman L (ed) Microscopic anatomy of invertebrates, vol 2. Placozoa, Porifera, Cnidaria and Ctenophora. Wiley-Liss, New York, pp 267–358
Garnett GP, Holmes EC (1996) The ecology of emergent infectious disease: infectious disease poses an ever-emerging threat to humanity. Bioscience 46(2):127–135
Gil-Turnes MS, Hay ME, Fenical W (1989) Symbiotic marine bacteria chemically defend crustacean embryos from a pathogenic fungus. Science 246:116–118
Glynn PW, Peters EC, Muscatine L (1986) Coral tissue microstructure and necrosis: relation to catastrophic coral mortality in Panama. Dis Aquat Org 1:29–37

Goldberg WM, Makemson JC, Colley SB (1984) *Entocladia endozoica* sp. nov., a pathogenic chlorophyte: Structure, life history, physiology, and effect on its coral host. Biol Bull 166:368–383

Harvell CD, Fenical W (1989) Chemical and structural defenses of Caribbean gorgonians (*Pseudopterogorgia* spp.): intracolony localization of defense. Limnol Oceanogr 34(2):382–389

Harvell CD, Tollrian R (1999) Why inducible defenses? In: Tollrian R, Harvell CD (eds) The ecology and evolution of inducible defenses. Princeton Univ Press, Princeton, pp 1–9

Harvell CD, Fenical W, Greene CH (1988) Chemical and structural defenses of Caribbean gorgonians (*Pseudoterogorgia* spp.) I. Development of an in situ feeding assay. Mar Ecol Prog Ser 49:287–294

Harvell CD, Kim K, Burkholder JM, Colwell RR, Epstein PR, Grimes DJ, Hoffman EE, Lipp EK, Osterhaus ADME, Overstreet RM, Porter JW, Smith GW, Vasta GR (1999) Emerging marine diseases: climate links and anthropogenic factors. Science 285:1505–1510

Harvell CD, Kim K, Quirolo C, Weir J, Smith GW (2001) Coral bleaching and disease: contributors to 1998 mass mortality in *Briarium asbestinum* (Octocorallia, Gorgonacea). Hydrobiologia 460:97–104

Harvell CD, Mitchell CE, Ward JR, Altizer S, Dobson AP, Ostfeld RS, Samuel MD (2002) Climate warming and disease risks for terrestrial and marine biota. Science 296:2158–2162

Hawkridge JM, Pipe RK, Brown BE (2000) Localization of antioxidant enzymes in the cnidarians *Anemonia viridis* and *Goniopora stokesi.* Mar Biol 137:1–9

Hayes RL, Goreau NI (1998) The significance of emerging diseases in the tropical coral reef ecosystem. Rev Biol Trop 46 [Suppl 5]:173–185

Hildemann WH, Raison RL, Cheung G, Hull CJ, Akaka L, Okamoto J (1977) Immunological specificity and memory in a scleractinian coral. Nature 270:219–223

Hoegh-Guldberg O (1999) Climate change, coral bleaching and the future of the world's coral reefs. Mar Freshwater Res 50:839–866

Hyman L (1940) The invertebrates, vol 1. Protozoa through ctenophora. McGraw-Hill, New York

Jensen PR, Harvell CD, Wirtz K, Fenical W (1996) The incidence of anti-microbial activity among Caribbean gorgonians. Mar Biol 125:411–420

Jolles AE, Sullivan P, Alker AP, Harvell CD (2002) Disease transmission of aspergillosis in sea fans: Inferring process from spatial pattern. Ecology 83(9):2373–2378

Jolles P, Muzzarelli RA (1999) Chitin and chitinase. Birkauser, Basel, Switzerland

Karban R, Baldwin IT (1997) Induced responses to herbivory. Univ Chicago Press, Chicago

Kim K, Harvell CD, Kim PD, Smith GW, Merkel SM (2000a) Fungal disease resistance of Caribbean sea fan corals (*Gorgonia* spp.). Mar Biol 136:259–267

Kim K, Kim PD, Alker AP, Harvell CD (2000b) Antifungal properties of gorgonian corals. Mar Biol 137:393–401

Koh EGL (1997) Do scleractinian corals engage in chemical warfare against microbes? J Chem Ecol 23(2):379–398

Kombrink E, Somssich IE (1995) Defense responses of plants to pathogens. In: Andrews JH, Tommerup IC (eds) Advances in botanical research (incorporating Advances in plant pathology), vol 21. Academic Press, London, pp 1–34

Kushmaro A, Rosenberg E, Fine M, Loya Y (1997) Bleaching of the coral *Oculina patagonica* by *Vibrio* AK-1. Mar Ecol Prog Ser 147:159–165

Lafferty KD, Kuris AM (1999) How environmental stress affects the impacts of parasites. Limnol Oceanogr 44(3):925–931

Lasker HR, Peters EC, Coffroth MA (1984) Bleaching of reef coelenterates in the San Blas Islands, Panama. Coral Reefs 3:183–190

Le Campion-Alsumard T, Golubic S, Priess K (1995) Fungi in corals: symbiosis or disease? Interaction between polyps and fungi cause pearl-like skeleton biomineralization. Mar Ecol Prog Ser 117:137–147

Lenihan HS, Micheli F, Shelton SW, Peterson CH(1999) The influence of multiple environmental stressors on susceptibility to parasites: an experimental determination with oysters. Limnol Oceanogr 44:910–924

Leonard C, Ratcliffe NA, Rowley AF (1985) The role of prophenoloxidase activation in non-self recognition and phagocytosis by insect blood cells. J Insect Physiol 31(10):789–800
Levin DA (1976) The chemical defenses of plants to pathogens and herbivores. Annu Rev Ecol Syst 7:121–159
Mattson P (1976) Regeneration. Merrill, Indianapolis
Meikle P, Richards GN, Yellowlees D (1988) Structural investigations on the mucous from six species of coral. Mar Biol 99(2):187–194
Meszaros A, Bigger C (1999) Qualitative and quantitative study of wound healing processes in the coelenterate, *Plexaurella fusifera*: spatial, temporal, and environmental (light attenuation) influences. J Invert Pathol 73:321–331
Metchnikoff E (1982) Leçons sur la pathologie comparée de l'Inflammation. Masson, Paris; reissued (1968) in English as: Lectures on the comparative pathology of inflammation. Dover, New York
Michalek-Wagner K, Willis BL (2001) Impacts of bleaching on the soft coral *Lobophytum compactum*. II. Biochemical changes in adults and their eggs. Coral Reefs 19:240–246
Morse DE, Morse ANC, Duncan H (1977) Algal "tumors" in the Caribbean sea fan, *Gorgonia ventalina*. Proc 3rd Int Coral Reef Symp 1:623–629
Muscatine L, Tambutte E, Allemand D (1997) Morphology of coral desmocytes, cells that anchor the calicoblastic epithelium to the skeleton. Coral Reefs 16:205–213
Nagelkerken I, Buchan K, Smith GW, Bonair K, Bush P, Garzon-Ferreira J, Botero L, Gayle P, Herberer C, Petrovic C, Pors L, Yoshioka P (1996) Widespread disease in Caribbean sea fans I. Spreading and general characteristics. Proc 8th Int Coral Reef Symp 1:679–682
Olano CT, Bigger CH (2000) Phagocytic activities of the gorgonian coral *Swiftia exserta*. J Invert Pathol 76:176–184
Overdijk B, van Stein GJ, Odds FC (1996) Chitinase levels in guinea pig blood are increased after systemic infection with *Aspergillus fumigatus*. Glycobiology 6(6):627–634
Patterson MJ, Landolt ML (1979) Cellular reaction to injury in the anthozoan *Anthoplexaura elegantissima*. J Invert Pathol 33:189–196
Paul JH, DeFlaun MF, Jeffrey WH (1986) Elevated levels of microbial activity in the coral surface microlayer. Mar Ecol Prog Ser 33:29–40
Peters EC (1984a) Comparative histology of selected tropical and temperate Atlantic scleractinian corals: an atlas. In: A survey of the normal and pathological histology of scleractinian corals with an emphasis on the effects of sedimentation stress, chap 1. Doctoral Dissertation, Graduate School of Oceanography, University of Rhode Island, Kingston, RI, pp 1–145
Peters EC (1984b) A survey of cellular reactions to environmental stress and disease in Caribbean scleractinian corals. Helgol Meeresunters 37:113–137
Peters EC, Pilson MEQ (1985) A comparative study of the effects of sedimentation stress on symbiotic and asymbiotic colonies of the coral *Astrangia danae*. J Exp Mar Biol Ecol 92:215–230
Petes L, Harvell CD, Peters E, Webb M, Mullen K (2003) Pathogens compromise reproduction and induce melanization in Caribbean sea fans. Mar Ecol Prog Ser 264:167–171
Phillips JH (1963) Immune mechanisms in the phylum Coelenterata. In: Dougherty EC, Brown ZN, Hanson ED, Hartman WD (eds) The lower Metazoa. Univ California Press, Berkeley, pp 425–431
Porter JW, Dustan P, Jaap WC, Patterson KL, Kosmynin V, Meier OW, Patterson ME, Parsons M (2001) Patterns of spread of coral disease in the Florida Keys. Hydrobiologia 460:1–24
Richardson LL (1998) Coral diseases: what is really known? Trends Ecol Evol 13:438–443
Richardson LL, Goldberg WM, Kuta K, Aronson RB, Smith GW, Ritchie KB, Halas JC, Feingold JS, Miller SL (1998) Florida's mystery coral-killer identified. Nature 392:557–558
Ringwood AH, Hameedi MJ, Lee RF, Brouwer M, Peters EC, Scott GI, Luoma SN, DiGuilio RT (1999) Bivalve biomarker workshop: overview and discussion group summaries. Biomarkers 4(6):391–399
Roberts WK, Selitrennikoff CP (1988) Plant and bacterial chitinases differ in antifungal activity. J Gen Microbiol 134:169–176

Rublee PA, Lasker HR, Gottfried M, Roman MR (1980) Production and bacterial colonization of mucus from the soft coral *Briareum asbestinum*. Bull Mar Sci 30:888–893
Salter-Cid L, Bigger CH (1991) Alloimmunity in the gorgonian coral *Swiftia exserta*. Biol Bull 181:127–134
Santavy DL (1995) The diversity of microorganisms associated with marine invertebrates and their roles in the maintenance of ecosystems. In: Allsopp D, Colwell RR, Hawksworth DL (eds) Microbial diversity and ecosystem function. CAB International in association with United Nations Environment Programme, Oxon, UK, pp 211–229
Santavy DL, Peters EC (1997) Microbial pests: coral disease research in the Western Atlantic. Proc 8th Int Coral Reef Symp 1:607–612
Schlichting CD, Pigliucci M (1998) Phenotypic evolution: a reaction norm perspective. Sinauer, Sunderland, MA
Smith GW, Ives LD, Nagelkerken IA, Ritchie KB (1996) Caribbean sea fan mortalities. Nature 383:487
Sparks AK (1972) Invertebrate pathology. Academic Press, New York
Stedman TL (1995) Stedman's medical dictionary, 26th edn. Williams and Wilkins, Baltimore
Szmant AM, Gassman NJ (1990) The effects of prolonged 'bleaching' on the tissue biomass and reproduction of the reef coral *Montastrea annularis*. Coral Reefs 8:217–224
Thaler JS, Fidantsef AL, Bostock RM (2002a) Antagonism between jasmonate- and salicylate-mediated induced plant resistance: effects of concentration and timing of elicitors on defense-related proteins, herbivore, and pathogen performance in tomato. J Chem Ecol 28:1143–1171
Thaler JS, Farag M, Pare P, Dicke M (2002b) Jasmonate-deficient tomato mutant has reduced direct and indirect defense. Ecol Lett 5:764–774
Thaler JS, Karban R, Ullman DE, Boege K, Bostock RM (2002c) Cross-talk between jasmonate and salicylate plant defense pathways: effects on several plant parasites. Oecologia 131:227–235
Thrall PH, Jarosz AM (1994) Host pathogen dynamics in experimental populations of *Silene alba* and *Ustilago violacea* II. Experimental tests of theoretical models. J Ecol 82:561–570
Thrall PH, Burdon JJ (2003) Evolution of virulence in a plant host-pathogen metapopulation. Science 299:1735–1737
Tjoelker LW, Gosting L, Frey S, Hunters CL, Trong HL, Steiner B, Brammer H, Gray PW (2000) Structural and functional definition of the human chitinase chitin-binding domain. J Biol Chem 275(1):514–520
Tomasi TB, Grey HM (1972) Structure and function of immunoglobin A. In: Kallas P, Walksman BH, de Weck A (eds) Progress in allergy, vol 16. Karger, New York, pp 81–213
Tronsmo A, Harman GE (1993) Detection and quantification of *N*-acetyl-β-D-glucosaminidase, chitobiosidase, and endochitinase in solutions and on gels. Anal Biochem 208:74–79
Tuzun S, Bent E (1999) The role of hydrolytic enzymes in multigenic and microbially-induced resistance in plants. In: Agrawal AA, Tuzun S, Bent E (eds) Induced plant defenses against pathogens and herbivores. APS Press, St Paul, MN, pp 95–115
Ward J, Lafferty K (2004) The elusive baseline for marine disease. Pub Libr Science (in press)
Weil E, Urreiztieta I, Garzón-Ferreira J (2000) Geographic variability in the incidence of coral and octocoral diseases in the wider Caribbean. Proc 9th Int Coral Reef Symp, Bali, Indonesia
Weil E, Smith GW, Mills M (2001) Spatial and temporal variability in coral and octocoral diseases in Bermuda. Abstract book, 30th Scientific Meeting of the AMLC, La Parguera Puerto Rico, 20 pp

23 Temperature Stress and Coral Bleaching

Paul L. Jokiel

23.1 Introduction

23.1.1 The Bleaching Phenomenon

Reef corals are symbioses that consist of an animal partner and dinoflagellate algae commonly known as zooxanthellae. This delicate association functions only within a very narrow range of environmental conditions in shallow tropical seas (Wells 1957). The symbiosis is thought to have developed as a means of coping with very low concentrations of nutrients in tropical marine waters (Muscatine and Porter 1977). This association provides benefits to both partners. The zooxanthellae are primary producers and supply their coral hosts with photosynthetic products vital to meeting their energetic requirements. In return, the zooxanthellae receive living space and nutrients in the form of waste metabolic products from the host. The tight cycling within the symbioses facilitates the high productivity and calcification rates observed in corals (Falkowski et al. 1984).

"Coral bleaching" occurs when environmental conditions disrupt the symbiosis, leading the degeneration and/or expulsion of zooxanthellae from the coral host. As a result of photosynthetic pigment loss, the white skeleton becomes visible through the transparent coral tissue, giving the organism a "bleached" white appearance. Bleaching is fatal to the coral unless the symbiotic relationship can be quickly re-established. Reef corals play a central role in coral reef communities, so their continued health is of critical importance.

Since the 1980s, spectacular regional bleaching events have occurred on coral reef areas throughout the world with increasing frequency and increasing geographic extent. These mass bleaching events correlate with elevated sea surface temperatures (SST). The mass bleaching phenomenon is believed to be associated with global warming due to anthropomorphic release of carbon dioxide and other gasses (reviewed by Jokiel and Coles 1990; Williams and Bunkley-Williams 1990; Glynn 1991, 1993; Goreau 1992; Pittock 1999; Fitt et al. 2001). The issue of global warming and mass coral bleaching will undoubtedly continue to be a major concern in the field of coral reef ecology.

23.1.2
Bleaching as a Disease Symptom

Coral bleaching is a condition that fits all the definitions of disease (Chap. 25, Rosenberg, this Vol.). The term bleaching describes a symptom that is produced by a variety of adverse physical, chemical and biological conditions in the environment. The term bleaching is quite subjective and has been applied to corals showing various degrees and patterns of pigment loss.

Bleaching can be caused by changes in numerous environmental parameters acting alone or in combination. These include increased sea surface temperatures (Jokiel and Coles 1990), low temperature (Steen and Muscatine 1987; Gates et al. 1992), increased irradiance (Lesser et al. 1990), decreased irradiance (Yonge and Nicholls 1931b; Franzisket 1970), altered spectral quality in the visible range (Kinzie et al. 1984), altered spectral quality in the ultraviolet radiation (UVR) range (Gleason and Wellington 1993), low salinity (Goreau 1964; Coles and Jokiel 1978), sedimentation (Meehan and Ostrander 1997; Philipp and Fabricius 2003), infectious disease (Kushmaro et al. 1996, 1998; Ben-Haim and Rosenberg 2002), exposure at low tide (Vaughan 1914; Yonge and Nicholls 1931b), oil contamination (Guzman et al. 1991) and exposure to toxic materials (Jones 1997; Jones and Steven 1997). Bleaching represents a single symptom that can be caused by a wide range of environmental stresses acting alone or in concert. Therefore, bleaching alone tells us little about the cause of the disease.

23.1.3
Coral Bleaching as an Environmental Indicator

Even the casual observer can easily recognize highly bleached corals. Thus, bleached corals have some value as an indicator of environmental stress. For example, documentation of severe bleaching events of 1998 depended on reports from untrained as well as trained observers on many remote reefs (Wilkinson et al. 1999). Establishing cause and effect of bleaching is a great challenge, because the term is highly subjective, imprecise and has many causes (Fitt et al. 2001). Conflicting reports arise because the "cause" has already passed before the conspicuous bleaching "effect" is noted. The damage has been done to the coral before bleaching becomes apparent. Observers generally are forced to conduct retrospective analysis based on limited environmental data. Measurements of the critically important environmental and physiological factors involved are usually not available. Thus, differences in the interpretation of causality arise.

23.1.4
Phenomenology Versus Mechanisms

The issue of coral bleaching will be discussed in two broad areas: (1) the phenomenology of bleaching and (2) the mechanisms involved in bleaching. The first category is descriptive and relatively straightforward. Coral bleaching events have happened, continue to occur with increasing frequency, and have been documented. Locations, species involved and environmental conditions that lead to bleaching have been described. The second category of mechanisms is far more uncertain and speculative and subject to extensive revision as new information is produced (see Chap. 25, Rosenberg, this Vol.). Interpretation of possible mechanisms might change, but the observed phenomena will remain. One of the main reasons for describing mechanisms is to provide predictive information on the possible future impact of continued global warming and the ability of reef corals to adjust to the increasing temperatures. For the purposes of this discussion, several terms will be defined. **Acclimation** refers to short-term physiological changes that lead to increased tolerance. In general, this term is applied in a situation where an organism is exposed to changes in a single parameter, generally under laboratory conditions. **Acclimatization** refers to longer-term phenotypic changes in response to changes in the natural environment. Such phenotypic responses are usually reversible. The process is limited by genotype, which determines the ultimate ability of a coral to tolerate stress in the environment. **Adaptation** is a long-term selective process that influences the genotype and thereby the ability of a coral to function in its environment and acclimatize to changes. It is useful to begin this discussion by describing what is known of the phenomenology of bleaching, because any proposed mechanisms must account for and be consistent with these observations.

23.2 Phenomenology Overview

Reef coral symbioses constantly adjust the population of zooxanthellae and/or the concentration of photosynthetic pigment per cell to meet changing environmental conditions. Under normal circumstances degraded zooxanthella cells are continually moved and released into the coelenteron (Boschma 1925, 1926; Yonge and Nicholls 1931a, b; Gates et al. 1992; Brown 1997; Stimson et al. 2002). Under severe environmental conditions most symbionts will be expelled, resulting in a bleached colony. Death results in reef corals if the association is not quickly re-established.

23.2.1 Bleaching and Coral Function

23.2.1.1 Photosynthesis and Respiration Phenomena

Photosynthesis (P) to respiration (R) ratio in corals decreases with increasing temperature above normal summer ambient temperature (Coles and Jokiel 1977). Tropical corals from Enewetak showed a higher P:R ratio at high temperature compared to subtropical varieties of the same species from Hawaii. Coles and Jokiel (1978) conducted controlled experiments involving corals acclimated for 2 months at different temperatures. Carbon fixation was subsequently measured in the corals held at those temperatures and showed that mean carbon fixation was substantially lower in the highest temperature treatment (28 °C) compared to the treatment approximating summer ambient (26 °C).

23.2.1.2
Growth and Regeneration Phenomena

Corals show optimum growth near the local summer temperature maxima (Jones and Randall 1973; Houck et al. 1977; Jokiel and Coles 1977; Coles and Jokiel 1978). A small prolonged increase of only 1–2 °C above this value will reduce skeletal growth (Jokiel and Coles 1977; Coles and Jokiel 1978). Short-term measurements of calcification rate in the coral *Pocillopora damicornis* show two calcification peaks, one at approximately 27 °C and a second at approximately 31 °C (Clausen and Roth 1975). The 27 °C peak is more prominent in subtropical Hawaiian corals and the 31 °C peak predominates in the more tropical Enewetak corals. Reduction of skeletal growth in corals at higher than normal summer temperatures has been shown in field experiments by Hudson (1981). During the 1987–1988 coral bleaching event in Jamaica, bleached colonies stopped growing during the bleaching period while unbleached corals grew at normal rates (Goreau and MacFarlane 1990). Tissue regeneration rates are lower when comparing bleached versus unbleached colonies (Meesters and Bak 1993).

23.2.1.3
Reproduction

Szmant and Gassman (1990) found that bleached corals are unable to complete gametogenesis, even in corals that have regained pigmentation. Highly bleached soft corals showed a significant reduction in reproductive output for two seasons after the corals regained full pigmentation (Michaelek-Wagner and Willis 2001a). Temperatures above normal summer maximum slow or prevent the development of gonads in corals (Hoegh-Guldberg 1999). Coral larvae exposed to 33 °C for 24 h (Edmunds et al. 2001) showed increased mortality, and reduced metamorphosis time compared to exposures at ambient temperatures (28 °C). Larval P:R ratios decreased with short-term exposures to elevated temperatures in these experiments, so presumably their longevity and competence are reduced. Settlement of *Pocillopora damicornis* coral larvae can be induced by brief exposure to high temperature (Edmondson 1946; Coles 1985). However, successful coral recruitment and growth rate of newly settled corals decreases dramatically at higher temperature, with tenfold reductions at 1 °C above the annual temperature maximum (Jokiel and Guinther 1978).

23.2.2
General Principles of Bleaching Phenomena

Complex ecological factors are involved in the bleaching phenomenon, but some generalizations are possible.

23.2.2.1
Regional Differences and Universal Response

Coles et al. (1976) concluded that corals throughout the world are living within 1–2 °C of their upper limit during the summer months in both tropical and subtropical environments. This statement has been substantiated by numerous other studies (Table 23.1). The 1–2 °C threshold has been observed universally, even though maximum seasonal temperature varies by 9 °C or more over the geographic range of these corals. Consequently, "bleaching hotspots" (reefs

Table 23.1. Estimated bleaching thresholds in relation to long-term mean summer maximum SST at various geographic locations

Location	Mean summer max. (°C)	Bleaching threshold (°C)	Increase above summer max. (°C)	Reference
Easter Island	25	27	2.0	Wellington et al. (2001)
Sodwana Bay, South Africa	26.5–27.0	27.5–28.8	1.0–1.8	Celliers and Schleyer (2002)
Hawaii	27–28	29–30	1.0–2.0	Jokiel and Coles (1990)
Gulf of Panama	28	–	–	Glynn (1977)
Gulf of Panama	–	30	2.0	Glynn and D'Croz (1990)
Gulf of Chiriqui	28	–	–	Glynn (1977)
Gulf of Chiriqui	–	30	2.0	Glynn and D'Croz (1990)
Bermuda	28	30	2.0	Cook et al. (1990)
Tahiti	28.2	29.5	1.3	Brown (1997)
Johnston Atoll	28.4	–	-	US Dept Commerce (1970)
Johnston Atoll	–	30.0	1.6	Cohen et al. (1997)
Puerto Rico	29	30	1.0	Goenaga and Canals (1990)
Enewetak	29–30	31–32	1.0–2.0	Coles et al. (1976)
Lizard Island, Australia	29	30	1.0	Hoegh-Guldberg and Smith (1989)
Papua, New Guinea	29	30.3	1.3	Davies et al. (1997)
Phuket, Thailand	29.5	30.1	0.7	Brown et al. (1996)
Belize	29.5	30.0–30.5	1.0–1.5	Aronson et al. (2000)
Palau	29.7–30	31	1.0–1.3	Bruno et al. (2001)
Maldives	30.3	31.4	1.1	Edwards et al. (2001)
Arabian Gulf	34	–	–	Coles (1988)
Arabian Gulf	–	35–36	1.0–2.0	Wilkinson et al. (1999)

likely to suffer from mass bleaching) have been defined as positive SST anomalies that exceed long-term mean maximum summer SST by 1–2 °C (Strong et al. 1997). The use of temperature increase above mean maximum summer SST rather than temperature yields a more predictive estimate. However, the satellite-derived temperature anomalies are based on offshore SST, which are generally 1–2 °C cooler than inshore waters during the summer months. Additional inshore heating of shallow restricted water masses by high solar irradiance during periods of low wind is needed to achieve bleaching temperatures.

The 1–2 °C margin of safety for reef corals identified by Coles et al. (1976) has already been eroded by global trends in warming on coral reefs. Corals are now living closer to their limit during summer months, and thus we are observing more frequent and more intense bleaching episodes. Since 1996, coral reefs throughout the world have been warming at a rate of approximately 0.01–0.02 °C per year or approximately 0.3–0.6 °C during that time interval (e.g., Brown et al. 1996; Hoegh-Guldberg 1999; Fitt et al. 2001; Winter et al. 2001). Climatic evidence suggests that the thermal tolerance thresholds of corals throughout the world will be exceeded every summer by the year 2030 (Hoegh-Guldberg 1999).

23.2.2.2 Temperature–Irradiance

A great deal of evidence indicates that the two primary factors involved in mass bleaching events are temperature and irradiance (reviewed in Fitt et al. 2001). High levels of solar irradiance interact with high temperature to produce coral bleaching (Jokiel and Coles 1977; Coles and Jokiel 1978; Hoegh-Guldberg and Smith 1989; Goenaga and Canals 1990; Fitt and Warner 1995; Brown et al. 1999a). A synergistic interaction occurs between temperature and irradiance (Coles and Jokiel 1978) that leads to chlorophyll reduction and lower rates of carbon fixation at temperatures above normal summer ambient. Individual corals usually show more pronounced bleaching and mortality on upper surfaces of the colony exposed to higher irradiance. Corals at shallow depths are exposed to high irradiance and are generally more prone to bleach at a given temperature than those at greater depths (Fisk and Done 1985; Oliver 1985; Lang et al. 1988; Wilkinson et al. 1999; Marshall and Baird 2000). Corals living at the same temperature and depth in turbid areas suffer less from bleaching than adjacent areas with higher turbidity (Phongsuwan 1998). High-temperature anomalies in the Society Islands during 1998 did not result in a bleaching episode, apparently because high cloud reduced solar radiation (Mumby et al. 2001).

High temperature and high irradiance are correlated. Obviously, this is because high temperature occurs during the summer season when irradiance is at its maximum. Further, high solar input combined with low wind conditions is the direct cause of high temperature anomalies. Two components of the solar spectrum are frequently implicated in bleaching. High photosynthetically active radiation (PAR) in conjunction with heat stress reduces the photosynthetic rate of the zooxanthellae (Coles and Jokiel 1977, 1978; Lesser et al. 1990). Ultraviolet radiation (UVR) may also be a detrimental factor that aggravates bleaching in corals (Jokiel 1980; Lesser et al. 1990; Gleason and Wellington 1993). Total solar energy, PAR and UVR are highly correlated (Jokiel et al. 1997). PAR and UVB are simply portions of the solar spectrum that are dependent variables on total solar radiation. PAR and UVR reaching the corals are further elevated during bleaching episodes because water transparency generally increases during the warming period due to low wind speed (Goenaga and Canals 1990; Rowan et al. 1997). The sudden and prolonged changes in water column clarity that occur during calm periods result in increased irradiance at all depths and may contribute substantially to coral bleaching (Gleason and Wellington 1993). Elevated levels of solar irradiance prior to the onset of high temperature may protect corals from subsequent bleaching (Dunne and Brown 2001).

23.2.2.3
Temperature–Irradiance–Duration

Duration of exposure to both temperature and irradiance is the third major factor controlling bleaching (Coles et al. 1976; Coles and Jokiel 1978). Prolonged exposure to temperatures slightly higher than the summer maximum can eventually produce bleaching in corals (Jokiel and Coles 1977). There is a strong relationship between severe bleaching and cumulative heat stress under conditions that exceed the normal summer maximum of 29 °C in Puerto Rico (Winter et al. 2001). Thus, the "hot spot" model (Strong et al. 1997) has recently been upgraded to include duration as well as intensity of heating. Bleaching thresholds are defined in degree heating weeks (DHWs), which are not allowed to accumulate unless the sea surface temperature is at least 1 °C above the maximum expected summertime temperature (Liu et al. 2003). One DHW is equivalent to 1 week of sea surface temperature at 1 °C above the expected summertime maximum. Two DHWs represent either 1 week at 2 °C above the expected summertime maximum or 2 weeks at 1 °C above the expected summertime maximum. Severe bleaching generally accompanies DHWs in excess of ten (http://www.osdpd.noaa.gov/PSB/EPS/method.html). The "hot spot model" does not include the effect of irradiance.

A conceptual model relating solar irradiance, heating above summer maximum, length of exposure and bleaching is presented as Fig. 23.1. The general model is based on published reports and 30 years of personal observation under experimental and field conditions. Extremely low irradiance situations are omitted for purposes of simplicity because darkness will inevitably lead to bleaching (Yonge and Nicholls 1931b; Franzisket 1970). The three-dimensional surface represents a bleaching tolerance "roof". Corals living at temperature-irradiance-duration conditions under the roof will not bleach. Those living under conditions at or slightly above the roof will bleach. Rapid mortality results from conditions higher above the roof. Note that at low irradiance, cor-

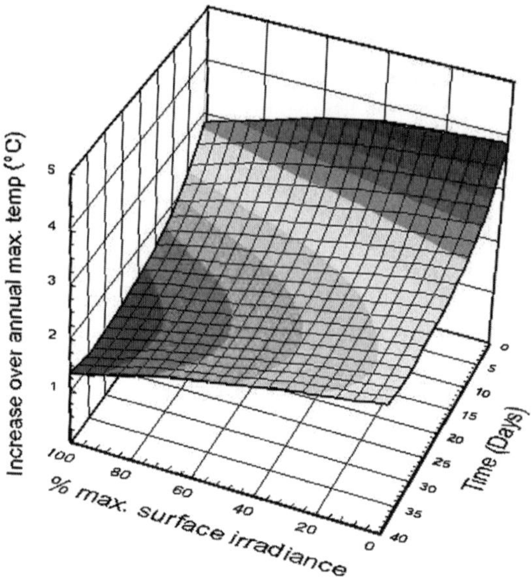

Fig. 23.1. The bleaching threshold for corals can be represented by a three-dimensional surface that is determined by a temperature increase above mean annual summer maximum, percent of peak annual surface irradiance (figure excludes very low levels of irradiance) and exposure time. Bleaching and eventual death occur under conditions at or above the ceiling

als can withstand exposure to high temperatures (3–4 °C over long-term mean summer maxima) for short time periods. At high irradiance, this capacity is reduced. In all cases, resistance to bleaching diminishes over time. At full solar irradiance, corals bleach at 1–2 °C above the mean annual maximum temperature within 2–4 weeks (Table 23.1). Shaded portions of colonies, colonies growing in deeper water, or corals under cloud cover at the same temperature will bleach at a slower rate. This is a generalized relative model assuming a more or less constant diurnal cycle of temperature and irradiance. The model describes conditions that on average lead to mass bleaching. Various species show higher or lower tolerance (e.g., Marshall and Baird 2000; Bruno et al. 2001). Another important point is that bleaching thresholds are defined in terms of mean temperature. Corals can tolerate considerable fluctuations (up to several degrees for limited periods) on reefs that can occur at low tide (e.g., Jokiel and Coles 1974; Craig et al. 2001), but the mean temperature ultimately determines the threshold.

This model assumes a relatively constant diurnal cycle of irradiance. Rapid increases in irradiance can occur in reef areas subjected to high terrigenous sedimentation. Prolonged periods of high wind create wave action that keeps fine sediments in suspension and greatly reduce irradiance penetration. In response, the corals photoacclimatize to the low light conditions. Onset of calm conditions under clear skies allows the sediments to settle with a consequent

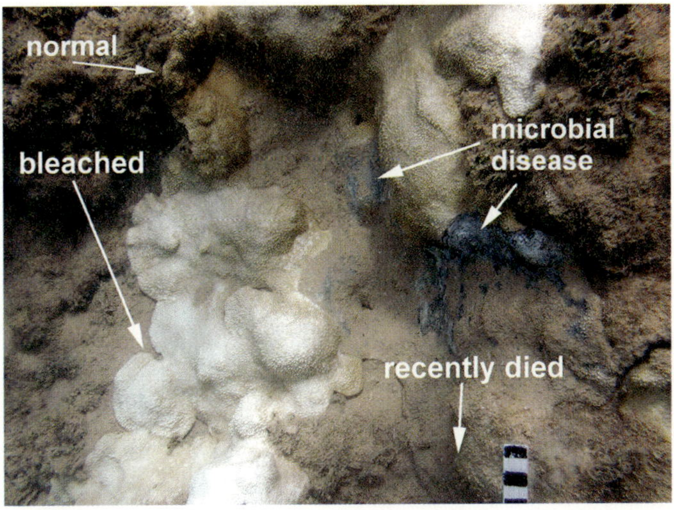

■ **Fig. 23.2.** Coral *Montipora capitata* growing on vertical wall was bleached by high turbidity and irradiance stress at Pilaa, Kauai. Microbial disease has invaded the bleached tissue. Unbleached tissues remain in area partially shaded by reef overhang. Scale in cm

rapid increase in irradiance reaching the corals. Such sudden increases in irradiance can result in bleaching (Fig. 23.2) at normal temperatures (Hoegh-Guldberg and Smith 1989).

23.2.2.4
Interactions Between Temperature–Irradiance and Other Ecological Factors

Water Motion. The relationship between duration, temperature and irradiance is further modified by water motion. Vigorous water motion has long been known to favor reef coral development. Water motion in the form of currents removes sediment and waste products and moderates factors such as temperature, salinity and oxygen concentration on the reef. Water motion influences growth, mortality and reproduction rate, and appears to influence corals by controlling the rate of material exchange across the interface between the seawater and the coral tissue (Jokiel 1978). *Acropora digitata* colonies held at 30 °C under low water-flow conditions suffered high bleaching mortality while colonies at the same temperature under high-flow conditions showed no bleaching effects (Nakamura and van Woesik 2001). Low wind conditions with resulting low water motion frequently occur during bleaching events, exacerbating the temperature–irradiance-induced stress (Goenaga and Canals 1990; Gleason and Wellington 1993).

Sedimentation/Turbidity. Turbid environments reduce high solar irradiance during periods of thermal stress and, consequently, can protect them from bleaching (Phongsuwan 1998). On the other hand, turbid environments can be the cause of bleaching, even at normal temperatures. For example, Fig. 23.2 shows a coral bleached at normal temperature in a high sediment environment off Pilaa, Kauai. Light intensity was greatly reduced in the inshore waters by high turbidity produced by massive mudflows from a grading project. Prolonged NE trade wind conditions produced onshore waves that kept the sediment in suspension. The corals became photoacclimatized to a very low irradiance. When the winds ceased, the sediment settled and the water became highly transparent. The corals were suddenly exposed to high irradiance, which caused them to bleach as described by Hoegh-Guldberg and Smith (1989). This type of bleaching has been observed in localized high sediment environments on the islands of Molokai, Kahoolawe, Lanai and Kauai (Jokiel, unpubl. observ.). Acevedo and Goenaga (1986) noted bleaching in south-western Puerto Rico that resulted from a turbidity plume that persisted for over 45 days.

Infectious Disease. Bleaching as a result of infectious disease agents is an emerging area of coral reef research (see Chap. 25, Rosenberg, this Vol.). Coral bleaching of the introduced Mediterranean coral *Oculina patagonica* has been linked to the infection and growth of the bacterium *Vibrio shiloi* at elevated temperature. These pathogens have been isolated and inoculated into unaffected coral colonies to cause bleaching of coral host tissues. Infection and bleaching are temperature-dependent. Most of the *O. patagonia* colonies bleach annually during the summer months as temperatures reach a maximum of 30–31 °C (Kushmaro et al. 1996, 1998). The corals recover during the winter months when SST falls to a minimum of 16 °C. The bacterium dies and lyses at the low temperature and cannot be detected in corals during the winter months. A similar bleaching relationship between the ubiquitous Pacific coral *Pocillipora damicornis* and the bacterium *Vibrio coralyticus* have also been demonstrated (Ben-Haim and Rosenberg 2002). It remains to be seen if such bacterial processes are involved in regional mass bleaching events (see Chap. 25, Rosenberg, this Vol.).

In some cases, increased disease caused by infectious agents may be the result of coral bleaching rather than the cause. In Hawaii, corals bleached by high sediment conditions on Molokai, Lanai, Kahoolawe and Oahu have shown evidence of microbial infections (Fig. 23.2). Coral bleaching could be a contributing factor in the outbreaks of various coral diseases that have occurred in Florida and Caribbean waters during the last decade (Cervino et al. 1998; Richardson et al. 1998).

Reduced Salinity. Even short-term exposure to salinity below 30 parts per thousand reduces the tolerance of coral to increased temperature, and these detrimental effects are further aggravated by exposure to high light intensities (Coles and Jokiel 1978). Salinity reductions from Hurricane Flora in 1963 resulted in bleaching of all reef corals to a depth of 3 m in the Port Royal area of Jamaica (Goreau 1964). On the other hand, a major reef kill in Kaneohe Bay, Oahu resulted in death without bleaching of corals to depths of 3 m (Jokiel et al. 1993).

23.2.3 Evidence for Acclimatization

Reef corals show a remarkable ability to acclimatize to changes in solar irradiance (e.g., Falkowski et al. 1990; Anthony and Hoegh-Guldberg 2003; see Chap. 9, Stambler and Dubinsky, this Vol.), but show little ability to acclimatize to elevated temperatures beyond their normal living range. Coles and Jokiel (1978) held corals under controlled laboratory conditions for 56 days at 28, 26, 24 and 20 °C and then exposed them to high temperature (31.5 °C). Mortality was inversely related to acclimation temperature. The increase in survival between the +2 °C temperature acclimation group (28 °C) and the summer ambient group (26 °C) was only 13%, so the effect was not substantial.

Marshall and Baird (2000) noted differences in the bleaching rate on different reefs and concluded that acclimatization due to previous thermal history was involved. Meesters and Bak (1993) concluded that acclimatization had occurred in corals growing in a thermal power plant plume during the 1990 bleaching event in Curaçao. Bleached corals from the heated plume showed higher rates of regeneration and regained their pigment at a higher temperature compared to bleached corals from outside the thermal plume. Berkelmans and Willis (1999) noted a difference in upper thermal limits of coral on a seasonal basis. Glynn and D'Croz (1990) found corals from an upwelling area in the Gulf of Panama to undergo greater bleaching at 30 °C in controlled experiments than the same species from the nonupwelling Gulf of Chiriqui, where ambient temperatures are higher and more stable.

Evidence of acclimatization to increased temperature in all of the above field studies is suspect, as many important environmental variables could not be controlled in the field observations. Differences in irradiance and water motion rather than temperature could easily account for the observed differences in bleaching tolerance. Acclimatization to higher temperature must be demonstrated under conditions with all other factors being held the same in the comparison.

23.2.4 Evidence for Adaptation

From the point of view of observed phenomena, the only evidence for genetic adaptation is the biogeographic observation that reef corals and/or their zooxanthellae show quite different bleaching thresholds that are related to maximum summer temperature regimes within their region. On a global scale, reported coral bleaching thresholds span a range of 9 °C or more (Table 23.1). On a local scale, different species and even different clones of the same species show differences in thermal tolerance (Marshall and Baird 2000; Stimson et al. 2002). During bleaching events, some colonies remain pigmented while adjacent colonies of the same species undergo bleaching, suggesting normal biological variation and genetic plasticity in response to high temperature.

23.3 Mechanisms Involved in Bleaching, Acclimatization and Adaptation

23.3.1 Primary Mechanisms of Cause

There is a widespread acceptance of the view that photoinhibition and damage to the photosynthetic apparatus of the zooxanthellae is a primary causal mechanism for bleaching (Chap. 9, Stambler and Zubinsky, this Vol.). However, this view is being challenged by the infectious disease hypothesis as discussed elsewhere in this volume (see Chap. 25, Rosenberg, this Vol.). According to the photoinhibition hypothesis, the early effects of heat stress on the photosynthesis of the zooxanthellae begin with strong nonphotochemical quenching of chlorophyll fluorescence, marked decrease in photosynthetic oxygen evolution and decreases in optimal quantum yield of photosystem II (PSII) and lead to damage of PSII and photoinhibition (e.g., Jones et al. 1998; Warner et al. 1999). In culture, photosynthesis in the zooxanthellae is impaired at temperatures above 30 °C and ceases completely at 34–36 °C (Iglesias-Prieto et al. 1992). Thus, any mechanism that prevents or reduces irradiance damage to photosystems of the zooxanthellae at high temperature will enhance the ability of the reef coral to resist bleaching.

Thus, one can argue that bleaching results from disruption of the delicate plant-animal association due to a breakdown of photosynthesis in the zooxanthellae, rather than a direct lethal effect on either partner. On an individual basis, the plant or animal components of the symbiosis appear to have slightly higher thermal limits and each can live independently at temperatures above the bleaching threshold. Once they have lost their symbionts, many corals will continue to live for a limited period of time under the same conditions of high temperature that led to their bleaching. Zooxanthellae isolated from the coral *Montipora verrucosa*, will grow well in culture at 31 °C (Kinzie et al. 2001), a temperature that will lead to rapid bleaching in the symbiosis (Jokiel and Coles 1977). Presumably, there is a mechanism in the coral that recognizes "healthy" zooxanthellae that are contributing fixed carbon to the animal host. Perhaps once a zooxanthella is metabolically unable to "pay the rent", it is recognized as dysfunctional by cellular processes and is evicted. Bhagooki and Hidaka (2002) exposed polyps of the coral *Galaea fasicularis* to temperatures of 24, 28 and 30 °C for 7 days. The corals held at 24 or 28 °C released low levels of degraded zooxanthellae particles. The temperature-stressed coral exposed to 30 °C released large numbers of healthy looking zooxanthellae, suggesting that high temperature caused a physiological dysfunction that lead to bleaching.

Bleached corals have lost most of their photosynthetic capacity and must depend on heterotrophic feeding and stored lipids to meet their energy needs. This is not possible, given the reliance of corals on photosynthesis (Mucatine 1990). Thus, depletion of energy reserves has been proposed as a mechanism to account for the failure of corals to complete gametogenesis even after regaining pigmentation (Szmant and Gassman 1990). Reduced survival, development, dispersal and recruitment of coral planula larvae at high temperature have been attributed to lower energy availability (Edmunds et al. 2001). Lipid and protein content remained lower in bleached colonies of the soft coral *Lobophytum compactum* and eggs produced by these colonies for at least 8 months following bleaching (Michalek-Wagner and Willis 2001b). Prolonged and repeated bleaching reduces energy reserves of the coral and may reduce the ability of coral to resist disease (Fig. 23.2).

23.3.2 Acclimatization Mechanisms

23.3.2.1 Production and Modification of Proteins and Enzyme Systems Involved in Metabolic Processes

The primary physiological challenge of increased temperature to reef corals and all other organisms is centered on the effect of temperature on enzyme reaction rates. The well-known concept of Q_{10} is defined as the ratio of enzyme reaction velocities at different temperatures and approximately doubles over a 10 °C range. Under conditions in their normal living range, organisms can acclimatize in order to offset the effects of temperature increases through quantitative changes in enzyme concentrations and/or by changing to variants with different kinetic properties (Hochachka and Somero 1984). As temperature increases, the coral symbiosis acclimatizes by modifying enzyme systems and other biochemical components in order to maintain homeostasis. The genotype of the coral and the genotype of the symbiotic algae set the ultimate limit on temperature acclimatization, because both the plant and the animal components of the symbiosis must act in unison to maintain the fragile nature of the relationship. Coral bleaching is a clear sign that the limits of coral acclimatization have been exceeded.

Heat shock proteins involved in acclimatization to temperature are found in reef corals (reviewed by Gates and Edmunds 1999). These serve to lessen the negative impact of higher temperature by reducing aggregation of structurally nonnative proteins, refolding reversibly heat-damaged proteins and aiding in the insertion of proteins into organelles. Both host and symbiotic algae produce heat shock proteins in response to elevated temperature (Brown et al. 2002b).

23.3.2.2 Mechanisms That Reduce Photoinhibition

Photoinhibition due to high temperature combined with high solar radiation is a primary factor influencing mass coral bleaching and is discussed in detail by Stambler and Dubinsky (Chap. 9, Stambler and Dubinsky, this Vol.). Coral bleaching starts with damage to the photosynthetic apparatus before there is a loss of zooxanthellae. Under normal conditions of temperature and irradiance, photosynthesis proceeds with uptake of dissolved carbon dioxide and water and transfer of protons through the photochemical systems of the light reaction. This is followed by release of oxygen and fixation of organic carbon in the dark reaction (reviewed by Fitt et al.2001; Chap. 9, Stambler and Dubinsky, this Vol.). At higher temperature and high irradiance, the rates of these processes can be increased to a level where more protons are produced in the light reaction than can be utilized to form organic carbon in the dark reaction (Brown 1997). A number of mechanisms serve to reduce such photosynthetic impairment.

Protective Enzymes. The production of oxygen free radicals or other toxic forms of oxygen in both the algal symbiont and coral host tissues increases at high temperature and high irradiance and can cause cellular damage, expulsion of symbionts and accelerate bleaching (Lesser et al. 1990; Lesser 1997). The enzymes superoxide dismutase, catalase and ascorbatae peroxidase act to detoxify active forms of oxygen. These enzymes inactivate harmful oxygen radicals and thereby prevent cellular damage. Activity and concentration of these protective enzymes increase in response to elevated temperature.

Production of Accessory Pigments. Ambarsari et al. (1997) and Brown et al. (1999b) have described the xanthophyll cycle in corals, which ameliorate the damaging effects of light at high levels of irradiance. Corals with a high capacity for dissipating excess excitation energy through this cycle appear to be less prone to temperature bleaching (Warner et al. 1996).

Production of Screening Pigments. Corals contain a group of UVR-absorbing compounds known as micosporine-like amino acids commonly known as MMAs (Dunlap et al. 1986) that show a maximum concentration in shallow water where high levels of UVR are present. MAAs increase bleaching resistance (Lesser et al. 1990) and counteract the effects of harmful UVR (Dunlap and Shick 1998). MAAs show high scavenging activities of the singlet oxygen and superoxide radical with very strong antioxidant activity on lipid peroxidation. Soft corals suffering MAA reduction caused by a previous bleaching event still amplified these compounds twofold in eggs, demonstrating the importance of MAAs in survival of the larvae (Michaek-Wagner and Willis 2001b). Coral larvae increase MAA in high UVR environments (Kuffner 2001). Reef corals often are not able to increase MAAs at a rate sufficient to counter the rapid UVR increases resulting from sudden increases in water clarity and can bleach as a result (Gleason and Wellington 1993). Corals growing in high water motion environments maintain high concentration of MAA protective compounds irrespective of irradiance (Jokiel et al. 1997). Thus, reef corals growing under conditions of high water motion are essentially pre-adapted to tolerate sudden increases in high solar radiation that accompany the onset of clear sky conditions associated with regional SST increases. Therefore, water motion can be an important environmental factor in bleaching resistance.

Corals contain other pigments that reduce irradiance reaching the zooxanthellae through reflection, absorption, fluorescence and/or scattering of excess or potentially damaging energy at different wavelengths. Green fluorescent pigment-like (GFP-like) proteins that fluoresce under UVR and/or visible light appear to play a role in protecting the photosynthetic machinery of the symbiotic dinoflagellates from the adverse effects of high irradiance (Dove et al. 2001; Labas et al. 2002). These pigments fluoresce the potentially damaging short wavelength radiation to longer green wavelengths. In the process they serve to reflect and scatter the energy away from the zooxanthellae. Corals containing GFP-like pigments are less prone to bleaching than nonfluorescent colonies of the same species (Salih et al. 2000). Nonfluorescent purple-blue pigments commonly found in corals may serve to reflect short wavelength radiation away from the zooxanthellae.

23.3.3 Behavioral Protective Mechanisms

Corals display limited behavioral responses at all life stages that may allow them to avoid bleaching and damage due to high temperature and high irradiance. Behavior related to reproduction, dispersal and settlement of larvae is largely directed at avoidance of unfavorable environmental conditions and unfavorable habitats that can lead to bleaching. Corals of the species *Pocillopora damicornis* extrude their brooded planula larvae into the water column when subjected to abnormally high temperature (Edmondson 1946). Presumably, this behavior reduces the metabolic demand on the coral polyps and allows the larvae to escape localized conditions of high temperature and high irradiance. Most broadcast spawning corals produce buoyant eggs and sperm bundles that rise to the surface layer of the ocean (Harrison et al. 1984). This mechanism serves to concentrate the gametes and insures higher rates of fertilization. During the day, however, the surface layer receives very high levels of irradiance and can be heated to temperatures 1–2 °C warmer than deeper water. Corals spawn in the evening, so the gametes and developing larvae are not subjected to the high damaging irradiance and higher surface temperatures during fertilization and early development of the planula larvae. Coral larvae will move vertically in the water column and appear to select depth of optimal light intensity (Kawaguti 1941; Raimondi and Morse 2000). Larvae will delay settling under conditions of high UVR (Kuffner 2001). Coral larvae actively select suitable settlement sites (Edmondson 1946; Harrigan 1972). After initial attachment to a substrate, the planula larvae will detach if conditions are unfavorable and presumably find another more suitable location (Harrigan 1972). In some cases, newly settled polyps will detach and undergo reverse metamorphosis even after initiating skeletal formation (Richmond 1985). Coral larvae of different species will select different light regimes of intensity and spectral quality before settling (Mundy and Babcock 1998). The larvae of many species tend to settle in shaded cryptic areas away from high light intensity (Morse et al. 1988; Raimondi and Morse 2000). Such behavior may allow them to avoid bleaching due to high irradiance as well as avoiding competition from macroalgae and providing protection from predation by grazers. Some species of coral selectively settle on specific crustose coralline algae, perhaps because these algae are indicators of a temperature-irradiance environment that is favorable to juvenile coral over the long run. In some cases, juvenile corals show less bleaching compared to adult colonies on the reef (Mumby 1999). A possible explanation is that the juveniles live in a micro-habitat with lower irradiance. However, mortality of juvenile corals during a bleaching event can be much higher than the mortality rate of mature colonies (Harriott 1985).

Polyp behavior among adult colonies may play a role in alleviating bleaching. Expansion of polyps can increase surface area to facilitate exchange of harmful metabolic products at high temperature (Abe 1939). Contraction of polyps into their calyces can reduce irradiance impact and reduce bleaching (Brown et al. 1994).

23.4 Adaptation Mechanisms

The primary mechanism controlling adaptation to higher temperature is natural selection (Darwin 1859). Variation in bleaching response within various species is known to occur. Genetic adaptation occurs when the less resistant members of a population are eliminated by the ecological conditions, leaving organisms with higher fitness to reproduce and recruit to an available habitat. To date, bleaching events have not resulted in complete mortality of any coral population, so the surviving genotypes presumably have high-temperature tolerance that will be imparted to their offspring. The process of natural selection for bleaching-tolerance at higher levels of temperature and irradiance is already underway. The time required for such genetic adaptation in reef corals is unknown, but presumably would take many coral generations and hundreds to thousands of years.

The time needed for adaptation could be decreased substantially if corals are able to substitute thermally resistant zooxanthellae for less resistant types as proposed by the "adaptive bleaching hypothesis" (Buddemeier and Fautin 1993; also Chap. 24, Buddemeier et al., this Vol.). Adaptation of reef corals to higher temperature by this mechanism remains to be demonstrated. Numerous species and types of zooxanthellae are now recognized (see Chap. 8, Baker, this Vol.), and their response to temperature and irradiance probably varies. For example, Jokiel and York (1982) showed the existence of sun-loving and shade-loving varieties of zooxanthellae. Differences in bleaching resistance that corresponded to zooxanthellae symbiont genotypes in the ubiquitous Pacific reef coral *Pocillopora damicornis* have been reported (Glynn et al. 2001). Further, some coral species have colonies that contain several types of zooxanthellae, which can undergo selective loss of the most sensitive type during thermal stress (Rowan et al. 1997). Reciprocal transplant experiments over a range of depths disrupted coral symbioses and showed that bleaching may provide an opportunity for replacement of the expelled zooxanthellae by other types of symbionts (Baker 2001). However, there may be costs involved, as shown by higher mortality in species that vary their algae with depth as compared to species that retain a single type of algae. Bleached anemones infected with a variety of zooxanthellae from other hosts often did not grow as well as anemones re-infected with their normal type of zooxanthellae (Kinzie and Chee 1979). Thus, new combinations of symbiont and host will be subject to the same process of natural selection as the other corals on the reef and may succeed or may be eliminated from the population.

23.5 Climate Change, Bleaching and the Future of Coral Reefs

The pattern of bleaching events that has increased in frequency and extent over the past 20 years will continue with increased severity in the future. Hoegh-Guldberg (1999) points out that reef corals have shown little ability to acclimatize or adapt to increasing global SST. Repeated bleaching has occurred at various locations over the past 20 years without any apparent increase in ability of the corals to acclimatize or adapt to higher temperature. The weight of available evidence is that coral reefs will be subjected to major environmental changes in future decades (Pittock 1999). Reef corals are already living at their thermal limit, and undergoing unprecedented mass bleaching due to higher summer temperatures (Wilkinson et al. 1999; Chap. 26, Hoegh-Guldberg, this Vol.). For example, the 1998 bleaching event in Belize central shelf lagoon areas produced catastrophic mortality (Aronson et al. 2002). Based on an analysis of reef cores, this was the first bleaching-induced mass mortality in at least the last 3000 years. Corals have continued to flourish despite major environmental and sea level fluctuations in the past (Buddemeier and Smith 1999). However, the ability of coral reefs to cope with such disturbances in the past is no guarantee of continued resilience in the face of unprecedented and much more rapid anthropogenic stresses (Hughes 1994). For example, in Jamaica, the effects of overfishing, hurricane damage and disease combined to destroy most corals. A dramatic phase shift has occurred, producing a system dominated by fleshy macroalgae (Hughes 1994). A study in the Bahamas from 1994 to 1998 documented the impact of a massive bleaching event that resulted in a significant decline in coral abundance and a rapid transition to an alga-dominated community (Ostrander et al. 2002). In future scenarios of projected climate change, it is likely that coral community decline will be influenced as much by increases in the frequency and intensity of extreme events as well as being impacted by increases in means of environmental parameters of temperature, atmospheric CO_2 and sea level (Done 1999).

Do reef corals possess mechanisms that may allow them to rapidly adjust to changing global conditions? Information on this question has been compiled by Buddemeier and Fautin (1993), Buddemeier and Smith (1999), Coles and

Brown (2003) and Buddemeier et al. (Chap. 24, this Vol.). Reef corals have the ability to acclimatize (modify phenotype) in response to the wide range of conditions normally encountered in their natural environment. For example, Hawaiian reef corals adjust readily to their normal seasonal temperature ranging from 21 to 28 °C without difficulty. However, prolonged exposure to temperatures of 29–30 °C at high irradiance typically exceeds the tolerance limits set by their genotype, and further acclimatization does not occur. The corals bleach. The ability of corals to acclimatize over a wide range of normal temperatures does not translate into an ability to acclimatize or adapt to higher than normal temperatures. The acclimatization limits are fixed by genotype and can only be increased through adaptation. The process of adaptation requires natural selection over many generations and could take hundreds to thousands of years in long-lived organisms such as corals. Conservationists and others concerned with the preservation of coral reefs have proposed a broad pragmatic approach to the bleaching problem that circumvents the need to understand mechanisms. The focus is on the identification and protection of those reef areas with the highest probability of enduring global changes (Salm and Coles 2001). These areas are to be integrated into a global network of coral reefs that would maintain high diversity and biomass and replenish the more impacted reef areas. Certain environmental factors favor survival or recovery of corals and other affected organisms during mass bleaching events. There are strong between-species and between-genera differences in susceptibility of corals to bleaching and mortality (Berkelmans and Willis 1999). Done (1999) used the term "tolerance" to describe intrinsic ability of species or colonies to withstand stressful events. West and Salm (2003) use the term "resistance" to expand the term "tolerance" to include extrinsic local environmental factors (upwelling, turbidity, strong currents, high cloud cover, topography, etc.) that allow coral reef communities to remain healthy during regional bleaching events. They also propose the term "resilience" to describe the ability of reef areas to recover after bleaching events. Some reefs have been observed to recover very quickly after an acute event (Coles and Brown, 2003) and would be said to have high resilience. Others recover slowly or not at all and would be said to have low resilience. The resilience of a reef is a function of abiotic as well as biotic factors. Coral reefs showing high resistance and high resilience would be set aside and protected from stresses such as sedimentation, fishing and eutrophication. Presumably, such protection would thereby reduce the anthropogenic stress load from fishing, pollution, sedimentation, etc. and increase the ability of these reefs to cope with global warming. These protected areas would represent a reservoir of reef corals that could repopulate impacted areas and thereby reduce, but not prevent, global reef decline. The proposed reserve system is contingent on the uncertain political willingness and economic ability of the international community to set aside and protect large reef areas.

23.6 Unresolved Questions and Future Research

The phenomenon of bleaching has now been well documented, yet the capacity of coral populations to adjust to global warming remains unknown. Further, the mechanisms involved are not fully understood and may even involve infectious agents (see Chap. 25, Rosenberg, this Vol.). On the level of phenomenology, coral bleaching has provided a clear early warning signal of future climatic change and more widespread impacts on coral reefs. There are considerable uncertainties in predictions of future rates of global change as well as uncertainties in biological response (e.g., bleaching) to such change. Humanity is rapidly moving forward with an unwitting experiment that involves massive anthropomorphic change to our planet. Perhaps the only way to answer questions on the fate of reefs will be to document the success or failure of these reefs to adapt to change over the next century. There will be increasing research opportunities to study the causes and consequences of coral bleaching and reef response, and to further document the decline of reef systems. In the future, descriptions of the causes of coral bleaching must be based on measurement of a wide range of interacting environmental factors, including the primary factors of temperature duration, solar irradiance intensity, spectral quality and water motion. Additional stresses that can result from global climate change include higher and more variable temperature, more frequent storms, changes in ocean currents, decreased salinity due to increased precipitation, and sea level rise. Global changes in ocean carbonate chemistry due to increased anthropogenic CO_2 will probably lead to reduced calcification rates in corals (Kleypas et al. 1999). Other anthropogenic stressors include nutrification, overfishing, increased sedimentation, increased levels of pollutants (pesticides, heavy metals, chlorinated hydrocarbons, etc.), shoreline construction and other factors. The impact of biological factors such as increased rate of disease, increased dominance of reefs by macroalgae, and impact of introduced species are unknowns that will contribute to the complexity of the problem. The bleaching response of corals to temperature varies with changes in other environmental parameters. These interactions are most important near the limits of tolerance for a given parameter (Coles and Jokiel 1978). Thus, synergistic effects due to a combination of the above climatic, oceanic, biological and anthropogenic factors are likely to have the greatest effect on coral reef ecosystems (Pittock 1999), and need to be described and understood in more detail.

References

Abe N (1939) On the expansion and contraction of the polyp of a reef-coral, *Caulastraea furcata* Dana. Palao Trop Biol Stn Stud 1:651–669

Acevedo R, Goenaga C (1986) Note on coral bleaching after a chronic flooding in southwestern Puerto Rico. Caribb J Sci 22:225

Ambarsari I, Brown BE, Britton G, Cummings D (1997) Fluctuations in algal chlorophylls and carotenoid pigments during solar bleaching in the coral *Goniastrea aspera* at Phuket, Thailand. Mar Ecol Prog Ser 159:303–307

Anthony KRN, Hoegh-Guldberg O (2003) Kinetics of photoacclimation in corals. Oecologia 134:23–31

Aronson RB, Precht WF, Macintyre IG, Murdoch TJT (2000) Coral bleach-out in Belize. Nature 405:36

Aronson RB, Precht WF, Toscano MA, Koltes KH (2002) The 1998 bleaching event and its aftermath on a coral reef in Belize. Mar Biol 141:435–447

Baker A (2001) Reef corals bleach to survive change. Nature 411:765–766

Ben-Haim Y, Rosenberg E (2002) A novel *Vibrio* sp. pathogen of the coral *Pocillopora damicornis*. Mar Biol 141:47–55

Berkelmans R, Willis BL (1999) Seasonal and local spatial patterns in the upper thermal limits of corals on the inshore Central Great Barrier Reef. Coral Reefs 18:219–228

Bhagooli R, Hidaka M (2002) Physiological responses of the coral *Galaxea fascicularis* and its algal symbiont to elevated temperatures. Galaxea JCRS 4:33–42

Boschma H (1925) The nature of the association between anthozoa and zooxanthellae. Proc Natl Acad 11:65–67

Boschma H (1926) On the food of reef corals. Proc Acad Sci Amsterdam 29:993–997

Brown BE (1997) Coral bleaching causes and consequences. Coral Reefs 16:S129–Sl38

Brown BE, Dunne RP, Scoffin TP, Le Tissier MDA (1994) Solar damage in intertidal corals. Mar Ecol Prog Ser 105:219–230

Brown BE, Dunne RP, Chansang H (1996) Coral bleaching relative to elevated seawater temperature in the Andaman Sea (Indian Ocean) over the last 50 years. Coral Reefs 15:151–152

Brown BE, Dunne RP, Ambarsari I, Le Tissier MDA, Satapoomin U (1999a) Seasonal fluctuations in environmental factors and variations in symbiotic algae and chlorophyll pigments in four Indo-Pacific coral species. Mar Ecol Prog Ser 191:53–69

Brown BE, Ambarsari I, Warner ME, Fitt WK, Dunne Rp, Gibb SW, Cummings DG (1999b) Diurnal changes in photochemical efficiency and xanthophyll concentrations in shallow water reef corals: evidence for photoinhibition and photoprotection. Coral Reefs 18:99–105

Brown BE, Clarke KR, Warwick RM (2002a) Serial patterns of biodiversity change in corals across shallow reef flats in Ko Phuket, Thailand, due to the effects of local sedimentation) and regional (climatic) perturbations. Mar Biol 141:21–29

Brown BE, Downs CA, Dunne RP, Gibb SW (2002b) Exploring the basis of thermotolerance in the reef coral *Goniastrea aspera*. Mar Ecol Prog Ser 242:119–129

Bruno JF, Siddon CE, Witman JD, Colin PL, Toscano MA (2001) El Niño related coral bleaching in Palau, Western Caroline Islands. Coral Reefs 20:127–136

Buddemeier RW, Fautin DG (1993) Coral bleaching as an adaptive mechanism – a testable hypothesis. Bioscience 43:320–326

Buddemeier RW, Smith SV (1999) Coral adaptation and acclimatization: a most ingenious paradox. Am Zool 39:1–9

Celliers L, Schleyer MH (2002) Coral bleaching on high-latitude marginal reefs at Sodwana Bay, South Africa. Mar Pollut Bull 44:1380–1387

Cervino JM, Goreau TJ, Hayes RI, Kaufman L, Nagelkerken I, Paterson K, Porter JW, Smith GW, Quirolo C (1998) Coral disease. Science 280:499

Clausen CD, Roth AA (1975) Effect of temperature and temperature adaptation on calcification rate in the hermatypic coral *Pocillopora damicornis*. Mar Biol 33:93–100

Cohen AL, Lobel PS, Tomasky GL (1997) Coral bleaching on Johnston Atoll, central Pacific Ocean. Biol Bull 193:276–279

Coles SL (1985) The effects of elevated temperature on reef coral planula settlement as related to power station entrainment. Proc 5th Int Coral Reef Congr 4:171–176

Coles SL (1988) Limitations on reef coral development in the Arabian Gulf: temperature or algal competition? Proc 6th Int Coral Reef Symp 3:211–216

Coles SL, Jokiel PL (1977) Effects of temperature on photosynthesis and respiration in hermatypic corals. Mar Biol 43:209–216

Coles SL, Jokiel PL (1978) Synergistic effects of temperature, salinity and light on the hermatypic coral *Montipora verrucosa*. Mar Biol 49:187–195

Coles SL, Brown BE (2003) Coral bleaching – capacity for acclimatization and adaptive selection. Adv Mar Biol 45:183–223

Coles SL, Jokiel PL, Lewis CR (1976) Thermal tolerance in tropical versus subtropical Pacific reef corals. Pac Sci 30:155–166

Cook CB, Logan A, Ward J, Luckhurst B, Berg CJ Jr (1990) Elevated temperatures and bleaching on a high latitude coral reef: the 1988 Bermuda event. Coral Reefs 9:45–49

Craig P, Birkeland C, Belliveau (2001) High temperatures tolerated by a diverse assemblage of shallow-water corals in American Samoa. Coral Reefs 20:185–189

Darwin C (1859) On the origin of species by means of natural selection. Murray, London, 502 pp

Davies JM, Dunne RP, Brown BE (1997) Coral bleaching an elevated sea-water temperature in Milne Bay Province, Papua New Guinea, 1996. Mar Freshwater Res 48:513–516

Done TJ (1999) Coral community adaptability to environmental change at the scales of regions, reefs and reef zones. Am Zool 39:66–79

Dove SG, Hoegh-Guldberg O, Ranganathan S (2001) Major colour patterns of reef-building corals are due to a family of GFP-like proteins. Coral Reefs 19:197–204

Dunlap WC, Shick JM (1998) Ultraviolet radiation absorbing mycosporine-like amino acids in coral reef organisms: a biochemical and environmental perspective. J Phycol 34:418–430

Dunlap WC, Chalker BE, Oliver JK (1986) Bathymetric adaptations of reef-building corals at Davies Reef, Great Barrier Reef, Australia. III. UV-B absorbing compounds. J Exp Mar Biol Ecol 104:239–248

Dunne RP, Brown BE (2001) The influence of solar radiation on bleaching of shallow water reef corals in the Andaman Sea, 1993–1998. Coral Reefs 20:201–210

Edmondson CH (1946) Behavior of coral planulae under altered saline and thermal conditions. Bernice P Bishop Mus Occ Pap 18:283–304

Edmunds PJ, Gates RD, Gleason MG (2001) The biology of larvae from the reef coral *Porites asteroides* and their response to temperature disturbances. Mar Biol 139:981–989

Edwards AJ, Clark S, Zahir H, Rajasuriya A, Naseer A, Rubens J (2001) Coral bleaching and mortality on artificial and natural reefs in the Maldives in 1998, sea surface temperature anomalies and initial recovery. Mar Pollut Bull 42:7–15

Falkowski PG, Dubinski Z, Muscatine L, Porter JW (1984) Light and the bioenergetics of a symbiotic coral. Bioscience 24:705–709

Falkowski PG, Jokiel PL, Kinzie RA III (1990) Irradiance and corals. In: Dubinsky Z (ed) Coral reefs. Ecosystems of the world, vol 25. Elsevier, Amsterdam, pp 89–107

Fisk D, Done T (1985) Taxonomic and bathymetric patterns of bleaching in corals, Myrmidon Reef (Queensland). Proc 5th Int Coral Reef Congr 6:149–154

Fitt WK, Warner ME (1995) Bleaching patterns of four species of Caribbean reef corals. Biol Bull 189:298–307

Fitt K, Brown BE, Warner ME, Dunne RP (2001) Coral bleaching interpretation of thermal tolerance limits and thermal thresholds in tropical corals. Coral Reefs 20:51–65

Franzisket L (1970) The atrophy of hermatypic reef corals maintained in darkness and their subsequent regeneration in light. Int Rev Hydrobiol 55:1–12

Gates RD, Edmunds PJ (1999) The physiological mechanisms of acclimatization in tropical reef corals. Am Zool 39:30–43

Gates RD, Baghdasarian G, Muscatine L (1992) Temperature stress causes host cell detachment in symbiotic cnidarians: implications for coral bleaching. Biol Bull 182:324–332

Gleason DF, Wellington GM (1993) Ultraviolet radiation and coral bleaching. Nature 365:836–838

Glynn PW (1977) Coral growth in upwelling and nonupwelling areas off the Pacific coast of Panama. J Mar Res 356:567–585

Glynn PW (1991) Coral reef bleaching in the 1980s and possible connections with global warming. Trends Ecol Evol 6:175–178
Glynn PW (1993) Coral reef bleaching: ecological perspectives. Coral Reefs 12:1–17
Glynn PW, D'Croz L (1990) Experimental evidence for high temperature stress as the cause of El Nino-coincident coral mortality. Coral Reefs 8:181–191
Glynn PW, Mate JL, Baker AC, Calderon MO (2001) Coral bleaching and mortality in Panama and Equador during the 1997–1998 El Niño Southern Oscillation event: spatial/temporal patterns and comparisons with the 1982–1983 event. Bull Mar Sci 69:79–109
Goenaga C, Canals M (1990) Island-wide coral bleaching in Puerto Rico. Caribb J Sci 26:171–175
Goreau TJ (1964) Mass expulsion of zooxanthellae from Jamaican reef communities after Hurricane Flora. Science 145:383–386
Goreau TJ (1992) Bleaching and reef community change in Jamaica: 1951–1991. Am Zool 32:683–695
Goreau TJ, MacFarlane AH (1990) Reduced growth rate of *Montastrea annularis* following the 1987–1988 coral bleaching event. Coral Reefs 8:211–215
Guzman HM, Jackson JBC, Weil E (1991) Short-term ecological consequences of a major oil spill on Panamanian subtidal reef corals. Coral Reefs 10:1–12
Harrigan JF (1972) The planula larva of *Pocillopora damicornis*. Lunar periodicity of swarming and substratum behavior. PhD Thesis, University of Hawaii, Honolulu, 213 pp
Harriott VJ (1985) Mortality rates of scleractinian corals before and during a mass bleaching event. Mar Ecol Prog Ser 21:81–88
Harrison PL, Babcock RC, Bull GD, Oliver JK, Walace CC, Willis BL (1984) Mass spawning in tropical reef corals. Science 223:1186–1189
Hochachka PW, Somero GN (1984) Biochemical adaptation. Princeton Univ Press, Princeton, 537 pp
Hoegh-Guldberg O (1999) Climate change, coral bleaching and the future of the world's coral reefs. Mar Freshwater Res 50:839–866
Hoegh-Guldberg O, Smith GJ (1989) The effect of sudden changes in temperature, light and salinity on the population density and export of zooxanthellae from the reef corals *Stylophora pistillata* Esper and *Seriatopora hystrix* Dana. J Exp Mar Biol Ecol 129:279–303
Houck JE, Buddemeier RW, Smith SV, Jokiel PL (1977) The response of coral growth rate and skeletal strontium content to light intensity and water temperature. Proc 3rd Int Coral Reef Symp 2:425–431
Hudson JH (1981) Response of *Montastrea annularis* to environmental change in the Florida Keys. Proc Fourth Int Coral Reef Symp 2:233–240
Hughes TP (1994) Catastrophes, phase shifts and large scale degradation of a Caribbean coral reef. Science 265:1547–1551
Iglesias-Prieto R, Matta J, Robins WA, Trench RK (1992) Photosynthetic response to elevated temperature in the symbiotic dinoflagellate *Symbiodinium microadriaticum* in culture. Proc Natl Acad Sci USA 89:10302–10305
Jokiel PL (1978) Effects of water motion on reef corals. J Exp Mar Biol Ecol 35:87–91
Jokiel PL (1980) Solar ultraviolet radiation and coral reef epifauna. Science 207:1069–1071
Jokiel PL, Coles SL (1974) Effects of heated effluent on hermatypic corals at Kahe Point, Oahu. Pac Sci 28:1–18
Jokiel PL, Coles SL (1977) Effects of temperature on the mortality and growth of Hawaiian reef corals. Mar Biol 43:201–208
Jokiel PL, Guinther EB (1978) Effects of temperature on reproduction in the hermatypic coral *Pocillopora damicornis*. Bull Mar Sci 28:786–789
Jokiel PL, York RH Jr (1982) Solar ultraviolet photobiology of the reef coral *Pocillopora damicornis* and symbiotic zooxanthellae. Bull Mar Sci 32:301–315
Jokiel PL, Coles SL (1990) Response of Hawaiian and other Indo-Pacific reef corals to elevated temperature. Coral Reefs 8:155–162
Jokiel PL, Hunter CL, Taguchi S, Watarai L (1993) Ecological impact of a fresh water "reef kill" in Kaneohe Bay, Oahu, Hawaii. Coral Reefs 12:177–184

Jokiel PL, Lesser MP, Ondrusek ME (1997) UV-absorbing compounds in the coral *Pocillopora damicornis*: interactive effects of UV radiation, photosynthetically active radiation, and water flow. Limnol Oceanogr 42:1468–1473

Jones RJ (1997) Zooxanthellae loss as a bioassay for assessing stress in corals. Mar Ecol Prog Ser 149:163–171

Jones RJ, Steven AL (1997) Effects of cyanide on corals in relation to cyanide fishing on reefs. Mar Freshwater Res 48:517–522

Jones RJ, Hoegh-Guldberg O, Larkum WD, Schreiber U (1998) Temperature-induced bleaching of corals begins with impairment of the carbon dioxide fixation mechanism in zooxanthellae. Plant Cell Environ 21:1219–1230

Jones RS, Randall RH (1973) A study of biological impact caused by natural and man-induced changes on a tropical reef. Tech Rep 7. Marine Laboratory, University of Guam

Kawaguti S (1941) On the physiology of reef corals V. Tropisms of coral planulae, considered as a factor of distribution of the reefs. Palao Trop Biol Stn Stud 2:319–328

Kim CS, Lim WA, Cho YC (2001) Mycosporine-like amino acids as the UV sunscreen with oxygen radical scavenging activity. Bull Natl Fish Res Dev Inst Korea 60:65–71

Kinzie RA III, Chee GS (1979) The effect of different zooxanthellae on the growth of experimentally reinfected hosts. Biol Bull 156:315–327

Kinzie RA III, Jokiel PL, York RH Jr (1984) Effects of light of altered spectral composition on coral zooxanthellae associations and on zooxanthellae in vitro. Mar Biol 78:239–248

Kinzie RA III, Takayama M, Santos SR, Coffroth MA (2001) The adaptive bleaching hypothesis: experimental tests of critical assumptions. Biol Bull 200:51–58

Kleypas JA, Buddemeier RW, Archer D, Gattuso J-P, Langdon C, Opdyke BN (1999) Geochemical consequences of increased atmospheric carbon dioxide on coral reefs. Science 284:118–120

Kuffner IB (2001) Effects of ultraviolet (UV) radiation on larval settlement of the reef coral *Pocillopora damicornis*. Mar Ecol Prog Ser 217:251–261

Kushmaro A, Loya Y, Fine M, Rosenberg E (1996) Bacterial infection and coral bleaching. Nature 380:396

Kushmaro A, Rosenberg E, Fine M, Haim Y, Loya Y (1998) Effect of temperature on bleaching of the coral *Oculina patagonica* by *Vibrio* AK-1. Mar Ecol Prog Ser 171:131–137

Lang JC, Wicklund RI, Dill RF (1988) Depth and related habitat bleaching of zooxanthellate reef organisms near Lee Stocking Island, Exuma Cays, Bahamas. Proc 6th Int Coral Reef Symp 3:313–318

Labas YA, Gurskaya NG, Yanushevich YG, Fradkov AF, Lukyanov KA, Lukyanov SA, Matz MV (2002) Diversity and evolution of the green fluorescent protein family. Proc Natl Acad Sci USA 99:4256–4261

Lesser MP (1997) Oxidative stress causes coral bleaching during exposure to elevated temperatures. Coral Reefs 16:187–192

Lesser MP, Stochaj WR, Tapley WR, Schick JM (1990) Bleaching in coral reef anthozoans: effects of irradiance, ultraviolet radiation, and temperature on the activities of protective enzymes against active oxygen. Coral Reefs 8:225–232

Liu G, Skirving W, Strong AE (2003) Remote sensing of sea surface temperatures during 2002 Barrier Reef coral bleaching. EOS 84:137–144

Marshall PA, Baird AH (2000) Bleaching of corals on the Great Barrier Reef: differential susceptibilities among taxa. Coral Reefs 19:155–163

Meesters EH, Bak RPM (1993) Effects of coral bleaching on tissue regeneration potential and colony survival. Mar Ecol Prog Ser 169:295–301

Meehan WJ, Ostrander GK (1997) Coral bleaching: a potential biomarker of environmental stress. J Toxicol Environ Health 50:529–552

Michalek-Wagner K, Willis BL (2001a) Impacts of bleaching on the soft coral *Lobophytum compactum*. I. Fecundity, fertilization and offspring viability. Coral Reefs 19:231–239

Michalek-Wagner K, Willis BL (2001b) Impacts of bleaching on the soft coral *Lobophytum compactum*. II. Biochemical changes in adults and their eggs. Coral Reefs 19:240–246

Morse DE, Hooker N, Morse ANC, Jensen RA (1988) Control of larval metamorphosis and recruitment in sympatric agariciid corals. J Exp Mar Biol Ecol 16:193–217

Mumby PJ (1999) Bleaching and hurricane disturbances to populations of coral recruits in Belize. Mar Ecol Prog Ser 190:27–35

Mumby PJ, Chisholm JRM, Edwards AJ, Andrefouet S, Jaubert J (2001) Cloudy weather may have saved Society Island reef corals during the 1998 ENSO event. Mar Ecol Prog Ser 222:209–216

Mundy CN, Babcock RC (1998) Role of light intensity and spectral quality in coral settlement: implications for depth dependent settlement? J Exp Mar Biol Ecol 223:235–255

Muscatine L (1990) The role of symbiotic algae in carbon and energy flux in reef corals. In: Dubinsky Z (ed) Coral reefs. Ecosystems of the world. vol 25. Elsevier, Amsterdam, pp 75–87

Muscatine L, Porter JW (1977) Reef corals: mutualistic symbioses adapted to nutrient-poor environments. Bioscience 27:454–464

Nakamura T, van Woesik R (2001) Water-flow rates and passive diffusion partially explain differential survival of corals during the 1998 blenching event. Mar Ecol Prog Ser 212:301–304

Oliver JK (1985) Recurrent seasonal bleaching and mortality of corals on the Great Barrier Reef. Proc 5th Int Coral Reef Congr 4:201–206

Ostrander GK, Armstrong KM, Knobbe ET, Gerace D, Scully EP (2002) Rapid transition in the structure of a coral reef community: the effects of coral bleaching and physical disturbance. Proc Natl Acad Sci USA 97:5297–5302

Philipp E, Fabricius K (2003) Photophysiological stress in scleractinian corals in response to short-term sedimentation. J Exp Mar Biol Ecol 287:57–78

Phongsuwan N (1998) Extensive coral mortality as a result of bleaching in the Andaman Sea in 1995. Coral Reefs 17:70

Pittock AB (1999) Coral reefs and environmental change: adaptation to what? Am Zool 39:10–29

Raimondi PT, Morse ANC (2000) The consequences of complex larval behavior in a coral. Ecology 81:3193–3211

Richmond RH (1985) Reversible metamorphosis in coral planula larvae. Mar Ecol Prog Ser 22:181–185

Richardson LL, Goldberg WM, Kuta KG, Aronson RB, Smith GW, Ritchie KB, Halan JC, Feingold JS, Miller SL (1998) Florida's mystery coral-killer identified. Nature 392:557–558

Rowan R, Knowlton N, Baker A, Jara J (1997) Landscape ecology of algal symbionts creates variation in episodes of bleaching. Nature 388:265–269

Salih A, Larkum A, Cox G, Kuhl M, Hoegh-Guldberg O (2000) Fluorescent pigments in corals are photoprotective. Nature 408:850–853

Salm RV, Coles SL (eds) (2001) Coral bleaching and marine protected areas. Report no 0102. The Nature Conservancy, Honolulu, Hawaii, 118 pp

Steen GR, Muscatine L (1987) Low temperature evokes rapid exocytosis of symbiotic algae by sea anemone. Biol Bull 172:246–263

Stimson J, Sakai K, Sembali H (2002) Interspecific comparison of the symbiotic relationship in corals with high and low rates of bleaching-induced mortality. Coral Reefs 21:409–421

Strong AE, Barrientos CS, Duda C, Sapper J (1997) Improved satellite techniques for monitoring coral reef bleaching. Proc 8th Int Coral Reef Symp 2:1495–1498

Szmant AM, Gassman NJ (1990) The effects of prolonged "bleaching" on the tissue biomass and reproduction of the reef coral *Montastrea annularis*. Coral Reefs 8:217–224

US Department of Commerce (1970) Surface water temperature and density. Pacific Coast. NOS Publ 31-3, 3rd edn. US Department of Commerce, Rockville, MD, 88 pp

Vaughan TW (1914) Reef corals of the Bahamas and of southern Florida. Carnegie Inst Wash Year Book 13:222–226

Warner ME, Fitt WK, Schmidt GW (1996) The effects of elevated temperature on the photosynthetic efficiency of zooxanthellae *in hospite* from four species of reef coral: a novel approach. Plant Cell Environ 19:291–299

Warner ME, Fitt WK, Schmidt GW (1999) Damage to photosystem II in symbiotic dinoflagellates: a determinant of coral bleaching. Proc Natl Acad Sci USA 96:8007–8012

Wellington GM, Glynn PW, Strong AE, Nauarrete SA, Wieters E, Hubbard D (2001) Crisis on coral reefs linked to climate change. EOS 82:1–7
Wells JW (1957) Corals. Mem Geol Soc Am Memoir 67:1087–1089
West JM, Salm RV (2003) Resistance and resilience to coral bleaching: implications for coral reef conservation and management. Conserv Biol 17:956-967
Wilkinson C, Linden O, Cesar H (1999) Ecological and socioeconomic impacts of 1998 coral mortality in the Indian Ocean: an ENSO impact and a warning of future change? Ambio 28:188–196
Williams EH Jr, Bunkley-Williams L (1990) The world-wide coral reef bleaching cycle and related sources of coral mortality. Atoll Res Bull 335:1–71
Winter A, Appeldorn RS, Bruckner A, Williams EH Jr, Goenaga C (2001) Sea surface temperatures and coral reef bleaching off La Parguera, Puerto Rico (northeastern Caribbean Sea). Coral Reefs 17:377–382
Yonge C M, Nicholls AG (1931a) Studies on the physiology of corals. IV. The structure, distribution and physiology of the zooxanthellae. Sci Rept Gr Barrier Reef Exp 1:152–176
Yonge C M, Nicholls AG (1931b) Studies on the physiology of corals. V. The effect of starvation in light and in darkness on the relationship between corals and zooxanthellae. Sci Rep Great Barrier Reef Exp 1:178–210

24 The Adaptive Hypothesis of Bleaching

Robert W. Buddemeier, Andrew C. Baker, Daphne G. Fautin, and J. Rebecca Jacobs

24.1 Introduction

24.1.1 Biological Background

Despite the perception that corals and coral reefs are limited to stable habitats distinguished by very narrow environmental parameters, the coral–algal symbiosis is capable of surviving under a variety of extreme conditions. Through the process of photoadaptation, corals and their algal symbionts adjust algal densities and pigment concentrations to function over a wide range of light levels ranging from direct exposure to full sunlight in intertidal corals to virtual darkness at the extreme limits of the photic zone (>200 m) on reef slopes (Zahl and McLaughlin 1959; Schlichter et al. 1986). Corals and reef communities in some areas (such as the Arabian Gulf) tolerate salinity and temperature conditions that are lethal when imposed rapidly on the same species in less extreme environments (Coles 1988; Sheppard 1988; Coles and Fadlallah 1991; Chap. 23, Jokiel, this Vol.). There are abundant reports of reef corals occurring in turbid, high nutrient, nearshore habitats (Larcombe et al. 2001). Coral reefs exist at the inherently variable interface between the sea, air and land (Smith and Buddemeier 1992), and reef communities have persisted over geological time through significant climate and sea-level fluctuations. Despite this, rates of speciation and extinction in scleractinian corals have been relatively low over the last 220 million years (Veron 1995).

Because the mechanisms by which reef corals adjust to these diverse environments are only poorly understood, we have little basis for estimating or predicting actual or potential rates of adaptive change. This issue is of particular interest because of the present high rates of coral mortality and reef degradation (e.g., Hoegh-Guldberg 1999; Wilkinson 2000) caused in substantial measure by the two large-scale threats of bleaching and other diseases.

24.1.2 Conceptual Background

Three lines of evidence concerning coral–algal symbioses converged in the early 1990s, concurrently with rising concern over the increasingly widespread phenomenon of frequent and severe coral bleaching. One line of evidence derived from molecular studies of the algal symbionts (often referred to as zooxanthellae) in scleractinian corals that revealed extraordinary diversity in the genus *Symbiodinium* (Rowan and Powers 1991). These findings reinforced the significance of the second line of evidence, consisting of earlier reports that various types of symbiotic algae could form metabolically different symbioses with the same host (e.g., Kinzie and Chee 1979). The third line of evidence arose from reports of variations in bleaching susceptibility, occurrence, and recovery (e.g., Oliver 1985; Jokiel and Coles 1990). Arguing for a consistent explanation for the combination of these observations, Buddemeier and Fautin (1993) proposed the "adaptive bleaching hypothesis" (ABH). A basic precept of the ABH is that changing combinations of hosts and symbiotic algae have the potential to create "new" ecospecies (see Sect. 24.2.1) that differ in environmental tolerances. Diversity and flexibility of partnerships within a host organism, occurring over small spatial scales and short (single generation) time scales, could explain differences from one population to another of corals (for example, in bleaching threshold temperature: e.g., Coles et al. 1976; Coles and Jokiel 1977). Buddemeier and Fautin (1993) interpreted these observations as evidence of differences in the ecospecies (see Sect. 24.2.1): what appeared in each case to be a single coral species was physiologically not the same.

Ware et al. (1996) used a simple competition model to explore the parameters underlying the ABH, and emphasized that disturbance frequency is an important factor in determining possible outcomes. Subsequently, Buddemeier et al. (1997) placed the ABH in the context of a more conventional, but still very rapid, form of adaptation, based on the potential for somatic mutation to produce genetic chimeras in colonial corals. In such chimeras, selection could potentially act on multiple animal genotypes within a single colony, with each animal genotype possibly interacting with a different type of photosymbiont. Beginning in 1999, the widespread application of molecular methods to identify algal symbionts, combined with increasingly intense bleaching events worldwide following the 1997–1998 El Niño, led to a relatively sudden proliferation of studies relating to the ABH, its mechanisms, or its underlying assumptions.

24.1.3 Aim and Scope

The purpose of this paper is to update the relevant literature, and to review some basic features of the ABH in order to clarify some confusion and misunderstanding that has become apparent since the original phrasing of the hypothesis.

The ABH has clear relevance for coral symbioses involving both eukaryotic and prokaryotic microbes. Accordingly, discussions of the relevance of the ABH to continued coral health can be framed in the context of both coral bleaching and microbial and viral pathogenic disease (which may initiate or be initiated by bleaching). The microbial community associated with corals is diverse and complex, having features in common with the coral-algal symbiosis (Rohwer et al. 2001, 2002; Knowlton and Rohwer 2003). As is the case with the bleaching of endosymbiotic algae, environmental change may elicit compensatory change in microbial symbiont communities that shift the degree to which each partner benefits from and contributes to the symbiosis.

24.2 Adaptive Bleaching: Characteristics and Definitions

24.2.1 Terminology

We use the term "ecospecies" to refer to any combination of host and symbiont(s). The introduction of this term implicitly assumes that conspecific hosts with different complements of symbionts (or conspecific symbionts in different hosts) may, and in at least some cases will, differ in environmental tolerances. It does not necessarily imply that each partner is a single entity; since multiple symbiont taxa may occur at different relative abundances within a single host, one functional ecospecies may differ from another either in kinds or in relative proportions of symbionts.

The ABH infers mechanisms from phenomena, rather than deducing outcomes from known mechanisms. One of the major phenomena under consideration is "adaptation" in the vernacular (non-Darwinian) sense: "...modification of an organism or its parts in a way that makes it more fit for existence under the conditions of its environment..." (Merriam-Webster 1997). Mechanistically, this may subsume any combination of what biologists define as acclimation, acclimatization, or adaptation. A change in ecospecies is a biological adaptation in the sense that the genetic composition of the holobiont is altered, but this adaptation involves neither speciation nor extinction, and occurs over very short timescales (within the lifetimes of the partners).

24.2.2 Diversity in Taxonomy, Scale, and Environment

The diversity of organisms involved, their differences in selectivity and mode of acquisition of symbiotic partners, and the degree to which their associations are obligate or facultative (Fig. 24.1) preclude easy generalizations from experimental observations in a limited number of organisms or locations. Experimental tests are necessary, but are of limited power individually.

Moreover, the symbiotic or physiological characteristics of any given ecospecies are not necessarily directly predictable from observations made on individual partners in isolation (e.g., in culture) or in other combinations. Distinguishing ecospecies from one another is further complicated by uncertainties in determining functional diversity of symbionts from (predominantly molecular)

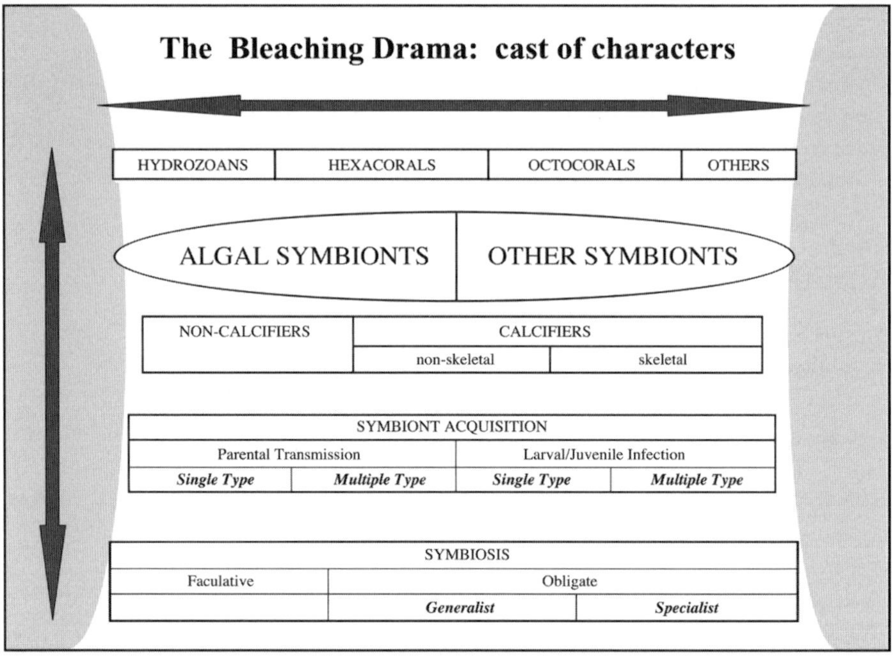

Fig. 24.1. *Symbiodinium* is found in a wide range of phylogenetically diverse host taxa, including "corals" (in the broadest sense) and their cnidarian relatives, both calcifying and noncalcifying. In addition, bivalves, sponges, and foraminifers are also hosts (Trench 1992; Baker 2003 for review). Various hosts may be facultatively or obligately symbiotic, and hosts and algal partners may exhibit varying degrees of selectivity. These diverse hosts differ not only in structure, habitat and life history, but also in their modes of symbiont acquisition

systematic data. Unambiguous identification of hosts can also be problematic. In the case of scleractinian corals, hybridization and reticulate evolution (e.g., Veron 1995; van Oppen et al. 2000, 2002), and potentially rapid rates of somatic mutation (Buddemeier et al. 1997; Fautin 1997) may result in functionally different, but taxonomically indistinguishable populations or individuals of the same apparent "species."

In addition to taxonomic uncertainties, there are uncertainties about the relationships among the environmental parameters that humans measure and the environment to which the ecospecies respond, and to which they are or are not adapted. Temperature and light (or proxies such as latitude and depth) are the variables most commonly investigated, but water motion (Nakamura and van Woesik 2001), water quality (including salinity and nutrients; e.g., Fagoonee et al. 1999), and possibly predation (Augustine and Muller-Parker 1998) all contribute to the environmental conditions experienced by intact associations. In addition, bacterial and viral communities present in these associations may constitute an important environmental component affecting the symbioses (see Chaps. 14, 15, 20 by Rohwer, Smith, and Bythell this Vol.). Interpreting environmental gradients in the context of single- or dual-variable changes could oversimplify the real picture.

The ABH can be envisioned as a process operating at either of two environmental end states, or along a continuum between them. One end state is a discrete change in environmental state, such as might occur with a climate-driven shift in major current patterns, or by alteration of lagoonal flushing and circulation due sea-level change crossing a threshold datum. Since the baseline condition is changed, a new and more favorable ecospecies formed in response to the stress at the onset of the new condition is likely to persist. By contrast, a stress and recombination event followed by return to the previous norm has a higher probability of reverting to the previously stable ecospecies. The transplantation experiments of Baker (2001) represent a step-function change, in contrast to many of the laboratory experiments that have compared the relative stability of homologous and heterologous infections under conditions known to be suitable for the homologous ecospecies.

The other end-state type of environmental change consists of rapid oscillations around a stable baseline of environmental conditions. Adaptation in this case is not to average conditions, or even to average conditions between stress (e.g., bleaching) events, but primarily to the frequency and intensity of stresses. This is exemplified in some of the model results of Ware et al. (1996).

In nature, environmental change can partake of both characteristics; at present, climate change is changing baseline conditions at rates that are relatively rapid on the scale of at least some host species' lifetimes, and concurrently, the present enhanced ENSO regime is imposing higher frequencies and intensities of extreme conditions than has been the case in the past. These phenomena are not the same, or even necessarily related, and they may put organisms and symbioses under different – but concurrent – selective pressures.

24.2.3 Stochasticity

Figure 24.2 illustrates the diverse mechanisms and processes that might be involved in adaptive bleaching. We use the analogy of a lottery to describe how chance may play an important role in determining the types of algae available to, and taken up by, potential hosts once the need or opportunity to re-equilibrate symbiotic relationships arises. Here, we consider only the odds of winning on a single lottery ticket; the question of how many tickets are purchased, or how often, is considered in the preceding section as part of the question of frequency and intensity of disturbance.

Some partners may be highly selective, or may be limited by modes of symbiont transmission or infection that make them ineligible to participate in the algal lottery at a given time or place. For those that do have a chance in the lottery, variations in abundance and condition of potential partners under the external (environmental) conditions determine the probabilities of successfully obtaining new algae that are either the same as or different from the pre-

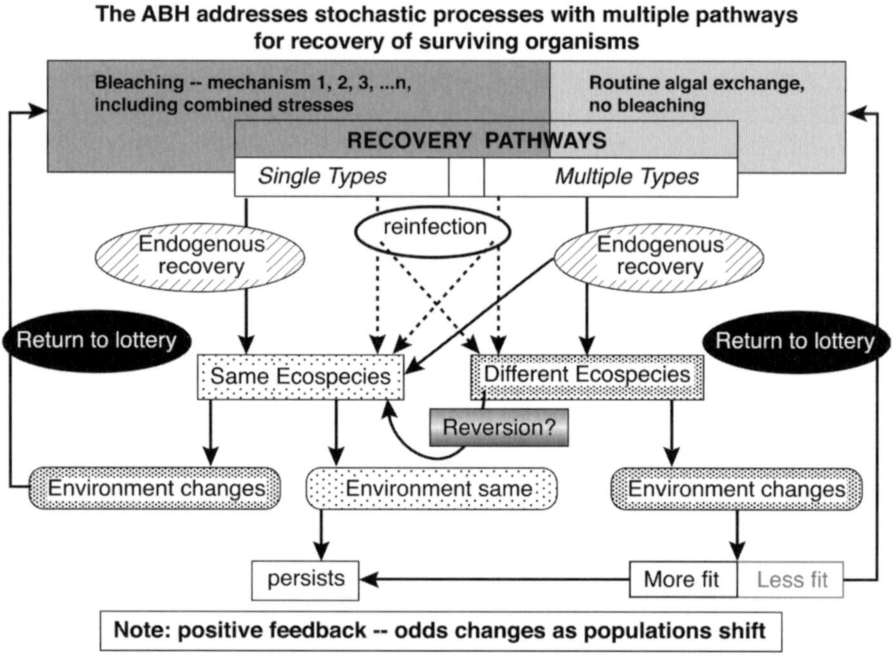

Fig. 24.2. Illustration of potential processes involved in transition from an initial to a final (different) ecospecies, depending on the characteristics and history of the original ecospecies, the local ecosystem (availability of symbionts), and the characteristics of the environment over time. The stochastic nature of the process is emphasized, and some of the complexity illustrated in Fig. 24.1 is omitted in the interests of relative clarity. The "reversion" process under stable environmental conditions interrupted only by occasional stress events is an example of an ecospecies transition that may occur without perceptible bleaching; it is discussed in Section 24.2.3

vious (dominant) type(s). However, distinguishing internal proliferation of existing symbiont populations from de novo acquisition of symbionts from the environment is difficult or impossible in most cases, and is immaterial to whether the ABH operates.

A particular bleaching event is not necessarily adaptive for any given individual because adaptive characteristics may be detectable only at the population scale. Adaptive change may occur over the course of multiple repetitive stress events, and may take several generations to become established at the community or ecosystem level. For this reason, the ABH is not readily falsifiable by limited individual observations – most specific cases will probably not document "adaptation" by recording change in symbiotic partnerships. Although unsatisfying for the experimental ecologist, this simply reflects the processes and scale of operation of this phenomenon.

Figure 24.2 also illustrates that significant change in the dominance of symbiotic algae types in hermatypic corals can routinely occur in the absence of observable bleaching or stress (e.g., Gates 1990; Brown et al. 1999; Fagoonee et al. 1999; Fitt et al. 2000; see also Sect. 24.4.2). Such shifts may occur in corals with multiple types of algal symbionts as a result of gradual environmental change, or of gradual competitive takeover by an alga that is slightly better suited to a stable environment than its co-symbionts. Similar fluctuations in symbiont populations occur in tropical sea anemones, but apparently not in their temperate counterparts (reviewed by Muller-Parker and Davy 2001). Although the mechanisms and outcomes would be those of the ABH, a focus on routine rather than extraordinary transitions between ecospecies would emphasize rather different ecological and evolutionary implications. The ABH was initially conceived to explain two obvious features of coral reefs and reef coral biology: the paradox of reefs seeming robust in the long term, but fragile in the short term; and what seems on its face a maladaptive response to trauma – the breakdown of a vital mutualism. Obvious bleaching may simply be an extreme manifestation of a mechanism that allows both partners to track the environment by continually making adjustments that shift ecospecies. Focus on the coral – and especially on the skeleton – gives the impression that change is slow and rare. Now that we can differentiate members of the other element in the symbiosis, we can assess the critical element, the ecospecies.

24.3 Accumulating Evidence

24.3.1 Taxonomic and Functional Diversity

24.3.1.1 Hosts

Particular attention has been paid to scleractinian corals in the context of bleaching and climate change because contemporary reefs are formed predominantly by about 750 species of this group (Veron 2000). However, sea anemones, soft corals, hydrozoans, bivalves, foraminifera and possibly sponges are also hosts to various types of *Symbiodinium* that are in many cases indistinguishable from those found in scleractinian hosts. Thus, a conservative estimate of the number of animal taxa that serve as hosts for endosymbiotic algae would be in the order of a few thousand; these occupy a wide range of habitats and form the reservoir of algal types for potential exchange and repopulation.

Multiple hosts, multiple types of algae (even within single host species) and the diverse habitats in which these organisms are found suggests a diverse reservoir of symbiotic algae types in almost every region where symbiotic organisms are common.

24.3.1.2
Symbiotic Algae

Since the original proposal of the ABH, the number of distinct types of *Symbiodinium* has expanded at an accelerating rate (see Baker 2003 and Chap. 8, Baker, this Vol., for review). Rowan and Powers (1991) initially documented five or six types in three clades identified as A, B, and C, with between-clade genetic distances comparable to those between some nonsymbiotic dinoflagellates in different orders. In less than 10 years, distinctions rose to 10–20 types (Rowan 1998; Baker 1999), and then, in less than another 5 years, to 50–60+ types (see Chap 8, Baker, this Vol.; Carlos et al. 1999; Pochon et al. 2001; LaJeunesse 2002; Santos et al. 2002a, b). Additional diversity clearly remains to be documented as both sampling and analytical techniques are further refined. Thus, an inventory in the order of hundreds of distinguishable taxa is probably not unrealistic in the near future. Rowan's (1998) remark that our estimates of diversity probably represent minima is as true now as when he wrote it: "By these methods [rDNA analysis], the observed number of taxa is a minimum estimate of the actual number and failure to reject the null hypothesis ['symbionts collected within one host species are not different'] cannot be regarded as evidence to support it."

Another challenge is distinguishing functionally different subpopulations. Some studies using relatively conserved molecular methods have failed to resolve algal taxa that clearly differ functionally in terms of their response to environmental parameters (Kinzie et al. 2001; Savage et al. 2002b). However, Santos et al. (2002a), using relatively variable chloroplast gene sequences to identify functional subpopulations described by Kinzie et al. (2001), found that cultures with the same responses had identical sequences, while distinctive responses were associated with unique sequences.

24.3.2
Existence of Distinguishable, Changeable Partnerships

24.3.2.1
Availability and Infectivity of Algal Symbionts

Current evidence indicates that ambient algae populations exist in seawater (Kinzie et al. 2001; Y. Benayahu, pers. comm., F. Rohwer, pers. comm.), that even recently expelled algae may remain sufficiently competent to participate in the algal lottery depicted in Fig. 24.2 (Ralph et al. 2001; Perez et al. 2001), and that there are vectors for transmission of algae to bleached hosts (Muller-Parker 1984).

Laboratory studies have demonstrated that aposymbiotic (bleached) hosts can become infected with "heterologous" symbionts isolated from hosts of other species (Kinzie 1974; Kinzie and Chee 1979; Schoenberg and Trench 1980; Colley and Trench 1983; Fitt 1985; Davy et al. 1997; Kinzie et al. 2001;

Toller et al. 2001b), thereby establishing that a single host can associate with various symbionts. Selectivity in these partnerships, however, may exist (Weis et al. 2001), and by analogy with other symbioses, it is likely that certain combinations of hosts and symbionts are impossible.

Field studies have detected multiple symbiont taxa in individual coral colonies as well as within single host species (Rowan and Knowlton 1995; Baker and Rowan 1997; Goulet and Coffroth 1997; Rowan et al. 1997; Baker 1999; Carlos et al. 2000; Darius et al. 2000; Toller et al. 2001b; LaJeunesse 2002). A single coral species may host different complements of algal symbionts in different parts of its geographic range (Loh et al. 2001; Rodriguez-Lanetty et al. 2001; van Oppen et al. 2001), or have the same suite of algae, but in different ratios. This diversity is often distributed over predictable environmental gradients, particularly light. Rowan (1998) suggested that "...polymorphic symbioses may be a common feature of holobiont biology."

An important caveat in discussing diversity is the potential for discrepancy between freshly separated and cultured algal symbionts (e.g., Rowan 1998; Carlos et al. 1999; Santos et al. 2001; Toller et al. 2001b; LaJeunesse 2002). LaJeunesse (2002) stated that "...in most instances the cultured type is not the symbiont that populates the host in nature ... hosts may contain low abundances of *Symbiodinium* spp. not detected by current molecular methods." The apparently low abundance of unusual types of *Symbiodinium* found in many hosts may be the result of laboratory parallels of the concept of adaptive bleaching during the culture process. Differential inoculation of *Symbiodinium* into culture or differential success of algal symbionts in culture (like the infection or acquisition of different symbionts by different hosts), combined with environmental conditions during culture that differ between experiments or laboratories (notably temperature and light: Rowan 1998; Kinzie et al. 2001), can control competitive interactions to allow the emergence of what had been minor symbionts in vivo as dominants in vitro.

The important question of the spatial distribution of multiple concurrent symbionts, especially for minor types, is not resolved. Rowan et al. (1997) provided dramatic illustration of niche partitioning by different algal symbionts within the same host. In this case, as in other studies of within-colony microhabitats, three distinct *Symbiodinium* taxa occupied distinct irradiance zones within the host colony. Macroscopic separation has also been studied in sea anemones which do not form colonies: in a species symbiotic with both a chlorophyte and *Symbiodinium*, proportions of the two phytosymbionts in an anemone differs with intertidal height (McCloskey et al. 1996; Bates 2000). Within individual anemones, proportions of the same two photosymbionts vary systematically as a function of location; the ratio of *Symbiodinium* to chlorophyte is systematically higher in the tips of tentacles than at their bases, and in the oral disc compared to the column (G. Muller-Parker, pers. comm.) Whether populations of algal symbionts are separated or intermingled has very different implications for algal competition and dominance shifts. Spatial patterning implies that the algae assort or are assorted in response to environmental parameters and the potential for direct competition is minimal, whereas microsympatry implies the opposite.

The widely held view that there is exclusive host and/or algal specificity is not representative of many symbioses, and the degree to which strictly specific symbioses may exist is not yet known. Most symbiotic systems appear to exhibit a range from specialists to generalists in both (all) partners, so it is reasonable to assume that some strictly specific relationships will be found. Studies that have reported a single symbiont taxon occurring in a given host species (Coffroth et al. 2001; Belda-Baillie et al. 2002; Diekmann et al. 2002) have often relied on relatively low-resolution rDNA analyses that are imprecise in their identifications.

In addition, there is evidence of hosts that can associate with several symbiont types as juveniles, but only one type as adults. Coffroth et al. (2001) found that juveniles of the octocoral *Plexaura kuna* could simultaneously associate with several symbionts, but individual adults appear to host just one, although that type varies from individual to individual (Goulet and Coffroth 2003). The evidence that hosts will phagocytize algae of at least several types suggests that if there is recognition of particular types by the host, it occurs after the symbiont enters a host cell (Schwarz et al. 2002).

24.3.3 Characteristics of Ecospecies

24.3.3.1 Existence of Environmental/Geographic Gradients

Environmental correlates of symbiont distributions and differences in sensitivity to stress-induced bleaching have been observed since the mid-1990s. Initial studies in the Caribbean suggested that in the *Montastrea* species complex, members of *Symbiodinium* in clades B and C were "sun" and "shade" specialists, respectively (Rowan and Knowlton 1995; Rowan et al. 1997), and members of clade A were stress-insensitive and weedy (Rowan 1998; Toller et al. 2001b). Extension of this work (Baker 1999; Toller et al. 2001a; LaJeunesse 2002) revealed additional symbiont taxa exhibiting ecological distributions in the Caribbean, although some site- and/or taxon-specific studies (Diekmann et al. 2002; Savage et al. 2002b), found no consistent environmental correlates.

In the tropical Indo-Pacific, initial observations of the ubiquity of clade C (Baker and Rowan 1997; Baker 1999) have been supported (Loh et al. 1998; Pochon et al. 2001; van Oppen et al. 2001; LaJeunesse et al. 2003; Baker 2003), but with increasing recognition of the occurrence of an additional clade (D) and of significant diversity within C. In the western Pacific, multiple geographic and systematic surveys have revealed both biogeographic and habitat- or host-specific distribution patterns for symbiont types (Loh et al. 2001; Rodriguez-Lanetty et al. 2001; van Oppen et al. 2001). In addition, emerging evidence suggests that clade D is relatively stress-resistant (Baker 1999; van

Oppen et al. 2001; LaJeunesse 2001), dominates in extreme environments, and tends to replace members of clade C in Indo-Pacific areas such as the eastern Pacific and the western Indian Ocean that have been affected by severe bleaching episodes (Baker 2003; see Chap. 8, Baker, this Vol.).

24.3.3.2
Evidence for Ecospecies Transitions

It has been shown that unusual host-algal symbiont combinations can become established following bleaching (e.g., Toller et al. 2001b), that both algae and the ecospecies they form with various hosts can be associated with specific environmental conditions or habitats, and that the individual mechanisms required for the ABH to function exist and operate in nature (this review, but see also Kinzie et al. 2001). A null hypothesis of the ABH could be expressed as: "hosts that recover from bleaching will always have the same type(s) of algal symbiont(s) (in the same proportions) as before the bleaching". This hypothesis has clearly been falsified. However, while data demonstrating the potential of the ABH and providing support for its existence have been reported, detailed demonstration of its importance or occurrence in the field is more difficult to obtain.

Field data indicate that El Niño-related bleaching on eastern Pacific reefs was much worse during the 1982–1983 event than the 1997–1998 event, even though temperature excursions during the two events were similar (Glynn et al. 2001; Guzmán and Cortés 2001; Podestá and Glynn 2001). Although factors other than temperature also affect bleaching, the difference in responses to these two events supports the general idea that corals can adapt to higher temperatures over decadal time scales, either through adaptive bleaching or through evolutionary selection for more heat/irradiance-tolerant individuals that survive bleaching events (Baker 2002, 2003). The fact that symbiont communities play a role in this adaptation is demonstrated by the observation that the regionally dominant algal clade had shifted from C to D over the course of the observations (Glynn et al. 2001).

In a field transplant experiment, Baker (2001) demonstrated that corals could bleach and recover with different *Symbiodinium* communities that were more suited to the new environmental conditions. There are also in progress some field studies and monitoring efforts (various pers. comm.; Fitt, unpubl.; Baker, unpubl.) which promise to provide colony-specific observations over time, but these have not yet reached the stage of publication.

24.4 Discussion and Development

24.4.1 The Adaptive Bleaching Hypothesis in Its Original Sense – Evidence for Existence

We assert that adaptive bleaching as originally proposed (Buddemeier and Fautin 1993) occurs to varying degrees and over varying time scales in various hosts. All of the most fundamental assumptions have been tested (Kinzie et al. 2001; Toller et al. 2001a, b; Santos et al. 2002a), several specific mechanisms have been individually demonstrated (Fitt et al. 2000, 2001), and some persuasive (but not rigorously conclusive) examples of field adaptation have been reported (Baker 2001; Glynn et al. 2001). We suggest that the level of available evidence justifies its acceptance as a working hypothesis, with the proviso that its stochastic nature must be recognized – it can be expected to function at the community and ecosystem level over multi-year scales, but it cannot provide deterministic predictions about the outcomes of specific events or the fate of individual specimens (or perhaps even taxa).

The immediate concern is the degree to which the ABH or its component features buffer reef systems against repeated stress episodes capable of inducing severe bleaching, and the absolute limits to ecospecies adaptation by recombination. Coupled with, and underlying, this immediate issue is the question of the larger evolutionary and ecological roles of flexible, multi-partner symbioses and the inevitable range of flexibility and specificity they exhibit.

24.4.2 Symbiotic Adaptation as a General Phenomenon

Certain aspects of the ABH that were implicit in its original formulation have become more central to the hypothesis with the discoveries of the last decade. Growing evidence for concurrent multiple infections, with a variety of viable symbiont types existing at very low levels within many hosts, shifts emphasis from a narrow window of reinfection opportunity following recovery from bleaching, and changes the odds in the lottery of Fig. 24.2 by suggesting that some hosts may be able to sample the environment and stockpile seed populations of alternative symbionts over time (the "moonlighting" symbionts of Baker 2003).

Catastrophic bleaching may merely represent the extreme end of a continuum that includes natural fluctuations in symbiont standing stocks over seasonal timescales, and catastrophic (or even visually detectable) bleaching (Fitt et al. 2000, 2001) may not be required for adaptive or acclimatory responses in ecospecies (Fitt et al. 2000, 2001).

The ABH has occasionally been misinterpreted and misapplied as being a deterministic process, or one that is relevant only to major bleaching events that result in prompt acquisition of heterologous algae from the external environment. Growing evidence suggests that the ABH is a synthetic hypothesis that is based on a number of relatively well established processes that occur with some frequency in photosymbiotic invertebrates. It is timely and appropriate to seek a set of terms and concepts that will further understanding of both the theoretical and applied aspects of the role of symbioses, and especially of changeable or multiple symbioses, in adaptive, acclimative, and evolutionary processes.

24.5 Conclusions and Recommendations

24.5.1 Summary

We offer the following conclusions from the foregoing review:

- There is abundant and rapidly expanding evidence for the existence of multiple types (taxa) of algal symbionts, with a wide range of host-systematic preferences and environmental optima,
- Many – perhaps most – hosts can support internal populations of multiple algal types, concurrently and/or serially. Transmission and/or infection mechanisms to establish these populations exist,
- The ecospecies concept, whereby different host–symbiont combinations exhibit different responses to environmental conditions, is strongly supported, and,
- The adaptive bleaching hypothesis, as a stochastic process operating over large population scales, can be regarded as confirmed in principle, but future tests and applications of the hypothesis can most profitably be viewed as an introduction to the larger issues of ecospecies roles in evolution and ecology over multiple scales.

24.5.2 Unresolved Questions and Future Research

24.5.2.1 Photosymbiont and Ecospecies Issues

Algal taxonomy, and especially understanding of the relationship between genetic and functional diversity, has greatly improved over the last decade, but remains a limiting factor in this research. Further surveys of the temporal stability of symbiotic associations in nature and of the taxonomic and environmental ranges of ecospecies and their component partners are needed.

Growing appreciation for the ecospecies concept suggests potential for a new generation of careful laboratory experiments to determine the physiological responses of different ecospecies and the underlying basis for documented examples from the literature (e.g., Coles et al. 1976; Coles and Jokiel 1977). Such experiments would be critical for understanding the fundamental environmental limits to symbiosis-based adaptation and acclimatization.

The actual nature, extent, distribution, and stability of multiple symbiosis with various types of photosymbionts need further investigation, in large part because ecological interpretations of selectivity and specificity in patterns of distribution (e.g., Loh et al. 2001; Rodriguez-Lanetty et al. 2001; van Oppen et al. 2001) may depend heavily on the relative importance of symbiont availability, partner preferences, and algal competitive advantages under various interacting combinations of environmental variables.

24.5.2.2 Disease and Microbial Symbiosis Issues

This volume provides an important linkage between the conceptually parallel and functionally interrelated fields of algal and microbial symbiosis in coral reef invertebrates. Initial demonstrations of the role of microbes as causative agents of bleaching (Kushmaro et al. 1996, 1998; Ben-Haim and Rosenberg 2002), and in the algal population responses to disease (Toller et al. 2001b) have provided some clues to the phenomenological links that may exist in the combined area of coral health, environmental stress, and disease.

Even more important for further development are the conceptual and process-based similarities in the acquisition, dynamics, and modulation of very different, but interactive symbiont communities by the same hosts. The disciplines involved have different conceptual approaches and experimental strengths; combining them effectively has the potential to yield wide-ranging integrative breakthroughs.

Acknowledgments. We are indebted to the organizers of the Coral Health and Disease Conference, E. Rosenberg and Y. Loya. The paper has benefited from helpful discussions with G. Smith, F. Rohwer, J. Bythell, P. Jokiel, Y. Benayahu, Y. Erez, and other conference participants. Portions of the research were supported by NSF grants OCE 00-03970 (DGF and RWB), OCE 0099301 (AB) and DEB 99-78106 (DGF), NOAA grant NURC 2002-15B (AB). M. Schoneweis assisted with preparation of the figures.

References

Augustine L, Muller-Parker G (1998) Selective predation by the mosshead sculpin *Clinocottus globiceps* on the sea anemone *Anthopleura elegantissima* and its two algal symbionts. Limnol Oceanogr 43:711–715

Baker AC (1999) The symbiosis ecology of reef-building corals. Doctoral dissertation in marine biology. University of Miami, Coral Gables, Florida, 120 pp

Baker AC (2001) Coral reefs bleach to survive change. Nature 411:765–766

Baker AC (2002) Is bleaching really adaptive? Reply to Hoegh-Guldberg et al. Nature 415:601–602

Baker AC (2003) Flexibility and specificity in coral–algal symbiosis: diversity, ecology and biogeography of *Symbiodinium*. Annu Rev Ecol Syst 34:661–689

Baker AC, Rowan R (1997) Diversity of symbiotic dinoflagellates (zooxanthellae) in scleractinian corals of the Caribbean and eastern Pacific. Proc 8th Int Coral Reef Symp 2:1301–1306

Bates A (2000) The intertidal distribution of two algal symbionts hosted by *Anthopleura xanthogrammica* (Brandt 1835). J Exp Mar Biol Ecol 249:249–262

Belda-Baillie CA, Baillie BK, Maruyama T (2002) Specificity of a model cnidarian–dinoflagellate symbiosis. Biol Bull 202:74–85

Ben-Haim Y, Rosenberg E (2002) A novel *Vibrio* sp. pathogen of the coral *Pocillopora damicornis*. Mar Biol 141:47–55

Brown BE, Dunne RP, Ambarsari I, Le Tissier MDA, Satapoomin U (1999) Seasonal fluctuations in environmental factors and variations in symbiotic algae in four Indo-Pacific coral species. Mar Ecol Prog Ser 191:53–69

Buddemeier RW, Fautin DG (1993) Coral bleaching as an adaptive mechanism: a testable hypothesis. BioScience 43:320–326

Buddemeier RW, Fautin DG, Ware JR (1997) Acclimation, adaptation, and algal symbiosis in reef-building scleractinian corals. In: den Hartog JC (ed) Proc 6th Int Conf Coelenterate Biology (16–21 July 1995, Noordwijkerhout, The Netherlands). National Museum of Natural History, Leiden, pp 71–76

Carlos AA, Baillie BK, Kawachi M, Maruyama T (1999) Phylogenetic position of *Symbiodinium* (Dinophyceae) isolates from tridacnids (Bivalvia), cardiids (Bivalvia), a sponge (Porifera), a soft coral (Anthozoa), and a free-living strain. J Phycol 35:1054–1062

Carlos AA, Baillie BK, Maruyama T (2000) Diversity of dinoflagellate symbionts (zooxanthellae) in a host individual. Mar Ecol Prog Ser 195: 93–100

Coffroth MA, Santos SR, Goulet TL (2001) Early ontogenetic expression of specificity in a cnidarian–algal symbiosis. Mar Ecol Prog Ser 222:85–96

Coles SL (1988) Limitations on reef coral development in the Arabian Gulf: temperature or algal competition? Proc 6th Int Coral Reef Symp 3:211–216

Coles SL, Jokiel PL (1977) Effects of temperature on photosynthesis and respiration in hermatypic corals. Mar Biol 43:209–216

Coles SL, Fadlallah YH (1991) Reef coral survival and mortality at low temperatures in the Arabian Gulf: new species-specific lower temperature limits. Coral Reefs 9:231–237

Coles SL, Jokiel PL, Lewis CR (1976) Thermal tolerance in tropical versus subtropical Pacific reef corals. Pac Sci 30:159–166

Colley NJ, Trench RK (1983) Selectivity in phagocytosis and persistence of symbiotic algae by the scyphistoma stage of the jellyfish *Cassiopeia xamachana*. Proc R Soc Lond B 219:61–82

Darius HT, Martin PMV, Grimont PAD, Dauga C (2000) Small subunit rDNA sequence analysis of symbiotic dinoflagellates from seven scleractinian corals in a Tahitian lagoon. J Phycol 36:951–959

Davy SK, Lucas IAN, Turner JR (1997) Uptake and persistence of homologous and heterologous zooxanthellae in the temperate sea anemone *Cereus pedunculatus* (Pennant). Biol Bull 192:208–216

Diekmann OE, Bak RPM, Tonk L, Stam WT, Olsen JL (2002) No habitat correlation of zooxanthellae in the coral genus *Madracis* on a Curaçao reef. Mar Ecol Prog Ser 227:221–232

Fagoonee I, Wilson HB, Hassell MP, Turner JR (1999) The dynamic of zooxanthellae populations: a long-term study in the field. Science 283:843–845

Fautin DG (1997) Cnidarian reproduction: assumptions and their implications. In: den Hartog JC (ed) Proc 6th Int Conf Coelenterate Biology. National Museum of Natural History, Leiden, pp 151–162

Fitt WK (1985) The effect of different strains of the zooxanthella *Symbiodinium microadriaticum* on growth and survival of their coelenterate and molluscan hosts. Proc 5th Int Coral Reef Symp 6: 131–136

Fitt WK, McFarland FK, Warner ME, Chilcoat GC (2000) Seasonal patterns of tissue biomass and densities of symbiotic dinoflagellates in reef corals and relation to bleaching. Limnol Oceanogr 45:677–685

Fitt WK, Brown BE, Warner ME, Dunne RP (2001) Coral bleaching: interpretation of thermal tolerance limits and thermal thresholds in tropical corals. Coral Reefs 20:51–65

Gates RD (1990) Seawater temperature and sublethal coral bleaching in Jamaica. Coral Reefs 8:193–197

Glynn PW, Maté JL, Baker AC, Calderón MO (2001) Coral bleaching and mortality in Panama and Ecuador during the 1997–1998 El Niño-Southern Oscillation event: spatial/temporal patterns and comparisons with the 1982–1983 event. Bull Mar Sci 69:79–109

Goulet TL, Coffroth MA (1997) A within colony comparison of zooxanthella genotypes in the Caribbean gorgonian *Plexaura kuna*. Proc 8th Int Coral Reef Symp 2:1331–1334

Goulet TL, Coffroth MA (2003) Genetic composition of zooxanthellae between and within colonies of the octocoral *Plexaura kuna*, based on small subunit and multilocus DNA fingerprinting. Mar Biol 142:233–239

Guzmán HM, Cortés J (2001) Changes in reef community structure after fifteen years of natural disturbances in the eastern Pacific (Costa Rica). Bull Mar Sci 69:133–149

Hoegh-Guldberg O (1999) Climate change, coral bleaching and the future of the world's coral reefs. Mar Freshwater Res 50:839–866

Jokiel PL, Coles SL (1990) Response of Hawaiian and other Indo-Pacific reef corals to elevated temperature. Coral Reefs 8:155–162

Kinzie RA III (1974) Experimental infection of aposymbiotic gorgonian polyps with zooxanthellae. J Exp Mar Biol Ecol 15:335–345

Kinzie RA III, Chee GS (1979) The effect of different zooxanthellae on the growth of experimentally re-infected hosts. Biol Bull 156:315–327

Kinzie RA III, Takayama M, Santos SR, Coffroth MA (2001) The adaptive bleaching hypothesis: experimental tests of critical assumptions. Biol Bull 200:51–58

Knowlton N, Rohwer F (2003) Multi-species microbial mutualisms on coral reefs: the host as a habitat. Am Nat (in press)

Kushmaro A, Loya Y, Fine M, Rosenberg E (1996) Bacterial infection and coral bleaching. Nature 380:396

Kushmaro A, Rosenberg E, Fine M, Haim Y, Loya Y (1998) Effect of temperature on bleaching of the coral *Oculina patagonica* by *Vibrio* AK-1. Mar Ecol Prog Ser 171:131–137

Larcombe P, Costen A, Woolfe KJ (2001) The hydrodynamic and sedimentary setting of nearshore coral reefs, central Great Barrier Reef shelf, Australia: Paluma Shoals, a case study. Sedimentology 48(4):811–835

LaJeuness TC (2001) Investigating the biodiversity, ecology and phylogeny of endosymbiotic dinoflagellates in the genus *Symbiodinium* using the ITS region: in search of a "species" level marker. J Phycol 37;866–880

LaJeunesse TC (2002) Diversity and community structure of symbiotic dinoflagellates from Caribbean reef corals. Mar Biol 141:387–400

LaJeunesse TC, Trench RK (2000) Biogeography of two species of *Symbiodinium* (Freudenthal) inhabiting the intertidal sea anemone *Anthopleura elegantissima* (Brandt). Biol Bull 199:126–134

LaJeunesse TC, Loh WKW, van Woesik R, Hoegh-Guldberg O, Schmidt GW, Fitt WK (2003) Low symbiont diversity in southern Great Barrier Reef corals relative to those of the Caribbean. Limnol Oceanogr 48:2046–2054

Loh W, Carter D, Hoegh-Guldberg O (1998) Diversity of zooxanthellae from scleractinian corals of One Tree Island (the Great Barrier Reef) Australian Coral Reef Society 75th Anniversary Conference, Heron Island, Australia, pp 141–151

Loh WK, Loi T, Carter D, Hoegh-Guldberg O (2001) Genetic variability of the symbiotic dinoflagellates from the wide ranging coral species *Seriatopora hystrix* and *Acropora longicyathus* in the Indo-West Pacific. Mar Ecol Prog Ser 222:97–107

Merriam-Webster (1997) Merriam Webster's Collegiate Dictionary, (10th edn. Merriam-Webster, Springfield, MA, p 13

McCloskey LR, Cove TG, Verde EA (1996) Symbiont expulsion from the anemone *Anthopleura elegantissima* (Brandt) (Cnidaria:Anthozoa). J Exp Mar Biol Ecol 195:173–186
Muller-Parker G (1984) Dispersal of zooxanthellae on coral reefs by predation on cnidarians. Biol Bull 167:159–167
Muller-Parker G, Davy SK (2001) Temperate and tropical algal–sea anemone symbioses. Invert Biol 120:104–123
Nakamura T, van Woesik R (2001) Water-flow rates and passive diffusion partially explain differential survival during the 1998 bleaching event. Mar Ecol Prog Ser 212:301–304
Oliver J (1985) Recurrent seasonal bleaching and mortality of corals on the Great Barrier Reef. Proc 5th Int Coral Reef Congr 4:201–206
Pawlowski J, Holzman M, Fahrni JF, Pochon X, Lee JJ (2001) Molecular identification of algal endosymbionts in large miliolid formaminifera 2. Dinoflagellates. J Eukaryotic Microbiol 48:368–373
Perez SF, Cook CB, Brooks WR (2001) The role of symbiotic dinoflagellates in the temperature-induced bleaching response of the subtropical sea anemone *Aiptasia pallida*. J Exp Mar Biol Ecol 256;1–14
Pochon X, Pawlowski J, Zaninetti L, Rowan R (2001) High genetic diversity and relative specificity among *Symbiodinium*-like endosymbiotic dinoflagellates in soritid foraminiferans. Mar Biol 139:1069–1078
Podestá GP, Glynn PW (2001) The 1997–98 El Niño event in Panama and Galápagos: an update of thermal stress indices relative to coral bleaching. Bull Mar Sci 69:43–59
Ralph PJ, Gademann R, Larkum AWD (2001) Zooxanthellae expelled from bleached corals at 33 °C are photosynthetically competent. Mar Ecol Prog Ser 220:163–168
Rodriguez-Lanetty M, Loh W, Carter D, Hoegh-Guldberg O (2001) Latitudinal variability in symbiont specificity within the widespread scleractinian coral *Plesiastrea versipora*. Mar Biol 138:1175–1181
Rohwer F, Breitbart M, Jara J, Azam F, Knowlton N (2001) Diversity of bacteria associated with the Caribbean coral *Montastraea franksi*. Coral Reefs 20:85–95
Rohwer F, Serigutan V, Azam F, Knowlton N (2002) Diversity and distribution of coral-associated bacteria. Mar Ecol Prog Ser 243:1–10
Rowan R (1998) Diversity and ecology of zooxanthellae on coral reefs. J Phycol 34:407–417
Rowan R, Powers DA (1991) A molecular genetic classification of zooxanthellae and the evolution of animal–algal symbioses. Science 251:1348–1351
Rowan R, Knowlton N (1995) Intraspecific diversity and ecological zonation in coral–algal symbiosis. Proc Natl Acad Sci USA 92:2850–2853
Rowan R, Knowlton N, Baker A, Jara J (1997) Landscape ecology of algal symbionts creates variation in episodes of coral bleaching. Nature 388:265–269
Santos SR, Taylor DJ, Coffroth MA (2001) Genetic comparisons of freshly isolated versus cultured symbiotic dinoflagellates: implications for extrapolating to the intact symbiosis. J Phycol 37:900–912
Santos SR, Taylor DJ, Kinzie RA III, Hidaka M, Sakai K, Coffroth MA (2002a) Molecular phylogeny of symbiotic dinoflagellates inferred from partial chloroplast large subunit (23S)-rDNA sequences. Mol Phylogenet Evol 23:97–111
Santos SR, Taylor DJ, Kinzie RA III, Sakai K, Coffroth MA (2002b) Evolution of length variation and heteroplasmy in the chloroplast rDNA of symbiotic dinoflagellates (*Symbiodinium*, Dinophyta) and a novel insertion in the universal core region of the large subunit rDNA. Phycologia 41:311–318
Savage AM, Goodson MS, Visram S, Trapido-Rosenthal H, Wiedenmann J, Douglas AE (2002a) Molecular diversity of symbiotic algae at the latitudinal margins of their distribution: dinoflagellates of the genus *Symbiodinium* in corals and sea anemones. Mar Ecol Prog Ser 244:17–26
Savage AM, Trapido-Rosenthal H, Douglas AE (2002b) On the functional significance of molecular variation in *Symbiodinium*, the symbiotic algae of Cnidaria: photosynthetic response to irradiance. Mar Ecol Prog Ser 244:27–37

Schlichter D, Fricke HW, Weber W (1986) Light harvesting by wavelength transformation in a symbiotic coral of the Red Sea twilight zone. Mar Biol 91:403–407

Schoenberg DA, Trench RK (1980) Genetic variation in *Symbiodinium* (=*Gymnodinium*) *microadriaticum* Freudenthal, and specificity in its symbiosis with marine invertebrates. 3. Specificity and infectivity of *Symbiodinium microadriaticum*. Proc R Soc Lond 207:445–461

Schwarz JA, Weis VM, Potts DC (2002) Feeding behavior and acquisition of zooxanthellae by planula larvae of the sea anemone *Anthopleura elegantissima*. Mar Biol 140:471–478

Sheppard CRC (1988) Similar trends, different causes: responses of corals to stressed environments in Arabian seas. Proc 6th Int Coral Reef Symp 3:297–302

Smith SV, Buddemeier RW (1992) Global change and coral reef ecosystems. Ann Rev Ecol Syst 23:89–118

Toller WW, Rowan R, Knowlton N (2001a) Zooxanthellae of the *Montastrea annularis* species complex: patterns of distribution of four taxa of *Symbiodinium* on different reefs and across depths. Biol Bull 201:348–359

Toller WW, Rowan R, Knowlton N (2001b) Repopulation of zooxanthellae in the Caribbean corals *Montastrea annularis* and *M. faveolata* following experimental and disease-associated bleaching. Biol Bull 201:360–373

Trench RK (1992) Microalgal–invertebrate symbiosis, current trends. Encyclopedia of microbiology. Academic Press, New York, pp 129–142

Van Oppen MJH, Willis BL, van Vugt HWJA, Miller DJ (2000) Examination of species boundaries in the *Acropora cervicornis* group (Scleractinia, Cnidaria) using nuclear DNA sequence analysis. Mol Ecol 9:1363–1373

Van Oppen MJH, Palstra FP, Piquet AM-T, Miller DJ (2001) Patterns of coral-dinoflagellate associations in *Acropora*: significance of local availability and physiology of *Symbiodinium* strains and host-symbiont selectivity. Proc R Soc B 268:1759–1767

Van Oppen MJH, Willis BL, Van Rheede T, Miller DJ (2002) Spawning times, reproductive compatibilities and genetic structuring in the *Acropora aspera* group: evidence for natural hybridization and semi-permeable species boundaries in corals. Mol Ecol 11:1363–1376

Veron JEN (1995) Corals in space and time. UNSW Press, Sydney

Veron JEN (2000) Corals of the world, 3 vols. Australian Institute of Marine Science, Townsville

Ware JR, Fautin DG, Buddemeier RW (1996) Patterns of coral bleaching: modeling the adaptive bleaching hypothesis. Ecol Model 84:199–214

Weis VM, Reynolds WS, de Boer MD, Krupp DA (2001) Host–symbiont specificity during onset of symbiosis between the dinoflagellates *Symbiodinium* spp. and planula larvae of the scleractinian coral *Fungia scutaria*. Coral Reefs 20:301–08

Wilkinson CR (ed) (2000) Status of coral reefs of the world: 2000. Australian Institute of Marine Science, Cape Ferguson, Queensland, 363 pp

Zahl PA, McLaughlin JJA (1959) Studies in marine biology. IV. On the role of algal cells in the tissues of marine invertebrates. J Protozool 6:344–352

25 The Bacterial Disease Hypothesis of Coral Bleaching

Eugene Rosenberg

25.1 Introduction

Before presenting and analyzing the evidence for the bacterial disease hypothesis of coral bleaching, it is logical to first ask the question: is coral bleaching a disease? The following definitions are instructive:

"Unhealthy conditions of an organism or part of an organism" (Oxford Dictionary)

"Any deviation from or interruption of the normal structure or function of a part, organ or system of a body as manifest by characteristics, symptoms and signs, the etiology, pathology and prognosis may be known or unknown" (Merck Medical Dictionary 2000)

"An illness of people, animals, plants, etc. caused by infection or a failure of health rather than by accident" (Cambridge Interactive Dictionary)

"A pathological condition of a part, organ, or system of an organism resulting from various causes, such as infection, genetic defect, or environmental stress, and characterized by an identifiable group of signs or symptoms" (American Heritage Dictionary of the English Language)

Clearly, coral bleaching fits precisely these definitions. It is an unhealthy condition which is, of course, why coral biologists and environmentalists are so concerned about it. In addition to reduced growth rate (Muscatine 1980, 1990), coral bleaching results in decreased coral sexual reproduction (Szmant and Gassman 1990; Fine et al. 2001) and increased coral mortality (Brown and Suharsono 1990; Glynn 1990). Coral bleaching has characteristic symptoms and signs. Like many diseases, such as cancer and heart disease, coral bleaching can have more than one cause. Published data demonstrate that coral bleaching can be induced by high seawater temperature (Jokiel and Coles 1977, 1990; Hoegh-Guldberg and Smith 1989; Glynn 1993; Fitt and Warner 1995), decreased seawater temperature (Coles and Fadlallah 1991; Kobluk and Lysenko 1994; Muscatine et al. 1991; Gates et al. 1992), too high or too low solar radiation (Hoegh-Guldberg and Smith 1989; Lesser et al. 1990; Gleason and Wellington 1993; Brown et al. 1994), reduced salinity (Goreau 1964; Fang et al. 1995; van Woesik et al. 1995), bacterial infection (Kushmaro et al. 1996, 1997; Ben-Haim et al. 2003a, b), a combination of elevated temperature and radiation (Brown 1997; Hoegh-Guldberg 1999) and a combination of elevated temperature and bacterial infection (Kushmaro et al. 1998).

Environmental triggers of coral bleaching have been referred to as stressors (Brown 1997). According to this generalization, bacterial infection is a biological stress. However, from the point of view of a microbiologist, an infectious agent is not a stressor, but a living organism with a comprehensive array of genes coding for virulence factors. Environmental stress factors, such as increased temperature, may make the host coral more susceptible to infection or may induce the infectious agent to become more virulent. The significance of distinguishing between a stress and an infectious agent becomes obvious when one considers the mechanism of a disease and potential technologies for preventing or curing the disease. For example, one would never ask what is the reservoir or mode of transmission of a stress, whereas studies on reservoirs and modes of transmission of infectious diseases have led to some of the most useful discoveries for combating diseases of plants, animals and man. Hopefully, these approaches will prove useful in preventing and/or curing infectious diseases of corals, including those coral bleaching diseases for which a microorganism (bacterium, fungus or virus) can be shown to be the causative agent.

Microbiologists make a distinction between "cause" and "causative agent" of a disease. For example, pneumonia can be caused by overexposure to cold or a weakened immunological system, but the causative agent is a bacterium or a virus. To demonstrate that a particular microorganism is the causative agent of a disease, Koch's postulates must be satisfied: (1) The organism is regularly found in lesions of the disease, (2) it can be isolated in pure culture on artificial media, (3) inoculation of this pure culture in healthy animals produces a similar disease and (4) the organism can be recovered from the lesions in these animals. Using these criteria, it was demonstrated that *Vibrio shiloi* is the causative agent of bleaching of the coral *Oculina patagonica* in the Mediterranean Sea (Kushmaro et al. 1996, 1997). It was further demonstrated in the field and in the laboratory that high seawater temperature was required for the infection to result in the coral bleaching disease (Kushmaro et al. 1998). Based on these results three questions were asked:

1. How does infection of *O. patagonica* with *V. shiloi* lead to the loss of the endosymbiotic zooxanthellae, i.e., what is the mechanism of the disease?
2. How does temperature affect the infectious process?
3. How general is bacterial bleaching of corals?

This chapter will summarize 5 years of a research effort dedicated to providing data relevant to these questions.

25.2 The Bleaching Disease of *Oculina patagonica*

Bleaching of the scleractinian coral *O. patagonica* was first observed along the Mediterranean coast of Israel in the summer of 1993 (Fine and Loya 1995). Subsequently, extensive bleaching has been recorded each summer when the seawater temperature reaches a maximum of 29–31 °C (Fig. 25.1). During the winter, most of the corals recover. Bleaching of colonies of *O. patagonica* is similar to bleaching of corals in reef systems in (1) correlation to high seawater

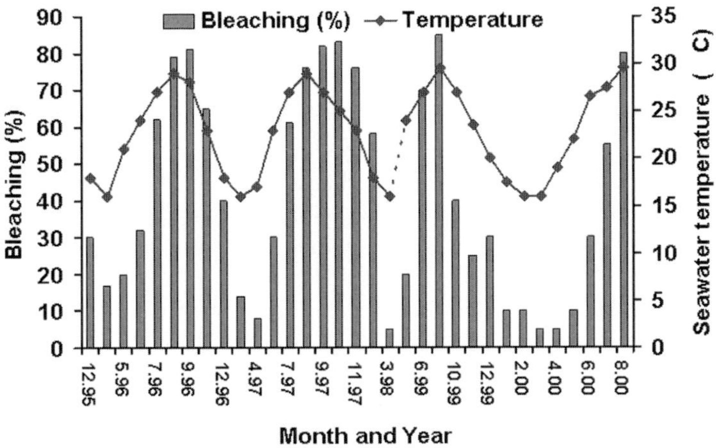

Fig. 25.1. Seasonal bleaching of *Oculina patagonica* in the Mediterranean Sea. The relationship between seawater temperature (*bars*) and percent of bleached colonies (*diamonds*) are plotted. (Adapted from Israely et al. 2001)

temperature, (2) loss of endosymbiotic zooxanthellae without any apparent tissue damage, (3) impairment of reproductive capability and (4) slow reversibility of the disease when temperature decreases. The ease of maintaining healthy *O. patagonica* in laboratory aquaria and the predictable cycle behavior of the disease along the coast of Israel make it an excellent model system to study coral bleaching. A colony of *O. patagonica* showing healthy (dark) and bleached (white) regions is shown in Fig. 25.2. The detailed ecology of *O. patagonica* is presented elsewhere in Chapter 6 (Fine and Loya, this Vol.).

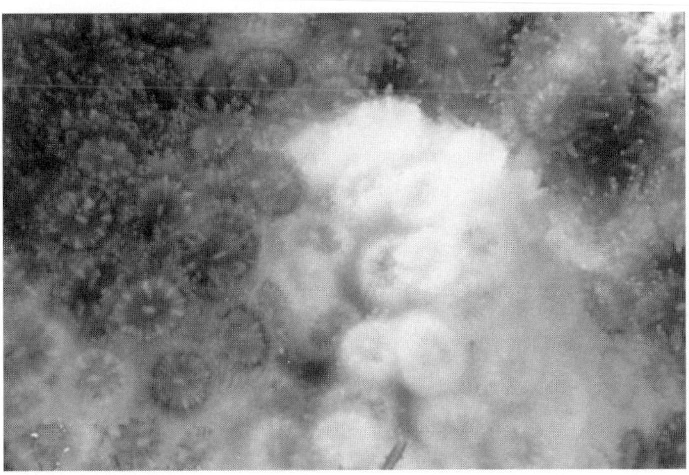

Fig. 25.2. A colony of *Oculina patagonica* showing bleached and healthy tissues

25.3
Vibrio shiloi: the Causative Agent of Bleaching of *O. patagonica*

As noted above, Koch's postulates were applied to demonstrate that the etiological agent of the bleaching disease of *O. patagonica* is *V. shiloi* (Kushmaro et al. 1996, 1997). The pathogen has been classified as a new species in the genus *Vibrio*, closely related to *Vibrio mediterranei* (Kushmaro et al. 2001). *V. shiloi* produces characteristic yellow colonies with serrated edges on TCBS agar. An electron micrograph of *V. shiloi* is seen in Fig. 25.3. In controlled aquaria experiments, it was shown that the infection of *O. patagonica* with *V. shiloi* and resulting bleaching were temperature-dependent, occurring only at elevated seawater temperatures (Kushmaro et al. 1998). At 28 °C, as few as 120 *V. shiloi* cells are sufficient to infect and bleach *O. patagonica*, whereas at 16 °C, even 1 million cells will not cause bleaching. During the last 5 years, data have been published outlining the sequential steps in the infection of *O. patagonica* by *V. shiloi* and how temperature affects various virulence factors.

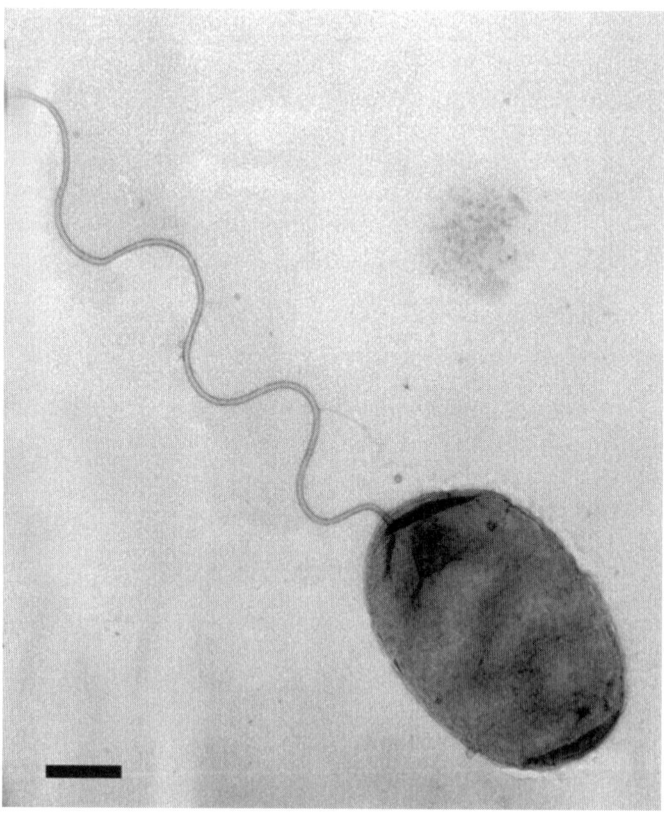

Fig. 25.3. Electron micrograph of *Vibrio shiloi*. *Bar* 0.2 μm

25.3.1
Virulence Mechanisms of *V. shiloi*

25.3.1.1
Chemotaxis

Using the capillary tube assay, it was shown that *V. shiloi* is attracted to the mucus obtained from *O. patagonica* (Banin et al. 2001a). Motility and chemotactic behavior were present when the bacteria were grown either at 16 or 25 °C. Chemotaxis of other *Vibrio* pathogens to the mucus of their host has been reported. For example, *Vibrio cholera* exerts a chemotactic response towards human intestinal mucus which has been correlated with the pathogen's virulence (Ferter and O'Brien 1981). *Vibrio anguillarum*, a fish pathogen, exhibits chemotaxis to different components in the fish mucus (O'Toole et al. 1999). In the case of *V. shiloi* chemotaxis to *O. patagonica* mucus, the attractive components in the mucus have not yet been identified.

25.3.1.2
Adhesion

The next step in the infectious process is the adhesion of *V. shiloi* to a β-galactoside-containing receptor on the coral surface (Toren et al. 1998). Adhesion is both coral- and bacteria-specific. The temperature of bacterial growth was critical for the adhesion of *V. shiloi* to the coral. When the bacteria were grown at the low temperature (16–20 °C), there was no adhesion to the coral, regardless of the temperature at which the coral had been maintained. However, bacteria grown at elevated temperatures (25–30 °C) adhered avidly to corals maintained either at low or high seawater temperatures. The important ecological ramification of these findings is that the environmental stress condition (high temperature) is necessary for the coral bleaching pathogen to initiate the infection and become virulent.

The β-galacoside-containing receptor that *V. shiloi* recognizes on the *O. patagonica* surface is present in the coral mucus. This was demonstrated by binding the coral mucus to ELISA plates and showing that the bacteria adhered avidly to the mucus-coated plates (Banin et al. 2001a). Adhesion was inhibited by methyl-β-D-galactopyranoside. Interestingly, the receptor is present only in corals that contain photosynthetically active zooxanthellae. *V. shiloi* does not adhere to bleached corals or to white (azooxanthellate) *O. patagonica* cave corals. When mucus was removed from healthy corals, adhesion of *V. shiloi* was inhibited. Furthermore, addition of 3-(3,4-dichlorophenyl)-1,1-dimethyl urea (DCMU), an inhibitor of algal photosynthesis, to mucus-depleted corals prevented the coral from synthesizing the receptor or transporting it to the mucus. Thus, *V. shiloi* is able to distinguish healthy corals from those which are photosynthetically impaired, and it only adheres to the healthy ones.

25.3.1.3
Penetration and Intracellular Multiplication

Electron micrographs of thin sections of *O. patagonica* following infection with *V. shiloi* demonstrated large numbers of bacteria in the epidermal layer of the coral (Banin et al. 2000). Using antibodies specific to *V. shiloi*, it was shown that the observed intracellular bacteria were, in fact, *V. shiloi*. The gentamycin invasion assay was used to measure the kinetics of *V. shiloi* penetration into the epidermal cells. The assay relies on the fact that the antibiotic gentamycin does not penetrate into eukaryotic cells (Isberg and Falkow 1985). Thus, only *V. shiloi* cells which have penetrated into the coral cells escape the killing action of gentamycin. The data are summarized in Table 25.1. After adhesion was complete (ca. 12 h), the bacteria began to penetrate into the coral as determined by both total counts and colony-forming units. By 24 h, 40–50% of the inoculated *V. shiloi* had penetrated into corals cells. From 24–72 h, the intracellular bacteria multiplied (based on total counts), reaching 3×10^8 bacteria per coral fragment. When the infected corals are maintained at the high summer temperatures, the bacteria remain at 10^8–10^9 cm^{-3} for at least 2 weeks.

25.3.1.4
Differentiation into the Viable, But Not Culturable State (Banin et al. 2000)

At the same time that *V. shiloi* multiplies inside the coral tissue (24–48 h after infection), the number of colony-forming units (cfu) decreases more than a 1000-fold (Table 25.1). Entry of bacteria into a state described as viable, but not culturable (VBNC) has been reported repeatedly with a large number of bacterial species, including several *Vibrio* species (Colwell et al. 1985; Oliver et al. 1991; Lee and Ruby 1995). A bacterium in the VBNC state has been defined as "a cell which can be demonstrated to be metabolically active, while being incapable of undergoing the sustained cellular division required for growth on a medium normally supporting growth of that cell" (Oliver 1993). Intracellular *V. shiloi* cells fit that definition, but unlike most cases of VBNC that have been

Table 25.1. Adhesion and penetration of *V. shiloi* onto and into *O. patagonica*[a]

Time after inoculation (h)	Bound *V. shiloi* (Total counts)	Intracellular *V. shiloi* (cfu)	Intracellular *V. shiloi* (Total counts)
4	4×10^6 (40%)	$<10^3$	$<10^6$
12	8×10^6 (80%	3×10^5	$<10^6$
24		4×10^6	5×10^6
48		2×10^3	8×10^7
72		$<10^3$	3×10^8

[a]Data derived from Banin et al. (2000).

studied, this is not brought about by starvation or low temperature. Rather, the entry of *V. shiloi* into the VBNC state occurs inside the coral epidermis, where nutrients are abundant. The fact that intracellular *V. shiloi* is viable, i.e., metabolically active, is apparent from the observation that they multiply intracellularly and they score as viable with the Live/Dead Baclight Bacterial Viability Kit (Molecular Probes, Eugene, Oregon). It is possible that the differentiated intracellular *V. shiloi* becomes dependent on one or more nutrients present in the coral cell.

The VBNC state of intracellular bacteria in corals could make it difficult to demonstrate that they are the causative agents of coral bleaching. In the case of *V. shiloi*, it was possible to isolate the bacterium from the coral mucus and then apply Koch's postulates to prove they are the causative agent of bleaching the coral *O. patagonica*. However, culturing potential pathogens from other corals may present a problem if they exist in the VBNC state. Therefore, it would be desirable to develop a general method for culturing intracellular coral bacteria. Further studies on the VBNC state of intracellular *V. shiloi* may be useful in this regard. Recently, it has been demonstrated that *V. shiloi* in the VBNC state is infectious, i.e., they can adhere, penetrate and multiply inside *O. patagonica* (Israely et al. 2001).

25.3.1.5
Toxin Production

V. shiloi produces extracellular toxins that block photosynthesis, bleach and lyse zooxanthellae (Ben-Haim et al. 1999). The toxin responsible for inhibition of photosynthesis, referred to as toxin P, is the following proline-rich peptide: PYPVYAPPPVVP (Banin et al. 2001b). In the presence of NH_4Cl, the toxin causes a rapid decrease in the photosynthetic quantum yield of zooxanthellae. Evidence has been presented that the toxin binds to algal membranes, forming a channel that allows NH_3 to rapidly pass, thereby destroying the pH gradient across the membrane and blocking photosynthesis (Banin et al. 2001b). This mode of action can help explain the mechanism of coral bleaching. *V. shiloi* also produces heat-sensitive, high molecular weight toxins that bleach and lyse isolated zooxanthellae. These toxins have not yet been isolated and characterized.

25.4
Effect of Temperature on *V. shiloi* Virulence Factors

Several of the virulence factors essential for a successful infection of *O. patagonica* by *V. shiloi* are only synthesized at elevated summer seawater temperatures, including:

1. An adhesin on the bacterial cell surface which recognizes a β-galactoside-containing receptor in the coral mucus (Toren et al. 1998; Banin et al. 2001a),

2. Superoxide dismutase (SOD) which allows the bacteria to survive in the oxygen-rich coral tissue (Banin et al. 2003),
3. Toxin P which binds to zooxanthellae membranes and inhibits photosynthesis (Ben-Haim et al. 1999; Banin et al. 2001b), and
4. Enzymes which lyse zooxanthellae (Ben-Haim et al. 1999).

The demonstration that the bacterial adhesin is not produced at temperatures below 20 °C provides an explanation for the failure of *V. shiloi* to infect *O. patagonica* in the field during the winter and in the laboratory at low temperatures. However, it does not explain why corals recover during the winter. When corals are infected with *V. shiloi* at the permissive temperature of 28 °C, the bacteria adhere, penetrate and begin to multiply intracellularly. If the infected corals are then shifted slowly (0.5 °C/day) to lower temperatures, the bacteria die and the infection is aborted (Israely et al. 2001). The failure of *V. shiloi* to survive inside coral tissue at temperatures below 20 °C is because it does not produce SOD at these low temperatures (Banin et al. 2003). This hypothesis was supported by constructing an SOD minus mutant. At 28 °C, the mutant adhered to the coral, penetrated into the tissue and then died. Death only occurred when the coral was exposed to light. The most reasonable explanation for these data is that the high concentration of oxygen and resulting oxygen radicals produced by the zooxanthellae during photosynthesis (Kuhl et al. 1995) is highly toxic to bacteria and is one of the mechanisms by which corals resist infection. At high temperatures, *V. shiloi* produces a potent SOD which helps it to survive in the coral tissue.

In a more general sense, the data on the importance of SOD for the survival of *V. shiloi* inside coral tissue provide strong evidence for the role of zooxanthellae in coral resistance to bacterial infection. Thus, zooxanthellae not only provide nutrients for coral growth, but also protect the coral against infection by pathogens.

25.5 Transmission of the Disease

The observation that *V. shiloi* could not be found inside *O. patagonica* during the winter (Israely et al. 2001) and that the bacterium could not survive in the coral below 16 °C (Banin et al. 2003) indicates that bleaching of *O. patagonica* requires a fresh infection each spring, rather than the activation of dormant intracellular bacteria. This poses a question: where does *V. shiloi* reside during the winter?

Using fluorescence in situ hybridization (FISH) with a *V. shiloi*-specific deoxyoligonucleotide probe, it was found that the marine fireworm *Hermodice caranculata* (Figs. 25.4, 25.5) is a winter reservoir for *V. shiloi* (Sussman et al. 2003). Worms taken directly from the sea during the winter contained $0.6–2.9\times10^8$ *V. shiloi* per worm by FISH analysis (Table 25.2). However, colony forming units (cfu) revealed only $0.4–1.8\times10^5$ *V. shiloi* per worm, indicating that ca. 99.9% of them were in the viable-but-not-culturable (VBNC) state.

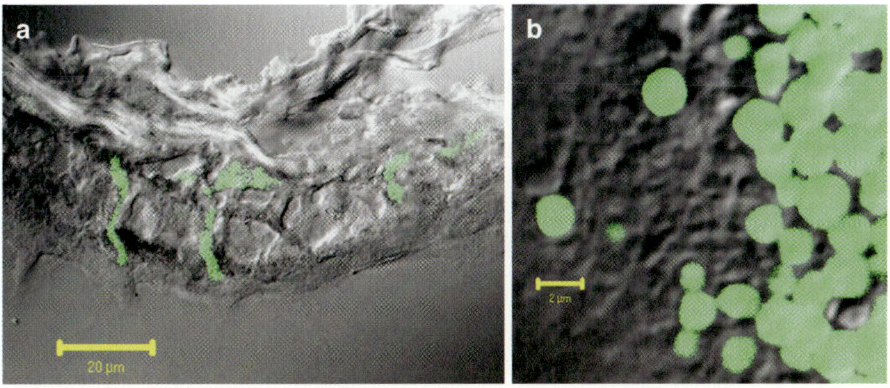

Fig. 25.4. Presence of *V. shiloi* in the fireworm *H. carunculata*. Using a *V. shiloi*-specific fluorescent oligonucleotide probe, the bacteria appear *purple* by fluorescence in situ hybridization (FISH) of tissue sections. **a** Low magnification, **b** higher magnification

Fig. 25.5. Photograph of the fireworm *H. carunculata* feeding on a coral colony

Table 25.2. Presence of *V. shiloi* inside the fireworm *H. carunculata* in a viable, but not culturable (VBNC) state. Worms were collected at night during the winter, homogenized and analyzed for total *V. shiloi* per worm by using a specific probe (see Sussman et al. 2003) and colony-forming units (cfu)

Worm length	*Vibrio shiloi* per worm	
(cm)	Total	cfu
5	0.7×10^8	0.4×10^5
6	1.1×10^8	0.4×10^5
7	0.6×10^8	0.8×10^5
7	1.3×10^8	0.9×10^5
13	2.9×10^8	1.8×10^5

When worms were infected with *V. shiloi* in the laboratory, most of the bacteria adhered to the worm within 24 h and then penetrated into epidermal cells. By 48 h, less than 10^{-4} of the intact *V. shiloi* in the worm gave rise to colonies, suggesting that they differentiated inside the worm into the VBNC state.

To test if worms carrying *V. shiloi* could serve as vectors for transmitting the pathogen to *O. patagonica*, worms infected with *V. shiloi* were placed in aquaria containing *O. patagonica*. Six out of the six corals that came into contact with the infected worms showed small patches of bleached tissue in 7–10 days and total leaching in 17 days. Uninfected worms did not cause bleaching. Thus, *H. carunculata* is not only a winter reservoir for *V. shiloi*, but also a potential vector for transmitting the bleaching disease to *O. patagonica*. It should be emphasized that knowledge about reservoirs and modes of transmission has proven useful in the past for developing technologies for controlling the spread of disease.

25.6 The Generality of the Bacterial Infection Hypothesis of Coral Bleaching

Although bleaching of *O. patagonica* by *V. shiloi* in the Mediterranean Sea is probably the best studied coral disease, there are several features about it that differ from coral bleaching in other parts of the world. To begin with, *O. patagonica* does not exist on coral reefs, but rather as colonies in temperate waters. Second, *O. patagonica* is a recent immigrant to the Mediterranean Sea (Fine et al. 2001) and thus, equilibrium between the coral and local marine organisms may not yet have been established. Third, there is a much larger seawater temperature difference between summer and winter in the Mediterranean Sea off the coast of Israel (ca. 14 °C) than is generally found in waters supporting coral reefs. These differences make it difficult to extrapolate from data obtained with the *V. shiloi/O. patagonica* model system to coral bleaching that takes place on coral reefs. In discussing the generality of the bacterial infection hypothesis of coral bleaching, it is useful to arbitrarily divide coral bleaching into "patchy bleaching" and "mass bleaching".

25.6.1 Patchy Bleaching

The recent discovery that bleaching of *Pocillopora damicornis* in the Indian Ocean and Red Sea is the result of an infection by *Vibrio coralliilyticus* strengthens the bacterial infection hypothesis of coral bleaching (Ben-Haim et al. 2003a, b). The data on the characteristics of *V. coralliilyticus*, the mode of infection and the effect of temperature on the process are presented in Chapter 17 (Ben-Haim and Rosenberg). The important points are the following: (1) Koch's postulates were satisfied, demonstrating that *V. coralliilyticus* is the causative agent of bleaching *P. damicornis*, (2) temperature determines the outcome of the infection, (3) *P. damicornis* is present on coral reefs throughout the world, and (4) the infection is host-specific, i.e., *V. coralliilyticus* only infects and bleaches *P. damicornis* and possibly closely related species.

We refer to the type of bleaching caused by *V. coralliilyticus* as "patchy" and "spreading". It is patchy in the sense that bleached corals on the reef are surrounded by apparently healthy corals of the same, as well as other, species. In the laboratory it was shown that the infection starts at one point and then spreads over the coral. Several investigators have reported on the patchy spatial distribution and spreading nature of coral bleaching (Fisk and Done 1985; Oliver 1985; Jokiel and Coles 1990; Lang et al. 1992; Edmunds 1994). It has been argued that the random mosaic patterns of bleaching observed in coral colonies are difficult to attribute solely to environmental stress, since neighboring regions of the colony must be exposed to the same extrinsic conditions (Hayes and Bush 1990). Even more significantly, the spreading nature of patchy coral bleaching is highly symptomatic of an infectious disease. We suggest that the patchy pattern of bleaching of *P. damicornis* observed on the Eilat coral reef, which was shown to be an infectious disease, may be a model for patchy bleaching diseases seen on other coral reefs. As discussed below, mass bleaching may have a different etiology.

25.6.2 Mass Bleaching

Unlike patchy bleaching, mass bleaching may involve entire reef systems and have a world-wide geographical distribution (Glynn 1984; Williams and Bunkley-Williams 1990; Lang et al. 1992; Hoegh-Guldberg 1999). For example, the mass bleaching event in 1998 that coincided with a prolonged El-Niño Southern Oscillation affected corals in many regions of the world (Wilkinson et al. 1999). As a result, significant coral mortality was reported in Australia, the Indian Ocean, Kenya, the Maldives, Seychelles and the Adamans (Harvell et al. 2002). As a result of global warming, it has been suggested that mass bleaching events will occur annually in most tropical oceans by the middle of this century (Hoegh-Guldberg 1999). Thus, it is urgent to understand the mechanism(s) of mass bleaching.

During the last few years, data have accumulated supporting the photoinhibition model of coral bleaching (Jones et al. 1998). Basically, the model states that at high solar radiation oxygen free radicals and other toxic products are produced in the chloroplast of zooxanthellae. Normally, free radicals are enzymatically detoxified; however, at higher-than-normal temperatures, the enzymes are denatured and repair is impaired, thereby leading to photoinhibition and cellular damage. Although there is, at present, more evidence for the photoinhibition model than any other mechanism to explain mass bleaching, the model clearly requires additional laboratory and field testing before it should be accepted as the primary explanation for mass bleaching events. For example, it fails to explain temperature-induced bleaching at moderate radiation that leads to the release of algae which are photosynthetically competent (Ralph et al. 2001).

The concept that microbial infection may be responsible for mass coral bleaching comes from the *V. shiloi/O. patagonica* studies. The annual mass bleaching that Hoegh-Guldberg (1999) predicts will occur annually on coral reefs in the near future has already occurred for the last 10 years with *O. patagonica* in the eastern Mediterranean Sea. As discussed above, bleaching of *O. patagonica* is the direct result of infection with *V. shiloi*, which becomes virulent at the high summer temperatures. It is interesting that toxin P, produced by *V. shiloi* at the high temperature, is an inhibitor of zooxanthellae photosynthesis (Banin et al. 2001b). The mode of photosynthesis inhibition by toxin P, disruption of the pH gradient across the thylakoid membrane of the chloroplast, is different to that resulting from high solar radiation, but the target is similar. It is also interesting that high solar radiation actually protects *O. patagonica* from bleaching by killing intracellular *V. shiloi* (Fine et al. 2002). This is consistent with the observation that there is much less bleaching of *O. patagonica* in shallow pools and reef flats than in deeper water (Fine et al. 2002). In general, bacteria are much more sensitive to UV radiation than algae or corals. Based on the above data, the occasional observation that bleaching is more severe in deeper than shallow water would tend to support the bacterial hypothesis of coral bleaching.

Consideration of the bacterial hypothesis of coral bleaching can lead to alternative explanations of existing data. For example, Dunne and Brown (2001) reported that exposure of corals to high solar radiation prior to sea temperature maxima protected them against subsequent bleaching. They conclude that "in this system, bleaching is shaped by experience, and that algae may acquire cross-protection against bleaching from solar to temperature stresses" (Brown et al. 2002). An alternate explanation based on the infectious disease hypothesis of coral bleaching is that the high UV radiation simply killed the pathogens, thereby protecting the corals from bleaching at the elevated temperatures. The latter hypothesis could be tested by carrying out microbial studies during the bleaching event.

Except for the *V. shiloi/O. patagonica* research, very few microbiological studies have been performed on bleached corals. Ritchie et al. (1994) compared culturable heterotrophic bacteria to healthy and bleached *Montastrea annularis* in the Bahamas. *Pseudomonas* which represented 13% of the isolates from nonbleached corals could not be isolated from bleached corals, whereas *Vibrio* were absent from healthy corals but represented 30% of the isolates from bleached corals. Although the authors suggest that a *Vibrio* sp. may be responsible for bleaching, the replacement of the *Pseudomonas* population with *Vibrio* in the bleached corals could be either the cause or the result of the disease. To distinguish between these two possibilities, infection experiments should be carried out under controlled conditions.

One of the apparent arguments against the bacterial disease hypothesis of coral bleaching is that mass bleaching involves many different species of corals, whereas bacterial infection is generally species-specific. In both cases where the causative agent of bleaching is known, *V. shiloi* and *V. coralliilyticus*, the pathogen only infects and bleaches its host and possibly closely related species. For bacteria to be responsible for mass bleaching, it would probably require that a pathogen for each coral species be present on the coral and that the environmental "trigger" induce each of them to become virulent almost simultaneously. I suggest (pure speculation) that this problem could be overcome in one of two ways. First, that bacteria normally present on or in the coral are opportunistic pathogens, and that the conditions that "trigger" mass bleaching either weaken the coral defense system or cause the indigenous bacterium to become virulent and cause bleaching. Although there are no data supporting this explanation for mass bleaching, there is strong evidence that it plays a key role in mortality following mass bleaching (Harvell et al. 2001). Bleached corals developed a high degree of necroses, and graft experiments revealed that the lesion-causing infections were transmissible.

The second speculation is that only one or a limited number of corals are infected by a host-specific pathogen, and that these infected corals release zooxanthellae which carry the infectious agent to other corals causing mass bleaching. The basis of this explanation is that the host specificity resides in the initial adhesion of the pathogen to its host. If the pathogen is transmitted by algae, the specificity of susceptible host corals is broadened. One of the advantages of this hypothesis is that it is testable.

25.7 Some Unresolved Questions and Future Research

At the outset of this chapter, three questions were presented:

1. How does infection with *V. shiloi* lead to the loss of the endosymbiotic zooxanthellae?
2. How does temperature affect the infectious process?
3. How general is bacterial bleaching of corals?

As often is the case in science, providing partial answers to questions raises additional questions, or as Albert Einstein wrote:

"*As the area of light increases, so does the circumference of darkness*".

With regard to the first question – although a considerable body of experimental data has been obtained outlining the infectious cycle, the following key questions remain unanswered:

1a. What is the chemical structure of the receptor in the coral mucus to which *V. shiloi* adheres to initiate the infection, and is it synthesized by the coral or the algae and then secreted into the mucus?

1b. What is the basis of the VBNC state of intracellular *V. shiloi*? Information that would lead to the ability to culture VBNC bacteria would be generally useful for the isolation of coral pathogens.

1c. What is the mode of action of the high molecular weight anti-algal toxin produced by *V. shiloi*? Does it play a role in the death or expulsion of the zooxanthellae?

In attempting to answer the second question, four *V. shiloi* virulence factors have been identified that are only produced at high summer temperatures: the adhesin that recognizes a β-galactoside-containing receptor in the coral mucus, superoxide dismutase, toxin P and the uncharacterized high molecular weight anti-algal toxin. These data lead to the following questions:

2a. Where are the genes coding for the synthesis of the virulence factors located? On plasmids? Pathogenicity islands?

2b. How are the temperature-dependent virulence genes regulated? For example, are they coordinately regulated by a common heat-sensitive repressor?

The generality of the bacterial disease hypothesis of coral bleaching remains the key question. So far, bacteria have been shown to be the causative agents of coral bleaching in two cases: extensive seasonal bleaching of *O. patagonica* in the eastern Mediterranean Sea and patchy bleaching of *P. damicornis* in the Red Sea and Indian Ocean. All other evidence for bacterial bleaching is circumstantial and indirect. The following questions should be addressed:

3a. Is patchy bleaching of different species of corals in different parts of the world caused by bacterial infection? If so, are different bacterial species responsible?

3b. Are bacterial infections responsible for mass bleaching? If so, how do they overcome the host-specificity problem?

3c. Are the recently discovered (Wilson et al. 2001) infectious viruses of zooxanthellae involved in coral bleaching? From a practical point of view, the most important future challenge is the development of technologies to cure and prevent the spread of coral disease, including the bleaching disease. Although there have been extensive reports on the distribution and dynamics of coral diseases and their relationship to environmental factors and the beginning of an understanding of the biochemistry and physiology of certain coral diseases, to this author's knowledge nothing has been done either experimentally or theoretically concerning methods for intervening in the process to cure or prevent diseases. For any infectious coral disease, answers to the following questions provide the basis for disease control:

4a. What is the infectious agent?

4b. Is the infectious agent present on healthy corals and induced to become virulent as a result of environmental factors (opportunistic pathogen)? If so, what are these environmental factors?

4c. If the infectious agent is not present on healthy corals, where does it reside when it is outside its host coral (i.e., its reservoir)?

4d. How does the pathogen move from its reservoir to its host? Does transmission involve a vector? Answers to these questions, coupled with over

100 years experience with human, animal and plant diseases, should provide the scientific basis for controlling coral disease. It is a daunting task that should not be ignored.

Acknowledgments. This work was supported by the Center for the Study of Emerging Diseases and the Pasha Gol Chair for Applied Microbiology.

References

Banin E, Israely T, Kushmaro A, Loya Y, Orr E, Rosenberg E (2000) Penetration of the coral-bleaching bacterium *Vibrio shiloi* into *Oculina patagonica*. Appl Environ Microbiol 66:3031–3036

Banin E, Israely T, Fine M, Loya Y, Rosenberg E (2001a) Role of endosymbiotic zooxanthellae and coral mucus in the adhesion of the coral-bleaching pathogen *Vibrio shiloi* to its host. FEMS Microbiol Lett 199:33–37

Banin E, Sanjay KH, Naider F, Rosenberg E (2001b) A proline-rich peptide from the coral pathogen *Vibrio shiloi* that inhibits photosynthesis of zooxanthellae. Appl Environ Microbiol 67:1536–1541

Banin E, Vassilakos D, Orr E, Martinez RJ, Rosenberg E (2003) Superoxide dismutase is a virulence factor produced by the coral bleaching pathogen *Vibrio shiloi*. Curr Microbiol 46(6):418–420

Ben-Haim Y, Banin E, Kushmaro A, Loya Y, Rosenberg E (1999) Inhibition of photosynthesis and bleaching of zooxanthellae by the coral pathogen *Vibrio shiloi*. Environ Microbiol 1:223–229

Ben-Haim Y, Thompson FL, Thompson CC, Cnockaert MC, Hoste B, Swings J, Rosenberg E (2003a) *Vibrio coralliilyticus* sp. nov., a temperature-dependent pathogen of the coral *Pocillopora damicornis*. Int J Syst Evol Microbiol 53:309–315

Ben-Haim Y, Zicherman-Keren M, Rosenberg E (2003b) Temperature-regulated bleaching and lysis of the coral *Pocillopora damicornis* by the novel pathogen *Vibrio coralliilyticus*. Appl Environ Microbiol 69(7):4236–4242

Brown BE (1997) Coral bleaching: causes and consequences. Proc 8th Int Coral Reef Symp 1:65–74

Brown BE, Suharsono (1990) Damage and recovery of coral reefs affected by El Niño related seawater warming in the Thousand Islands, Indonesia. Coral Reefs 8:163–170

Brown BE, Dunne RP, Scoffin RP, Le Tissier MDA (1994) Solar damage in intertidal corals. Mar Ecol Prog Ser 105:219–230

Brown BE, Dunne RP, Goodson MS, Douglas AE (2002) Experience shapes the susceptibility of a reef coral to bleaching. Coral Reefs 21:119–126

Coles SL, Fadlallah YH (1991) Coral reef survival and mortality at low temperatures in the Arabian Gulf: new species-specific lower temperature limits. Coral Reefs 9:231–237

Colwell RR, Brayton PR, Grimes DJ, Roszak DB, Huq SA, Palmer LM (1985) Viable but non-culturable *Vibrio cholerae* and related pathogens in the environment: implications for release of genetically engineered microorganisms. Bio/Technology 3:817–820

Dunne RP, Brown BE (2001) The influence of solar radiation on bleaching of shallow water reef corals in the Andaman Sea, 1993–1998. Coral Reefs 20:201–210

Edmunds PJ (1994) Evidence that reef-wide patterns of coral bleaching may be the result of the distribution of bleaching-susceptible clones. Mar Biol 121:137–142

Fang LS, Liao CW, Liu MC (1995) Pigment composition in different-colored scleractinian corals before and during the bleaching process. Zool Stud 34:10–17

Ferter R, O'Brien CM (1981) Role of chemotaxis in the association of motilic bacteria in intestinal mucosa: fitness and virulence of nonchemotactic *Vibrio cholera* in infant mice. Infect Immun 34:222–223

Fine M, Loya Y (1995) The coral *Oculina patagonica*: a new immigrant to the Mediterranean coast of Israel. Isr J Zool 41:81

Fine M, Zibrowius H, Loya Y (2001) *Oculina patagonica*, a non lessepsian scleractinian coral invading the Mediterranean Sea. Mar Biol 138:1195–1203

Fine M, Banin E, Israely T, Rosenberg E, Loya Y (2002) Ultraviolet (UV) radiation prevents bacterial bleaching of the Mediterranean coral *Oculina patagonica*. Mar Ecol Prog Ser 226:249–252

Fisk DA, Done TJ (1985) Taxonomic and bathymetric patterns of bleaching in corals, Myrmidon Reef (Queensland). Proc 5th Int Coral Reef Symp 6:149–154

Fitt WK, Warner ME (1995) Bleaching patterns of four species of Caribbean reef corals. Biol Bull (Woods Hole) 187:298–307

Gates RD, Baghdasarian G, Muscatine L (1992) Temperature stress causes host cell detachment in symbiotic cnidarians: implications for coral bleaching. Biol Bull (Woods Hole) 182:324–332

Gleason DF, Wellington GM (1993) Ultraviolet radiation and coral bleaching. Nature 365:836–838

Glynn PW (1984) Widespread coral morality and the 1982–83 El Niño warming event. Environ Conserv 11:133–146

Glynn PW (ed) (1990) Coral mortality and disturbances to coral reefs in the tropical eastern Pacific. In: Global ecological consequences of the 1982–83 El Niño-Southern Oscillation. Elsevier, Amsterdam, pp 55–126

Glynn PW (1993) Coral reef bleaching ecological perspectives. Coral Reefs 12:1–17

Goreau TE (1964) Mass expulsion of zooxanthellae from Jamaican reef communities after hurricane Flora. Science 145:383–386

Harvell CD, Mitchell CE, War JR, Altizer S, Dobson AP, Ostfeld RS, Samuel MD (2002) Climate warming and disease risks for terrestrial and marine biota. Science 296:2158–2162

Harvell D, Kiho K, Quirlo C, Weir J, Smith G (2001) Coral bleaching and disease: contributors to 1998 mass mortality in *Briareum asbestinum* (Octocorallia, Gorgonacea). Hydrobiologia 460:97–104

Hayes RL, Bush PG (1990) Microscopic observations in recovery in the reef-building scleractinian coral, *Monastrea annularis*, after bleaching on a Cayman reef. Coral Reefs 8:203–209

Hoegh-Guldberg O (1999) Climate change, coral bleaching and the future of the world's coral reefs. Mar Freshwater Res 50:839–866

Hoegh-Guldberg O, Smith GJ (1989) The effect of sudden changes in temperature, irradiance and salinity on the population density and export of zooxanthellae from the reef corals *Stylophora pistillata* (Esper 1797) and *Seriatopora hystrix* (Dana 1846). J Exp Mar Biol Ecol 129:279–303

Isberg RR, Falklow S (1985) A single genetic locus encoded by *Yersinia pseudotuberculosis* permits invasion of cultured animal cells by *Escherichia coli* K-12. Nature 317:262–264

Israely T, Banin E, Rosenberg E (2001) Growth, differentiation and death of *Vibrio shiloi* in coral tissue as a function of seawater temperature. Aquat Microb Ecol 24:1–8

Jokiel PL, Coles SL (1977) Effects of temperature on the mortality and growth of Hawaiian reef corals. Mar Biol 43:201–208

Jokiel PL, Coles SL (1990) Response of Hawaiian and other Indo Pacific reef corals to elevated temperatures. Coral Reefs 8:155–162

Jones R, Hoegh-Guldberg O, Larkum AWL, Schreiber U (1998) Temperature induced bleaching of corals begins with impairment of dark metabolism in zooxanthellae. Plant Cell Environ 21:1219–1230

Kobluk DR, Lysenko MA (1994) Ring bleaching in Southern Caribbean *Agaricia agaricites* during rapid water cooling. Bull Mar Sci 54:142–150

Kuhl M, Cohen Y, Dalsgard T, Jorgensen BB, Revsbech NP (1995) Microenvironment and photosynthesis of zooxanthellae in scleractinian corals studied with microsensors for O_2, pH and light. Mar Ecol Prog Ser 117:159–172

Kushmaro A, Loya Y, Fine M, Rosenberg E (1996) Bacterial infection and coral bleaching. Nature 380:396

Kushmaro A, Rosenberg E, Fine M, Loya Y (1997) Bleaching of the coral *Oculina patagonica* by *Vibrio* AK-1. Mar Ecol Prog Ser 147:159–165

Kushmaro A, Rosenberg E, Fine M, Ben-Haim Y, Loya Y (1998) Effect of temperature on bleaching of the coral *Oculina patagonica* by *Vibrio shiloi* AK-1. Mar Ecol Prog Ser 171:131–137
Kushmaro A, Banin E, Stackebrandt E, Rosenberg E (2001) *Vibrio shiloi* sp. nov: the causative agent of bleaching of the coral *Oculina patagonica*. Int J Sys Evol Microbiol 51:1383–1388
Lang JC, Lasker HR, Gladfelter EH, Hallock P, Jaap WC, da Losa FJ, Muller RC (1992) Spatial and temporal variability during periods of 'recovery' after mass bleaching on Western Atlantic coral reefs. Am Zool 32:696–706
Lee KH, Ruby EG (1995) Symbiotic role of the viable but nonculturable state of *Vibrio fischeri* in Hawaiian coastal seawater. Appl Environ Microbiol 61:278–283
Lesser MP, Stochaj WR, Tapley DW, Shick JM (1990) Bleaching in coral reef anthozoans: effects of irradiance, ultraviolet radiation, and temperature on the activities of protective enzymes against active oxygen. Coral Reefs 8:225–232
Muscatine L (1980) Productivity of zooxanthellae. In: Falkowski PG (ed) Primary productivity in the Sea. Plenum Press, New York, pp 381–402
Muscatine L (1990) The role of symbiotic algae in carbon and energy flux in reef corals. Coral Reefs 25:1–29
Muscatine L, Grossman D, Doino J (1991) Release of symbiotic algae by tropical sea-anemones and corals after cold shock. Mar Ecol Prog Ser 77:233–243
Oliver J (1985) Recurrent seasonal bleaching and mortality of corals on the Great Barrier Reef. Proc 5th Int Coral Reef Symp 4:201–206
Oliver JD (1993) Formation of viable but nonculturable cells. In: Kjelleberg S (ed), Starvation in bacteria. Plenum Press, New York, pp 239–272
Oliver JD, Nilsson, L, Kjelleberg S (1991) Formation of nonculturable *Vibrio vulnificus* cells and its relationship to the starvation state. Appl Environ Microbiol 57:2640–2644
O'Toole R, Lundberg S, Fredriksson S, Jansson A, Nillson B, Wolf-Watz H (1999) The chemotactic response of *Vibrio anguillarum* to fish intestinal mucus is mediated by a combination of multiple mucus components. J Bacteriol 181:4308–4317
Ralph PJ, Gademann R, Larkum AWD (2001) Zooxanthellae expelled from bleached corals at 33 °C are photosynthetically competent. Mar Ecol Prog Ser 220:163–168
Ritchie KB, Dennis JH, McGrath T, Smith GW (1994) Bacteria associated with bleached and nonbleached areas of *Monastrea annularis*. Proc Symp Nat Hist Bahamas 5:75–80
Sussman M, Loya Y, Fine M, Rosenberg E (2003) The marine fireworm *Hermodice carunculata* is a winter reservoir and spring–summer vector for the coral-bleaching pathogen *Vibrio shiloi*. Environ Microbiol 5(4):250–255
Szmant A, Gassman NJ (1990) The effects of prolonged bleaching on the tissue biomass and reproduction of the reef coral *Monastrea annularis*. Coral Reefs 8:217–224
Toren A, Landau L, Kushmaro A, Loya Y, Rosenberg E (1998) Effect of temperature on adhesion of *Vibrio* strain AK-1 to *Oculina patagonica* and on coral bleaching. Appl Environ Microbiol 64:1379–1384
Van Woesik R, De Vantier LM, Glazebrook JS (1995) Effects of cyclone Joy on nearshore coral communities of the Great Barrier Reef. Mar Ecol Prog Ser 128:261–270
Wilkinson CR, Linden O, Cesar H, Hodgson G, Rubens J, Strong AE (1999) Ecological and socioeconomic impacts of 1998 coral mortality in the Indian Ocean: an ENSO impact and a warning of future change? Ambio 28:188–196
Williams EH Jr, Bunkley-Williams L (1990) The world-wide coral reef bleaching cycle and related sources of coral mortality. Attol Res Bull 355:1–72
Wilson HW, Francis I, Ryan K, Davy SK (2001) Temperature induction of viruses in symbiotic dinoflagellates. Aquat Microb Ecol 25:99–102

26 Coral Reefs and Projections of Future Change

Ove Hoegh-Guldberg

Any attempt to project the future is inherently daunting. The importance of answering this question, however, has been emphasized by the observation that coral reefs are in an accelerating pattern of decline. Given the support that coral reefs provide to critical industries and human societies that line tropical coastlines, the rapid decline in the health of coral reefs could potentially cost billions of dollars per year and displace tens of millions of people. The current decline is being attributed to a wide array of causes including coastal development, overexploitation of fish stocks, destructive fishing, agricultural runoff, marine-based pollution and climate change. These influences are subsequently manifesting themselves as an increased incidence of syndromes arising from stress and disease. In previous chapters, this volume has dealt in detail with the range of syndromes and disease that are currently affecting coral reefs. The intention of the current chapter is to attempt to integrate this information in the context of how reefs might change over the next 20–50 years. Much of the direction of this chapter will be set by climate change, which, in concert with other stresses is a major threat to the distribution of healthy coral reefs and to the people and societies that depend on them.

26.1 Introduction

26.1.1 The State of the World's Reefs

Exquisite and productive, coral reefs contain the greatest biodiversity of any other marine ecosystem. Reef-building corals and their symbiotic dinoflagellates provide the physical framework and much of the primary productivity of coral reefs. These activities lend habitat and resources to millions of species of protists, plants and animals, and critical services and industry to at least 100 million people (Bryant et al. 1998). Over longer time scales, coral reefs are critical to the stabilization of coastlines and the conditions necessary for associated ecosystems such as sea grass and mangroves.

Reefs are being influenced by a wide array of human activities. These all originate from the fact that the earth's population (6.3 billion people) is growing at 1.2% per annum (United Nations 2002) or roughly 77 million people per year. Most of these people (>95%) are being added to developing nations along coastlines, many of which are bordered by large proportions of the world's coral reefs. With increased densities of people along coastlines, and falling agricultural resources per capita, people are increasing exploiting resources associated with coral reefs. This pressure has skyrocketed over the last 50 years. This has led to increased stress and disease within tropical ecosystems and, as a result, major changes in the state of these ecosystems. A growing body of evidence has suggested that changes to coastal ecosystems like coral reefs may have begun as long ago as centuries or millennia (Jackson et al. 2001). This perspective matches growing evidence from terrestrial ecosystems that the increase in human populations thousands of years ago led to large-scale changes in species and ecosystems.

Assessing the current state of the world's coral reefs is complicated by two factors. The first is associated with the fact that a truly objective sampling of the world's coral reefs is a gargantuan, if not impossible, job. Coral reefs are distributed across at least 100 countries and territories, and vary from reefs in muddy coastal waters to those associated with clear oceanic environments. They vary from coral reefs formed from a handful of species growing on rocky substrates to those that involve a complex limestone meshwork made from hundreds of species (Spalding et al. 2001). Reefs are influenced by a background of natural factors that may explain their ecological state at any one point in time. Variability in space and time creates uncertainty about the reasons for the particular state of a reef system at any point in time.

The second factor that complicates the assessment of the state of coral reef health is the complications caused by the "shifting baseline" syndrome (Sheppard 1995). Any assessment of the health of a coral reef is invariably relative to a previous state. Several authors have highlighted the difficulty of assessing the current state of a coral reef without knowing the original components and states of complex ecosystems like coral reefs.

These objections aside, there is little doubt that the decline of coral reefs has accelerated over the past 30 years. Characteristics that were prominent on coral reefs as little as 10–20 years ago (e.g. dominant *Acropora* communities; Patterson et al. 2002) have disappeared from some areas (e.g. Caribbean). Long-term assessments of coral cover across the globe reveal a general decline across a diversity of sites. Programs such as Reef Check (Hodgson and Liebeler 2002) and the Global Coral Reef Monitoring Network (Wilkinson 2002) are now adding the assessment of reef health at even wider scales.

The conclusion from these global assessments is far from encouraging. Wilkinson and Buddemeier (1994), for example, concluded that 10% of the world's reefs had already been degraded and that another 30% are likely to be lost within 10–20 years. This assessment was undertaken prior to the large-scale El Niño-driven impacts of 1997–1998. During the single episode of mass coral bleaching during 1997–1998, both Reef Check (Hodgson and Liebeler 2002) and the Global Coral Reef Monitoring Network (Wilkinson 2002) reported alarming losses of living coral across all sites of 10 and 16%, respectively. Coral bleaching and subsequent death of corals due to the unusually warm conditions of 1997–1998 were identified as the sole reason for the decline in coral cover. The loss of coral when examined on a regional basis was much higher. Reefs in the Indian Ocean, which experienced much higher sea temperatures than other sites, lost an estimated 46% of their living coral cover.

26.2 Human Impacts on Reefs

26.2.1 Anthropogenic Changes to Reef Health

Today, coral reefs are highly threatened by a large range of human activities (Table 26.1). These activities cross a spectra of pressures on reefs that stem from increasing population and resource use. The first five sources of stress are relatively more direct and local, being associated with human activities that are being pursued in the vicinity of coral reefs. The last, climate change, is indirect and based on a change to global climate under which corals reefs grow. Direct or indirect, most of these factors link to the current incidence of disease on coral reefs. Changing water quality via intensive coastal development or agricultural activity can affect the bacterial diversity and loading of the water column with direct impacts on the incidence of disease. Patterson et al. (2002), for example, describe how *Serratia marcescens*, which is a component of human sewage, causes the 'white pox' disease. The expansion of coastal communities throughout the Caribbean (and hence the amount of sewage entering the waters above coral reefs in the Caribbean) is seen as a major factor underlying the widespread incidence of this particular disease. These issues have been dealt with by several authors in this volume.

Other influences of the activities listed in Table 26.1 may have subtle and unexpected ramifications for incidence of disease. The sudden loss of the black sea urchin, *Diadema antillarum*, across the Caribbean due to a virus between 1982 and 1984 may have had its origins in population explosions of urchins resulting from the over-exploitation of predatory fish stocks that previously kept sea urchin densities in check (Hughes 1994). Greater transmission of the viral agent due to higher population densities may have lead to the catastrophic decline of sea urchins. Major changes resulted from the loss of the sea urchin. With no other grazer to take the place of *D. antillarum*, macroalgae (seaweeds) out-competed corals for space on Caribbean reefs, and eventually dominate the substrate.

Most recent assessments place climate change as the primary threat to coral reefs due to the geographical scale and intensity of its impacts (Hughes et al. 2003). The rest of this chapter will outline this threat and projections of future impacts. With this said, it is important to realize that the synergistic action of the factors listed in Table 26.1 with climate change represents the greatest challenge for coral reefs.

Table 26.1. Principal sources of stress for coral reefs worldwide. References are meant as samples of key literature and are not intended to be exhaustive. (See Bryant et al. 1998; Spalding and Jarvis 2002; Wilkinson (2002) for further details)

Activity	Description	Reference
Coastal development	Intense coastal development (urban centres, ports, tourism) results in the removal of coastal vegetation and a change in flow patterns off the land, the amount of sediments and nutrients entering the water column tends to increase. This has been implicated in stimulating algal overgrowth on inshore coral reefs and in setting off outbreaks of potential coral predators (e.g. Crown-of-Thorns starfish, *Acanthaster planci*). Other major impacts are more physical in nature such as that associated with increased visitation by tourists leading to anchor damage, diver damage, boat strikes and trampling.	Bryant et al. (1998); McClanahan (1999)
Overexploitation	Both local subsistence fishers and fishing industries are putting large pressures on fish stocks associated with coral reefs. As a result, most fish stocks in tropical as well as temperate oceans are in major decline and are at 10% of the previous unexploited state. The removal of fish leads to a loss of functional groups (herbivores, key predators) that are important to controlling other organisms within the ecosystem. Widespread ecological changes may result.	Hughes (1994); Boersma and Parrish (1999); Jackson et al. (2001); Meyers and Worm (2003)
Destructive fishing	The methods employed to catch fish have also had major impacts on coral reefs. Blast fishing (use of dynamite) and cyanide fishing (the use of cyanide to stun fish) are widespread. This has quite serious impacts on surrounding reef communities, both through the direct physical destruction of reefs and the impact on the health of corals and associated organisms.	Wilkinson (1998, 2002); Jones et al. (1999); McClannahan et al. (2002)
Agricultural runoff	The growth of agriculture along tropical shorelines has led to major increases in the amount of sediments, nutrients, pesticides and herbicides entering adjacent coastal water columns. These have a range of effects, including algal overgrowth and potential coral kills from the associated toxins.	Cortes (1994); Koop et al. (2001); McClanahan et al. (2002)
Marine-based pollution	Chemicals dumped by shipping or coastal developments leads to a buildup of compounds that poison corals and associated organisms. Associated with this is physical pollution or trash. This can have major impacts on coral reef fauna when dumped from shipping or coastal communities. Plastic bags can choke turtles and fish, or cover coral reefs, blocking off the sunlight needed to keep the reefs alive. Another form of significant trash is that of discarded fishing nets ("ghost nets"), which can continue to kill fish and other marine life long after decommissioning.	Edinger et al. (1998); Wilkinson (1998, 2002); Abelson et al. (1999); Bastidas et al. (1999)
Climate change	Rising sea temperatures and sea levels, and decreasing calcium carbonate saturation states are changing the conditions under which coral reefs have prospered for at least 400,000 years. In 1998, along with a worldwide episode of warmer than normal water temperatures, an estimated 16% of the world's corals died. Seen by many as the primary threat to reefs now as oceans undergo sustained warming over this century. In addition to coral bleaching, thermal stress is also probably also responsible for an increase in other coral diseases.	Glynn (1991); Brown (1997); Hoegh-Guldberg (1999); Done et al. (2003)

26.2.2
Climate Change: Past and Current

The earth is undergoing accelerating climate change that is being driven by rapidly increasing greenhouse gas concentrations. Concentrations of gases such as carbon dioxide are much higher than at any other time in the last 400,000 years (Fig. 26.1) if not 20 million years (Hoegh-Guldberg 2001). These changes are drastically modifying the conditions under which the earth's fauna and flora have flourished for many millions of years. There is now extensive evidence of changes to the distribution, abundance and health of earth's terrestrial and aquatic ecosystems (Walther et al. 2002; Parmesan and Yohe 2003). Species are migrating towards the poles, stationary ecosystems like coral reefs are experiencing increasingly stressful conditions and populations of organisms are in decline as a result of a combination of climate change and other anthropogenic impacts.

Evidence that rapid changes in global climate are occurring is now overwhelming (IPCC 2001). Since the beginning of instrumental records around

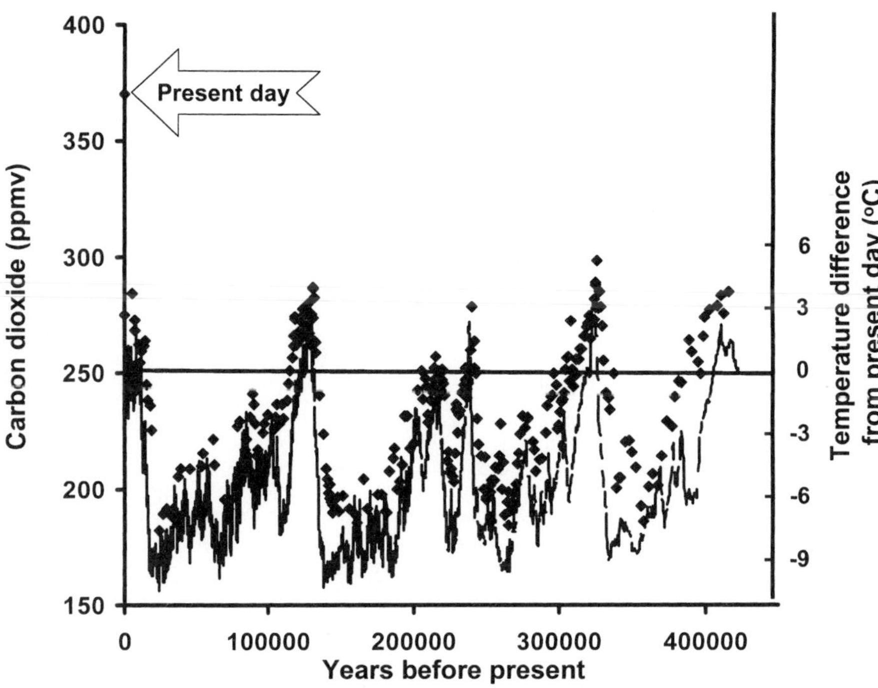

Fig. 26.1. Variation in carbon dioxide (*symbols*) and temperature variation relative to recent temperatures (*line*) in Antarctica from ice core data drilled at the Vostok station. Carbon dioxide data are replotted from Barnola et al. (1999) and temperature data from Petit et al. (1999). Present-day carbon dioxide levels are indicated by the *arrow*

1880, global temperature has increased by 0.6±0.2 °C, with the 1990s being the warmest decade on record (Jones et al. 1999). Within this decade, January–May 1998 was also the warmest period (IPCC 2001) for more than a century. The role of humans in these changes is equally undeniable. After reviewing vast amounts of information, the Intergovernmental Panel on Climate Change (IPCC 2001) concluded: "There is new and stronger evidence that most of the warming observed over the last 50 years is attributable to human activities." The Third Assessment Report of the Intergovernmental Panel on Climate Change is unambiguous in its conclusions. Anthropogenic climate change will cause major modifications to both the atmosphere and biosphere.

26.2.3
Coral Bleaching and Climate Change

Reef-building corals live within 1–2 °C of their upper temperature limits (Coles et al. 1976), which vary geographically with water temperature. They undergo coral bleaching if this upper temperature is exceeded for extended periods of time, which impacts their physiology, life expectancy and ecology (Chap. 23, Jokiel, this Vol.). Mass coral bleaching events were unknown in the scientific literature prior to the mid-1970s, but the volume of reports has expanded rapidly since that time (Fig. 26.2; Hoegh-Guldberg 1999). As has been

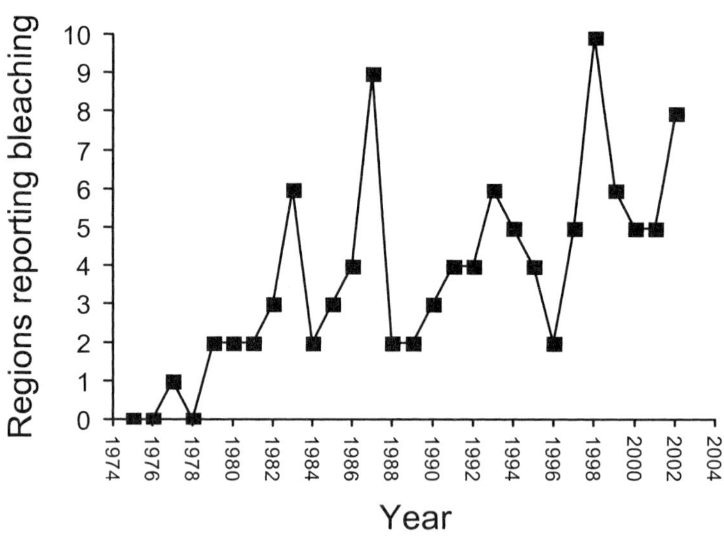

Fig. 26.2. The number of coral reef regions reporting bleaching since 1975. Regions were: Caribbean, Atlantic coast of South America, western Pacific, central Pacific, eastern Pacific, northern Pacific, South-East Asia, North-East Asia, western Indian Ocean and eastern Indian Ocean. Data collected from over 3800 bleaching records from nearly 100 countries in the ReefBase bleaching report database (http://www.reefbase.org, 23 Dec. 2002)

discussed previously in this volume (Chaps. 6 and 23, Fine and Loya, Jokiel, this Vol.), the frequency and intensity of coral bleaching have been expanding rapidly over the past two decades. Events coincide with periods of warmer than normal water and relate to changes in the physiology of the symbiont and symbiosis as thermal stress increases. The relationship between bleaching and warm seas is causal as has been demonstrated experimentally in many studies (Coles and Jokiel 1977; Hoegh-Guldberg and Smith 1989). Previous chapters in this volume (Chaps. 3, 4, 5, 9, 23–25) have thoroughly dealt with the various aspects that underpin the physiology and ecology of coral bleaching.

Increasing thermal stress is also probably a major factor in the rise of other coral diseases, as discussed in previous chapters. These stresses either cause opportunistic infections or a breakdown in the symbiotic milieu of surface bacteria and viruses. The last area of great importance is how the frequency and intensity of coral bleaching will increase over the rest of this century, and how changes in the fundamental physiological well-being of reef-building corals will potentially influence many other aspects of coral reefs. These aspects will be the focus of this final chapter.

26.3 Projections of Future Change

26.3.1 Projected Conditions over this Century

Anthropogenic activities such as expanding agriculture, clearing forests and burning fossil fuels have and are continuing to change the composition of the atmosphere and global climate (IPCC 2001). Future changes in climate are to some extent already determined due to the long residence times of gases in the atmosphere. Effective residence times (time for removal of 63% of the anthropogenic excess of a greenhouse constituent in the atmosphere if anthropogenic production falls to zero) of gases like carbon dioxide are of the order of approximately 230 years or more (Fuglestvedt et al. 1999). Many other greenhouse constituents like methane have more potent greenhouse warming potentials (ten times CO_2), but shorter residence times in the atmosphere. The overall conclusion, however, is that global temperature will increase over the next 25–50 years no matter how much we reduce greenhouse emissions. As will be discussed, reducing greenhouse emissions drastically today is highly desirable due to the huge benefits delivered later this century, especially in terms of achieving stabilization or not.

General circulation models or GCMs provide projections of future conditions by simulating future additions and removals of greenhouse gases and the resulting heat-trapping behaviour of the atmosphere. They also increasingly take into account behaviour and interaction of components of the climate system such as ice, clouds, currents and land-mass effects. Greenhouse gas concentrations and climate change projected by these models are subject to large uncertainties in

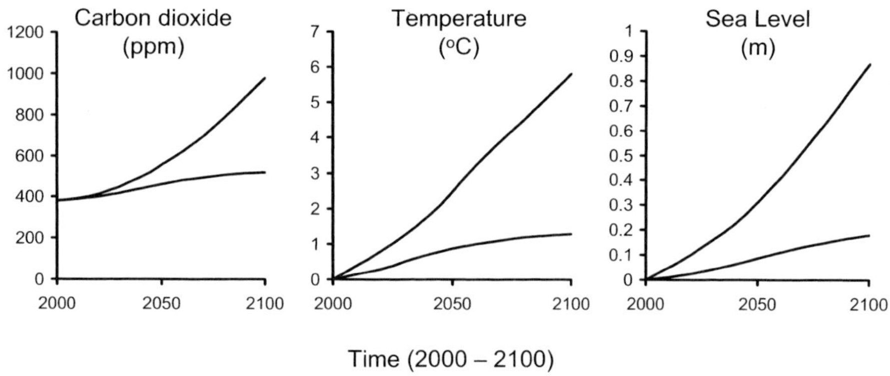

Fig. 26.3. Projected envelope for atmospheric carbon dioxide concentration, global temperature and sea level for all SRES scenarios over the period from 2000–2100. (Modified from IPCC 2001)

the effects of both natural processes and human activities. This has led to scenario-building exercises that take into account different sets of conditions and assumptions. The Intergovernmental Panel on Climate Change has extensively reviewed the outputs of the major GCMs for 40 quantitative scenario variations as part of its Special Report on Emission Scenarios (IPCC 2001). These scenarios take into account a full range of technological, social and economic possibilities, as well as the uncertainties inherent in the GCM components. The full range of scenarios has been published in the recent IPCC Third Assessment report (IPCC 2001). These scenarios give a range of future global responses that include ranges of two- to three-fold increases in GHG concentrations, 1.4–5.8 °C and 0.3–0.5 m increases in temperature and sea level by 2100, respectively (Fig. 26.3, all 40 SRES quantitative scenario variations, IPCC 2001).

The expected changes in global temperature will not be distributed evenly and will have a diverse influence on a range of climate variables. According to the IPCC (2001), it is likely to very likely that the following changes will be seen in the latter half of the twenty-first century in terrestrial environments:

- Higher maximum temperatures and more hot days over nearly all land areas.
- Higher minimum temperatures, fewer cold and frost days over nearly all land areas.
- Reduced diurnal temperature range over most land areas.
- Increase in heat index over land areas.
- More intense precipitation events in some areas.
- Increased summer continental drying and associated risk of drought.
- Increase in tropical cyclone peak wind intensities.
- Increase in tropical cyclone mean and peak precipitation intensities.

In addition to the changes above, oceanic environments are likely to change as the calcium carbonate saturation state (Ω) is reduced by 30–60% (Gattuso et al. 1998; Kleypas et al. 1999), global sea level rises by 28–98 cm and sea temperature increases by 2–5.8 °C by the end of this century (IPCC 2001).

Given the causal relationship between sea temperatures and mass coral bleaching and mortality, projections of change within this chapter focus mostly on those associated with sea temperature. This is not to downplay the importance of factors such as the calcium carbonate saturation state. These factors, as well as the other factors listed in Table 26.1, will be considered in the discussion on the important interaction of factors to determining how coral reefs will fare under a rapidly changing climate. Involving their influence is likely to heighten the effects of temperature, and hence the projections of change presented within this chapter are likely to be conservative.

26.3.2 Responses of Coral Populations to Global Climate Change

As described in previous chapters, thermal stress induces bleaching which either leads to the death of affected corals or a series of sublethal impacts such as reduced growth, reproduction and competitive ability (see Chaps. 6, 9 and 23). As also described, coral species and, hence, communities have different thermal tolerances which may or may not be related to characteristics of the coral or symbiont genotype or history (acclimation). Thermal tolerances vary with latitude, species, clone, other physical factors (e.g. light) and history (Edmunds 1994; Jones et al. 1998; Berkelmans and Willis 1999; Hoegh-Guldberg 1999; Brown et al. 2002). Despite additional influences, satellite measurements of sea surface temperature anomalies can be used to predict bleaching events several weeks in advance with greater than 90% accuracy (Hoegh-Guldberg 1999). Explanations of addition influences essentially explain a small, but significant part of the variance around the highly reliable primary variable sea temperature.

An individual coral colony will have a specific tolerance threshold above which it will bleach. A population of corals will have bands of tolerance, each due to local factors, which cause differences between individuals. A community of coral species, on the other hand, will have a broader range of tolerances that are driven by inherent interspecific differences in thermal tolerance (e.g. Hoegh-Guldberg and Salvat 1995; Loya et al. 2001). Against a rapidly rising sea temperature, whether during an individual bleaching event or over decades of climate change, the response of the community of corals will depend on the diversity of behaviour exhibited by a particular coral community within a particular thermal habitat (Fig. 26.4).

An integration of the response of corals to the projected changes in sea temperature under climate change was explored by Hoegh-Guldberg (1999). Mean thermal thresholds for coral bleaching on reefs from six parts of the world were compared to GCM outputs for the seas surrounding these reef systems under what is now understood to be a lower to mid-range climate scenario (IS92a scenario or doubling of carbon dioxide; IPCC 2001). The results of this analysis were startling. The known thresholds for coral bleaching were ex-

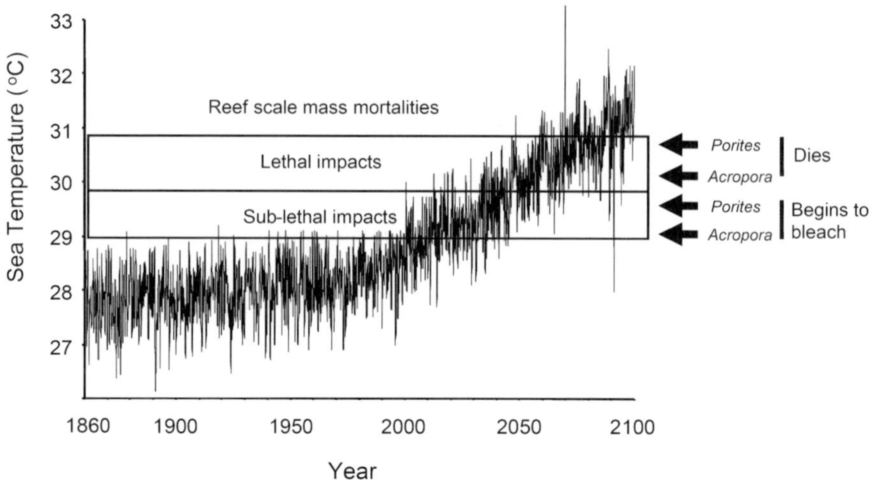

Fig. 26.4. The impact on a hypothetical coral community of differences in bleaching and mortality tolerances among species as average sea temperature changes under mild climate change (IS92a or a doubling of carbon dioxide). Predictions of sea temperature are based on the ECHAM4-OPC3 model from the Max Planck Institute described by Hoegh-Guldberg (1999). At any one point, differences between coral species will produce different responses. *Acropora*, a thermally sensitive genus has been compared to *Porites*, a relatively thermally insensitive species to illustrate this effect. Differences in thermal tolerance dictate that coral community composition will change prior to mass mortalities in which the tolerances of all species within a community are exceeded and coral cover rapidly declines

ceeded on an annual basis by 2030–2050 for all six coral reef regions. Given the scale of impacts seen during 1997–1998, and the absence of compelling evidence for mechanisms such as adaptation (sensu stricto; see definitions outlined in Chap. 23, this Vol.) that might modify the thermal tolerances of reefs, it was concluded that coral reefs may have only remnant populations of reef-building corals by the middle of this century.

Another way to project how coral reefs might change under the extreme temperatures being projected is to refine the use of thermal thresholds by including the important effect of time of exposure. A combination of elevated sea temperature and exposure time predicts mass coral bleaching with great accuracy (Strong et al. 1996, 2000; Hoegh-Guldberg 1999, 2001). Table 26.2 outlines information from the global event in 1998 in which anomaly size and exposure were multiplied together to give a degree heating month (akin to degree heating weeks of Strong et al. 2000). The four sites that experienced major post-bleaching mortalities had three-fold higher degree heating month indices than those that bleached and recovered. While some fine tuning needs to be done with regard to the influence of other factors (e.g. irradiance, Berkelmans and Willis 1999; Brown et al. 1999; Mumby et al. 2001), the relationship between SST anomalies, exposure time and coral bleaching and mortality gives strong indications of how reefs might change as heat stress increases over the next cen-

Table 26.2. Comparison of recent degree heating months and mass bleaching mortality estimates from incidents of bleaching within the 1998 mass bleaching event. (Adapted from Hoegh-Guldberg 2001)

Location	Degree heating months	Mortality (%)	Source
Severe events (mortality >80%)			
Palau	3.9	70–90	J. Bruno (unpubl. data)
Seychelles	3.1	Up to 75	Spencer et al. (2000)
Okinawa	3	90–95	Loya et al. (2001)
Scott Reef	2.6	75–90	L. Smith and A. Heyward (unpubl. data)
Mean ± 95% CI	3.2±0.47		
Mild events (mortality <10%)			
Southern GBR (reef crest)	1.7	10–30	Jones et al. (2000)
Central GBR (inner reefs)	1.4	1–16	Marshall and Baird (2000)
Moorea (outer reef crest)	0.9	0 mortality	Pers. observ. (10% bleached)
Cook Is. (Southern; reef crest)	0.4	0 mortality	Pers. observ. (5% bleached)
Mean ± 95% CI	1.1±0.49		

tury. A doubling of CO_2 (IS92a scenario) will lead to degree heating months (DHM) for most tropical regions that will be greater than three-fold higher than those which caused large-scale mortality events in Palau, Okinawa, Seychelles and Scott Reef (Hoegh-Guldberg 2001).

The next step in this type of analysis is to relate the DHM-related responses listed in Table 26.2 to the future DHM values calculated in Fig. 26.5. Operating objectively on the behaviour of coral reefs over the past two decades, two categories of response can be defined:

1. Reefs that bleach, but 'recover': reefs that experience 0.5 degree heating months (DHM) during the summer months will experience mass bleaching. They will recover if stress levels return to previous levels.
2. Reefs that experience near total coral mortality: reefs that are exposed to 3.2 DHM/year or more will experience almost complete mortality of their coral populations. This is a conservative estimate as reefs probably experience major mortality events at lower DHM values (e.g. Scott Reef, 2.6 DHM in 1998; Table 26.2).

In between these two DHM values, reefs will vary from 'recovery' to greater extents of mortality as DHM values seen each year increase. To understand how reefs will respond to increasing thermal stress, we need two further assumptions. The first is that reefs that bleach every second year will experience a decrease in reef quality. This is logical given that bleaching has strong sublethal effects on both growth and reproduction and coral bleaching over the past decade (which has a periodicity of 3–4 years) has already started to de-

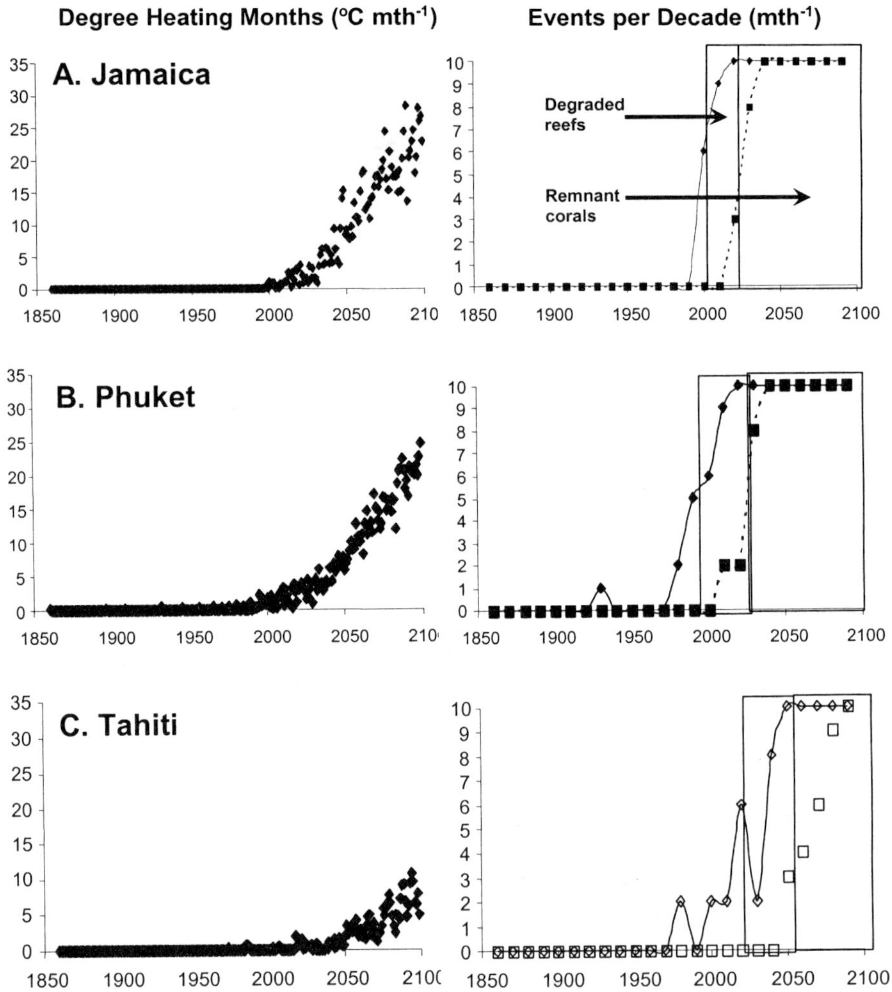

Fig. 26.5. Projected patterns of bleaching projected for reefs along the southern coast of **A** Jamaica (17.5°N, 76.5°W), **B** Phuket (7.5°N, 98.5°E) and **C** Tahiti (17.5°S,149.5°W) from 1860 up to 2100. Sea temperature data from the ECHAM4/OPYC3 (Roeckner et al. 1996) model (details in Hoegh-Guldberg 1999). *Leftside* in each case shows the accumulated DHM values once they had exceeded 0.5 DHM. *Right side* shows the frequency of bleaching events per decade over the next century (DHM >0.5, *solid line*) and severe events (DHM >3.2, *dotted line*). *Left column* shows period in which reefs are in decline (coral cover beginning to decrease). *Right column* is period in which coral is remnant (<5%) based on the assumption that coral communities will be unable to survive three severe bleaching events per decade (severe as those seen in Okinawa, Palau, Seychelles and Scott Reef in 1998)

grade coral cover in many parts of the world. Equally, total mortality events that occur three times per decade will no longer have coral-dominated reefs. This is clearly supported by the observation that reefs like those of Palau, NW Australia and Okinawa have not recovered fully from the 1997–1998 mass bleaching event.Obviously, if 90% of the coral cover on a reef is lost, it will not grow back completely within 3 years, especially given that recovery after a mass mortality of coral may take 10–40 years.

Wilkinson (2002) sums this issue up in the Executive summary to the latest Global Coral Reef Monitoring Network assessment of coral reefs worldwide in relationship to the reefs showing the fastest signs of recovery since the last global mass bleaching event: "There has been considerable recovery in the unstressed reefs of Southeast and East Asia and Palau, and also along the Great Barrier Reef of Australia, but it will take several decades before reefs return to pre-1998 status. There is broad concern that another Climate Change/El Niño event could arrest the recovery." The overwhelming conclusion, however, is that mass mortality events like those of Palau, Scott Reef and Okinawa cannot occur every 3–4 years without eventually bringing coral cover close to zero. This assumption is probably highly conservative given that the anomaly size (as well as frequency) also grows (far beyond a DHM value of 3.2, the worst case seen in 1998).

If the basic definitions listed above are accepted and integrated with even the mildest rates of climate change (a doubling of greenhouse gas concentrations), reefs will soon progress through a degraded state to one in which corals are rare, remnant organisms. Reefs will be largely devoid of reef-building corals. While it is true that reefs will change rather than disappear, these new ecological states will be far distant relatives of the coral-dominated ecosystems that we currently associate with coral reefs.

These conclusions should evoke great concern. While the weight of evidence supports these perspectives, there is a need to consider alternative views. A large and rapid change in the thermal threshold of coral species, for example, would modify the conclusions of this type of analysis (Hoegh-Guldberg 2000; Hughes et al. 2003). There is very little, if any, evidence that suggests that the thermal threshold of corals and their zooxanthellae is moving or adapting to changes in sea temperature. Mortality appears to be rapidly increasing, not decreasing (see Hoegh-Guldberg 1999; Wellington et al. 2001 for recent reviews; but see Glynn et al. 2001), and thermal thresholds of coral populations appear to be similar to those 20 years ago (e.g. Brown 1997; Hoegh-Guldberg et al. 1997). The balance of evidence, therefore, favours the conclusion that rates of change are exceeding the rates at which reef-building coral populations can adapt. This is not surprising given the slow-growing, largely asexual nature of corals and their complex intracellular symbiosis (with a unicellular dinoflagellate protist).

Similar issues underpin the reason for hosts not being able to switch to completely new symbionts in the 'ecological' time frame of a bleaching event (see discussion in Hoegh-Guldberg 1999; Hoegh-Guldberg et al. 2002, see also Chap. 24, this Vol., for an alternative view). The establishment of new symbioses with higher thermal tolerances, like adaptation, is inherently chancy. The successful recombination of a host, for example, with a "beneficial" symbiont (as opposed to the more likely suboptimal or even harmful one) is likely to be extremely rare in ecological time scales. This may explain why unambiguous evidence of a completely new externally derived symbiont type (genotype) replacing another through the process of mass bleaching is non-existent at this point of time.

It is important to realize that adjusting the ratio of two or more pre-existing algal genotypes within polycladal symbioses does not provide the necessary new

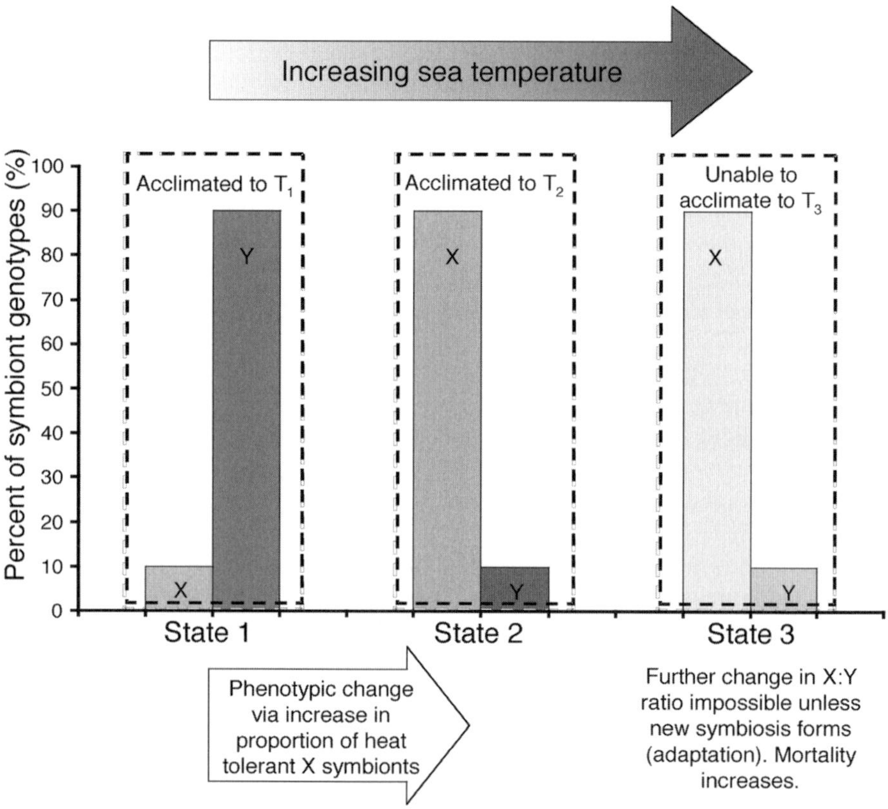

Fig. 26.6. Acclimation to sea temperature may involve polycladal symbioses shifting the ratio of different algal genotypes to match new sea temperatures. In the hypothetical example above, two algal genotypes (*X* and *Y*) are shifted from Y being dominant in *State 1* to X being dominant in *State 2*. As genotype X is more thermally tolerant, the shift in the ratio to one dominated by X allows the symbiosis to shift its thermal tolerance to match the warmer conditions. As with any phenotypic response, there is a limit to the temperatures to which the symbiosis can acclimate to. In *State 3*, sea temperatures have increased to T_3. The symbiosis, however, does not have a third alga with which to further shift its thermal tolerance to match the higher temperature. Over long periods of time (decades to centuries), another alga with the required thermal tolerance may form a symbiosis with higher thermal tolerance. This, however, is unlikely in ecological time

genetic combinations of host and symbiont to keep pace with the rapidly changing selective pressures associated with climate change (Fig. 26.6). As pointed out in Chapter 24, there is ample evidence that shifts in the ratio of algal genotypes in polycladal symbioses do occur (see Chap. 24; Verde and McClosky 1996; Baker 2001; Muller-Parker and Davy 2001). However, they are, like all genotypes, restricted ultimately by an upper thermal tolerance set by the host and its existing partners. To extend the thermal tolerance of the symbiosis, the host must form a new symbiosis with a more thermally tolerant alga. This is a complex process (especially for endosymbiotic associations) that is unlikely to happen in the ecological time frame of a bleaching event or at a rate fast enough to keep pace with the current increases in sea temperature.

In any discussion of adaptation, it is important to note that the rate of climate change that has and is currently occurring is approximately 100–200 times faster than that seen during the shift from the glacial to interglacial period over the past 400,000 years (Fig. 26.1; Barnola et al. 1987; Petit et al. 1999; see calculations in Hoegh-Guldberg and Hoegh-Guldberg 2003). Given the huge changes in the distribution and abundance of organisms that occurred during these comparatively slower shifts in climate, it is unlikely that adaptation of a complex trait like thermal tolerance (possibly involving scores of required changes) will keep up with the rapid pace of climate change facing us. Reefs are changing and will continue to change, with the spectre of very low coral abundance being more likely than not.

26.3.3
Implications for Reef Distribution and Biodiversity

From the preceding discussion, even the mildest climate change scenarios project substantial decreases in the amount of coral cover. Community changes like those seen by Loya et al. (2001) in Okinawa are broad sweeping, with the loss of sensitive coral species and the retention of more tolerant genera and species. This is despite the fact that global temperature has only increased by 0.6–0.8 °C over the past century. While a reef may become more "tolerant" due to the loss of thermally sensitive species, biodiversity has decreased with expected impacts on other more global aspects such as reef resilience and recovery rates (Hughes et al. 2003). Our understanding of how climate change will affect coral reefs is also extremely limited. In order to understand how quickly coral reefs will migrate in a poleward direction, for example, an in-depth understanding is needed of the rates of migration and recruitment between coral reef regions. Our understanding of these processes, however, is limited and is hindered currently by the precision of available techniques (Ridgway and Hoegh-Guldberg 2001). It is also apparent that the conditions under which corals prosper are not necessarily limited to sea temperature. The provision of suitable carbonate alkalinities (which decline in a poleward direction and under climate change), light and appropriate substrate are also very important determinants of coral growth. Consequently, the prospect of "coral reefs off Sydney or New York" as climate change develops may never eventuate.

Ultimately, successful adaptation and re-assortment of the world's coral populations to our changing climate will depend on when and where stabilization of sea temperatures occurs. Without stabilization, selection will continue to operate on each generation of corals, with coral populations continually being thrown through the selective bottleneck and population densities remaining low. Only when stabilization occurs, is it possible for genes to re-assort and tolerant coral populations to expand to fill the reefs at their new stable sea temperatures. It is important to note that stabilization is only possible under the lower range of scenarios. At the higher end of possible scenarios (equivalent to the pathway that global society is currently on), global temperature including sea temperatures will continue to increase at high rates for several centuries (IPCC 2001). This is the main reason why immediate action on greenhouse emissions is required.

Corals, while forming key elements of the framework of coral reefs, only represent part of the ecosystem at risk from climate change. How the changes in coral cover and diversity will affect the hundreds of thousands of other organisms on coral reefs is still being examined. Organisms that depend on corals for food or shelter and which reproduce via external fertilization might be predicted to face extinction as their primary habitat corals become extinct. The response of fish communities over the short term has yielded some surprises. In the Seychelles, for example, Spalding and Jarvis (2002) found that the overall structure of fish communities had changed very little despite massive decreases (3–20-fold) in living coral cover after the 1997–1998 bleaching event. Contrary to this is the observation of rapid decreases in the abundance of species that are obligate corallivores. These species are directly dependent on the presence of coral for their existence and disappear quickly if coral is removed. The orange-spotted filefish (*Oxymonacanthus longirostris*), a coral obligate, rapidly disappeared from Okinawan reefs after the 1998 bleaching event (Kokita and Nakazona 2001). Abundances of some fish also appear to increase following the loss of reef-building corals from reef communities. Lindahl et al. (2001), for example, showed an overall increase in fish abundance after the 1998 mass bleaching event on Tanzanian reef systems. This was largely linked to an increase in herbivores. Similar conclusions have been made in studies at other sites by Chabanet (2002).

Fish populations are a small, but important sector of the ecosystems under threat. Our understanding of the impacts of climate change on the rest of the biodiversity of coral reefs is also in its infancy and must be a future priority. While the pathway and time course of this change are relatively undefined, few experts are suggesting that biodiversity will be unaffected by a rapid loss of reef-building corals from the system. The irony is that corals, by way of being asexual for at least part of their life cycle, may be the least prone to extinction, being able to persist at low density until conditions improve. On the other hand, the hundreds of thousands of coral-dependent species, many being highly dependent on sexual reproduction, may not. In these cases, close proximity of individuals for sexual reproduction may be critical to species survival. Corals that are smaller and more distantly separated on the world's future reefs may well contribute to the rapid extinction of coral-associated species.

26.3.4 Integrated Impacts: the Importance of Reef Resilience

Given that greenhouse gases like carbon dioxide have a long residence time in the atmosphere, many of the impacts of climate change on coral reefs are inevitable. Jackson et al. (2001) and, most recently, Hughes et al. (2003) also highlight the loss of resilience of reefs (the ability to return to a previous state after a perturbation) as another important way that reef systems have been changed by humans (see Folke 2004 on the impact of human intervention on ecosystem resilience). For example, the loss of herbivores through over-exploitation appears to have been responsible for leading to the large-scale losses of coral cover from coral reefs around Jamaica and other sites (Hughes 1994). In this case, herbivory controls algal population growth and ensures open spaces for the recruitment of juvenile corals. On unperturbed reefs both fish and invertebrate grazers take over this critical role. However, after fish grazers were removed during the twentieth century by sustained fishing, the only grazer left was the sea urchin *Diadema*. This left Caribbean reefs with a reduced capacity to absorb shocks, a fact that was demonstrated when a Caribbean-wide virus decimated sea urchin populations in the 1980s. The result of the loss of the only remaining herbivore was that Caribbean reefs rapidly turned from coral reefs to algal-dominated ecosystems.

The preceding example is directly relevant to the impacts of climate change on coral reefs. As has been argued in this and many of the other chapters of this volume, catastrophic events like some of those seen to occur on reefs during 1998 may see vastly reduced coral populations, which may strive to repopulate reefs through the dispersal of larvae. However, if high rates of herbivory by fishes and invertebrates are not present, spaces for new coral recruits will not be created and coral recruitment may be blocked by macroalgae and other benthic species. Protection of critical fish stocks through no-take zones and other management techniques may ensure that these critical aspects of reef resilience are preserved. Protected reefs off Tanga in Tanzania showed almost double the recovery of reefs that were unprotected (Lindahl et al. 2001). Other examples can be found in the most recent report of the Global Coral Reef Monitoring Network report (Wilkinson 2002). Factors other than fishing pressure (sedimentation, nutrient loading) would be expected to interact to produce effects on the ability of reef systems to recover from coral bleaching and mortality.

The concept of reef resilience is critically important to how society responds to potential impacts of climate change on coral reefs. While reducing greenhouse gas emissions is an urgent priority, equally important is the need to ensure that reefs have the highest ecological resilience possible as they face these challenges. The threats listed in Table 26.1 are all contributing to the rapid decline in both the health and resilience of coral reefs to withstand the impacts of climate change. Management options in which problems like the over-exploitation of fish species (e.g. marine protected areas, MPAs) and degradation of coastal water quality are dealt with decisively will all contribute to helping coral reefs through the extremely stressful conditions of the next 100 years. The more effectively these options are put in place, the faster any return in historical time and the lower the ultimate extinction rate and loss of coral reef biodiversity. They may also be critical to the outcome that is produced when the earth's climate stabilizes.

26.4 Conclusion: Socio-Economic Costs and the Precautionary Principle

Coral reefs are one of the most beautiful ecosystems on the planet. Few experiences compare to that of swimming through the rich splendour of a coral-dominated reef resplendent with tropical reef fish. In an ever hungry world, however, the utility of coral reefs to the hundreds of million of people as resources of food, coastal protection, building materials and industry may provide the greatest drive to protect them. It is instructive to understand what economic benefits are at stake if reefs continue to degrade. Bryant et al. (1998) estimated that almost half a billion people live within 100 km of a coral reef, with the implication that these people benefit from goods and services that these ecosystems provide. The value of the resources (such as fish) and services (such as tourism and coastal protection) derived from coral reefs is about $375 billion each year (Constanza et al. 1997). The cost of destroying just 1 km of reef ranges from about $137,000 to almost $1.2 million in lost goods and services over a 25-year period (Cesar 1996). A recent study done for the Great Barrier Reef has estimated that the decline in the coral reefs that are associated with mild (+2 °C under the B2 scenario of the IPCC 2001 report) and high levels of climate change (+5 °C under the A2 scenario) will shave an estimated Au$ 3 and 8 billion off the Australian economy over the period 2003–2023, respectively (Hoegh-Guldberg and Hoegh-Guldberg 2003).

The serious nature of the decline of coral reefs must be addressed at the highest levels of government and the agencies assigned to protect coral reef resources. Current management practices in most parts of the world are not adequate to arrest the decline of the world's coral reefs. Rapid anthropogenic climate change has added another major stress to the mix of stresses that are devastating coral reefs, making their prognosis even bleaker. These factors drive home the conclusion that management practices have to be further altered in order to take into account the continuing, rapid decline of the world's coral reefs. With this in mind, the protection of coral reefs has to involve a new phase in which management and protective measures take into account current threats as well as those that are being projected. The precautionary principle of developing management practices that take into account the risk of worst-case scenarios (as opposed to those assuming the least or the most optimistic), has to play an even larger role in an uncertain world that faces a rate of environmental change perhaps not seen for at least 20 million years.

References

Abelson A, Shteinman B, Fine M, Kaganovsky S (1999) Mass transport from pollution sources to remote coral reefs in Eilat (Gulf of Aqaba, Red Sea). Mar Pollut Bull 38:25–29

Baker AC (2001) Reef corals bleach to survive change. Nature 411:765–766

Barnola JM, Raynaud D, Korokevich YS, Lorius C (1987) Vostok ice core provides 160,000 year record of atmospheric CO_2. Nature 329:408–414

Bastidas C, Bone D, García EM (1999) Sedimentation rates and metal content of sediments in a Venezuelan coral reef. Mar Pollut Bull 38:16–24

Berkelmans R, Willis BL (1999) Seasonal and local spatial patterns in the upper thermal limits of corals on the inshore Central Great Barrier Reef. Coral Reefs 18(3):219–228

Boersma PD, Parrish JK (1999) Limiting abuse: marine protected areas, a limited solution. Ecol Econ 31:287–304

Brown BE (1997) Coral bleaching: causes and consequences. Coral Reefs 16 [Suppl]:S129–S138

Brown BE, Ambarsari I, Warner ME, Fitt WK, Dunne RP, Gibb SW, Cummings DG (1999) Diurnal changes in photochemical efficiency and xanthophyll concentrations in shallow water reef corals: evidence for photoinhibition and photoprotection. Coral Reefs 18:99–105

Brown BE, Dunne RP, Goodson MS, Douglas AE (2002) Experience shapes the susceptibility of a reef coral to bleaching. Coral Reefs 21:119–126

Bryant D, Burke L, McManus J, Spalding M (1998) Reefs at risk: a map-based indicator of threats to the worlds coral reefs. World Resources Institute, Washington, DC, 56 pp

Cesar H (1996) Economic analysis of Indonesian coral reefs. World Bank, Washington, DC

Chabanet P (2002) Coral reef fish communities of Mayotte (Western Indian Ocean) two years after the impact of the 1998 bleaching event. Mar Freshwater Res 53:107–113

Coles SL, Jokiel PL (1977) Effects of temperature on photosynthesis and respiration in hermatypic corals. Mar Biol 43:209–216

Coles SL, Jokiel PL, Lewis CR (1976) Thermal tolerance in tropical versus subtropical Pacific coral reefs. Pac Sci 30:159–166

Cortes J (1994) A reef under siltation stress: a decade of degradation. In: Ginsburg RN (compiler) Proceedings of the colloquium on global aspects of coral reefs: health, hazards and history. Univ Miami, Florida, pp 240–246

Constanza R, d'Arge R, de Groot R, Farber S, Grasso M, Hannon B, Limburg K, Naeem S, ONeill RV, Paruelo J, Rashin RG, Sutton P, van den Belt M (1997) The value of the worlds ecosystem services and natural capital. Nature 387:253–260

Done TJ, Whetton P, Jones R, Berkelmans R, Lough J, Skirving W, Wooldridge S (2003) Global climate change and coral bleaching on the great barrier reef. Final report to the State of Queensland Greenhouse Taskforce through the Department of Natural Resources and Mining, Townsville

Edmunds PJ (1994) Evidence that reef-wide patterns of coral bleaching may be the result of the distribution of bleaching-susceptible clones. Mar Biol 121:127–142

Folke C (2004) Social-ecological resilience and behavioural responses. In: Biel A, Hansson B, Mårtensson M (eds) Individual and structural determinants of environmental practice. Ashgate Publ, London (in press)

Fuglestvedt JS, Berntsen TK, Isaksen ISA, Mao H, Liang X-Z, Wang W-C (1999) Climatic forcing of nitrogen oxides through changes in tropospheric ozone and methane; global 3-D model studies. Atmos Environ 33:961–977

Gattuso J-P, Frankignoulle M, Bourge I, Romaine S, Buddemeier RW (1998) Effect of calcium carbonate saturation of seawater on coral calcification. Global Planet Change 18:37–47

Glynn PW (1991) Coral reef bleaching in the 1980s and possible connections with global warming. Trends Ecol Evol 6:175–179

Glynn PW, Mate JL, Baker AC, Calderon MO (2001) Coral bleaching and mortality in Panama and Ecuador during the 1997–1998 El Niño-Southern Oscillation event: spatial/temporal patterns and comparisons with the 1982–1983 event. Bull Mar Sci 69:79–109

Hodgson G, Liebeler J (2002) The global coral reef crisis: trends and solutions 1997–2001. Reef Check Foundation, USA, http://www.reefcheck.org

Hoegh-Guldberg O (1999) Coral bleaching, climate change and the future of the world's coral reefs. Mar Freshwater Res 50:839–866

Hoegh-Guldberg O (2000) Global climate change and the thermal tolerance of corals. Galaxea JCRS 2:1–11

Hoegh-Guldberg O (2001) The future of coral reefs: integrating climate model projections and the recent behaviour of corals and their dinoflagellates. Proc 9th Int Coral Reef Symp 23–27 Oct 2000, Bali, Indonesia

Hoegh-Guldberg H, Hoegh-Guldberg O (2003) Biological, economic and social impacts of climate change on the Great Barrier Reef. World Wide Fund for Nature, special report, 400 pp

Hoegh-Guldberg O, Salvat B (1995) Periodic mass-bleaching and elevated sea temperatures: bleaching of outer reef slope communities in Moorea, French Polynesia. Mar Ecol Prog Ser 121:181–190

Hoegh-Guldberg O, Smith GJ (1989) The effect of sudden changes in temperature, light and salinity on the population density and export of zooxanthallae from the reef corals *Stylophora pistillata* Esper and *Seriatopora hystrix* Dana. J Exp Mar Biol Ecol 129:279–303

Hoegh-Guldberg O, Berkelmans R, Oliver J (1997) Coral bleaching: implications for the Great Barrier Reef Marine Park, CRC January 1997 conference. Research and Reef Management Proceedings, pp 21–43

Hoegh-Guldberg O, Hoegh-Guldberg H, Stout DK, Cesar H, Timmermann A (2000) Pacific in peril: biological, economic and social impacts of climate change on Pacific coral reefs. Greenpeace International, Amsterdam, 72 pp

Hoegh-Guldberg O, Jones RJ, Ward S, Loh WK (2002) Is coral bleaching really adaptive? Nature 415:601–602

Hughes TP (1994) Catastrophes, phase-shifts and large-scale degradation of a Caribbean coral reef. Science 265:1547–1551

Hughes TP, Baird AH, Bellwood DR, Card M, Connolly SR, Folke C, Grosberg R, Hoegh-Guldberg O, Jackson JBC, Kleypas J, Lough JM, Marshall P, Nyström M, Palumbi SR, Pandolfi, JM, Rosen B, Roughgarden J (2003) Climate change, human impacts, and the resilience of coral reefs. Science 301:929–933

Intergovernmental Panel on Climate Change (IPCC 2000) Special report on emissions scenarios. www.grida.no/climate/ipcc/emission

Intergovernmental Panel on Climate Change (IPCC 2001) Technical summary of the Working Group 1 Report. www.grida.no/climate/ (Same general website for subsequent IPCC listings)

Jackson JB, C, Kirby MX, Berger WH, Bjorndal KA, Botsford LW, Bourque BJ, Bradbury RH, Cooke R, Erlandson J, Estes JA, Hughes TP, Kidwell S, Lange CB, Lenihan HS, Pandolfi JM, Peterson CH, Steneck RS, Tegner MJ, Warner RR (2001) Historical overfishing and the recent collapse of coastal ecosystems. Science 293:629–638

Jones R, Hoegh-Guldberg O, Larkum AWD, Schreiber U (1998) Temperature induced bleaching of corals begins with impairment of dark metabolism in zooxanthellae. Plant Cell Environ 21:1219–1230

Jones RJ, Kildea T, Hoegh-Guldberg O (1999) PAM chlorophyll fluorometry: a new in situ technique for stress assessment in scleractinian corals, used to examine the effects of cyanide from cyanide fishing. Mar Pollut Bull 38(10):864–874

Jones RJ, Ward S, Yang Amri Abin, Hoegh-Guldberg O (2000) Changes in quantum efficiency of Photosystem II of symbiotic dinoflagellates of corals after heat stress and during the 1998 Great Barrier Reef mass bleaching. Mar Freshwater Res 51(1):63–71
Kleypas JA, Buddemeier RW, Archer D, Gattuso J-P, Langdon C, Opdyke BN (1999) Geochemical consequences of increased atmospheric carbon dioxide on coral reefs. Science 284:118–120
Kokita T, Nakazono A (2001) Rapid response of an obligately corallivorous filefish *Oxymonacanthus longirostris* (Monacanthidae) to a mass coral bleaching event. Coral Reefs 20:155–158
Koop K, Booth D, Broadbent A, Brodie J, Bucher D, Capone D, Coll J, Dennison W, Erdmann M, Harrison P, Hoegh-Guldberg O, Hutchings P, Jones GB, Larkum AWD, O'Neil J, Steven A, Tentori T, Ward S, Williamson J, Yellowlees D (2001) ENCORE: the effect of nutrient enrichment on coral reefs: synthesis of results and conclusions. Mar Pollut Bull 42:91–120
Lindahl U, Ohman MC, Schelten CK (2001) The 1997/1998 mass mortality of corals: effects on fish communities on a Tanzanian coral reef. Mar Pollut Bull 42:127–131
Loya Y, Sakai K, Nakano Y, Sambali H, van Woesik R (2001) Coral bleaching: the winners and the losers. Ecol Lett 4:122–131
Marshall PA, Baird AH (2000) Bleaching of corals on the Central Great Barrier Reef: differential susceptibility among taxa. Coral Reefs 19:155–163
McClanahan TR (1999) Is there a future for coral reef parks in poor tropical countries? Coral Reefs 18:321–325
McClanahan TR, Cokos B, Sala E (2002) Algal growth and species composition under experimental control of herbivory, phosphorus and coral abundance in Glovers Reef, Belize. Mar Pollut Bull 44:441–451
Meyers RA, Worm B (2003) Rapid worldwide depletion of predatory fish communities. Nature 423:280–283
Muller-Parker G, Davy SK (2001) Temperate and tropical algal-sea anemone symbioses. Invert Biol 120:104–123
Mumby PJ, Chisholm JRM, Edwards AJ, Andrefouet S, Jaubert J (2001) Cloudy weather may have saved Society Island reef corals during the 1998 ENSO. Mar Ecol Prog Ser 222:209–216
Parmesan C, Yohe G (2003) A globally coherent fingerprint of climate change impacts across natural systems. Nature 421:37–43
Patterson KL, Porter JW, Ritchie KB, Polson SW, Mueller E, Peters EC, Santavy DL, Smith GW (2002) The etiology of white pox, a lethal disease of the Caribbean elkhorn coral, *Acropora palmata*. Proc Natl Acad Sci USA 99:8725–8730
Petit JR, Jouzel J, Raynaud D, Barkov NI, Barnola J-M, Basile I, Benders M, Chappellaz J, Davis M, Delayque G, Delmotte M, Kotlyakov VM, Legrand M, Lipenkov VY, Lorius C, Pépin L, Ritz C, Saltzman E, Stievenard M (1999) Climate and atmospheric history of the past 420,000 years from the Vostok ice core, Antarctica. Nature 399:429–436
Ridgway T, Hoegh-Guldberg O (2003) Reef recovery in disturbed coral reef ecosystems. Proc 9th Int Symp for Reef Stud 2:1117–1122
Roeckner E, Oberhuber JM, Bacher A, Christoph M, Kirchner I (1996) ENSO variability and atmospheric response in a global atmosphere ocean GCM. Climate Dynam 12:737–754
Sheppard C (1995) The shifting baseline syndrome. Mar Poll Bull 30:766–767
Spalding MD, Jarvis GE (2002) The impact of the 1998 coral mortality on reef fish communities in the Seychelles. Mar Pollut Bull 44:309–321
Spalding MD, Ravilious C, Green EP (2001) World atlas of coral reefs prepared at the UNEP world conservation monitoring centre. Univ California Press, Berkeley
Spencer T, Teleki TA, Bradshaw C, Spalding M (2000) Coral bleaching in the Southern Seychelles During the 1997–1998 Indian Ocean warm event. Mar Pollut Bull 40:569–586
Strong AE, Barrientos CS, Duda C, Sapper J (1996) Improved satellite techniques for monitoring coral reef bleaching. Proc 8th Int Coral Reef Soc Symp (Panama)
Strong AE, Gjovig KK, Kearns E (2000) Sea surface temperature signals from satellites – an update. Geophys Res Lett 27(11):1667–1670
United Nations (2002) World population prospects: the 2002 revision I. Comprehensive tables. UN, Geneva

Verde EA, McClosky LR (1996) Photosynthesis and respiration of two species of algal symbionts in the anemone *Anthopleura elegantissima* (Brandt) (Cnidaria; Anthozoa). J Exp Mar Biol Ecol 195:187–202

Walther G-R, Post E, Convey P, Menzel A, Parmesan C, Beebee TJC, Fromentin J-M, Hoegh-Guldberg O, Bairlein F (2002) Ecological responses to recent climate change. Nature 416:389–395

Wellington GM, Glynn PW, Strong AE, Navarrete SA, Wieters E, Hubbard D (2001) Crisis on coral reefs linked to climate change. EOS 82(1):1

Wilkinson C (ed) (1998), Status of coral reefs of the world: 1998. Australian Institute of Marine Science, Townsville

Wilkinson C (ed) (2000) Status of coral reefs of the world: 2000. Global Coral Reef Monitoring Network, Australian Institute of Marine Science, Townsville

Wilkinson C (ed) (2002) Status of coral reefs of the world: 2002. Global Coral Reef Monitoring Network, Australian Institute of Marine Science, Townsville

Wilkinson CR, Buddemeier RW (1994) Global climate change and coral reefs: implications for people and reefs. Report of the UNEP-IOC-ASPEI-IUCN Global Task Team on the implications of climate change on coral reefs. IUCN, Gland, Switzerland, 124 pp

Subject Index

Printing: Saladruck, Berlin
Binding: Stein+Lehmann, Berlin